NANOBIOTECHNOLOGY FOR AGRICULTURAL SCIENCES

Nano-Agri-Input Products for Crop Production and Environmental Protection

NANOBIOTECHNOLOGY FOR AGRICULTURAL SCIENCES

Nano-Agri-Input Products for Crop Production and Environmental Protection

J. C. Tarafdar, PhD
Deepak Kumar, PhD
Sanjay K. Singh, PhD
Krishna Pal Singh, PhD
Editors

AAP | APPLE ACADEMIC PRESS

First edition published 2025

Apple Academic Press Inc.
1265 Goldenrod Circle, NE,
Palm Bay, FL 32905 USA

760 Laurentian Drive, Unit 19,
Burlington, ON L7N 0A4, Canada

CRC Press
2385 NW Executive Center Drive,
Suite 320, Boca Raton FL 33431

4 Park Square, Milton Park,
Abingdon, Oxon, OX14 4RN UK

Library and Archives Canada Cataloguing in Publication

Title: Nanobiotechnology for agricultural science : nano-agri-input products for crop production and environmental protection / J.C. Tarafdar, PhD, Deepak Kumar, PhD, Sanjay K. Singh, PhD, Krishna Pal Singh, PhD, editors.
Names: Tarafdar, J. C., editor | Kumar, Deepak (Research scientist), editor | Singh, Sanjay K. (Mycologist), editor. | Singh, Krishna Pal, editor.
Description: First edition. | Includes bibliographical references and index.
Identifiers: Canadiana (print) 20250133229 | Canadiana (ebook) 20250133296 | ISBN 9781774919682 (hardcover) | ISBN 9781774919699 (softcover) | ISBN 9781003620273 (ebook)
Subjects: LCSH: Agricultural innovations. | LCSH: Nanostructured materials. | LCSH: Nanotechnology.
Classification: LCC S494.5.I5 N36 2025 | DDC 338.1/6—dc23

Library of Congress Cataloging-in-Publication Data

CIP data on file with US Library of Congress

ISBN: 978-1-77491-968-2 (hbk)
ISBN: 978-1-77491-969-9 (pbk)
ISBN: 978-1-00362-027-3 (ebk)

About the Editors

J. C. Tarafdar, PhD
Former Emeritus Scientist and National Fellow, ICAR, Central Arid Zone Research Institute, Jodhpur, Rajasthan, India

J. C. Tarafdar, PhD, received his MSc and PhD degrees in soil science and agricultural chemistry from the Indian Agricultural Research Institute, New Delhi, India, and his postdoctoral degree from the Institute of Agricultural Chemistry, Göttingen, Germany. He was awarded merit scholarships throughout his academic career. Dr. Tarafdar has made an original and well-recognized contribution to the mobilization of native phosphorus. He has developed an in vivo filter paper technique for phosphatase estimation. He developed biological phosphorus (bio-phos) fertilizer, which is now being promoted by the state government of Rajasthan. He is the first in the world to have successfully developed biosynthesized nanonutrients and nano-induced polysaccharide powder for agricultural use. He also developed a sequential P fractionation scheme and a rapid method for assessing plant residue quality. Dr. Tarafdar has pioneered finding out the origin of different phosphatase fractions and also developed a potential biological indicator. He has developed many new techniques that are now used in soil biology. The most notable of them are the visual demonstration technique of germinating AM spores; the soil solarization technique for mass-scale production of AM fungi; the freeze-drying technique to understand nutrient movement; and the electrofocusing technique to demonstrate the origin of enzymes. Dr. Tarafdar has traveled and lectured widely in many countries and published 384 research articles in national and international journals, which include 42 book chapters and five books. He has four patents and 73 new organisms to his credit. In recognition of his professional contribution, Dr. Tarafdar has received many awards, most notable being the Sukumar Basu Memorial Award for the biennium 1993–1994, IMPHOS-FAI Award (2006), Prof. S. K. Mukherjee Memorial Award (2008), Prof. R. S. Murthy Memorial Award (2008), the Bharat Jyoti Award 2011, Glory of India Gold Medal 2012, Prof. R. V. Tamhane Memorial Award 2013, Dr. Kartik Oaron Memorial Award 2014, Dr. N. S. Randhawa Memorial Award 2014, ISSS

Platinum Jubilee Commemoration Award, 2016, Prof. N. P. Datta Memorial Award, 2020, INSO Outstanding Scientist Award, 2022. Dr. Tarafdar is a fellow of the most prestigious Alexander von Humboldt (AvH) Germany and DAAD Germany. He is also a fellow of the National Academy of Agricultural Sciences (2003), the Indian Society of Soil Science (2002), the Indian Society of Salinity Research Scientists (2003), and the Society of Fertilizer and Environment (2022). He has more than 9200 citations for his publications and 45 h-index. Dr. Tarafdar has been placed among the Top 2% of Scientists across all sciences in the world by Stanford University, USA. He has guided seven PhD and 18 MSc students.

Deepak Kumar, PhD
Executive Director, Nextnode Bioscience Pvt. Ltd., Kadi, Gujarat, India; Director (R&D), Shri Ram Solvent Extractions Pvt. Ltd., Jaspur, Uttarakhand, India

Deepak Kumar, PhD, presently works as the Executive Director (R&D) at Nextnode Bioscience Pvt. Ltd., Kadi, and Director (R&D) of the DSIR-recognized laboratory of Shri Ram Solvent Extractions Pvt. Ltd. He is an author and coauthor of 20 peer-reviewed journal articles, three books, one patent, and 20 book chapters. He has been involved in the quality production of biofertilizers and biopesticides for the last 12 years. Dr. Kumar is a Certified Farm Advisor (organic farming, certified by MANAGE, Government of India) and a Food Safety Mitra (certified by FSSAI, GOI). He also worked as a technical expert member (biofertilizer) to start the ICAR-IISR Agribusiness Incubation Centre (ABI), Indore, Madhya Pradesh. Additionally, Dr. Kumar was selected as an AYE mentor by IFCI Venture Capital Funds Limited (an Indian Government undertaking) for SC/ST mentorship under the Ministry of Social Justice and Empowerment, Government of India (from February 2022 to date). Moreover, Dr. Kumar also works as a guest faculty for the course, PG Diploma in Biotechnology Law and Policy (PGDBLP), at Gujarat National Law University. He obtained his MSc (Biotechnology) from C.C.S. University, Meerut; his MBA (Human Resource Management) from Punjab Technical University, Jalandhar; and his PhD (Ag Biotechnology) from Acharya Narendra Deva University of Agriculture & Technology, Ayodhya, Uttar Pradesh, India. He has received many prestigious awards from different scientific and market research organizations, like the Indo-Global Award for Innovation in Biotech (2021) from the Indian Economic Development & Research Association (IEDRA), New Delhi; the Distinguished Industrial Microbiologist Award (2020) from the Microbiologists Society of India (MSI); the India Leadership Awards (2019) in the Best Agribiotech Entrepreneur category from Blindwink Bangalore; the Science Entrepreneur Award (2018) from Biologix Research and Innovation Centre (BRICPL, India); and the Young Scientist Award (2018) from Doctor's Agricultural and Horticultural Development Society, Lucknow, Lucknow, India. He also completed and guided six indigenous and externally funded (DSIR and DST) research projects. Dr. Kumar has recently developed and licensed a neem endophytic mycoflora-based mycofertilizer, namely "AsperPhos," which showed potential in mineral solubilization. Moreover, he is a technical consultant for more than a dozen companies, providing customized technical support. He is an active member of several associations and bodies like the Fertilizer Associations of India (FAI), the Society for Technology Management (STEM), the International Competence Centre for Organic Agriculture (ICCOA), and the Association of Microbiologists of India (AMI), respectively.

Sanjay K. Singh, PhD
Scientist 'G' and In-Charge, Biodiversity and Paleobiology Group, Agharkar Research Institute, Pune, Maharashtra, India

Sanjay K. Singh, PhD, obtained his postgraduate degree in botany from Kanpur University, India, and PhD in botany (mycology) from Gorakhpur University, Gorakhpur, UP. He was a postdoctoral fellow at CSIR, New Delhi (1996), received the Young Scientist Award of DST, New Delhi (1997–1999), and a Research Associateship of a Planning Commission-sponsored project at Goa University (2000–2001). He received training in the fields of fungal biology, biodiversity, conventional and genomic identification, systematics, germplasm conservation, and secondary metabolite profiling and their applications in agriculture, medicine, and industry. With vast experience in fungal biology and plant pathology, Dr. Singh has made a major contribution by establishing a state-of-the-art National Facility for Culture Collection of Fungi with financial support from DST, Govt. of India, in 2008. Under this flagship program, he established an exclusive repository for fungi as the National Fungal Culture Collection of India (NFCCI), affiliated with the World Federation for Culture Collection (WFCC). More than 700 academic research institutions and about 100 industries spread across 28 states in India are beneficiaries of this facility. Dr. Singh, under the DBT Associate program for young scientists (under age 40) in the niche area of agri-biotechnology, received specialized training with world-renowned researcher Prof. Gary A. Strobel, Montana State University (MSU), USA, on the biology of endophytes and their bioactive metabolites. He attended International Advance Training on Genetic Resources and Intellectual Property Rights (GRIP), sponsored by the Swedish International Development Authority (SIDA), at SLU, Sweden, and Johannesburg, South Africa. Dr. Singh supervised Dr. Frank Ackiah, from Ghana, who was selected for the CV Raman International Fellowship for African Researchers sponsored by DST, Government of India, in 2018. He has also served as a member of the Research Advisory Committee of the CSIR-Institute of Microbial Technology (IMTECH), Chandigarh, and the ICAR-NBAIM, Mau, by the ICARI, New Delhi. Dr. Singh is an elected fellow of the Maharashtra Academy of Sciences (FMASc), the Indian Phytopathological Society (FIPS), and an honorary fellow of the Society for Biotechnology (India). He was elected President, West Zone, Indian Phytopathological Society (IPS-2008–2009), and Council Member of IPS and MSI. He is a recipient of the A. K. Sarbhoy Memorial Award of the Indian Phytopathological Society (ICAR), New Delhi, and the Dr. S. K. Shome Memorial Award of the Mycological Society of India (MSI). He is a life member of several prestigious societies, such as ISCA, IPS, MSI, and the Association of Fungal Biologists (AFB). Dr. Singh has operated 14 projects, published more than 115 research papers, three books, 20 book chapters, and delivered more than 30 invited lectures in symposiums, seminars, and workshops. He has organized three conferences and more than 15 capacity-building national-level workshops and served as a reviewer and editorial board member for prestigious journals. He has visited Nepal, Fiji, Brazil, South Africa, Sweden, and USA for various research-related activities

Krishna Pal Singh, PhD

Vice-Chancellor, Mahatma Jyotiba Phule Rohilkhand University, Bareilly, India
Professor, Biophysics Unit, College of Basic Sciences & Humanities, G.B. Pant University of Agriculture & Technology, Pantnagar, Uttarakhand, India

Krishna Pal Singh, PhD, obtained his MSc., MPhil, and PhD degrees in Chemistry (area of research: membrane biophysics) from Aligarh Muslim University, Aligarh, Uttar Pradesh. He went for further research on structure elucidation of biomolecules, nanosignaling, targeted drug delivery, and nanobiosensors at the Technical University Bratislava (Slovakia), Ruder Boskovic Institute Croatia (Europe), and Cranfield University, UK. He joined as an associate professor of Biophysics at the G. B. Pant University of Agriculture and Technology, Pantnagar, in 2005. He is an author and coauthor of 80 peer-reviewed journal articles, 10 book chapters, four conference papers, and 10 books. His area of research is nanobiosensors, molecularly imprinted polymers as plastic antibodies, targeted drug delivery through nanomaterials, nano-signaling through biomembranes, nanotoxicity, and molecular diagnostics. He has served as Director, Uttarakhand Council for Biotechnology, Government of Uttarakhand, India. He has also served as Vice Chancellor of Chaudhary Charan Singh Haryana Agricultural University, Hisar, Haryana, India, since 2016, and additionally, he has served as Vice Chancellor of Maharana Pratap Horticultural University, Karnal, Haryana, India, since 2017. Currently, he has been working as Vice Chancellor of Mahatma Jyotiba Phule Rohilkhand University, Bareilly, India, since 2020. His research interests include nano-biotechnology, nanobiosensors, nanomedicine, targeted drug delivery, nanofertilizers, and bioenergy, where he has authored several publications in terms of articles, books, patents, policy papers, and international scientific and strategic documents. Currently, he is working on sustainable development through scientific innovation and research encompassing applied agriculture, applied science, and nanotechnology. Dr. Singh's emphasis is on the creation of an eco-hub for startups and entrepreneurship using cutting-edge technologies.

Contents

Contributors

Krishnendu Acharya
Molecular and Applied Mycology and Plant Pathology Laboratory, Centre of Advanced Study, Department of Botany, University of Calcutta, Kolkata, India

Jishnu Banerjee
Department of Botany, Ramakrishna Mission Vivekananda Centenary College, Rahara, Khardaha, West Bengal, India

Yatin Batra
Department of Chemistry, Guru Nanak Dev University Amritsar, Punjab, India

Vaidehi Bhagwat
Nanobioscience Group, Agharkar Research Institute, G. G. Agarkar Road, Pune, Maharashtra, India

Prithu Bhattacharyya
Department of Botany, Ramakrishna Mission Vivekananda Centenary College, Rahara, Khardaha, West Bengal, India

Nivedita Chaudhri
Department of Chemistry, Guru Nanak Dev University Amritsar, Punjab, India

Vladimir Chistyakov
Southern Federal University, Academy of Biology and Biotechnology named by D. I. Ivanovsky, Rostov on Done, Russia

Deepali Choudhary
Nanobioscience Group, Agharkar Research Institute, G. G. Agarkar Road, Pune, Maharashtra, India

Himanki Dabral
Department of Genetics & Plant Breeding, School of Agricultural Sciences, Shri Guru Ram Rai University, Pathribagh, Dehradun, Uttarakhand, India.

Sunandana Das
Department of Botany, Scottish Church College, Kolkata, West Bengal, India

Tamal Das
Department of Botany, Ramakrishna Mission Vivekananda Centenary College, Rahara, Khardaha, West Bengal, India

Sarita Dhaka
Department of Chemistry, Sanatan Dharm (PG) College, Muzaffarnagar, Affiliated to Maa Shakumbhari University, Saharanpur, Uttar Pradesh, India

Rahul Kumar Dhaka
Department of Chemistry (COBS&H), CCS Haryana Agricultural University, Hisar, India

Amita Gharu Dhariwal
Department of Chemistry, Jai Narain Vyas University, Jodhpur, Rajasthan, India

Saurabh Dubey
Formulation Development and Consultant, Bhiwadi, Rajasthan, India

Shreshta Dubey
Formulation Development and Consultant, Bhiwadi, Rajasthan, India

Maheshwari G.
Nanobioscience Group, Agharkar Research Institute, Pune, Maharashtra, India

Akanksha Gautam
School of Biotechnology, Jawaharlal Nehru University, New Delhi, India

Vandana Ghormade
Nanobioscience Group, Agharkar Research Institute, Pune, Maharashtra, India

Sandipta Ghosh
School of Biological and Environmental Sciences, Shoolini University of Biotechnology and Management Sciences, Solan, India

Raja Husain
Department of Agriculture, Himalayan University, Itanagar, Arunachal Pradesh, India

Nusrat Iqbal
Formulation Division, Institute of Pesticide Formulation Technology, Udyog Vihar Gurugram, Haryana, India

Pallavi Joshi
Graphic Era Hill University, Dehradun, Uttarakhand, India

Ashish Kalkal
Department of Mechanical Engineering, University College London, London, UK

Somanjana Khatua
Department of Botany, Faculty of Science, University of Allahabad, Prayagraj, Uttar Pradesh, India

Sergey Kolesnikov
Southern Federal University, Academy of Biology and Biotechnology Named By D. I. Ivanovsky, Rostov on Done, Russia

Deepak Kumar
Nextnode Bioscience Pvt Ltd., Opposite GEB Office, Kadi-Kalol Road, Kadi, Gujarat, India
Shri Ram Solvent Extraction Pvt Ltd., Kashipur Road, Jaspur, Uttarakhand, India

Hemant Ritturaj Kushwaha
School of Biotechnology, Jawaharlal Nehru University, New Delhi, India

Saglara Mandjieva
Southern Federal University, Academy of Biology and Biotechnology named by D. I. Ivanovsky, Rostov on Done, Russia

Nikita Minin
Southern Federal University, Academy of Biology and Biotechnology named by D. I. Ivanovsky, Rostov on Done, Russia

Tatiana Minnikova
Southern Federal University, Academy of Biology and Biotechnology named by D. I. Ivanovsky, Rostov on Done, Russia

Tatiana Minkina
Southern Federal University, Academy of Biology and Biotechnology named by D. I. Ivanovsky, Rostov on Done, Russia

Tatiana Minkina
Soil Science and Land Evaluation Department, Southern Federal University, Russia

Arkadeb Mukhopadhyay
Division of Agricultural Chemicals, ICAR-Indian Agricultural Research Institute, New Delhi, India

Md Abu Nayyer
Department of Agriculture, Integral Institute of Agricultural Science and Technology (IIAST) Integral University, Dasauli, Uttar Pradesh, India

Pramila Pandey
Department of Botany, BSNVPG College, Charbagh, Lucknow, India

Vishnu D. Rajput
Southern Federal University, Academy of Biology and Biotechnology named by D. I. Ivanovsky, Rostov on Done, Russia

Anuj Rana
Department of Microbiology (COBS&H), CCS Haryana Agricultural University, Hisar, Haryana, India

Joy Sarkar
Department of Botany, Dinabandhu Andrews College, Kolkata, India

Md. Shamim
Department of Molecular Biology and Genetic Engineering, Dr. Kalam Agricultural College, Kishanganj, Bihar Agricultural University, Sabour, Bhagalpur, Bihar, India

S. Shanthakumar
Gujarat National Law University (GNLU), Gandhinagar, Gujarat, India

Manu Sharma
School of Nano Sciences, Central University of Gujarat, Gandhinagar, Gujarat, India

Mahaveer P. Sharma
Agribusiness Incubation Centre, ICAR-Indian Institute of Soybean Research, Indore, Madhya Pradesh, India

Prernaa Sharma
Graphic Era Hill University, Dehradun, Uttarakhand, India

Mohammed Haris Siddiqui
Department of Agriculture, Integral Institute of Agricultural Science and Technology (IIAST) Integral University, Kursi Road, Dasauli, Uttar Pradesh, India

Saba Siddiqui
Department of Agriculture, Integral Institute of Agricultural Science and Technology (IIAST) Integral University, Kursi Road, Dasauli, Uttar Pradesh, India

Anu Singh
School of Biotechnology, Jawaharlal Nehru University, New Delhi, India
Samvet Bharat Under the Aegis of Samvetan Society for Social & Scientific Research, Dehradun, Uttarakhand, India

Awantika Singh
School of Biotechnology, Jawaharlal Nehru University, New Delhi, India

Sanjay K. Singh
Biodiversity and Palaeobiology Group, MACS' Agharkar Research Institute, Pune, India

Ramesh Kumar Singh
Samvet Bharat Under the Aegis of Samvetan Society for Social & Scientific Research, Dehradun, Uttarakhand, India

Krishna Pal Singh
Biophysics Unit, College of Basic Sciences & Humanities, G. B. Pant University of Agriculture & Technology, Pantnagar, Uttarakhand, India
Mahatma Jyotiba Phule Rohilkhand University, Bareilly, India

Shivangi Singh
Nanobioscience Group, Agharkar Research Institute, Pune, Maharashtra, India

Deepti Srivastava
Department of Agriculture, Integral Institute of Agricultural Science and Technology (IIAST) Integral University, Kursi Road, Dasauli, Uttar Pradesh, India

Pratibha Srivastava
Bioprospecting Group, Agharkar Research Institute, Pune, Maharashtra, India

Ayon Tarafdar
Livestock Production and Management Section, ICAR-Indian Veterinary Research Institute, Izatnagar, Bareilly, Uttar Pradesh, India

J. C. Tarafdar
ICAR—Central Arid Zone Research Institute, Jodhpur, Rajasthan, India

Sourabh Tyagi
Special Centre for Systems Medicine, Jawaharlal Nehru University, New Delhi, India

Priyank Vyas
Graphic Era (Deemed to Be) University, Dehradun, Uttarakhand, India

Abbreviations

AI	active ingredient
ABA	abscisic acid
AFM	atomic force microscopy
AMF	arbuscular mycorrhizal fungi
APX	ascorbate peroxidase
APVMA	Australian Pesticides and Veterinary Medicines Authority
BIS	Bureau of Indian Standards
BRs	brassinosteroids
CAT	chloramphenicol acetyltransferase
CEPA	Canadian Environmental Protection Act
CEC	cation exchange capacity
CAPA	Canada Agricultural Products Act
CPP	cell-penetrating peptide
CIB&RC	Central Insecticide Board and Registration Committee
CAGR	compound annual growth rate
CRISPR	clustered regularly interspaced short palindromic repeats
CODEX	Codex Alimentarius Commission
CKs	cytokinins
CNTs	carbon nanotubes
CNAP	cyantraniliprole
CFRP	carbon fiber-reinforced polymers
DAI	Daphnia Acute Immobilization
DEFRA	Department for Environment, Food, and Rural Affairs
DBT	Department of Biotechnology
DiSTAP	Disruptive and Sustainable Technologies for Agricultural Precision
DSC	dynamic light scattering
DHAR	dehydroascorbate reductase
DDVP	dichlorovinyl dimethyl phosphate
ET	ethylene
EC	electrical conductivity
ECA	European Chemical Agency
EDX	energy-dispersive X-ray spectroscopy
ENMs	engineered nanomaterials
EFSA	European Food Safety Authority
EPA	Environmental Protection Agency
EM	electron microscopy

EU	European Union
EIA	Environmental Impact Assessment
ECHA	European Chemicals Agency
EFSA	European Food Safety Authority
FAO	Food and Agricultural Organization
FAI	Fertilizer Association of India
FCO	Fertilizer (Control) Order
FDO	Food and Drug Organisation
FSANZ	Food Standards Australia New Zealand
FDA	Food and Drug Administration
FRET	Förster Resonance Energy Transfer
FSA	Food Standards Agency
FFDCA	Federal Food, Drug, and Cosmetic Act
FIFRA	Federal Insecticide, Fungicide, and Rodenticide Act
FE-SEM	field emission scanning electron microscopy
FET	fish acute toxicity
FSSAI	Food Safety and Standards Authority of India
GM	genetically modified
GAs	gibberellins
GST	glutathione-S-transferase
HSE	Health and Safety Executive
ICH	International Council for Harmonisation of Technical Requirements for Pharmaceuticals for Human Use
ISO	International Standard Organisation
INIC	Iran Nanotechnology Initiative Council
IAA	indole-3-acetic acid
MARA	Ministry of Agriculture and Rural Affairs
MDAR	monodehydroascorbate reductase
MEE	Ministry of Ecology and Environment
MoEFCC	Ministry of Environment, Forest, and Climate Change
MDA	malondialdehyde
MOFs	metalloporphyrinic metal-organic frameworks
NAIPs	nanoagri-input products
NAPs	nanoagri products
NMs	nanomaterials
NPs	nanoparticles
NPD	nanotechnology products database
NCNST	National Centre for Nanoscience and Technology
NHC	National Health Commission
OECD	Organization for Economic Cooperation and Development
PA	precision agriculture
PHAC	Public Health Agency of Canada
PCPA	Pest Control Products Act

POD	peroxidase
PPO	polyphenol oxidase
REACH	Registration, Evaluation, Authorisation, and Restriction of Chemicals
RWC	relative water content
ROS	reactive oxygen species
R&D	research and development
RGO	reduced graphene oxide
SANS	small-angle neutron scattering
SA	salicylic acid
SAMR	State Administration for Market Regulation
SL	strigolactones
SEM	scanning electron microscopy
SPR	surface plasmon resonance
SOD	superoxide dismutase
SOC	soil organic carbon
TEM	transmission electron microscopy
TAT	tyrosine aminotransferase
TCPP	tetra-carboxy-phenylporphyrin
THPP	tetra-hydroxy-phenyporphyrin
TG	test guidelines
TANIDA	Taiwan Nanotechnology Industry Development Association
TSCA	Toxic Substances Control Act
TMV	tobacco mosaic virus
US-FDA	United States Food and Drug Administration
US EPA	United States Environmental Protection Agency
WHO	World Health Organization
XRD	X-ray powder diffraction

Preface

Agriculture and allied sectors around the world are experiencing significant problems in improving agricultural output in order to feed the world's growing population. Environmental pollution, climate change, development of resistant pests, ecosystem imbalance, toxicity for humans and nontarget organisms, and so on are all consequences of indiscriminate pesticide usage. Precision agriculture approaches are required in this scenario to ensure agriculture's long-term viability. As a result, nanotechnology may provide effective answers to present agricultural difficulties.

Today, nanotechnology is a prominent area of research and growth in technology. The size of the global agricultural nanotechnology market was estimated at 321.13 billion in 2022 and is expected to grow at a compound annual growth rate (CAGR) of 11.94% from 2023 to 2031 to reach 868.98 billion (Insight Ace Analytic, 2023). The advancement in nanotechnology provides new opportunities for agricultural productivity, plant growth regulators, biostimulants, fertilizers, pesticides, biofertilizers, plant growth regulators, antitranspiration agents, insect-pest/disease prevention, and crop genetic modification using nanomaterials as carrier systems. Innovative nano-interventions in the food and agricultural sectors have a significant impact on product processes and can lead to the creation of low cost, extremely potent products.

Research efforts from various nations are now being made to transition from chemical-based agriculture to green agriculture, in which biopesticides and biological nanomaterials play an important role in pest management. However, information on their environmental behavior is limited, making it difficult to estimate the new risk and benefits in comparison to already utilized agrochemicals. For crop-pest management, a wide range of metal nanoparticles such as gold (Au), silver (Ag), silica (Si), aluminum (Al), and metal-oxide-based polymers titanium dioxide (TiO_2), selenium dioxide (SeO_2), and zinc oxide (ZnO) that are eco-friendly and synthesized using plants and microbes are being developed. However, practical studies conducted in the subject of nanomaterials and pest management are few till date.

In addition, excessive chemical fertilizer use lingers in the soil and generates ammonia, causing significant crop harm as well as environmental issues. Nontoxic nanoparticles can be utilized as additions in fertilizers for agricultural development to boost food production while reducing pollution. Many academics believe that the creation of macronutrient nano-based fertilizers could contribute to a new green revolution, helping to feed the world's ever-growing population while also improving agriculture's environmental sustainability.

A number of good books on nanotechnology have recently become available in the market. However, no single book has been completely dedicated to agricultural nano-input products or environmental protection. This book assesses the state of nanotechnology for agricultural applications while also emphasizing the biogenic synthesis of NAIPs and technology's economic potential in the establishment of a sustainable agriculture. This aspect is absolutely important for global farming as nano-input has the potential to replace chemical inputs, now mostly used in agriculture, so that we may overcome the problems of nutrient use efficiency, native nutrient mobilization, utilization of plant efficiency, controlled release of nutrient for better utilization, salt accumulation in soil, increase in soil carbon sequestration, more stress tolerance in plants, soil and plant health improvement, and improvement of yield in comparison to all generations of fertilizer.

Nanoagri-input management can dictate precise nutrient and health management, have a significant commercial impact, and has the potential to make high-quality products at a very low cost and at a faster pace. It has the bright future of revolutionizing agricultural farming in the world.

This book consists of four parts, in which 16 chapters cover all aspects of nanomaterials in the context of synthesis and their application in crop production and environmental protection. The first part (Chapters 1–5) of the book includes discussions related to the global scenario of nanomaterials, their synthesis, characterization, applications, and regulatory framework for commercialization. Green-synthesized nanomaterials show promise in sustainable agriculture, with potential benefits for soil fertility enhancement and management of biotic and abiotic stresses.

The second part (Chapters 6–12) of the book is dedicated to the green synthesis of nanomaterials and their impacts on soil fertility and biotic and abiotic stress management. Nanomaterials play a significant role in the production and applications of biochar and nanobiochar, enhancing their properties and expanding their potential uses. In this context, the third part (Chapters 13–14) of the book includes dedicated chapters related to biochar and nano-biochar and their potential benefits and challenges.

The last or fourth part (Chapters 15–16) of the book comprises chapters that explore the application of nanomaterials in soil remediation and environmental protection. Nanomaterials have shown great potential in soil remediation and environmental protection due to their unique properties and applications. They offer innovative solutions for addressing various environmental challenges, including soil contamination, water pollution, and air quality issues.

The present book is arranged in such a way that readers will acquire recent knowledge on nanoagri-input technologies and their uses and feel more comfortable learning the subject. We hope readers find this book very inspiring and motivating. Our purpose is to provide up-to-date knowledge about various nanoagri-products and their importance to commercialization. We have outlined this book in such a way that both researchers and industrialists may acquire a lot of ideas for the research and commercialization of the products to bring global farming into a golden age. The main purpose of this book is to equip students, researchers, scientists, and industries with extensive knowledge on agricultural nano-inputs and the ethical issues involved in their future implementation.

We welcome creative suggestions from the readers and hope that this book will certainly find a place in the libraries around the world. The editors sincerely wish to acknowledge the chapter authors for their significant contributions to the development of this volume. The editors recognize the assistance and support received from numerous books, experts, journals, and publishers whose works were used to prepare the manuscripts. We express our gratitude to Apple Academic Press for publishing this book and considering our proposal.

—***Editors***

PART I

Global Presence, Synthesis, Characterization, Applications, and Regulatory Framework of Nanomaterials

CHAPTER 1

The Global Status of Commercial Nano-Agri-Input-Products (NAIPs) and Their Techno-Commercial Viability

DEEPAK KUMAR[1,2], VISHNU D. RAJPUT[3], TATIANA MINKINA[4], J.C. TARAFDAR[5], and SANJAY K. SINGH[6]

[1]*Nextnode Bioscience Pvt Ltd., Kadi, Gujarat, India*

[2]*Shri Ram Solvent Extraction Pvt Ltd., Jaspur, Uttarakhand, India*

[3]*Southern Federal University, Rostov-on-Don, Rostov Oblast, Russia*

[4]*Soil Science and Land Evaluation Department, Southern Federal University, Rostov Oblast, Russia*

[5]*ICAR – Central Arid Zone Research Institute, Jodhpur, Rajasthan, India*

[6]*Biodiversity and Palaeobiology Group, MACS' Agharkar Research Institute, Pune, India*

ABSTRACT

The application of nanotechnology has opened up new possibilities for agricultural productivity, plant growth regulators, biostimulants, fertilizers, pesticides, biofertilizers, antitranspiration agents, insect-pest/disease prevention, and genetic modification of crops using nanomaterials (NMs) as carrier systems for crop benefit applications. Innovative nano-interventions in the food and agriculture industries have a prominent impact on product processes and result in low-cost, highly effective products being developed. Globally, a number of novel nano-based agricultural inputs and food products are released each year. However, a small number of nano-agri products

Nanobiotechnology for Agricultural Sciences: Nano-Agri-Input Products for Crop Production and Environmental Protection. J. C. Tarafdar, Deepak Kumar, Sanjay K. Singh, and Krishna Pal Singh (Eds.)

(NAPs) are available in the market as a result of stringent regulatory restrictions, new technology, the high cost of formulation development, and unpredicted breakeven points or returns that prevent the commercialization of nanoformulations. Moreover, some major concerns are also associated with nano-enabled products, such as the potential concentration of nano-agri-input products (NAIPs) used, particle size, synthesis methods, and delivery systems for applications. This chapter explored the use of NAIPs, market potential, regulatory framework, techno-commercial viability for mass production, limitations, and some other features based on recent findings.

1.1 INTRODUCTION

Nanotechnology is a rapidly evolving field that involves the manipulation and control of matter at the nanoscale, typically at the level of individual atoms or molecules. It has the potential to revolutionize various industries by enabling the development of novel materials, devices, and systems with unique properties and functionalities. Several commercial nanoproducts (NPs) based on nanotechnology have been rapidly developed and promoted over the past two decades. One needs to comprehend the complexity involved in the creation of these items, the range of their applications, and their marketing tactics as more and more products are added every year. Since these NPs have been found to have a significant negative influence on both the environment and human health, regulations have been put in place for their production, transportation, regular use, and disposal (Antony et al., 2021).

The term "nanomaterial" (NM) refers to a material with at least one dimension between 1 and 100 nm in size, as well as any material with enhanced characteristics or phenomena, even if these dimensions are outside the nanoscale range (up to 1000 nm). There are different ways that various nations and their regulatory agencies define "NMs" in terms of size, shape, and related efficacy.

Due to the high surface area-to-volume ratio of NMs, they exhibit special features. Innovative nano-interventions in the food and agriculture industries may lead to the development of low-cost, highly effective goods and procedures. On the other hand, a nano-agri-input product (NAIP) is an agricultural input composition containing NMs that is designed for applications on crops for farming purposes using soil, seed, foliar, and drip treatments, as well as other methods. Materials with any of the three dimensions—zero, one, or two—or with an internal or surface structure on the nanoscale make up NAIPs. Whereas, an agricultural preparation including NMs intended for consumption or use in food or feed and their supplements, as well as nutraceutical delivery, is referred to as a nano-agri product (NAP). NAPs incorporate nanotechnology to enhance their performance and efficiency (DBT India, 2020, https://dbtindia.gov.in/).

However, it is important to consider the potential risks and ethical implications associated with the use of nanotechnology in agriculture. Safety assessments, regulations, and responsible use of NAIPs and NAPs should be prioritized to ensure environmental and human health protection. Adopting sustainable nanotechnology principles can help ease the burden of unsafe NAIPs or NAPs and their commercialization. It is an excellent opportunity to make safe NAPs at this time because many NPs are still in the early stages of development. It is essential to have a comprehensive regulatory framework and a strong nanotech policy that can aid start-ups in this industry in order to accurately assess the sustainability of NPs and their manufacturing and commercialization (Cinelli et al., 2016; Subramanian et al., 2015; Malsch et al., 2015).

1.2 GLOBAL MARKET TRENDS

The size of the global agricultural nanotechnology market was estimated at 321.13 billion in 2022 and is expected to grow at a CAGR of 11.94% from 2023 to 2031 to reach 868.98 billion (Insight Ace Analytic, 2023). Agriculture is currently faced with a number of issues and challenges, such as population growth, environmental deterioration, and a limited supply of plant nutrients such as phosphorus and potassium. The fusion of enhanced agricultural practices with nanotechnologies over the following 5 years will also accelerate the growth of the worldwide market for agricultural nanotechnology. Agricultural nanotechnology's advantages also lessen concerns such as eutrophication, soil acidification, groundwater pollution, decreased soil fertility, and biodiversity loss. Over the next 5 years, the increasing worldwide population will contribute to the growth of the agricultural nanotechnology business. The exorbitant cost of agricultural machinery developed with cutting-edge nanotechnology severely restricts the market's expansion. The rate of market expansion would be further slowed by changes in raw material prices. The strong competition for customers among vendors would once again stifle the market's expansion. Market segments for agricultural nanotechnology include type, application, and end users. The market is divided into categories based on type, including NMs, nano-biosensors, nanopesticides, nanofertilizers, and others. The market is divided into six application groups: soil enhancement, plant breeding, crop production and protection, water purification, recycling farm waste, and diagnostics. Farmers, R&D facilities, governmental agencies, and other entities are some of the market's end consumers (Insight Ace Analytic, 2023).

In 2021, the market for agricultural nanotechnology was dominated by the category of nanofertilizers. Due to the rapidly growing global population and the necessity to feed everyone, the global market is expanding. Because of their outstanding capacity to boost nutrient absorption efficiency, nanofertilizers provide advantages in nutrition management. Applied nutrients are bonded to nano-dimensional adsorbents as opposed to conventional fertilizers, allowing for a far more gradual release of nutrients. This strategy improves fertilizer effectiveness while decreasing fertilizer loss into groundwater. Nano-nutrients can also increase a plant's resistance to abiotic stress. There are further advantages to these so-called nano-biofertilizers, which are nanofertilizers blended with bacteria. The development of novel NMs as a result of ongoing research into new techniques has also helped to support the development of the sub-segment and drive the expansion of the global market, particularly in nations like the United States, Germany, the United Kingdom, China, and India (Insight Ace Analytic, 2023).

TABLE 1.1 An Overview of the Global Market Scope of Agricultural Nanoproducts.

Market Attribute	Specifications
Market size value in 2022	USD 321.13 billion
Revenue forecast in 2031	USD 868.98 billion
Compound annual growth rate (CAGR)	11.94% from 2023 to 2031
Prominent countries	United States, Germany, United Kingdom, China, India, Canada, Japan, South Korea, Brazil, Mexico, France, and South East Asia
Prominent manufacturers	Nanosys Inc., Nanoco Group Plc, NanoScale Corporation, Indian Farmers Fertiliser Cooperative Limited (IFFCO), Agrilife, Nanoshell LLC, Xinglu Chemical, Hyperion Catalysis International, Chemat Technology Inc., DuPont de Nemours, Inc, Cabot Microelectronics Corp., Applied Materials, Inc., Lazuriton Nano Biotechnology Co., Ltd.

Source: Insight Ace Analytic (2023), https://www.insightaceanalytic.com/.

1.3 TYPES OF NANO-AGRI PRODUCTS (NAPS)

NAPs refer to agricultural products or technologies that incorporate nanotechnology for various purposes, such as enhancing crop productivity, improving nutrient delivery systems, pest control, and water management. Some examples of NAPs and their potential applicationsare listed in the following sections.

1.3.1 NANOFERTILIZERS

Nanofertilizers are designed to deliver nutrients more efficiently to plants, reducing nutrient waste and increasing nutrient uptake. Nanofertilizers can potentially enhance crop yields while reducing the amount of fertilizer required. In order to improve their characteristics and composition, nanofertilizers are modified fertilizers created by physical, chemical, or biological means (Meena et al., 2017). The use of chemical fertilizers is increasingly being cited as one of the main causes of air and water pollution. Examples of key macronutrients that are given to crops in a variety of chemical forms are nitrogen (N) and phosphorus (P). The majority of the provided nutrients are lost rather than being completely absorbed by plants (Mujtaba et al., 2021). Nanofertilizers have the ability to lessen the loss of nutrients through runoff and leaching, in addition to focusing on utilizing smaller amounts. Nanofertilizers are more resistant to abrupt changes in chemical nature, which improves the effectiveness of nutrient utilization and makes them more environmentally benign (Mujtaba et al., 2021; Suttiponparnit et al., 2011). Moreover, in comparison to conventional fertilizers, nanofertilizers have a number of advantages that improve farming's quality standards (López-Valdez et al., 2018). The production of large-scale NMs with precise physical and chemical properties at a reasonable cost requires a compositional strategy for nanoparticles intended for usage as nanofertilizers (Ahmad et al., 2020). Different chemical synthesis techniques are used to create NMs derived from mineral nutrients. "Wet" techniques include hydrothermal, homogeneous precipitation, sol-gel, and reversed micelle methods, while "dry" techniques include aerosol-based processes. Current chemical synthesis methods could be used to quickly produce nanofertilizers in huge quantities by going directly into massive reactors (Mujtaba et al. 2021; Dimkpa et al., 2017; Tarafdar et al., 2014, 2012).

1.3.2 NANOPESTICIDES

Nanotechnology can be used to develop targeted delivery systems for pesticides, allowing for precise and controlled release. This approach may minimize the environmental impact of conventional pesticides by reducing the amount of chemicals required and minimizing off-target effects. A pesticide formulation that is smaller in size than an active composition or is a designed structure with potent pesticidal properties is referred to as a nanopesticide (Bergeson, 2013). In other words, nanopesticides are a type of pesticide that utilizes nanotechnology to enhance their effectiveness in controlling pests and managing agricultural or environmental challenges. Nanotechnology involves the manipulation and utilization of materials at the nanoscale, which is the scale of atoms and molecules (Pandey et al., 2021). Nanopesticides offer several potential benefits over traditional pesticides. They can provide more targeted and efficient pest control, as the nanoscale materials can be designed to specifically interact with the pests or pathogens while minimizing their impact on nontarget organisms. The increased surface area of nanoparticles also allows for better adhesion to plants or target surfaces, improving their efficacy. Furthermore, nanopesticides can potentially reduce the amount of chemicals required for pest control, leading to lower environmental contamination and reduced health risks for humans and other organisms. They can also enhance the stability and controlled release of active ingredients, improving their longevity and reducing the need for frequent reapplication (Jogaiah et al., 2020; Sreenivasa et al., 2020; Joshi et al., 2019; Kah, 2015; Ali et al., 2017).

Research and development (R&D) in the field of nanopesticides are ongoing, focusing on various aspects such as nanoparticle synthesis, formulation techniques, delivery mechanisms, and understanding their environmental impact. However, it's important to note that the long-term effects and potential risks associated with the use of nanopesticides are still being studied and evaluated. Regulatory authorities in different countries have varying approaches to the approval and regulation of nanopesticides. They typically assess their safety, efficacy, and potential environmental impact before granting approvals for their commercial use (DBT India, 2020, https://dbtindia.gov.in/).

1.3.3 NANO-ENCAPSULATION OF BIOACTIVE COMPOUNDS

Encapsulation is the process of enclosing a bioactive molecule in its liquid, solid, or gaseous state inside an inert matrix or substance, typically a polymer. The coated substance (food or flavor molecules or ingredients) can be preserved by an enclosed wall, enhancing its stability in the appropriate environment (Malekhosseini et al., 2019; Gómez et al., 2018). Nanostructures can be used to encapsulate bioactive compounds, such as vitamins, antioxidants, and growth regulators. This technique improves the stability, controlled release, and bioavailability of these compounds in agricultural applications. Nano-encapsulation specifically produces particles of less than 100 nm, and the encapsulated particles can be divided into three categories: macrocapsules (>5000 μm), microcapsules (0.2–5000 μm), and nanocapsules (<0.2 μm), and the homogeneity that nano-encapsulation gives, leading to improved encapsulation efficiency and acceptable physical and chemical properties, which is one of the key advantages of using it (Human et al., 2019; Jafari, 2017; Nikmaram et al., 2017).

Encapsulation has drawn interest because it allows for the regulated release of the majority of biologically active substances and because the biomaterials employed as coatings are environmentally acceptable (Li et al., 2020). Encapsulation provides particles with high lipophilicity and hydrophilicity, which improves their capacity to enter plant tissues. This procedure produces capsules with improved biological properties by coating or packaging a bioactive substance or active agent, which is referred to as the core material, in a carrier (protective material) (Mandaji et al., 2022). Phytohormones such as indole-3-acetic acid (IAA), gibberellins (GAs), brassinosteroids (BRs), salicylic acid (SA), jasmonic acid (JA), abscisic acid (ABA), ethylene (ET), cytokinins (CKs), and strigolactones (SL) can be nano-encapsulated to provide controlled distribution of the active ingredients. New ecologically sound and financially viable methods are constantly being developed, and new biomaterials with a great affinity for transporting and encapsulating bioactive chemicals are also being developed, which all contribute to ongoing changes in encapsulation. The tolerance of the plant to various stimuli is modulated by phytohormones (Sampedro-Guerrero et al., 2023). Due to their capacity to foster plant growth and mitigate pathogen effects, encapsulated phytohormones have become more often used in plant treatments in recent years (Godoy et al., 2021). In the encapsulation process, a variety of materials are employed, including waxes, lipids, proteins that are synthetic or inorganic, and polysaccharides from various sources, including plants, marine algae, and fungi. For the encapsulation of plant extracts (flavonoids, fatty acids, and primary phytohormones), recent agricultural developments seek to use inert or biodegradable matrices (Szopa et al., 2022).

1.3.4 NANOSENSORS

The science of nanotechnology has advanced significantly over the past few decades, and this advancement has greatly aided in the creation of inexpensive, efficient, and economically viable bio- and nanosensors for identifying various contaminants in natural agroecosystems that have the advantage of being safe for human health. The swift growth of nanosensors for the detection of various composites, such as multiple metal ions, proteins, insecticides, and entire microbes, has been facilitated by the expansion of nanotechnology (Sharma et al., 2021). The potential of nanosensors in agriculture has also been established. Nanotechnology-enabled sensors can monitor various parameters, including soil moisture, nutrient levels, and pest infestations. These sensors provide real-time data, enabling farmers to make informed decisions about irrigation, fertilization, and pest management (Chhipa, 2019; Chen and Yada, 2011). The creation of such nanosensors has played a vital role in the growth of agriculture and will likely continue to do so. The overuse of pesticides and fertilizers has been curbed thanks to real-time monitoring, which helps lower product costs and environmental damage. Nanosensor applications transform traditional agricultural methods into smart agriculture, which uses more environmentally friendly and energy-efficient methods for sustainable agricultural practices. In order to increase nutrient dispersion and wettability, smart agriculture practices include (a) nano-formulation-based fertilizers or pesticide delivery systems; (b) nano-detectors for pesticide or fertilizer residues; and (c) remote-sensing-based crop growth and disease-monitoring systems. The high sensitivity and low detection limit of nanosensors make them more advantageous for intelligent agriculture (Fraceto et al., 2016; Mohanraj et al., 2022).

Nanosensors are elemental, tiny devices designed to detect a specific molecule, environmental condition, or biological component. In comparison to their macroscale predecessors, these sensors detect at a level that is much lower, are portable, and are cost-effective. There are three essential parts to a typical nanosensor device operation: (i) sample preparation, (ii) recognition, and (iii) signal transduction.

1. **Sample preparation:** It could consist of a homogeneous or intricate mixture of solid, liquid, and gaseous phases. Agroecosystem sample preparation is highly difficult because of contaminants and interferences. The sample includes specific chemicals, molecular functional groups, or living things that the sensors can detect. The term "analyte" refers to the organisms or molecules that are being studied, such as

ions (metals, surfactants, halogens), dyes, colors, pesticides, volatile compounds, gas or vapor (carbon dioxide, oxygen, water vapors), microorganisms (fungi, bacteria, viruses), and environmental factors (temperature, pH, weather, humidity, etc.).

2. **Recognition:** The analytes present in the sample are recognized by specific molecules or components. These recognition molecules, which include enzymes, aptamers, chemical legends, and antibodies, have high affinities, specificities, and selectivity for their analytes that enable quantification to acceptable levels.

3. **Signal transduction:** These simple devices have been divided into various sorts by specific signal transduction techniques, including optical, electrochemical, pyroelectric, piezoelectric, gravimetric, and electronic biosensors. They turn recognition events into signals that can be analyzed further to produce data.

The development of nanosensors for pesticide detection has made use of a variety of metal NMs, such as gold nanoparticles (Au NP), graphene, carbon nanotubes (CNT), quantum dots (QD), nanorods, nanowires, nanofibers, nanofilms, nanoplates, and different nanocomposites with polymers (Arduini et al., 2020; Chhipa, 2019). The nanosensors outperformed conventional sensors in many ways, including a rapid response time of only a few seconds, high sensitivity, more consistent and reliable results, the ability to detect small amounts of substance (up to nanograms or less), the ability to be applied in various matrixes, and support for quick electron transfer kinetics (Scognamiglio et al., 2013).

1.3.5 NANOREMEDIATION

NMs can be used to remediate contaminated soil or water by adsorbing or degrading pollutants, such as heavy metals or organic pollutants. This technology has the potential to mitigate soil and water pollution, thus improving agricultural sustainability. Nanoremediation is currently one of the most significant methods for cleaning up various environmental toxins. By utilizing nanoparticles or NMs, nanoremediation purifies contaminated soil in industries including sustainable agriculture production, increased crop yield, decreased nutrient loss, and disease management. Rhizospheric engineering, agro-biotechnological applications, and molecular biological methods based on nanotechnology are a few of the techniques. Additionally, it has uses in cleaning up groundwater, sediment, air, and wastewater to make it possible to use these contaminated areas for agricultural production and close the gap between the world's food needs and supply (Roychoudhury and Bhowmik, 2023).

Organic exudates, bacterial or viral particles, colors, or heavy metals such as chromium (Cr), lead (Pb), Zinc (Zn), arsenic (As), cadmium (Cd), nickel (Ni), or mercury (Hg) that are nonbiodegradable in nature and pose a major risk to human health are some examples of contaminants in the environment. An environmentally benign and frequently economical way of removing heavy metal contamination from places is nanoremediation. This frequently entails the inclusion of reactive substances, which cleanse the contaminated area by dissolving the contaminants or immobilizing them. Nanoparticles efficiently clean up better than larger-sized particles because of their special characteristics of strong reactivity, small size, and huge surface area (Karn et al., 2009; Roychoudhury and Bhowmik, 2023). Effective treatment of waste water and contaminated groundwater may be achieved by using NMs such as nanomembranes, nanomotors, and nanosorbents. In order to effectively decontaminate heavy metals, nanophotocatalysis uses visible light/UV light/ sunshine to activate or stimulate NMs (Jhaveri and Murthy, 2016). Carbon nanotube and graphene oxide membranes are examples of carbon polymer-based membranes that can be loaded with NMs such as titanium oxide, alumina, or zeolites (Waduge et al., 2015; Del Prado-Audelo et al., 2021).

Crop output must be adequate to feed the world's population in light of a growing population. An adequate amount of food crops must also be produced in addition to crops for horticulture, biomass for biofuels, and other uses. Utilizing the polluted regions, which make up a startling one-third of arable land, is one method to boost output. The need for more space to produce enough work in the modern world might be alleviated by remediating this sizable region. The agro-biotechnological industry would be greatly impacted by innovations like nanoparticles that can remove heavy metal pollution, ensuring food safety, food quality, and sustainability (Roychoudhury and Bhowmik, 2023; Del Prado-Audelo et al., 2021; El-Ramady et al., 2017).

1.4 REGULATORY FRAMEWORK FOR COMMERCIALIZATION OF NANO-AGRI PRODUCTS (NAPS) GLOBALLY

The practice of agriculture is used to produce food through the breeding of livestock and the growing of a variety of crops. It is regarded as the foundation of the economy in the majority of emerging nations and is essential to progress and development. Scientists and engineers are currently experimenting with novel techniques to boost agricultural production as a result of the increasing need for food due to the world's growing population (Baruah and Dutta, 2009). Agriculture nanotechnology has recently concentrated on research and application to address difficulties with sustainability, crop enhancement, and increased production in agriculture and the environment. Regarding the reduction of hunger, undernourishment, and child mortality rates, agricultural nanotechnology appears to be very intriguing for poor countries (Neme et al., 2021; Googs et al., 2012). According to a higher number of publications and patents, developed and developing nations including Germany, the United States of America (USA), Brazil, China, India, France, and Korea have a greater interest in employing NMs for agricultural purposes (Neme et al., 2021; Adholeya et al., 2017; Googs et al., 2012).

However, there are concerns regarding the potential environmental and health risks associated with the use of nanotechnology in agricultural goods. In response to these worries, regulatory organizations from all over the world have created standards and rules that promise the secure application of nanotechnology in agricultural products. Here are some of the most important laws and rules pertaining to agricultural products using nanotechnology (Kumari et al., 2023; Mishra and Singh, 2021).

1.4.1 UNITED STATES (US)

Several governmental agencies, including the Food and Drug Administration (FDA), the Environmental Protection Agency (EPA), and the US Department of Agriculture (USDA), regulate agricultural products based on nanotechnology in the US. Under the Federal Food, Drug, and Cosmetic Act (FFDCA) (US-FDA), the Food and Drug Administration (FDA) is in charge of the regulatory aspects of a product intended for agricultural application. The FFDCA does not, however, include any standards for goods based on nanotechnology. NMs do not yet have a regulatory definition from the FDA. The FDA has made explicit in various guidelines (FDA-regulated product involves the application of nanotechnology) that "a case-by-case approach" is required in determining the finished NM product's safety. On the other hand, the Federal Insecticide, Fungicide, and Rodenticide Act (FIFRA) gives the US Environmental Protection Agency (US EPA) the power to regulate pesticides. However, at this time, this regulation does not offer any explicit protections for NMs. The US EPA has published a notice asking for feedback from the public on the regulation and inclusion of nano-enabled pesticides under FIFRA. It is highly advised for businesses planning to register nanopesticides to get in touch with the "US-EPA's Pesticide Registration Ombudsmen." Recently, a pesticide product with nano-silver as a new active ingredient received conditional FIFRA registration (Adholeya et al., 2017; Kumari et al., 2023). In general, these organizations share the objective of ensuring the efficacy and safety of nano-agri products, as well as their conformity with relevant laws. Moreover, the USFDA has cleared certain NMs for use in food-related applications. The following are many noted examples:

1. **Zinc oxide (ZnO):** Nanoscale zinc oxide has received authorization for use as a food coloring and dietary supplement (USFDA, 2020).
2. **Silica:** Some culinary products, such as coffee creamer that is powdered, use nanoscale silica as an anticaking agent (USFDA, 2020).
3. **Iron oxide (Fe_2O_3):** The use of nanoscale iron oxide as a food coloring agent has been approved (USFDA, 2020).
4. **Titanium oxide (TiO_2):** Titanium dioxide is a typical food ingredient that is used to brighten and whiten a variety of foods, including candy, gum, and powdered sugar. Titanium dioxide in nanoscale form has been permitted for use in food products (Powell et al., 2016).

1.4.2 EUROPEAN UNION (EU)

There are several legal frameworks in the European Union (EU) that deal with nano-enabled items used in the food, agriculture, and feed industries. One of them is the Regulation on the Provision of Food Information to Consumers (1169/20119). Others include the Regulation on Plastic Food Contact Materials and Articles (10/2011), the Regulation on Active and Intelligent Materials and Articles (450/2009), the Regulation on Biocidal Products (528/2012), the Regulation on Cosmetic Products (1223/2009), and the Registration, Evaluation, Authorization, and Restriction of Chemicals (REACH) Regulation (1907/2006). Despite the fact that the current plant protection products (PPP) Regulation (EC) (1107/2009) applies to nano-enabled pesticides, the EU views them as a distinct class of pesticides and, as such, requires a separate risk assessment and regulatory authority permission (Adholeya et al., 2017; Kumari et al., 2023).

1. **Regulation (EU) No 2019/1009:** The guidelines for putting fertilizers with the CE mark on the market are laid out in this regulation. Additionally, it mandates that all fertilizers be safe for both the environment and human health. All fertilizsers that incorporate nanoparticles are subject to the legislation.
2. **Regulation (EC) No 1107/2009:** In accordance with this law, PPPs may only be sold within the European Union (EU). Before being sold or used, all PPPs must first be licensed. Aside from that, the law outlines the information needed for approving PPPs, including those that use NMs.
3. **Regulation (EC) No 396/2005:** This rule defines the maximum residual levels (MRLs) for pesticides in or on food and feed derived from plants and animals. It also holds true for pesticides made with nanoparticles.
4. **Regulation (EC) No 1935/2004:** This law establishes the general safety criteria for goods and materials that come into contact with food. It applies to all types of NMs, including those used in agricultural products, that come into contact with food.
5. **Regulation (EC) No 767/2009:** The EU's licensing and marketing policies for feed additives are established by this legislation. All feed additives must also be safe for both the environment and animals. Any feed additive that contains NMs is subject to the rule.

Moreover, various European agencies and organizations, including the European Chemicals Agency (ECHA) and the European Food Safety Authority (EFSA), have also issued guidelines and recommendations on the safety evaluation of NMs used in agri products in addition to these rules.

1.4.3 CANADA

Agri products based on nanotechnology are subject to a number of laws and regulations to guarantee both consumer and environmental safety. In Canada, the Public Health Agency of Canada (PHAC) and the Canadian Food Inspection Agency (CFIA) regulate aspects of NPs use in agriculture, food, and feed sectors jointly. Moreover, several other important laws, together with the dates and references that match, are as follows:

1. **Canadian Environmental Protection Act, 1999 (CEPA):** The main federal law governing how nanotechnology-based products, especially agricultural products, affect the environment and people is this one. Under the New Substances Notification Regulations, the CEPA provides the framework for the evaluation and management of NMs (CEPA, 1999; Chemicals and Polymers, 2015).
2. **Pest Control Products Act (PCPA):** This act is the main piece of legislation in Canada that governs pest control products, including those that make use of nanotechnology. The PCPA specifies the criteria for pesticide product safety and efficacy, as well as the registration and labeling regulations (PCPA, 2002).
3. **Canada Agricultural Products Act (CAPA):** This law governs how Canadian agricultural products are marketed and inspected. This law places inspection and quality control requirements on agri products that make use of nanotechnology in order to guarantee consumer safety (CAPA, 1985)
4. **Food and Drugs Act (FDA):** This is the federal law governing consumer health and food safety in Canada. The legal framework for guaranteeing the security, quality, and effectiveness of food items, including those that make use of nanotechnology, is provided by the FDA. The FDA also specifies the labeling rules for foods that contain NMs (FDA, 2018).

1.4.4 UNITED KINGDOM (UK)

Several governmental organizations in the United Kingdom, notably the Department for Environment, Food, and Rural Affairs (DEFRA), the Food Standards Agency (FSA), and the Health and Safety Executive (HSE), are in charge of regulating agri products based on nanotechnology. The Nanotechnology Safety Guidance, created by the Health and Safety Executive (HSE) in 2011, is one of the key rules controlling the safety of nanotechnology-based agricultural products in the United Kingdom. This advice offers details on how NMs should be handled and applied safely in many industrial contexts, including agriculture.

The Food Standards Agency (FSA) is in charge of making sure that all food items, including those made using nanotechnology, are safe and of high quality. A report on the safety of nanoparticles in food was released by the FSA in 2014, and it suggested that any use of nanotechnology in food products be subjected to risk analysis and evaluation. The Department for Environment, Food, and Rural Affairs (DEFRA) has also released guidelines on how to handle and dispose of NMs in agricultural settings, as well as on their application in agriculture. In order to reflect the most recent scientific understanding of the potential dangers related to NMs, this advice was updated in 2018. Overall, the regulation of nanotechnology-based agri products in the United Kingdom is a sector that is quickly expanding, with new laws and advice being released on a regular basis to reflect the most recent advances in scientific knowledge of the potential risks and advantages of nanotechnology (Kumari et al., 2023).

1.4.5 CHINA

National standards in the field of nanotechnology are established by the National Center for Nanoscience and Technology (NCNST) and the Commission on Nanotechnology Standardization. However, Chinese regulatory authorities have not yet approved nano-enabled items designed for use in food and agricultural applications. Moreover, several government organizations, including the Ministry of Agriculture and Rural Affairs (MARA), the State Administration for Market Regulation (SAMR), the National Health Commission (NHC), and the Ministry of Ecology and Environment (MEE), are in charge of regulating agricultural products based on nanotechnology in China. The important laws, safety requirements, and rules pertaining to nanotechnology-based agricultural goods in China are listed here, along with their references and the years they were enacted (Kumari et al., 2023, http://en.nim.ac.cn/).

1. The 2001 MARA Order No. 7 regulations on the safety assessment of agricultural genetically modified organisms: The guidelines for approving genetically modified agricultural goods, including those that make use of nanotechnology, are outlined in this regulation.
2. Nanomaterial Safety Requirements for Food and Food Additives (NHC No. 13)—2011: In accordance with this recommendation, food and food additives using NMs, including those used in agriculture, will need to meet certain safety standards and go through evaluation processes.

3. Technical Guidelines for Safety Assessment of Nanoscale Agricultural Products, 2014 (MARA No. 198): This standard offers a framework for evaluating the safety of agricultural products made with nanotechnology, including how they are made, processed, and used.
4. Regulatory Measures for Safety Assessment of New Varieties of Agricultural Genetically Modified Organisms (SAMR Order No. 8)—2020: This rule defines the processes for evaluating the safety of novel genetically modified agri products, including those that make use of nanotechnology.

1.4.6 INDIA

The Department of Biotechnology (DBT), the Ministry of Environment, Forest and Climate Change (MoEFCC), the Food Safety and Standards Authority of India (FSSAI), and the Indian Council of Agricultural Research (ICAR) are among the agencies and laws in India that have jurisdiction over the regulation of nanotechnology-based agriproducts. In 2020, the DBT issued detailed guidelines for the evaluation of nano-based agri-inputs and food products in India for commercialization. Both NAIPs and NAPs are covered under these regulations. These rules also apply to nanocomposites, NM-based sensors, and sensors that need to come into contact with crops, food, or feed in order to collect data. These recommendations do not apply to traditional products or formulations that contain natural NMs incidentally. The current regulations were created to support the already-existing national regulatory laws of the Central Insecticide Board and Registration Committee (CIB&RC), Fertilizer Control Order (FCO, 1985), FSSAI, and Bureau of Indian Standards (BIS), with particular needs and adaptations for NAIPs and NAPs when deemed relevant. The applicable provisions for NAIPs and NAPs under the European law to protect human health and the environment, that is, regulation on the registration, evaluation, authorization, and restriction of chemicals (REACH), the Organization for Economic Co-operation and Development (OECD), the United States Environmental Protection Agency (USEPA), the US law Toxic Substances Control Act (TSCA) of 1976, Australian Pesticides and Veterinary Medicines Authority (APVMA), FAO/ WHO, the United States Food and Drug Administration (USFDA), the EFSA, the Food Standards Australia New Zealand (FSANZ), and Codex, as well as the International Council for Harmonisation of Technical Requirements for Pharmaceuticals for Human Use (ICH) principles, are likewise aligned with these guidelines (DBT India, 2020, https://dbtindia.gov.in/).

Major nanofertilizers and nano-micronutrients must be registered under the current FCO, 1985, and nanopesticides must be registered under the current Insecticides Act, 1968, and Rules, through the nodal agency of the Government of India and the Agriculture Department of State Governments. The requirements for these regulatory frameworks' datasets for pesticide and fertilizer registration have already been established. As per the guidelines, there are the following requirements for registration and licensing of NAIPs and NAPs:

1. **Nanofertilizers (with or without carriers [nano or non-nano]):** For registration of any nanofertilizers, safety, efficacy, functionality, and other quality data should be collected in accordance with FCO, 1985, with additional requirements for the inclusion of nanofertilizers. The Essential Commodities Act of 1955, which establishes registration procedures for fertilizers, governs the issuance of FCO.
2. **Nanopesticides (with or without carriers [nano or non-nano]):** For registration (for manufacture or import), studies on the chemistry, bio-efficacy and residues, toxicity, packaging, and processing of nanopesticide products should be carried out in accordance with the regulatory aspect provisions under Section 9 of the Insecticides Act, 1968 (Act 46 of 1968), with additional criteria for inclusion of nanopesticides as per the requirement of CIB&RC of the Directorate of Plant Protection, Quarantine, and Storage (DPPQS).
3. **Nanofood:** The Food Safety and Standards Act, 2006, with new criteria for the inclusion of nanofood, and the FDA guidelines implemented by the FSSAI are applied for registration of any nanofood.
4. **Nanofeed:** The Cattle Feed (Regulation of Manufacture and Sale) Order, 2009, should be followed, along with any new criteria for the inclusion of nanofeed that may be implemented by FSSAI. This includes safety, assessment, and other quality studies of nanofeed.

Moreover, standards should be implemented in accordance with the BIS, with extra requirements for NAIPs and NAPs inclusion. The BIS is a national Standards Body that falls under the purview of the Ministry of Consumer Affairs, Food, and Public Distribution of the Government of India. It is responsible for developing and enforcing standards, running certification programs for both systems and products, organizing and managing testing labs, promoting consumer awareness, and keeping close contact with international standards.

1.4.7 TAIWAN

In 2003, the Ministry of Economic Affairs' Industry Development Bureau developed the nanoMark. It is the world's first mechanism for certifying NPs. The goal is to safeguard consumer rights and interests, support outstanding businesses' sustainable growth, and increase the nanotechnology sector's global competitiveness. After 13 years of operation, the nanoMark has created a favorable environment for the growth of the nanotechnology sector by coordinating the efforts of the public sector, private sector, and general public. The national nanotechnology science and technology program, which was in charge of research development and technical transfer to industry, was responsible for the establishment and promotion of the nanoMark. After 12 years of implementation, the program has successfully completed various industrialization missions, with nanoMark being one of the noteworthy successes. The Industry Development Bureau developed the industrial development transformation plan to industrialize nanotechnology applications in order to meet the rapid advancement of international standards and establish a sustainable development of NPs. In order to continue the operation for a new milestone, it also passed the nanoMark verification system to the Taiwan Nanotechnology Industry Development Association (TANIDA, 2024, http://www.tanida.org.tw/nanomark_e.php?nm=markIntroduction_e).

FIGURE 1.1 nanoMark: A nanoproduct certification system in Taiwan.

Source: TANIDA, 2024, http://www.tanida.org.tw/nanomark_e.php?nm=markIntroduction_e

1.4.8 OTHER COUNTRIES

The Iran Nanotechnology Initiative Council (INIC) oversees the establishment of policies for NP R&D. Guidelines for nano-enabled items (pharmaceuticals, medical equipment, cosmetics, food, beverages, and pharmaceutical supplements) have recently been established by the Food and Drug Organisation (FDO). However, FDO did not include the agricultural sector. In Russia, the Russian Corporation of Nanotechnologies is in charge of Russian legislation and policy pertaining to industries based on nanotechnology. This regulatory body has started several federal programs that essentially incorporate the safe reference guidelines established by the FAO/WHO (Kumari et al., 2023; Adholeya et al., 2017). The regulatory frameworks for NPs in other countries such as Japan, South Korea, Switzerland, Australia, and New Zealand are briefly discussed in Table 1.2.

1.5 COMMERCIAL NANO-AGRI-INPUT PRODUCTS (NAIPS) AVAILABILITY IN THE MARKET AND LIMITATIONS

Money spent on R&D in the field of nanotechnology reflects the fact that this discipline has advanced significantly. The National Nanotechnology Initiative (NNI) of the United States has proposed a budget of $27 billion for the year 2019; Saudi Aramco Energy Ventures (SAEV) has given $10 million to Nano-Mech, a leading manufacturer of NPs; and the Massachusetts Institute of Technology (MIT) has received $350 million to build a cutting-edge nanoscale center dubbed "MIT nano" (NNI-Budget, 2021, https://www.nano.gov/about-nni/what/funding). Comprehensive changes to traditional agricultural practices are urgently needed to achieve the Zero Hunger goal, which is the United Nations' target for sustainable development to be attained by 2030. The use of Nanotechnological inventions can bring about these developments (Rajput et al., 2021a; Khan et al., 2021). Researchers have developed a variety of tactics over the past few decades to replace traditional agricultural practices. One of them, nanotechnology (NT), is a developing technique that is anticipated to enhance modern agriculture in sustainable ways, including crop productivity through mitigating effects by improving plant nutrition, precision farming, water use efficiency, crop protection against pests and diseases by transforming nano-enabled formulations, and restoration of degraded sites through nano-bioremediation (Rajput et al., 2021b; Chauhan et al., 2020). It has been reported that nano-enabled items, such as smart and controlled delivery materials, nanofertilizers, growth stimulators, soil-improving agents, and nano-sensors, are already available on the market and used in sustainable agriculture (Rajput et al., 2021a, 2021c; Raliya et al., 2018).

TABLE 1.2 Status of Nanoproduct Regulation in Agri-Food Systems Globally.

Country	Responsible Organization/ Regulatory Body	Concerned Legislature/ Laws and Documents	Existing Provisions and Remarks
United States (US)	US-Environmental Protection Agency (US EPA), and Food and Drug Administration (FDA)	Federal Insecticide, Fungicide and Rodenticide Act (FIFRA), Federal Food, Drug and Cosmetic Act (FFDCA)	The FFDCA of FIFRA contains no specifications for nanoproducts
European Union (EU)	European Parliament and Council, European Commission (EC)	Regulation (EU) No. 1169/2011(8) and Regulation (EU) No. 2015/2283	Specifies that materials that, among other things, have particle sizes in the specified nanoscale (1–100 nm) meet the criteria for designed nanomaterials (NMs) in Novel Foods on the Provision of Food Information to Consumers, gives instructions on how to conduct risk analyses of novel foods, FCMs, food/feed additives, and pesticides, among other NMs in the food and feed sector
		Regulation (EC) No. 1332/2008 on food enzyme	Says that a food enzyme that is already on the community list but was produced using different techniques or starting materials that are significantly different from those considered in the authority's risk assessment should be submitted for re-evaluation (it is specified that "significantly different" could mean a change in particle size)
		Regulation (EC) No. 1333/2008	Explains that before being marketed, a food additive that has already been approved but was obtained via nanotechnology must be reevaluated
		Regulation (EC) No. 450/2009a	While NMs are not specifically mentioned, there is a reference to "substances intentionally engineered to particle size that exhibit functional physical and chemical properties that significantly differ from those at a larger scale." As a result, a case-by-case analysis must be used for active and intelligent materials and items containing NMs

TABLE 1.2 *(Continued)*

Country	Responsible Organization/ Regulatory Body	Concerned Legislature/ Laws and Documents	Existing Provisions and Remarks
		Directive 2002/46/ EC	Dietary supplements: the use of food supplements (minerals or vitamins) that are listed by the EC is allowed. Due to the differences in manufacture and probable variations in nutritional content and bioavailability compared to their macro-scale equivalents, the use of nanoforms of minerals and vitamins necessitates a safety review prior to marketing, which will be carried out under the Novel Food Regulation
		Food contact material regulation 1935/2004	Supervises the passage of nanocomponents into food from food contact materials and articles. This regulation mandates that these NMs may not transfer their constituents into food under usual and foreseeable usage conditions in quantities that could endanger human health or bring about an unacceptable change in the composition of food
		(EU) No 528/2012	NMs-based biocidal products are currently ineligible for a streamlined authorization process. The test methods used on the NMs must be complemented and standardized by an explanation addressing their precise suitability, taking into account each NM's unique characteristics for subsequent NM-based product authorization and approving NMs as active substances
		Regulation (EU) No. 10/2011a	Explains that compounds in nanoform should only be used if they are specified in this regulation
		EFSA's Novel Food Regulation 258/97	Pre-market permission for products, components (other than food additives), and procedures that are not used in the EU, It places a focus on the nutritional value, composition, metabolism, degree of microbiological and chemical component research on toxicity and allergenicity, and intended usage (it does not take particle size into account).

TABLE 1.2 *(Continued)*

Country	Responsible Organization/ Regulatory Body	Concerned Legislature/ Laws and Documents	Existing Provisions and Remarks
		Registration, Evaluation, Authorization and Restriction of Chemicals (REACH) regulation (1907/2006)	There are nano-specific provisions in the law, but no separate NM legislation
Canada	Public Health Agency of Canada (PHAC), and Canadian Food Inspection Agency (CFIA)	Food and Drugs Act (7)	Nanoproducts are subject to current legislative regulations
Australia and New Zealand	Food Standards Australia New Zealand (FSANZ)	Australian New Zealand Food Standards Code	New food standards to manage risks from nanoproducts are supported by an updated FSANZ application handbook
Switzerland	Swiss Federal office of public health (FOPH), and Federal Office for the Environment (FOEN)	--	Existing laws assure the safety of NMs
South Korea	Korean Agency for Technology and Science (KATS),	(2012/2016) and Food Sanitation Act	No NM specification
	Korean Food and Drug Administration (KFDA)		
	Ministry of Food and Drug Safety (MFDS)	National Nano-Safety Strategies Plan	No NM specification
China	Ministry of Health, Ministry of Agriculture	Food Safety Law of China	The commission on nanotechnology standardization and the National Center for Nanoscience and Technology (NCNST). The authority has not yet approved the use of nanoproducts in agri-food systems
Japan	Ministry of Health, Labour and Welfare	Food Sanitation Law	No NM specification

TABLE 1.2 *(Continued)*

Country	Responsible Organization/ Regulatory Body	Concerned Legislature/ Laws and Documents	Existing Provisions and Remarks
Taiwan	Ministry of Economic Affairs' Industry Development Bureau, Taiwan Nanotechnology Industry Development Association (TANIDA)	nanoMark verification system	Nanoproducts are approved
Iran	Nanotechnology Committee of Food and Drug Organization (FDO), Iran Nanotechnology Initiative Council (INIC)	--	FDO recommendations regarding the use of nanoproducts in the food, drink, pharmaceutical, and medical device sectors, However, agriculture is still not included
India	Department of Biotechnology, Ministry of Science & Technology, Ministry of Agriculture and Farmers' Welfare	Guidelines for the evaluation of Nano-Based Agri-Inputs (NAIPs) and Food Products (2020), Central Insecticide Board and Registration Committee (CIB&RC), Fertiliser Control Order (FCO, 1985), Food Safety and Standards Authority of India (FSSAI), and Bureau of Indian Standards (BIS)	Both nano-agri-input products (NAIPs) and nano-agri products (NAPs) are covered under these regulations
Thailand	Food & Drug Administration of the Ministry of Public Health	Approved by the Nanotechnology Association of Thailand	The Nanotechnology Association of Thailand developed the NanoQ mark for nanoproducts that have been certified but not for agricultural or food products

Source: Mishra et al. (2020); Subramanian and Rajkishore (2018); DBT India (2020), https://dbtindia.gov.in/.

Although it is still in its early stages, the practise of using NPs in agriculture looks to have a promising future (Ali et al., 2021). It is very possible that the production of modified NPs) for agricultural use will change how people view traditional farming (Yadav et al., 2020). Utilizing cutting-edge materials and technologies aids in resolving enduring issues in agriculture. The majority of these issues typically include the proper use of pesticides, herbicides, fertilizers, etc., NPs are distributed in plants through a unique path that has many benefits (Rajput et al., 2021a; Usman et al., 2020; Schwab et al., 2016). According to the website for the Nanotechnology Products Database (NPD), which is used to maintain current information on items utilizing nanotechnology, there are currently 10,860 commercially available NPs made by 68 different nations and 3674 different firms (NNI-Budget, 2021). These nano-enabled products are categorized into different fields such as agriculture, environment, renewable energy, cosmetics, food, medicine, automotive, sports, home appliances, fitness, and other nano-enabled products. As per the NPD, 244 different types of agricultural NPs in 88 companies, with head offices spread across 28 nations, are responsible for introducing these products into international markets. The products are divided into the sub-industrial sectors of fertilizers, animal husbandry, soil improvement, plant breeding, and plant protection that are shown in Table 1.3 (NNI-Budget, 2021).

TABLE 1.3 An Overview of the Commercial Agricultural Nanoproducts Available in the Global Market.

Agricultural Industrial Sector	Products	Companies	Countries
Fertilizer	107	46	17
Plant protection	46	23	12
Soil improvement	8	3	3
Plant breeding	27	8	7
Animal husbandry	56	25	14

Source: Nanotechnology Products Database (NPD), NNI-Budget (2021).

Moreover, as per the NPD database, manufacturers frequently release nutrients, chelates, and fertilizers as products. Other commercially available product types include precursors, chemicals, and boosters (such as controlled-release nanocomposites) used in the production of fertilizers. Additionally, soil enhancers with nanomaterials have received a lot of interest recently. The principal goods in the plant protection division include pesticides, herbicides, algicides, fungicides, biocides, and plant cleaning agents. In the agriculture sector, there are hundreds of nano-based manufacturers in operation, but only a few of them can be referred to as industry titans. The following companies are some examples: Neufarm GmbH, IFFCO Limited, Kanak Biotech, Plant Vitality Ltd., FRAmelco, Organic Fertilizing, Aqua-Yield Hub, Reed Mariculture Inc., Kimitec Group, Richfield Fertilisers Pvt., Prodotti Arca S.r.l., Danaflex Nano, Blue Planet Environmental, AgriLife, Bioteksa, NanoLandBaltic, DVS BioLife Ltd., and Vive Crop Protection (https://product.statnano.com/). The most frequent NMs reported in the agricultural sector include silver, silicon dioxide, titanium dioxide, and zinc oxide. These materials have special physical, chemical, and mechanical properties that improve plant foliage, increase stress tolerance, lengthen product shelf lives, and improve nutrient uptake from the soil. Manganese, iron, potassium, magnesium, and iron nanocapsules are also crucial in this field. Additionally, porous substances like zeolites can promote immunity and increase crop yields (Campos, 2021; Rajput et al., 2021a, 2021b). The top countries in nanotechnology promotion include India, Germany, the UK, the USA, Vietnam, Taiwan, Brazil, China, Malaysia, and the Netherlands. These nations are the most active in encouraging manufacturers to take advantage of nanotechnology's benefits for the success of agriculture (Campos, 2021). Some of the commercial NAPs are available on the market and are popular among farmers for nutrient management and crop protection (Figure 1.2; Table 1.4). However, very limited NAPs have a presence in the market because of production and strict regulatory limitations that hinder the commercialization of nanoformulations, such as the fact that new technology is still little used, the high cost of formulation development, and unpredicted breakeven points or returns that have not yet been achieved.

FIGURE 1.2 Some of the commercial nano-agri-input products (NAIPs) available in the market.

TABLE 1.4 The Availability of Selected Commercial Nano-Agri-Input Products (NAIPs) on the Global Market for Agriculture Applications.

Commercial Brand Name	Product Category	Manufacturer	Type of Nanomaterial	Applications
HydroBlool	Oxygen-rich nano-bubble technology (NBT)	Blue Plant Environmental (USA)	Plant nutrient	Increase plant root system uptake of nutrients in hydroponic systems to regulate and accelerate plant growth
Nano-Ag Answer®	Fertilizer	Urth Agriculture (USA)	Ag-based nano-powder/ nanoparticles	Balances and promotes the growth of beneficial organisms for soil fertilization
NanoZn™	Fertilizer	Aqua-YeldHuB (USA)	Unspecified nanomaterial	Regulation of plant development by delivering zinc to certain crop tissues
Nano-Gro™	Fertilizer	Agro Nanotechnology Crop (USA)	Zeolite	Boosts yield and growth evenly. Increases resistance to adverse weather and climatic conditions
Root Zone Mass	Seed treatment nutrient	Plant Vitality Ltd. (UK)	Unspecified nanomaterial	Helps awaken latent seeds and encourages faster seed hatching
Nano Silicate+	Plant growth regulator	Plant Vitality Ltd. (UK)	Unspecified nanomaterial	Enhances plant seed germination and aids in reviving dormant seeds. Increases resistance to illness and aridity
Hydro: Grow	Seed treatment nutrient	Plant Vitality Ltd. (UK)	Unspecified nanomaterial	Increases early growth and root development by a large amount
Hydro: Flower/Coco: Flower/Coco: grow/ Soil: Flower/Soil: grow	Soil enhancer	Plant Vitality Ltd (UK)	Unspecified nanomaterial content	A blend of nanoparticles that enhances root activity, seed germination, and plant growth
Nano Zinc Chelate Fertilizer	Fertilizer	AFME Trading Group (UK)	Zinc (Zn) nano-powder/ nanoparticles	Chelated fertilizer that gives plants vital micronutrients (Zn), accelerating or regulating plant growth

TABLE 1.4 *(Continued)*

Commercial Brand Name	Product Category	Manufacturer	Type of Nanomaterial	Applications
Nano Potassium Chelate Fertilizer	Fertilizer	AFME Trading Group (UK)	Potassium (K) nano-powder /nanoparticles	Chelated fertilizer that supplies plants with vital micronutrients (Fe), accelerating or regulating plant growth
Nano Iron Chelate Fertilizer	Fertilizer	AFME Trading Group (UK)	Iron (Fe) nanopowder/ nanoparticles	Chelated fertilizer that gives plants vital micronutrients (Fe) results in the acceleration or restriction of plant growth
Nano Calcium Chelate Fertilizer	Fertilizer	AFME Trading Group (UK)	Calcium (Ca) nanopowder/ nanoparticles	Chelated fertilizer accelerates or controls plant growth by giving plants the necessary micronutrients (Ca)
AZteroid Fungicide	Fertilizer/Fungicide	Vive Crop Protection (Canada)	Allosperse™ (polymer nanomaterials shuttles)	Promotion of seed germination and broad-spectrum disease management for several seed and seedling illnesses
SomGuardsilcop	Plant nutrient	AgriLife (Greece)	Silver (Ag) nanopowder/ nanoparticles	Nontoxic disinfectant that is extremely effective against pathogenic microorganisms
Nano SiO_2 Powders (Water Soluble)	Fertilizer	Hongwu International Group Ltd. (China)	Silicon dioxide (SiO_2) nanoparticles	Enhances photosynthesis, root activity, and plant pest resistance when applied to plants
Nano Plant Growth Promoter	Fertilizer	Nanjing High Technology Nano Material (China)	Unspecified nano-oxides	Increases plant yield, encourages development, and strengthens its defences against pests and diseases
Pesticide Synergistic	Pesticide	Nanjing High Technology Nano Material (China)	Unspecified nanomaterial	Nanoscale catalysts that increase biological activity while lowering toxicity

TABLE 1.4 *(Continued)*

Commercial Brand Name	Product Category	Manufacturer	Type of Nanomaterial	Applications
NovaLand Nano- Mn, Zn, N, Fe, Cu, Mo)	Fertilizer	Land Green & Technology Co. (Taiwan)	Unspecified nanoparticles of minerals (Mn, Zn, N, Fe, Cu, Mo)	Accelerating and regulating plant development by increasing chlorophyll production and strengthening plant defences against harmful situations
NovaLand-F	Fertilizer	Land Green & Technology Co. (Taiwan)	Unspecified nanomagnet	Accelerating plant metabolism helps regulate and accelerate plant growth
Groagro 1: Super Nano-catalyst Kalium+ TE	Fertilizer	Microwell Bio Solutions SdnBhd (Malaysia)	Unspecified nanocomposite of potassium	Improved plant growth and boosted nutritional value
Nanofertilizer	Fertilizer	NANOPAC (M) SdnBhd (Malaysia)	Unspecified nanomaterial content	Accelerated plant development and increased protection against germs, fungi, and illnesses
FOSVIT K30	Fertilizer	Kimitec Group (Spain)	Potassium (K) and (P) nanopowder/ nanoparticles	Stimulates the production of phytoalexins, which are poisonous to harmful organisms and hence help plants mount defences
Nanosept®Aqua Super/ Nanosepts® Aqua/ Nanosept®	Plant disinfect formulation	Nanosept (Hungary)	Silver (Ag) nanoparticles	Fungicide, bactericide, and virucide
NeuCytokin®	Plant growth regulator	Neufarm GmbH (Germany)	Unspecified nanomaterial content	A natural foliar fertilizer made of potassium and seaweed that promotes crop growth
Nano nutrients for Crop	Fertilizer	NanoLandBaltic (Lithuania)	Unspecified nanonutrient	Supply nutrients to chlorophyll directly when sprayed on leaves
Hyper Feed	Fertilizer	Bio Nano Technology (Egypt)	Unspecified nanomaterial content	A balanced combination of major and minor nutrients

TABLE 1.4 *(Continued)*

Commercial Brand Name	Product Category	Manufacturer	Type of Nanomaterial	Applications
NANOCU®	Pesticide	Bio Nano Technology (Egypt)	Copper nanoparticles	Accelerating plant growth and possessing fungicide and bactericide characteristics
NUBIOTEC® ULTRA Ca	Fertilizer	Bioteska (Mexico)	Calcium (Ca) nanopowder/ nanoparticles	Acceleration and control of plant growth as well as increased resilience to harmful situations
Nonovec TSS 80	Fertilizer	Laboratorios Bio-Me´dicin (Brazil)	Molybdenum, Manganese, and Zinc	Accelerates plant growth, enhances the germination of seeds, and strengthens the root system and absorbent radicles
FERTIL CALMAG	Fertilizer	HPL Agronegocios (Brazil)	Magnesium (Mg) nanocapsules	Due to magnesium (Mg) and calcium (Ca) deficiency, plant growth is regulated, increasing plant resistance to fungus, bacteria, and diseases
NUBIOTEK® Hyper Fe+ Mg	Fertilizer	Bioteska (Mexico)	Iron (Fe) nanopowder/ nanoparticles and magnesium (Mg) nanocapsules	Acceleration/regulation of plant growth and improvement of plant resistance
NANO FERTILIZERS	Fertilizer	Silvertech Chemical Industry Trade Ltd. (Turkey)	TiO_2 and ZnO nanoparticles/ nanopowder	By producing more chlorophyll, plants grow faster, and produce more food
Lithocal	Fertilizer	Litho Plant (Brazil)	Manganese (Mn) and Calcium nanoparticles/ nanopowder	Enhances plant tolerance to drought and increases soil nutrient and water absorption
Nano Chelated Zinc	Fertilizer	KHAZRA (Iran)	Unspecified nanomaterial content	This chelated slow-release fertilizer helps plants absorb nutrients to their fullest potential and is stable in both alkaline and acidic soils

TABLE 1.4 *(Continued)*

Commercial Brand Name	Product Category	Manufacturer	Type of Nanomaterial	Applications
Nano Chelated Magnesium	Fertilizer	KHAZRA (Iran)	Unspecified nanomaterial content	This chelated fertilizer has a delayed release to ensure that plants can use its nutrients to their full potential
Nano Calcium	Fertilizer	KHAZRA (Iran)	Unspecified nanomaterial content	Plants use nutrients optimally, an this slow-release chelated fertilizer is used in both acidic and alkaline soils, the nanoproduct is resilient and stable
Nano Chelated NPK (12-12-36)/ Nano Chelated NPK (20-20-20)	Fertilizer	KHAZRA (Iran)	Unspecified nanomaterials content	These chelated slow-release fertilizers help plants absorb nutrients to their fullest potential. Both alkaline and acidic soils do not affect the nanoproduct's resistance or stability
IFFCO NANO DAP (liquid)	Fertilizer	Indian Farmers Fertiliser Cooperative Limited (IFFCO), (India)	Nanoclusters of Phosphorus (P: 16.0% as P_2O_5 w/v) and Nitrogen (N: 8.0% w/v) functionalized with biopolymers (particle size: ≥100 nm)	The plant's system assimilates NANO DAP, which results in stronger seeds, more chlorophyll, improved photosynthetic efficiency, better quality, and higher crop output
IFFCO NANO UREA (liquid)	Fertilizer	Indian Farmers Fertiliser Cooperative Limited (IFFCO), (India)	It contains 4% N (as nitrogen analogues) encapsulated in an organic matrix (nanoparticle size: 20–50 nm)	With NANO UREA applications, crop availability is increased by more than 80%. It minimises gaseous emissions and nutrient loss
Somguard®	Nano disinfectant	AgriLife™, (India)	Nano silver (AgNP: 100 ppm) encapsulated by a polymer	A universal disinfectant of natural origin with low environmental impact, Somguard® AgNP is used to eliminate any kind of pathogenic microorganism

TABLE 1.4 *(Continued)*

Commercial Brand Name	Product Category	Manufacturer	Type of Nanomaterial	Applications
Somguard® SilCop	Nano disinfectant	AgriLife™, (India)	Nano silver (AgNP: 100 ppm) and nano copper (CuNP: 100 ppm), encapsulated by a biopolymer	An all-purpose, natural-origin disinfectant is used to control all kinds of harmful pathogens (viz., bacteria, fungi, and viruses).
BioCopp™	Nano disinfectant	AgriLife™, (India)	Nano copper (CuNP: 500 ppm), encapsulated by a biopolymer	A universal disinfectant made from natural sources that is eco-friendly and effective against all kinds of phytopathogens It is quite efficient against diseases that affect plants, such as the downy mildew that affects grapes
Nano Chitosan	Nano disinfectant	AgriLife™, (India)	Nano Chitosan (1.0% w/v)	Nano chitosan (a linear polysaccharide) extracted from natural sources shows phytotonic, fungistatic, and bacteriostatic properties
Microtrace	Nano nutrient/ fertilizer	AgriLife™, (India)	Nanoparticles of Mg, Zn, Mo, Fe, Mn, and Boron (each of 100 ppm), encapsulated by chitosan-based biopolymer, particle size >100 nm	Supplied all trace elements or minerals to crops for faster growth, more chlorophyll, and more food production
NanoCal	Nano nutrient/ fertilizer	AgriLife™, (India)	Nano calcium particles (size: 100 nm, potency: 10,000 ppm) are encapsulated by a chitosan-based biopolymer	NanoCal provides bioavailable calcium to plants

TABLE 1.4 *(Continued)*

Commercial Brand Name	Product Category	Manufacturer	Type of Nanomaterial	Applications
NanoPhos	Nano nutrient/ fertilizer	AgriLife™, (India)	Nano Phosphorus particles (size: 100 nm, potency: 10,000 ppm) are encapsulated by a chitosan-based biopolymer	NanoPhos provides bioavailable phosphorus to the plant. For many biological processes, including plant respiration and photosynthesis, phosphorus is an essential macronutrient
NanoPUFA	Nano nutrient/ fertilizer	AgriLife™, (India)	Nano polyunsaturated particles (potency: 30%, size: 100 nm) are encapsulated by a chitosan-based biopolymer	Lipids, a significant source of metabolic energy and a source of accessible, necessary polyunsaturated fatty acids for plants, are found in nano-PUFA
NanoGold	Nano disinfectant	AgriLife™, (India)	Nano gold particles (size: 100 nm, potency: 10 ppm) are encapsulated by a chitosan-based biopolymer	All kinds of phtopathogens can be disinfected with the help of NanoGold, a universal disinfectant with natural origins
NanoIron	Nano nutrient/ fertilizer	AgriLife™, (India)	Nano iron particles (size: 100 nm, potency: 10,000 ppm) are encapsulated by a chitosan-based biopolymer	The plant receives bioavailable iron from nano iron. Nano iron applications directly and indirectly affect photosynthesis and respiration in plants.

TABLE 1.4 *(Continued)*

Commercial Brand Name	Product Category	Manufacturer	Type of Nanomaterial	Applications
Nano Silica	Nano nutrient/ fertilizer	AgriLife^{TM}, (India)	Nano silica particles (size: 100 nm, potency: 10,000 ppm) are encapsulated by a chitosan-based biopolymer	The plant receives bioavailable silica, increasing its ability to withstand biotic and abiotic stresses, tolerating drought, and defending itself against pest and disease attack
NANO-SILVER	Nano disinfectant	Kanak Biotech, (India)	Silver nanopowder/ nanoparticles	Defend plants from phytopathogens and diseases
NANO-PHOSPHORUS	Plant growth regulator	Kanak Biotech, (India)	Phosphorus nanopowder/ nanoparticles	Nano phosphorus improves plant growth, reproduction, and overall yield
NANO-IRON	Plant growth regulator	Kanak Biotech, (India)	Iron nanopowder/ nanoparticles	Increases the iron that is available to plants and aids in the processes of photosynthesis and respiration
ANO-ZINC	Plant growth regulator	Kanak Biotech (India)	Zinc oxide (ZnO) nanopowder/ nanoparticles	Since zinc is a crucial structural component or regulatory co-factor of a variety of enzymes, it promotes the growth and health of plants
Nano Green	Fertilizer	Nano Green Sciences, Inc (India)	Unspecified nano material content	Increases plant photosynthesis, increasing carbon dioxide absorption in a proportional manner

Source: Adapted from Campos (2021).

Moreover, the costs associated with registering new products and the absence of a specific regulation are impeding the commercialization of nano-based products (Lade et al., 2019; Deshpande, 2019). In order to expand the market for agricultural NPs, five critical factors must be addressed: (i) the development of harmonized international legislation; (ii) the establishment of a network of nano-based product developers to enable communication between public and private organizations; (iii) the standardization of the definition of nanotechnology and NMs; (iv) the standardization of methods for determining the risk-benefit of NMs; and (v) the development of standardized methods for characterization, quantification, and tracking these NMs in the environment and evaluating their impact on human health and ecosystems (Campos, 2021; Kah et al., 2018a, 2018b; Parisi et al., 2015).

1.6 TECHNO-COMMERCIAL VIABILITY

The techno-commercial viability of NPs depends on several factors, including the maturity of the technology, manufacturing scalability, cost-effectiveness, regulatory considerations, and consumer acceptance. While some NPs have already reached the market and gained commercial success, others are still in the R&D phase or facing challenges scaling up production. NAPs that incorporate nanotechnology for enhanced performance, efficiency, or functionality. Evaluating the techno-commercial viability of NAPs involves assessing both the technological aspects and the economic feasibility. Here are some key considerations for evaluating the techno-commercial viability:

1. **Technological advancements:** Analyze the scientific and technological advances that NAPs offer. This includes determining if nanotechnology has a positive impact on plant health in general, crop productivity, disease resistance, nutrient absorption, and water use. Think about the accuracy and repeatability of findings from studies and field tests.
2. **Cost-benefit analysis:** To assess whether the advantages of nano-agricultural products outweigh the disadvantages, perform a thorough cost-benefit analysis. Analyze the whole cost of manufacturing, taking into account the costs of putting the solutions based on nanotechnology into use and keeping them up-to-date. Comparing these expenses against the possible gains—such as higher crop

yields, a need for fewer chemical inputs, better resource management, and better-quality products—is important.

3. **Market demand and potential:** Analyze the current market demand for nano-agricultural products and determine how much room there is for expansion. Take into account elements like market size, regulatory environment, and customer acceptance. Find out if there is a specific crop or specialized market where nano-agri products can provide a competitive edge.
4. **Environmental impact:** Over the course of their life cycles, evaluate the environmental impact of nano-agri products. Think about things like trash production, energy usage, and any hazards related to using NMs. Analyze whether the advantages of using nano-agricultural products outweigh any potential risks to the environment.
5. **Regulatory and safety considerations:** Examine the safety issues and legal requirements related to nano-agricultural products. Ensure adherence to national and international laws governing the application of NMs in agriculture. Analyze the potential dangers brought on by the discharge of NMs into the environment as well as their effects on ecosystems and human health.
6. **Scalability and practical implementation:** Determine whether deploying NAPs on a large scale is feasible and scalable. Think about things like distribution channels, manufacturing capacity, and suitability with current agricultural techniques. Consider whether the technology can be simply integrated into current farming systems without causing any disruption.
7. **Long-term viability:** Think about the sustainability and long-term viability of NAPs. Determine whether the technology has the ability to change and advance in response to shifting market dynamics, consumer preferences, and regulatory needs. Examine whether the advantages provided by NAPs are long-lasting and able to survive any future difficulties or restrictions.

It's crucial to remember that determining the techno-commercial feasibility of NAPs is a continuous process that calls for continuing monitoring, investigation, and evaluation as the technology develops and new information becomes available. Additionally, other elements such as R&D, quality assurance, prototyping, regulatory compliance, marketing, and product positioning have a big impact on how economically viable any new NP production is from a technological standpoint. The following are the general procedures required for creating and marketing NPs:

1. **R&D:** Businesses spend money on R&D to create novel NMs, nanocomponents, and nanosystems. In order to do this, it is necessary to investigate the special characteristics and behavior of nanoscale materials and pinpoint prospective applications.
2. **Prototyping:** Prototypes of NPs are developed after promising NMs or nanocomponents have been found to test their usability, performance, and viability for mass manufacturing.
3. **Manufacturing:** The manufacturing procedure is established to make NPs on a wider scale after successful prototyping. In order to maintain consistency in quality and regulate the materials' nanoscale properties, specialized production methods may be necessary.
5. **Quality control:** To assure the performance, dependability, and safety of NPs, stringent quality control procedures are put in place. Testing for these characteristics at the nanoscale may entail physical, chemical, and biological components.
6. **Regulatory compliance:** Depending on the location and the intended use, certain laws may apply to NPs. Regarding safety, labeling, and environmental impact, businesses must make sure they are in conformity with all applicable laws.
7. **Marketing and distribution:** To raise awareness and sell NPs to specific consumers, efficient marketing tactics are devised. To make the products accessible to consumers, distribution channels are formed. These channels may include direct sales, partnerships with shops, and internet marketplaces.
8. **Commercialization:** NPs are introduced to the market, and sales efforts are directed at earning income and gaining market share. In order to increase the company's reach and clientele, this may entail partnerships, licensing deals, or corporate collaborations.
9. **Post-market monitoring:** After NPs are commercialized, businesses continue to track their performance and user reviews. This makes it easier to find areas that need improvement, deal with problems as they develop, and keep customers happy.

1.7 CONCLUSION AND FUTURE PERSPECTIVES

In the current scenario, nanotechnology is a prominent area of growing technology. The critical analysis of the scientific literature revealed that agricultural nanotechnology is constantly evolving, but its marketability as NAIPs is not assured because of huge investments in R&D and mass production. Although nano-based formulations have benefits, it is important to ensure that using them won't have a negative impact on the environment or people's health. The main issues with nanotechnology-enabled products are related to the potential concentration of NPs used since NPs exhibit hazardous effects to varying degrees at increasing concentrations. Continuous use of nanotechnology-enabled products, particularly in farming, could raise the amount of these materials in the soil and in crops. Therefore, green nanotechnology is a multidisciplinary area in developing clean, secure, and ecologically acceptable substitutes for chemically synthesized NPs. Green-synthesized NPs are produced utilizing a variety of environmentally friendly methods, including the use of plant components, microbes, amino acids, glucose, and so on. Biosynthesized NPs are less toxic and safer for the environment as well as human health. In conclusion, in order to overcome the potential bottleneck of nanotechnology applications that prevent the widespread adoption of this technology in sustainable agriculture, actions should be taken. Improvements in technology are also necessary for the mass production of effective NPs that encourage low startup costs and higher returns on investment.

KEYWORDS

- **nano-agri-input products (NAIPs)**
- **nano-encapsulation of bioactive compounds**
- **nanofertilizers**
- **nanomaterials**
- **nanopesticides**
- **nanoremediation**
- **nanosensors**
- **nanotechnology**
- **techno-commercial viability**

REFERENCES

Adholeya, A., Das, R. K., Dubey, M. K., Kochar, M., & Singh, R. (2017). *Zero draft policy on regulation of nanoproducts in agriculture*. The Energy and Resources Institute (TERI), 1–7. https://www.teriin.org/policy-brief/zero-draft-policy-regulation-nanoproducts-agriculture

Ahmad, H., Venugopal, K., Rajagopal, K., De Britto, S., Nandini, B., Pushpalatha, H. G., Konappa, N., Udayashankar, A. C., Geetha, N., & Jogaiah, S. (2020). Green synthesis and characterization of zinc oxide nanoparticles using *Eucalyptus globules* and their fungicidal ability against pathogenic fungi of apple orchards. *Biomolecules, 10*(3), 425. https://doi.org/10.3390/biom10030425

Ali, S. S., Al-Tohamy, R., Koutra, E., Moawad, M. S., Kornaros, M., Mustafa, A. M., Mahmoud, Y. A. G., Badr, A., Osman, M. E. H., Elsamahy, T., Jiao, H., & Sun, J. (2021). Nanobiotechnological advancements in agriculture and food industry: Applications, nanotoxicity, and future perspectives. *Science of the Total Environment, 792*, 148359. https://doi.org/10.1016/j.scitotenv.2021.148359

Ali, E. O. M., Shakil, N. A., Rana, V. S., Sarkar, D. J., Majumder, S., Kaushik, P., Singh, B. B., & Kumar, J. (2017). Antifungal activity of nano emulsions of neem and citronella oils against phytopathogenic fungi, *Rhizoctonia solani* and *Sclerotium rolfsii*. *Industrial crops and products, 108*, 379–387.

Antony, S., Rathinam, P., Reshmy, R., Sindhu, R., Binod, P., & Pandey, A. (2021). Consumer nanoproducts: A new viewpoint. In *Handbook of consumer nanoproducts*. Springer, Singapore. https://doi.org/10.1007/978-981-15-6453-6_106-1

Arduini, F., Cinti, S., Scognamiglio, V., & Moscone, D. (2020). Nanomaterial-based sensors. In *Handbook of nanomaterials in analytical chemistry* (pp. 329–359). DOI: 10.1016/B978-0-12-816699-4.00013-X. Epub 2019 Oct 25. PMCID: PMC7153334.

Baruah, S., & Dutta, J. (2009). Nanotechnology application in polluting sensing and degradation in agriculture: A review. *Environmental Chemistry Letters, 7*, 191–204.

Bergeson, L. L. (2013). Sustainable nanomaterials: emerging governance systems. *ACS Sustainable Chemistry & Engineering, 1*(7), 724–730. https://doi.org/10.1021/sc4000863

Campos, E. V. R. (2021). Commercial nanoproducts available in world market and its economic viability. In S. Jogaiah, L. F. Fraceto, H. B. Singh, & R. Lima (Eds), *In Advances in Nano-Fertilizers and Nano-Pesticides in Agriculture: A Smart Delivery System for Crop Improvement*. Woodhead Publishing. https://doi.org/10.1016/B978-0-12-820092-6.00004-5

CAPA. (1985). Canada agricultural products act (CAPA) R.S.C c 20 (4th supp). https://laws-lois.justice.gc.ca/eng/acts/A-6.01/

CEPA. (1999). Canadian environmental protection act, 1999 (CEPA, C 33). https://laws-lois.justice.gc.ca/eng/acts/C-15.31/.

Chauhan, R., Yadav, H. O. S., & Sehrawat, N. (2020). Nanobioremediation: A new and a versatile tool for sustainable environmental clean-up-Overview. *Journal of Materials and Environmental Science, 11*, 564–573.

Chemicals and Polymers (New Substances Notification Regulations). (2015). Chemicals and polymers (new substances notification regulations) SOR/2015-81. https://laws-lois.justice.gc.ca/PDF/SOR-2015-81.pdf.

Chen, H., & Yada, R. (2011). Nanotechnologies in agriculture: New tools for sustainable development. *Trends in Food Science & Technology, 22*(11), 585–594.

Chhipa, H. (2019). Applications of nanotechnology in agriculture. *Methods in Microbiology, 46*, 115–142. https://doi.org/10.1016/bs.mim.2019.01.002.

Cinelli, M., Coles, S. R., Sadik, O., Karn, B., & Kirwan, K. (2016). A framework of criteria for the sustainability assessment of nanoproducts. *Journal of Cleaner Production, 126*, 277–286. https://doi.org/10.1016/j.jclepro.2016.02.118

Del Prado-Audelo, M. L., García Kerdan, I., Escutia-Guadarrama, L., Reyna-González, J. M., Magaña, J. J., & Leyva-Gómez, G. (2021). Nanoremediation: Nanomaterials and nanotechnologies for environmental cleanup. *Frontiers in Environmental Science, 9*, 793765. DOI: 10.3389/fenvs.2021.793765

Department of Biotechnology, India (DBT, India). (2020). *Guidelines for evaluation of nano-based agri-input (NAIP) and food products in India.* https://dbtindia.gov.in/sites/default/files/uploadfiles/Guidlines%20for%20Evaluation%20of%20Nano%20Based%20Agri%20Input%20and%20Food%20Products%20in%20India.pdf

Deshpande, M. V. (2019). Nanobiopesticide perspectives for protection and nutrition of plants. In *Nano-Biopesticides Today and Future Perspectives* (pp. 47–68). Elsevier. https://doi. org/10.1016/B978-0-12-815829-6.00003-6

Dimkpa, C. O., Bindraban, P. S., Fugice, J., Agyin-Birikorang, S., Singh, U., & Hellumus, D. (2017). Composite micronutrient nanoparticles and salts decrease drought stress in soybean. *Agronmy for Sustainable Developement, 37*(1), 5.

El-Ramady, H., Alshaal, T., Abowaly, M., Abdalla, N., Taha, H. S., Al-Saeedi, A. H., Shalaby, T., Amer, M., MiklósFári, É. D.-S., Sztrik, A., Prokisch, J., Selmar, D., Elizabeth, A. H., Smits, P., & Pilon, M. (2017). Nanoremediation for sustainable crop production. In *Nanoscience in food and agriculture, sustainable agriculture reviews* (Vol. 5, pp. 335–363). Springer, Cham.

FDA. (2018). U.S. Food and drug administration, nanotechnology. https://www.fda.gov/science-research/nanotechnology-programs-fda

Fraceto, L. F., Grillo, R., de Medeiros, G. A., Scognamiglio, V., Rea, G., & Bartolucci, C. (2016) Nanotechnology in agriculture: Which innovation potential does it have? *Frontiers in Environmental Science*, *4*(20). DOI: 10.3389/fenvs.2016.00020

Godoy, F., Olivos-Hernández, K., Stange, C., & Handford, M. (2021). Abiotic stress in crop species: Improving tolerance by applying plant metabolites. *Plants*, *10*(2), 186. https://doi.org/10.3390/plants10020186.

Gómez, B., Barba, F. J., Domínguez, R., Putnik, P., Bursać Kovačević, D., Pateiro, M., Toldrá, F., & Lorenzo, J. M. (2018). Micro-encapsulation of antioxidant compounds through innovative technologies and its specific application in meat processing. *Trends in Food Science and Technology*, *82*, 135–147. DOI: 10.1016/j.tifs.2018.10.006.

Googs, A., Knauer, K., & Bucheli, T. D. (2012). Nanomaterials in plant protection and fertilization: Current state, foreseen applications, and research priorities. *Journal of Agricultural and Food Chem*istry, *60*, 9781–9792.

Human, C., de Beer, D., van der Rijst, M., Aucamp, M., & Joubert, E. (2019). Electrospraying as a suitable method for nanoencapsulation of the hydrophilic bioactive dihydrochalcone, aspalathin. *Food Chemistry*, *276*, 467–474.

Insight Ace Analytic. (2023). *Agricultural nanotechnology market size, share, trends and future scope analysis to 2031.* https://www.insightaceanalytic.com/report/agricultural-nanotechnology-market/1530

Jafari, S. M. (2017). An overview of nanoencapsulation techniques and their classification. In S. M. Jafari (Ed.), *Nanoencapsulation technologies for the food and nutraceutical industries* (pp. 1–34). Academic Press.

Jhaveri, J. H., & Murthy, Z. V. P. (2016). A comprehensive review on anti-fouling nanocomposite membranes for pressure driven membrane separation processes. *Desalination, 379*,137–154.

Jogaiah, S., Fraceto, L. F., Singh, H. B., & Lima, R. (2020). *Advances in Nano-fertilizers and nano-pesticides in agriculture: A smart delivery system for crop improvement*. Woodhead Publishing. ISBN: 9780128200926. https://doi.org/10.1016/B978-0-12-820092-6.00004-5

Joshi, S. M., De Britto, S., Jogaiah, S., & Ito, S. I. (2019). Mycogenic selenium nanoparticles as potential new generation broad spectrum antifungal molecules. *Biomolecules, 9*(9) 419. https://doi.org/10.3390/biom9090419

Kah, M. (2015). Nanopesticides and nanofertilizers: Emerging contaminants or opportunities for risk mitigation? *Frontiers in Chem*istry, *3*, 64. DOI: 10.3389/fchem.2015.00064

Kah, M., Kookana, R. S., Gogos, A., & Bucheli, T. D. (2018a). A critical evaluation of nanopesticides and nanofertilizers against their conventional analogues. *Nature Nanotechnology*, *13*, 677–684. https://doi.org/10.1038/s41565-018-0131-1

Kah, M., Walch, H., & Hofmann, T. (2018b). Environmental fate of nanopesticides: Durability, sorption and photodegradation of nanoformulated clothianidin. *Environment Science: Nano*, *5*, 882–889. https://doi.org/10.1039/C8EN00038G

Karn, B., Kuiken, T., & Otto, M. (2009). Nanotechnology and in site remediation: A review of the benefits and potential risks. *Environmental Health Perspectives*, *117*(12), 1813–1831.

Khan, M. K., Pandey, A., Hamurcu, M., Gezgin, S., Athar, T., Rajput, V. D., Gupta, O. P., & Minkina, T. (2021). Insight into the prospects for nanotechnology in wheat biofortification. *Biology*, *10*, 1123.

Kumari, R., Suman, K., Karmakar, S., Mishra, V., Lakra, S. G., Saurav, G. K., & Mahto, B. K. (2023). Regulation and safety measures for nanotechnology-based agri-products. *Frontiers in Genome Editing, 5*, 1200987. DOI: 10.3389/fgeed.2023.1200987

Lade, B. D., Gogle, D. P., Lade, D. B., Moon, G. M., Nandeshwar, S. B., & Kumbhare, S. D. (2019). Nanobiopesticide formulations: Application strategies today and future perspectives. In *Nano-Biopesticides Today and Future Perspectives* (pp. 179–206). Elsevier. https:// doi.org/10.1016/B978-0-12-815829-6.00007-3

Li, Q., Li, X., & Zhao, C. (2020). Strategies to obtain encapsulation and controlled release of small hydrophilic molecules. *Frontiers in Bioengineering and Biotechnology*, *8*, 437. https://doi.org/10.3389/fbioe.2020.00437.

López-Valdez, F., Miranda-Arámbula, M., Ríos-Cortés, A. M., Fernández-Luqueño, F., & de-la-Luz, V. (2018). Nanofertilizers and their controlled delivery of nutrients. In F. López-Valdez & F. Fernández-Luqueño (Eds.), *Agricultural nanobiotechnology*. Springer, Cham. https://doi.org/10.1007/978-3-319-96719-6_3

Malekhosseini, P., Alami, M., Khomeiri, M., Esteghlal, S., Nekoei, A., & Hosseini, S. M. H. (2019). Development of casein-based nanoencapsulation systems for delivery of epigallocatechin gallate and folic acid. *Food Science & Nutrition*, 7, 519–527. DOI: 10.1002/fsn3.827.

Malsch, I., Subramanian, V., Semenzin, E., Hristozov, D., & Marcomini, A. (2015). Supporting decision-making for sustainable nanotechnology. *Environment Systems Decisions*, *35*, 54–75.

Mandaji, C. M., da Silva Pena, R., & Campos Chisté, R. (2022). Encapsulation of bioactive compounds extracted from plants of genus *Hibiscus*: A review of selected techniques and applications. *Food Research International*, *151*, 110820. https://doi.org/10.1016/j.foodres.2021.110820.

Meena, S. D., Gautam, C., Patidar, O. P., Meena, H. M., & Prakasha, V. G. (2017). Nano fertilizer is a new way to increase nutrients use efficiency in crop production. *International Journal of Agricultural Science*, *9*, 3831–3833.

Mishra, A., & Singh, H. B. (2021). Nanoparticles in agro-products: Benefits, applications and concerns. In *Nanomaterials for agriculture and forestry applications* (pp. 73–88). Springer, Singapore.

Mohanraj, J., Subramanian, K. S., Rahale, S., & Priyadharshini, M. (2022). Application of biosensors and nanosensors in agriculture. V*igyan Varta*, *3*(8), 13–15.

Mujtaba, M., Sharif, R., Ali, Q., Rehman, R., & Khawar, K. M. (2021). Biopolymer based nanofertilizers applications in abiotic stress (drought and salinity) control. In S. Jogaiah, L. F. Fraceto, H. B. Singh & R. Lima (Eds), *Advances in nano-fertilizers and nano-pesticides in agriculture: A smart delivery system for crop improvement.* Woodhead Publishing. https://doi.org/10.1016/B978-0-12-820092-6.00004-5

Neme, K., Nafady, A., Uddin, S., & Tola, Y. B. (2021). Application of nanotechnology in agriculture, postharvest loss reduction and food processing: Food security implication and challenges. *Heliyon*, *7*, e08539. https://doi.org/10.1016/j.heliyon.2021.e08539

Nikmaram, N., Roohinejad, S., Hashemi, S., Koubaa, M., Barba, F. J., Abbaspourrad, A., & Greiner, R. (2017). Emulsion-based systems for fabrication of electrospun nanofibers: Food, pharmaceutical and biomedical applications. *RSC Advances*, *7*, 28951–28964.

NNI-Budget. (2021). National Nanotechnology Initiative. https://www.nano.gov/about-nni/what/funding (accessed on 7 July 2023).

Pandey, S., Giri, V. P., Tripathi, A., Bajpai, R., Sharma, D., Bahadur, L., Mishra, A. (2021). Interaction, fate and risks associated with nanomaterials as fertilizers and pesticides. In S. Jogaiah, L. F. Fraceto, H. B. Singh & R. Lima (Eds), *Advances in nano-fertilizers and nano-pesticides in agriculture: A smart delivery system for crop improvement.* Woodhead Publishing. https://doi.org/10.1016/B978-0-12-820092-6.00004-5

Parisi, C., Vigani, M., & Rodrı´guez-Cerezo, E. (2015). Agricultural nanotechnologies: What are the current possibilities? *Nano Today*, *10*, 124–127. https://doi.org/ 10.1016/j.nantod.2014.09.009.

PCPA. (2002). Pest control products act (PCPA) (SC 2002, c 28). https:// laws-lois.justice.gc.ca/eng/acts/P-9.01/

Powell, J. J., Ainley, K., Harvey, R. S., & Mason, A. S. (2016). The role of oral processing in nanoparticle uptake and relevance to consumer exposure: A review. *British Journal of Nutrition*, *116*(11), 1886–1897.

Rajput, V. D., Minkina, T., Kumari, A., Harish, Singh, V. K., Verma, K. K., Mandzhieva, S., Sushkova, S., Srivastava, S., & Keswani, C. (2021a). Coping with the challenges of abiotic stress in plants: New dimensions in the field application of nanoparticles. *Plants*, *10*, 1221.

Rajput, V. D., Singh, A., Minkina, T., Rawat, S., Mandzhieva, S., Sushkova, S., Shuvaeva, V., Nazarenko, O., Rajput, P., Komariah, A., Verma, K. K., Singh, A. K., Rao, M., & Upadhya, K. (2021b). Nano-enabled products: Challenges and opportunities for sustainable agriculture. *Plants*, *10*, 2727. https://doi.org/10.3390/ plants10122727

Rajput, V. D., Singh, A., Singh, V. K., Minkina, T. M., & Sushkova, S. (2021c). Impact of nanoparticles on soil resource. In A. Amrane, D. Mohan, T. A. Nguyen, A. A. Assadi & G. Yasin (Eds.), *Nanomaterials for soil remediation* (pp. 65–85). Elsevier.

Raliya, R., Saharan, V., Dimkpa, C., & Biswas, P. (2018). Nanofertilizer for precision and sustainable agriculture: Current state and future perspectives. *Journal of Agricultural and Food Chemistry*, *66*(26), 6487–6503. https://pubs.acs.org/doi/10.1021/acs.jafc.7b02178

Roychoudhury, A., & Bhowmik, R. (2023). Nanoremediation of heavy metals in agricultural soil. In F. Fernandez-Luqueno & J. K. Patra (Eds.), *Agricultural and environmental nanotechnology. Interdisciplinary biotechnological advances*. Springer, Singapore. https://doi.org/10.1007/978-981-19-5454-2_16

Sampedro-Guerrero, J., Vives-Peris, V., Gomez-Cadenas, A., & Clausell-Terol, C. (2023). Efficient strategies for controlled release of nanoencapsulated phytohormones to improve plant stress tolerance. *Plant Methods*, 19, 47. https://doi.org/10.1186/s13007-023-01025-x

Schwab, F., Zhai, G., Kern, M., Turner, A., Schnoor, J. L., & Wiesner, M. R. (2016). Barriers, pathways and processes for uptake, translocation and accumulation of nanomaterials in plants—Critical review. *Nanotoxicology*, *10*, 257–278.

Scognamiglio, V., Stano, P., Polticelli. F., Antonacci, A., Lambreva, M. D., Pochetti, G., Giardi, M. T., & Rea, G. (2013). Design and biophysical characterization of atrazine-sensing peptides mimicking the *Chlamydomonas reinhardtii* plastoquinone binding niche. *Physical Chemistry Chemical Physics*, *15*(31), 13108–13115. DOI: 10.1039/c3cp51955d. PMID: 23824019.

Sreenivasa, N., Meghashyama, P. B., Udayashankar, A. C., Lakshmeesha, T. R., Geetha, N., & Jogaiah, S. (2020). Biosynthesis and characterization of *Dillenia indica*-mediated silver nanoparticles and their biological activity. *Applied Organometallic Chemistry*, *34*. https://doi.org/10.1002/aoc.e5567e5567

Subramanian, K. S., & Rajkishore, S. K. (2018). Regulatory framework for nanomaterials in agri-food systems. In M. Rai & J. K. Biswas (Eds.), *Nanomaterials: Ecotoxicity, safety, and public perception* (pp. 319–342). Springer International Publishing, Cham https://doi.org/10.1007/978-3-030-05144-0_16

Subramanian, V., Semenzin, E., Hristozov, D., Zondervan-van den Beuken, E., Linkov, I., & Marcomini, A. (2015). Review of decision analytic tools for sustainable nanotechnology. *Environment Systems and Decisions*, *35*, 29–41.

Suttiponparnit, K., Jiang, J., Sahu, M., Suvachittanont, S., Charinpanitkul, T., & Biswas, P. (2011). Role of surface area, primary particle size, and crystal phase on titanium dioxide nanoparticle dispersion properties. *Nanoscale Research Letters*, *6* (1), 27.

Szopa, D., Mielczarek, M., Skrzypczak, D., Izydorczyk, G., Mikula, K., Chojnacka, K., & Witek-Krowiak, A. (2022). Encapsulation efficiency and survival of plant growth-promoting microorganisms in an alginate-based matrix—A systematic review and protocol for a practical approach. *Industrial Crops and Products*, *181*, 114846.

Taiwan Nanotechnology Industry Development Association (TANIDA). (2024). Origin of nanoMark. http://www.tanida.org.tw/nanomark_e.php?nm=markIntroduction_e

Tarafdar, J., Raliya, R., Mahawar, H., & Rathore, I. (2014). Development of zinc, nanofertilizer to enhance crop production in pearl millet (*Pennisetum americanum*). *Agricultural Research*, *3*(3), 257–262.

Tarafdar, J., Raliya, R., & Rathore, I. (2012). Microbial synthesis of phosphorus nanoparticle from tri-calcium phosphate using *Aspergillus tubigenesis* TFR-5. *Journal of Bionanoscience*, *6*(2), 84–89.

US Food and Drug Administration. (2020). CFR - code of federal regulations title 21. https://www.accessdata.fda.gov/scripts/cdrh/cfdocs/cfcfr/CFRSearch. cfm?fr=172.480

Usman, M., Farooq, M., Wakeel, A., Nawaz, A., Cheema, S. A., Rehman, H. U., Ashraf, I., & Sanaullah, M. (2020). Nanotechnology in agriculture: Current status, challenges and future opportunities. *Science of the Total Environment*, *721*, 137778.

Waduge, P., Larkin, J., Upmanyu, M., Kar, S., & Wanunu, M. (2015). Programmed synthesis of freestanding graphene nanomembrane arrays. *Small, 11*, 597–603.

Yadav, R. K., Singh, N. B., Singh, A., Yadav, V., Bano, C., Khare, S., & Vegetas, N. (2020). Expanding the horizons of nanotechnology in agriculture: Recent advances, challenges and future perspectives. *Vegetos*, *33*, 203–221.

CHAPTER 2

Role of Nano-Agri-Input Products in Agriculture 4.0: Futuristic Farming Technology

DEEPAK KUMAR[1,2], MANU SHARMA[3], S. SHANTHAKUMAR[4], MAHAVEER P. SHARMA[5], and RAJA HUSAIN[6]

[1]*Nextnode Bioscience Pvt Ltd., Kadi, Gujarat, India*

[2]*Shri Ram Solvent Extraction Pvt Ltd., Jaspur, Uttarakhand, India*

[3]*School of Nano Sciences, Central University of Gujarat, Gandhinagar, Gujarat, India*

[4]*Gujarat National Law University (GNLU), Gandhinagar, Gujarat, India*

[5]*Agribusiness Incubation Centre, ICAR-Indian Institute of Soybean Research, Indore, Madhya Pradesh, India*

[6]*Department of Agriculture, Himalayan University, Itanagar, Arunachal Pradesh, India*

ABSTRACT

In the current scenario, it is very necessary to meet the dietary needs of the constantly expanding world population. Due to the misuse of chemical pesticides and fertilizers in traditional farming over the past three decades, the soil has become toxic and degraded, along with other environmental problems. More cutting-edge technology is therefore urgently needed to solve these problems. In this way, the agrotechnological revolution—also known as Agriculture 4.0, which is currently in the process of reforming the global agriculture system—has been made possible by nanotechnology.

Nanobiotechnology for Agricultural Sciences: Nano-Agri-Input Products for Crop Production and Environmental Protection. J. C. Tarafdar, Deepak Kumar, Sanjay K. Singh, and Krishna Pal Singh (Eds.)

Nanotechnology offers novel nano-agri-inputs and innovative delivery systems to boost crop yields and minimize chemical pesticide and fertilizer usage in agriculture systems. Additionally, nanotechnology has the potential to improve agriculture with novel nanotools (viz., nanosensors) for rapid disease diagnostics, moisture detection in soil, and enhancing the capacity of plants to absorb nutrients to track the pathways of macro- and micronutrient utilization. This chapter covers all aspects of nano-agri-input products (NAIPs) and their roles in Agriculture 4.0, along with potential risks associated with the use of nanomaterials (NMs) in agriculture.

2.1 AGRICULTURE 4.0: INTRODUCTION

The agriculture industry is today confronted with significant problems in sustainably feeding a growing global population while addressing major issues such as resource depletion and climate change. Major technical developments are occurring concurrently in a variety of industries, including nanotechnology, robotics, gene technology, energy production, machine learning, artificial intelligence, and many more (NFU, 2023; Firbank et al., 2018). The emergence of these new technologies will bring about the "fourth agricultural revolution," also known as "Agriculture 4.0," which analysts have dubbed the emergence of a wide range of potential "future agricultures" or "future food systems" that are characterized by cutting-edge, potentially game-changing technologies. Of course, earlier agricultural revolutions were innovative at the time: the first one saw nomadic life transition to cultivated agriculture (Agriculture 1.0); the following one was marked by innovation as a result of the British Agricultural Revolution, which noticed new machines such as Jethro Tull's seed drill (Agriculture 2.0); and the third revolution involved changes in production in developing countries with the rise of the Green Revolution (Agriculture 3.0) (Klerkx and Rose, 2020; Rose and Chilvers, 2018).

Currently, the phrase "Agriculture 4.0" is unclear and is used to refer to a variety of technologies and concepts that are linked to theories about the Industrial Revolution 4.0 (Zambon et al., 2019). Future agricultural and food systems under the new agricultural 4.0 include ideas such as circular agriculture, vertical farming, bioeconomy, digital agriculture, and aquaponics. These systems are frequently being created and put into practise, although they have not yet reached their full potential (Klerkx and Rose, 2020; Hermans, 2018). The phrase is used in this context to describe potentially revolutionary technologies that have the potential to fundamentally alter how food is processed, produced, exchanged, and eaten. These include innovations in nanotechnology, bioreactors, sensors associated with precision farming technology, artificial intelligence, 3D printing, gene editing, robotics, blockchain management, machine learning, internet of things (IoT), cultured meat, and synthetic food production (i.e., synthetic protein) (NFU, 2023; Klerkx and Rose, 2020; Grieve et al., 2019; Stephens et al., 2019). Moreover, Agriculture 4.0 has the capacity to be revolutionary and disruptive in a variety of ways. Food and nutrition security, as well as the structure and operation of agricultural production systems, nanotechnology in the agri-tech revolution, and digital solutions, may be affected on a biophysical, economic, and social level. It will also have an impact on how agriculture is incorporated into ecosystems and landscapes (Lowry et al., 2019; Brandt and Barrangou, 2019; Watanabe et al., 2018). According to a traditional agricultural and food system concept, Agriculture 4.0 is frequently associated with and seen to benefit large-scale, technologically advanced, and specialized farms (Bronson, 2019).

The use of nanomaterials (NMs) in agriculture is a current area of research, with new uses including the targeted delivery of nutrients or pharmacological capsids for the diagnosis and treatment of diseases, the delivery of bioactive compounds to specific regions to improve crop development, and other cutting-edge methods (Javid et al., 2022; Cicek and Nadaroglu, 2015). Applying NPs can enhance plant growth and productivity, according to research reports. Numerous NPs, including those made through emulsification, polymerization, ionic gelation, oxide reduction, and others, significantly increase crop yields (Fraceto et al., 2016). NMs have a larger surface area for aiding several plant metabolic pathways, boosting the rate of photosynthesis to provide more dry matter and crop output. NMs are different from bulk materials in terms of their physical and chemical characteristics. For instance, when applied in nanoform, rock phosphate increases the amount of phosphorus that is available to plants because the direct application of nano-rock phosphate prevents soil fixation (Adhikari et al., 2014). Therefore, in the agricultural scientific community, NMs are receiving a lot of interest primarily because of their excellent penetration capability, compact size, and large surface area. These substances differ from other bulk materials in a certain way, making them special (Yadav et al., 2023). Moreover, to better understand the effects of NMs on soil microorganisms and the fate of given nanoproducts (NPs), more research on NM deliveries, optimum dosage of concentrations in plant systems, and precision farming is required (Figure 2.1).

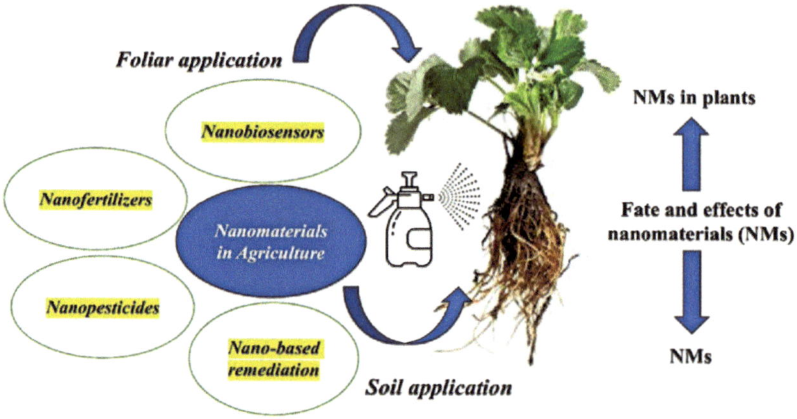

FIGURE 2.1 Fate and effects of nanomaterials (NMs) on plants.

2.2 NANOTECHNOLOGY IN AGRICULTURE 4.0

Agriculture 4.0 refers to the integration of modern technologies, data-driven decision-making, and smart systems in agriculture to improve efficiency, productivity, and sustainability (Lowry et al., 2019). Nanotechnology, which involves working with materials at the nanoscale (between 1 and 100 nanometers), has the potential to be a key component of Agriculture 4.0 by presenting cutting-edge solutions to different agricultural problems. The production of sophisticated NMs for improved seed germination and micronutrient supply, as well as the reduction of biotic and abiotic stress, are among the advantages of nanotechnology in Agriculture 4.0. Additionally, because nanotechnology-based fertilizers and insecticides may be applied in smaller doses, they pose fewer risks to the environment and public health. A further important benefit is the introduction of state-of-the-art nanobiosensors and nanodiagnostic systems that track soil quality indicators, plant illnesses, and stress—all essential for precision agriculture (PA) (i.e., Agriculture 4.0). Additionally, nanotechnology has shown promise in reducing agricultural waste, creating high-value goods, and utilizing techniques and tools for tagging, monitoring, and tracking agricultural products. In addition to these advancements, cloud computing, and smartphone-based biosensors have become essential tools for data collection and analysis (Yadav et al., 2023; Javid et al., 2022). The controlled release of agrochemicals on targets for effective nutrient uptake and disease resistance is part of this technology. Nanoscale transporters, nanofertilizers (NFs), nanopesticides, nanosensors, and nanoherbicides are examples of nanotechnology-based products. The farming community may eliminate agrochemicals while maintaining good crop output, safeguarding the health of the soil and water, and contributing to a cleaner environment by implementing PA practises based on nanotechnology (Klerkx and Rose, 2020). Here are some instances in which Agriculture 4.0 might use nanotechnology:

1. **Nanopesticides and NMs:** Nanotechnology enables the synthesis of tiny particles that can more efficiently fertilize and apply pesticides to crops. These nano-agents can improve the targeted delivery of nutrients or pesticides, lowering the amount of chemicals required and lowering the negative effects on the environment (Yadav et al., 2023; Javid et al., 2022).
2. **Precision farming and sensing:** Systems for monitoring soil or plants can use nanosensors to deliver real-time information on the state of the soil, the availability of water, the levels of nutrients, and the health of the plants. With the use of this knowledge, farmers can make decisions based on facts, maximizing resource use and minimizing waste (Yadav et al., 2023).
3 **NMs for soil remediation**: In order to improve the removal of contaminants and support efforts to remediate soil, nanoparticles can be made to bind with pollutants in the soil. In places impacted by industrial pollution or chemical spills, this is especially beneficial (Del Prado-Audelo et al., 2021).
4. **Smart delivery systems:** Nanotechnology can help with the regulated release of agrochemicals and nutrients over time, guaranteeing a consistent supply to plants as needed. This regulated delivery mechanism aids in the reduction of overapplication and environmental damage (Yadav et al., 2023; Hussain, 2017; Solanki et al., 2015).
5. **Disease and pest management:** The use of nanotechnology can help create biopesticides with nanoscale formulations that are more effective in controlling diseases and pests while causing the least amount of harm to the environment and beneficial insects (An et al., 2022; Javid et al., 2022).
6. **Improved water management:** The installation of nanoscale water purification systems, the reduction of water waste, and the assurance of the availability of clean water for crops and livestock are all ways that nanotechnology can help improve water efficiency in agriculture (Singh et al., 2022; Lu et al., 2016).
7. **Plant genetic improvement:** For the purpose of enhancing the expression of desirable features in plants and enhancing their resilience to

diseases and environmental challenges, nanoparticles can be utilized for targeted delivery of genetic material, such as RNA interference (RNAi) (Squire et al., 2023; Yan et al., 2022; Ahmar et al., 2021).

8. **Food quality and safety:** In food packaging, nanoscale sensors can be used to spot contamination or rotting, ensuring the quality and safety of agricultural products along the supply chain (Onyeaka et al., 2022; Ashfaq et al., 2022).

2.3 NANOMATERIALS AS NANO-AGRI-INPUT PRODUCTS (NAIPS)

The agriculture sector, which is a nation's primary source of revenue, steadily uses more nanotechnology products to address issues that arise. The use of agricultural inputs (such as fertilizers, pesticides, herbicides, and so on) can be reduced by increasing their efficiency through the use of nano-carriers; nanosensors can simultaneously detect environmental conditions and crop development in the field; nanoarrays can reduce sample volume and analyte usage; nanofilters can effectively treat water resources; and nanoparticles can speed up crop development (Samra and Hayrunnisa, 2015). The use of nanoparticles in agriculture to improve plant growth, crop protection, and overall agricultural output is referred to as "nano-agri-input products (NAIPs)." In 2020, the Department of Biotechnology (DBT), Ministry of Science and Technology, Government of India, released guidelines for the evaluation of nano-based agricultural inputs (NAIP) and food products. As per the NAIP guidelines, "a NAIP is a preparation for agricultural inputs that contains NMs and is designed for external and internal applications (via foliar, drip, soil, and seed in crops, as well as through other means) on crops for the goal of agricultural farming" (DBT India, 2020, https://dbtindia.gov.in/). NMs are substances with dimensions between 1 and 100 nanometers, and because of their small size and high surface area-to-volume ratio, they have special features. Furthermore, nanoparticles can be classified based on their shape (sphericity, flatness, and aspect ratio), composition, dimensions (0, 1, 2, and 3), agglomeration qualities, and core material. It can take the form of an infused, aggregated, or agglomerated material and hase a tubular, spherical, or irregular shape. Nanoparticle toxicity is influenced by their shape and aggregation characteristics, in addition to their individual chemical makeup. They may be beneficial in a variety of agricultural applications due to these characteristics. The definition of NM application in agriculture and the food sector in different countries is comparatively different in context to size, solubility, distribution threshold, novel properties, intentionally manufactured or engineered, as per the regulatory framework of each country (DBT India, 2020; Subramanian and Rajkishore, 2018; Boverhof et al., 2015). The comparison of the NM definition in context to size in selected countries as per their regulatory frameworks is illustrated in Table 2.1.

The use of NMs in agriculture includes the production of nanopesticides, nanoherbicides, NFs, nanosensors, and nanotracers that can contain a variety of particles, including titanium, iron, silver, zinc, gold, magnesium, carbon, and silica (Sharma et al., 2022; Okey-Onyesolu et al., 2021). Agri-input products in nanoform have active ingredient(s) of novel material (organic, inorganic, or composite) with altered novel properties that are to be used in agriculture and allied sectors. The NAIPs are not limited to NFs and nanopesticides but are also used as carrier systems for disease prevention, anti-transpiration agents, biostimulants, pheromone delivery, and plant growth regulators. Here are a few examples of NMs used in agricultural inputs.

2.3.1 NANOFERTILIZERS

A sort of agricultural product based on nanotechnology called a "NM" is intended to increase the effectiveness of nutrient delivery to plants. Traditional fertilizers support plant growth by providing vital minerals such as nitrogen, phosphorous, and potassium, but they can be ineffective and lead to nutrient losses to the environment. NMs, usually between 1 and 100 nanometers in size, are added to fertilizer formulations to produce NFs. Due to their small size and high surface area, these nanoparticles can be manufactured from a variety of substances, including metals, metal oxides, or carbon-based compounds (Yadav et al., 2023). The use of NFs is a really cutting-edge and innovative approach. In NFs, micro- or macronutrients can be given as emulsions or nanoparticles, enclosed in NMs, coated with a thin layer of protection, or encapsulated by NMs (DeRosa et al., 2010). By using NFs, it is possible to release nutrients into the soil gradually and under controlled conditions, preventing eutrophication and water resource pollution. NFs increase crop output and quality while increasing the efficiency with which nutrients are used, lowering production costs, and promoting agricultural sustainability (Bratovcic et al., 2021, Usman et al., 2020). Regarding the advancement of NF technology, the market entry strategy should focus on the creation of high-efficiency NAIPs for the major and secondary elements N, P, K, S, Ca, and Mg, which are used in greater quantities but are currently less efficient and have a greater negative environmental impact. The total expense of fertilization per unit of fruit output will be greatly decreased

TABLE 2.1 Comparison of the Definitions of Nanomaterials in the Current Regulatory Frameworks in Selected Countries That are Relevant to Agriculture and the Food Industry.

Country and Regulation	Solubility	Size	Distribution Threshold	Aggregates and Agglomerates	Novel Properties	Intentionally Engineered/ Manufactured
United States Environmental Protection Agency	No	1–100 nm	10% by weight	Yes	Yes	Yes
Norwegian Environmental Agency	No	1–100 nm and larger	50% by number	Yes	No	No
Danish Ministry of the Environment (2014)	No	1–100 nm	50% by number	Yes	No	No
Belgian Federal Public Service Health, Food Chain Safety and Environment (Belgium 2014)	No	1–100 nm	50% by number	Yes	No	Yes
Swiss Federal Office of Public Health and Federal Office for the Environment	No	1–100 nm and larger	50% by number	Yes	No	No
Health Canada	No	1–100 nm and large	No specific mention	Yes	Yes	Yes
Australian Government Department of Health and Ageing	No	1–100 nm	10% by number	Yes	Yes	Yes
Taiwan Council of Labor Affairs	No	1–100 nm and larger	No specific mention	No specific mention	Yes	Yes
United States Food and Drug Administration	No	1–100 nm and larger	No specific mention	No specific mention	Yes	Yes
European Commission Biocides Directive	No	1–100 nm	50% by number	Yes	No	No
European Commission Cosmetics Directive	Yes	1–100 nm	No specific mention	Yes	No	Yes
European Commission Recommendation for a Definition	No	1–100 nm	50% by number	Yes	No	No
European Parliament and the Council of the European Union on the Provision of Food Information to Consumers	No	1–100 nm and larger	No specific mention	Yes	Yes	Yes

Source: DBT India (2020); Subramanian and Rajkishore (2018); Boverhof et al. (2015).

if it is possible to use less kg per hectare and fewer treatments per season with the use of NFs (Ojeda-Barrios et al., 2020). Due to redox potential, pH-dependent phenomena, and the availability of ligands in the soil, NFs have lower rates of nutrient release than conventional fertilizers while, on the other end, increasing their stability by adsorption or aggregation without resulting in changes in chemical variation or speciation (Raliya et al., 2018). SiO_2 and TiO_2 nanoparticles enhanced the capacity of plants to absorb nutrients and water, increasing the activity of nitrate reductase in soybeans (Lu et al., 2002). On a 2-week-old pearl millet plant, a concentration of nano-nitrogen (80 ppm) application led to a noticeably better root length, root area, root perimeter, average root diameter, number of tips, and total biomass. The outcome unequivocally showed that it is possible to biosynthesize nano-N for effective application as a nanofertilizer (Thomas et al., 2016).

The commercial production of NMs as NFs and plant growth biostimulants is regulated under Schedule VII of the Fertilizer (Inorganic, Organic, or Mixed) Control Order, 1985 (as amended up to July 2021) in India. The general specification and parameters of NF are defined in Part D: Methodology of Testing of Schedule VII of the FCO, 1985 (The Fertilizer Association of India, 2021, https://www.faidelhi.org/). The detailed general specifications and parameters of NF are mentioned in Table 2.2.

TABLE 2.2 General Specification of Nanofertilizer as per Schedule VII of Fertilizer (Inorganic, Organic or Mixed) Control Order (FCO), 1985.

Sr. No.	Parameters	Specifications
1.	Nitrogen concentration (%)	1–5
2.	Particle size in nano-meter (nm) in one dimension	20–50
	Physical particle size (as per TEM analysis)	20–80
	Hydrodynamic particle size (as per DLS analysis)	
	At least 50% of the particles in the tested samples should be in the above range, fulfilling the stipulated DBT guidelines	
3.	Zeta potential in mV (± scale)	>30
	Liquid nanofertilizer sample (measured directly)	
	Solid nanofertilizer sample (measured in 1:10 fertilizer: H_2O suspension)	
4.	Viscosity in cps	5–30
	Liquid nanofertilizer sample (measured directly)	
	Solid nanofertilizer sample (measured in 1:10 fertilizer: H_2O suspension)	
5.	pH	4.5–6.0
	Liquid nanofertilizer sample (measured directly)	
	Solid nanofertilizer sample (measured in 1:10 fertilizer: H_2O suspension)	

Source: The Fertilizer Association of India (2021), https://www.faidelhi.org/.

2.3.2 NANOPESTICIDES

Nanopesticides, which are nanoscale versions of conventional pesticides, can be made using NMs. Due to their specific delivery to pests or diseases, these nanopesticides have higher dispersibility, a greater surface area for better adherence to plant surfaces, and increased efficacy. They can reduce the number of pesticides used and their negative effects on the environment (Bratovcic et al., 2021). With the help of nanopesticides, it is possible to regulate the delivery of pesticides and get better results from a smaller chemical dose. It is possible to envision the first class of nanopesticides as having an active ingredient(s) that are made of manufactured NMs, such as metal nanoparticles (MNP) (such as silver and copper) or metal oxide nanoparticles (such as copper oxide, zinc oxide, titanium dioxide, manganese dioxide, and silicon dioxide) (Bratovcic et al., 2021; Lakshmeesha et al., 2020; Bratovcic, 2020; Shukla et al., 2020; Guilger-Casagrande et al., 2019). The production and utilization of nanopesticides aim to increase a pesticide's effectiveness and longevity, lessen the quantity of active components used, and minimize or completely remove any potential risks.

The term "encapsulation" is widely used in the production of nanopesticides and NFs, and it refers to the process of entrapping one substance inside another to create particles with diameters ranging from a few nm to a few mm. The fundamental justification for utilizing encapsulation is that some nutrients don't stay in the soil for very long or may interact with other soil elements in ways that have unfavorable impacts (Bratovcic and Suljagic, 2019). Both pesticides and fertilizers are processed via nanoencapsulation for commercial formulations. In order to increase stability, improve dispersion in aqueous media, and enable a controlled release of the active ingredient, nanoencapsulation is also used for pesticides, particularly hydrophobic ones. This technique improves stability, improves dispersion in aqueous media, and permits a controlled release of the active ingredient, which increases its effectiveness in the control of pests and lowers treatment doses for crops and human exposure (Grillo et al., 2016; Nuruzzaman et al., 2016). The majority of the polymers employed in nanoencapsulation for nanopesticides are of natural origin. Examples include polysaccharides like xanthan gum, alginate, chitosan, carrageenan, and modified starch, as well as proteins like gelatine, egg albumin, zein, and collagen (Bratovcic et al., 2021; Pestovsky and Martinez-Antonio, 2017). In order to create an adhesive nanopesticide (CNAP-HMS-PDAAM), Gao et al. (2020) used an adhesive hollow mesoporous silica hybrid with a well-specified spherical form as a nanocarrier of cyantraniliprole (CNAP). The findings showed that CNAP-HMS-PDAAM demonstrated long-term control efficacy against *Cnaphalocrocis medinalis* (Guenee) and *Chilo suppressalis* (Walker), mostly because of its potent adhesion to rice leaves and prolonged release capabilities. Due to their broad antimicrobial spectral range, copper-based (nano) insecticides in agroecosystems may have unexpected effects on soil microbial communities that are not their intended targets. Peixoto et al. (2021) investigated the effects of a commercial $Cu(OH)_2$-nanopesticide on the function, structure, and abundance of microbial communities over a period of 90 days at levels for both single and seasonal agricultural applications. This study identified that the soil microbial population is affected by $Cu(OH)_2$-nanopesticide, potentially changing its ecological function. In the current scenario, experts are focusing on the use of silica nanoparticles in agriculture. Crop development and productivity can be enhanced by using insecticides/ herbicides/fertilizers based on nanosilica. Plants can simply be injected with silica nanoparticles. It might alter the anabolism and catabolism of the plant (Rajiv et al., 2020).

Different regulatory agencies take different approaches to approving and regulating nanopesticides. Before approving them for commercial use, they normally evaluate their safety, efficacy, and potential environmental impact (DBT India, 2020, https://dbtindia.gov.in/). In India, nanopesticides (with or without nano- or non-nano-carriers) are registered for commercialization in accordance with the regulatory aspect requirements under Section 9 of the Insecticides Act, 1968 (Act 46 of 1968), with additional criteria for inclusion of nanopesticides as per the requirement of CIB&RC of the Directorate of Plant Protection, Quarantine, and Storage (DPPQS).

There are hundreds of nano-based manufacturers working in the agriculture market, but only a select few of them can be called industry titans. Some important manufacturers include Neufarm GmbH, IFFCO Limited, Kanak Biotech, Plant Vitality Ltd., Reed Mariculture Inc., Kimitec Group, Richfield Fertilisers Pvt., Danaflex Nano, Blue Planet Environmental, AgriLife, Bioteksa, BioLife Ltd., and Vive Crop Protection (https://product.statnano). Silver, silicon dioxide, titanium dioxide, and zinc oxide are the NMs that are most frequently reported in the agriculture industry. For nutrient control and crop protection, some commercial Nano-Agri-Products (NAPs) are offered on the market and are well-liked by farmers. Despite this, there are only a handful of NAPs on the market due to production and stringent regulatory restrictions that prevent the commercialization of nanoformulations. Some selected approved nano-agri-inputs (i.e., nanopesticides and NFs) that are available on the global market are mentioned in Table 2.3.

TABLE 2.3 Some of the Approved Nano-Agri-Input Products (NAIPs) That are Used in Agriculture Globally.

Nano-Agri-Input Product (NAIP) Name	Product Category	Manufacturer Name	Country of Origin	Ingredients Specification
BioCopp™	Nano disinfectant	AgriLife™	India	Nano copper (CuNP: 500 ppm), encapsulated by a biopolymer
Nano Chitosan	Nano disinfectant	AgriLife™	India	Nano Chitosan (1.0% w/v)
Somguard® SilCop	Nano disinfectant	AgriLife™	India	Nano silver (AgNP: 100 ppm) and nano copper (CuNP: 100 ppm), encapsulated by a biopolymer
NANO-SILVER	Nano disinfectant	Kanak Biotech	India	Unspecified silver nanopowder/nanoparticles
NANOCU®	Pesticide	Bio Nano Technology	Egypt	Unspecified copper nanoparticles
IFFCO NANO DAP (liquid)	Fertilizer	Indian Farmers Fertilizer Cooperative Limited (IFFCO)	India	Nanoclusters of phosphorus (P: 16.0% as P_2O_5 w/v) and Nitrogen (N: 8.0% w/v) functionalized with biopolymers (particle size: $\geq$100 nm)
IFFCO NANO UREA (liquid)	Fertilizer	Indian Farmers Fertilizer Cooperative Limited (IFFCO)	India	It contains 4% N (as nitrogen analogues) encapsulated in an organic matrix (nanoparticle size: 20–50 nm)
Nano ultra-fertilizer	Fertilizer	SMTET Eco-technologies Co., Ltd	Taiwan	5.5% organic matter; 10% T-N; 9% T-P_2O_5; 14% T-K_2O; 8% AC-P_2O_5; CA-K_2O; 3% CA-MgO
Green nano	Fertilizer	Green Organic world Co., Ltd	Thailand	0.5% N; 0.7% P_2O_5; 3.9% K_2O; 2% Ca; 0.2% Mg; 0.8% S; 1% Fe; 49 ppm Mn; 17 ppm Cu; 12 ppm Zn
PPC Nano	Fertilizer	WAI International Development Co., Ltd	Malaysia	19.6% M protein; 0.3% Na_2O; 2.1% K_2O; 1.7% $(NH_4)2$ SO_4; 76% diluent
Hero Super Nano	Fertilizer	World Connet Plus Myanmar Co., Ltd	Thailand	0.7% N; 2.3% P_2O_5; 8.9% K2O; 0.5% Ca; 0.2% Mg; 0.4% S; 12.08 pH
Nano calcium (magic green)	Fertilizer	PAC International Co., Ltd	Germany	77.9% $CaCO_3$; 7.4% $MgCO_3$; 7.47% SiO_2; 0.2% K; 0.03% Na; 0.02% P; 7.4 ppm Fe; 6.3 ppm Al_2O_3; 804 ppm Sr; 278 ppm sulphate; 174 ppm Ba; 172 ppm Mn; 10 ppm Zn
Nano Capsule (The Best)	Fertilizer	The Best International Co., Ltd	Thailand	0.5% N; 0.7% P_2O_5; 3.9% K_2O; 2% Ca; 0.2% Mg; 0.8% S; 2% Fe; 0.004% Mn; 0.007% Cu; 0.004% Zn

2.3.3 NANOHERBICIDES

Herbicides play a big part in protecting crops by controlling weeds. Its widespread use, though, has led to financial and environmental problems. Herbicides are administered in large quantities to crops since plants absorb them at rates below 1% (Schnoor et al., 2018). In order to encourage crop development, farmers frequently use herbicides in higher quantities than advised (Sookhtanlou et al., 2022). These actions encourage the growth of weeds that are resistant to herbicides. In agriculture, herbicide resistance is a serious issue that requires the development of new drugs and management techniques. The reduction of the use of pesticides and the harm they do to the environment is one of the main objectives of PA based on nanotechnology. The use of nanotechnology in the agricultural pesticide industry could provide a sustainable solution to the issue of chemical residues that doesn't pollute the environment. Nano-based particles are used to synthesize herbicides, while, on the other side, effective application of herbicide synthesis using nanoparticle-based delivery systems results in nanoherbicides. Nanoherbicides have the following benefits: they use fewer synthetic chemicals, have an efficient delivery mechanism, protect people, and have a quick response time. Nanoherbicides are a cutting-edge technology that corrects all of the problems with conventional herbicides. The development of weed resistance, which necessitates strategic planning for its ultimate eradication, is one of the main issues that conventional herbicides are currently confronting. Such issues may be resolved with the use of nanoherbicides with qualities such as high penetration capability and an efficient delivery system (Patil-Patankar and Sanghvi , 2022).

Herbicides (active ingredients) are charged with NMs before being applied in nanoherbicide formulations to promote plant bioavailability and improve weed control. The development of nanoherbicides depends on the choice of NM. The herbicidal substance must be able to fit inside the dimensions of the NM that is employed and, ideally, interact with the NM through chemical bonding (Yadav et al., 2023). In one study, the application of poly (ε-caprolactone) (PCL) nanocapsules containing atrazine that were 10 times diluted was found to be more effective than a commercial formulation that contained conventional atrazine doses at inhibiting fungal growth in *Amaranthus viridis* (slender amaranth) and *Bidens pilosa* (hairy beggarticks) (Sousa et al., 2018). In another study, the toxicity of the herbicides imazapic and imazapyr was successfully lowered using nanoencapsulation, potentially reducing the effect on unintended targets and the larger environment

(Maruyama et al., 2016). Biodegradable polymer-based nanoparticles have been proven to be effective herbicide transporters, enhancing weed control and safeguarding the active ingredient in crop fields. Takeshita et al. (2022) studied the control effects of *nano*MTZ (i.e., nanoencapsulation formulation of herbicide metribuzin) on the physiological parameters of *Ipomoea grandifolia* plants. It offered a substitute that would increase effectiveness and lessen environmental losses (Takeshita et al., 2022).

2.3.4 NANOFUNGICIDES

Around $45 billion in agricultural losses globally are attributed to phytopathogenic fungi each year. Pesticide consumption around the world totals 2.5 million metric tonnes (MMT) per year, costing around $100 billion. Since 90% of applied agrochemicals vanish in open fields due to overland flow, harming the ecosystem and increasing farmers' costs, chemical treatments for fungal control have negatively impacted the environment and slowed economic progress (Yadav et al., 2023; Kamel et al., 2015; Pimente, 2009). Nanotechnology provides a clever answer to environmental problems at the nanoscale. It is a new area of research, and there are numerous ways to synthesize nanoparticles. Nanofungicides are safe, eco-friendly compounds or materials that can be used to treat plant diseases. Because of their solubility, permeability, low dose, low dose–dependent toxicity, increased bioavailability, targeted delivery, increased bioavailability, and controlled release, nanofungicides are effective. Numerous metallic substances, including Cu, Ag, Zn, and TiO_2, are accessible for use as nanofungicides. The effectiveness and antifungal potential of nanoparticles are inversely correlated with their size (Periakaruppan et al., 2023; Kumari and Khan, 2017). According to reports, hydroponically or in irrigated areas, nanofungicides completely eradicate fungal infections from crops while posing no environmental dangers (Yadav et al., 2023; Sharma et al., 2012; Park et al., 2006). In order to get rid of fungal phytopathogens that adhere to the cytosolic membrane's S protein groups, nanoparticles (NPs) alter cell permeability, damage DNA, interfere with protein oxidation, and disrupt the cell's electron transport chain. NPs also produce reactive oxygen species and prevent nutrient uptake. To fight phytopathogens, which can also aid in plant growth, they are used as foliar sprays on the leaves. When used as nanofungicides, metallic NP-containing agrochemicals are widely used (Paul and Roychoudhury, 2021; Agrawal and Rathore, 2014). Taha et al. (2023) studied the effectiveness of nano-silica (nano-SiO_2) and nano-selenium (nano-Se) against the *Alternaria alternata* leaf spot disease in common beans (*Phaseolus vulgaris* L.). The *in vitro* studies revealed that the combined applications of Se and SiO_2 at half dosages had an efficacy rate of 77.8%, and nano-Se (100 ppm) had an effectiveness rate of 85.1% on *A. alternata* mycelial growth. The field study revealed that the application of nano-Se, nano-Se, and nano-SiO_2 considerably reduced the disease severity of *A. alternata*. The opportunity associated with copper and copper oxide nanoparticles (CuNPs and CuONPs, respectively) as nano-fungicides is brought about by their antimicrobial properties, in addition to the fact that Cu is an important metal micronutrient that serves as a cofactor for many enzymatic activities in plants. CuNPs and CuONPs have been used as nanofungicides against toxigenic fungi, and Thipe et al. in 2022 studied these compounds' modes of action. The ecotoxicology and safety of CuNPs and CuONPs as nanofungicides demonstrated that nanoCu-enabled products were declared safe for agricultural practises.

2.3.5 NANOSENSORS

Nanosensors are mechanical or chemical sensors that can be used to measure physical parameters like temperature or to detect the presence of chemical species and nanoparticles. Environmental pollution and a severe loss of soil and water quality are the results of intensive farming and unsustainable agricultural practises. Nanosensors are one of several technological advancements that have been produced to address the difficulties and limitations of modern agriculture (Beegum and Das, 2022). Recent developments in nanotechnology have produced tiny sensors with exceptional sensitivity and adaptability (Sagadevan and Periasamy, 2014). A variety of nanosensors, such as Förster resonance energy transfer (FRET)-based nanosensors, plasmonic nanosensors, electrochemical nanosensors based on carbon, antibody nanosensors, and nanowire nanosensors, have been used in plants. Nanosensors have made it possible to examine cellular processes and metabolic flux, keep track of the spatiotemporal dynamics of analytes, and find pathogens like viruses and fungi (Shaw and Honeychurch, 2022; Kosaka et al., 2014). PA depends on nanosensors in many ways. The soil's humidity, pesticide traces, nutritional needs, and plant diseases can all be detected with nanosensors in a variety of environmental settings. The advantages of nanosensors over traditional sensors have been demonstrated by their smaller detection limits, rapid response times, high surface-to-volume ratios, and more trustworthy results.

According to studies, the detection thresholds for atrazine, acetamiprid, and glyphosate are all between nanograms and micrograms, nanomoles, and micromoles, respectively (Beegum and Das, 2022).

Fertilizer application is crucial to raising agricultural output. But excessive fertilizer use can change the chemical ecology of the soil and constrict the amount of land that can be used for crop production. The development of direct connections between growers and plants is made possible by non-destructive nanosensors that can convert plant signals into digital signals, enabling controlled fertilizer release while reducing fertilizer use. Electrochemical nanosensors may also measure the concentration of different ions in the soil and provide information on the proper rates of fertilizer application (Shang et al., 2019). Electrodes that are ion-selective have been used to track the sap of broccoli and potatoes (Errebhi et al., 1998; Kubota et al., 1997). Heavy metal ions as well as ions necessary for plant growth, such as H^+, K^+, and Na^+, can be found using electrochemical nanosensors (Li et al., 2010; Esser et al., 2012). These ion-selective electrodes can be used in systems for the greenhouse industry to control liquid fertilization techniques (Shaw and Honeychurch, 2022; Gutiérrez et al., 2007). Likewise, as chemical fertilizers, recent decades have also seen an increase in concern about the harmful effects of pesticides on the agricultural ecosystem, which has led to the requirement for monitoring programs to ascertain the destiny and accumulation of pesticides in the soil (Sabzevari and Hofman, 2022). For efficient pesticide application and to prevent pesticide overexposure, it is important to comprehend the behavior of pesticide translocation. Real-time pesticide translocation in tomato plant tissues, including the flowers and leaves, has been seen using surface-enhanced Raman scattering (SERS) using gold nanoparticles (Yang et al., 2019). Additionally, self-assembling uniform SERS sensor films to detect pesticides have been made quickly using flame aerosol technology. In a simple, low-cost procedure, this method combines particle synthesis with easy film fabrication (Li et al., 2022).

2.4 ROLE OF NANO-AGRI-INPUT PRODUCTS IN AGRICULTURE 4.0

Nanotechnology has a significant potential to improve agricultural sustainability. It has an important role in the agricultural industry to raise crop yields, increase the generation of secondary metabolites, be employed as fertilizer and pesticide, identify crop infections quickly and accurately, and monitor soil conditions. Agriculture 4.0 (i.e., precision farming) based on nanotechnology depends on measuring crop-based and environmental characteristics using computers, GPS, other remote sensing equipment, and the application of nano-agri-inputs. Agricultural inputs that employ nanotechnology to improve their efficiency and efficacy in farming techniques are referred to as NAIPs (Yadav et al., 2023; Klerkx and Rose, 2020). Agriculture 4.0, also referred to as smart farming or futuristic farming technology, refers to the incorporation of cutting-edge technologies to enhance agricultural operations. Products made from nano-agri-inputs contribute significantly to Agriculture 4.0 by improving different facets of farming and providing a number of benefits. The following are some crucial functions played by nano-agri-input items in Agriculture 4.0:

2.4.1 ENHANCED NUTRIENT DELIVERY

The need for increasing food production on a worldwide scale is growing, demanding novel agricultural strategies that could raise crop output while reducing environmental impacts. Utilizing nanoscale materials, in particular NAIP, has been shown to increase the efficiency with which crops use nutrients while minimizing the negative impacts of over-fertilization. By encapsulating nutrients in nanoscale particles, nanotechnology provides tailored delivery of nutrients to particular plant tissues. This boosts nutrient utilization effectiveness, decreases waste, and improves nutrient absorption (Cicek and Nadaroglu, 2015). The process to produce NFs often involves encasing nutrients inside nano-scale carriers, such as metal oxide nanoparticles (NPs) or polymeric nanocapsules, which enable controlled nutrient release while minimizing nutrient losses from leaching or volatilization (Muhammad, 2019). Studies have demonstrated that the use of NFs can significantly increase the efficiency of nitrogen uptake by plants. In one study, the application of ZnO-NP-coated urea resulted in a 51% increase in wheat grain production compared to the control group (Dimkpa et al., 2020). Similarly, zinc oxide NPs have been shown to boost rice plants' ability to absorb zinc, which leads to higher zinc content in the grain and better plant development (Ali et al., 2019). Additionally, PA based on nanotechnology has the potential to improve nutrient utilization efficiency as well as foster the development of more specialized and long-term nutrient management techniques. For instance, farmers may now improve the rates and timing of nutrient administration based on real-time data thanks to nanosensors that have been developed to monitor soil nutrient levels (Senapaty et al., 2023). This strategy lessens the negative effects of agriculture on the environment, like greenhouse gas emissions and eutrophication, while also improving nitrogen uptake in plants (Yadav et al., 2023; Kaushal and Wani, 2017).

Moreover, another important aspect of mineral absorption is directly and indirectly associated with hidden hunger in human beings. For a healthy lifestyle, humans need over 20 minerals and 40 nutrients from their food. Malnutrition results from human diets, which, regrettably, frequently lack the critical nutrients that are needed (WHO, 2000). The likelihood of developing anemia, heart disease, cancer, blindness, and early mortality is decreased by a fortified diet. Fortifying the crops is necessary to combat this global threat. According to Smith (2000), and Bouis et al. (2017), there are numerous ways to address micronutrient deficiencies, including supplementation, dietary modification, industrial fortification, and biofortification of staple crops. Staples and other food sources can be changed or improved to increase the micronutrient content or bioavailability (Elemike et al., 2019). A quick and simple solution to the deficiency of these vital elements in soils and plants is the agronomic technique of biofortifying crops with micronutrients. It entails growing types that are high in vitamins and minerals. With this technique, the micronutrient content of farmed crops like cereals and legumes is increased. It is important to underline that developing nations might benefit more from the agronomic biofortification process (Velu et al., 2014; Elemike et al., 2019). In this context, NMs like fullerene (C_{70}), fullerol ($C_{60}(OH)_{20}$), carbon nanotubes (CNTs), ZnO, Fe_3O_4, Cu, TiO_2, CeO_2, Au, Ag, and Fe are among the most often investigated ones (Rico et al., 2011). They have an impact on soil fertility and plant nutrition in a variety of ways. These NM-based nutrients may also be helpful to encounter the issue of micronutrient deficiency in plant produce, resulting in hidden mineral deficiencies in humans, but the toxicity of these NMs should also be considered before their application in crop production systems for agronomic biofortification (Elemike et al., 2019; Chen et al., 2010).

2.4.2 PRECISION AGRICULTURE

PA is the science of enhancing agricultural yields and supporting management choices with cutting-edge sensor and analysis capabilities. The goal of PA is to increase the sustainability of agricultural production through observation, measurement, and response to temporal and spatial variability. Using information technology (IT), PA makes sure that soil and crops receive the exact nutrients they require for optimum health and yield. Additionally, this ensures environmental protection, sustainability, and financial success (Shaw and Honeychurch, 2022; Yadav et al., 2023). Nanotechnology is now widely used in precision farming as well as contemporary agriculture. Due to their small sizes, large surface areas, and unique optical properties, NMs are used in farming, plant protection, and the management of feeding and farming techniques. Regarding development and metabolic processes, plants respond to nanoparticles in various ways. Nanoencapsulation is essential for environmental protection because it prevents dangerous compounds from evaporating and leaching into the environment. By delivering precise and regulated agricultural input applications, nano-agri-input solutions support precision farming approaches. Farmers may make data-driven decisions for the best resource management by using nanosensors and NMs to monitor soil conditions, plant health, and environmental elements in real-time (Samreen et al., 2022). Nanosensors and wireless field networking are used in PA to monitor and manage farming techniques. It controls site-specific crops as well as features of pre- and post-harvesting. Investigating the intriguing qualities of the functional materials used to make nanobiosensors could aid in the precise analysis of soil moisture, nutrients, water, and phytopathogens in PA (Burrell et al., 2004; Antonacci et al., 2014). A typical biosensor is made up of four components: (i) a sensor, (ii) a signal conditioning unit, (iii) a microprocessor chip, and (iv) a radio module enabling wireless connection with the monitoring station (Dasgupta et al., 2017). Recent advances in nanotechnology have made it possible to analyse the biochemical interactions that different diseases cause in plant cells and tissues. A probe is introduced into the xylem conduit at the root base as part of the procedure. Ionic activity, radial electrical gradients, and xylem pressure are all measured by the probe (Wegner, 2012; Bandyopadhyay et al., 2013). Moreover, the term "satellite farming" also applies to PA. Precision farming aims to build a database or information system that will allow access to all of the farm's data. It makes use of numerous 21st-century technological advancements, some of which are listed below: Unmanned aerial vehicles (UAV), including drones, sensors, robots, remote sensing, smartphone apps, GPS (global positioning system), and GNSS (global navigation system) technologies, as well as phytogeomorphological methods and satellite images, are examples of UAVs (Shaikh et al., 2021, Figure 2.2).

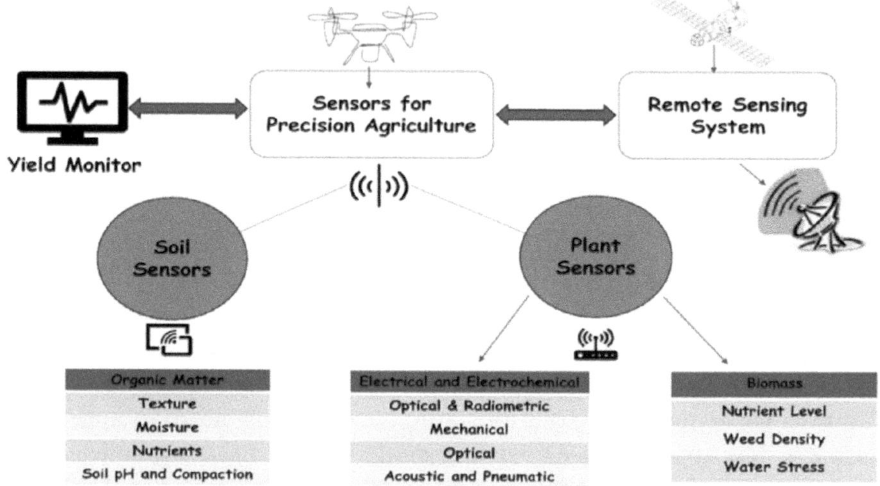

FIGURE 2.2 Graphical representation of nanosensors in precision agriculture.
Source: Adapted from Yadav et al. (2023).

Researchers from the Disruptive and Sustainable Technologies for Agricultural Precision (DiSTAP) in Singapore have created a novel kind of plant nanobionic optical sensor that can detect and monitor levels of the extremely toxic heavy metal arsenic in the underground environment in real-time. As arsenic is a contaminant in many common agricultural products such as rice, vegetables, and tea leaves, this development offers significant advantages over conventional methods used to measure arsenic in the environment. It could be crucial for both environmental monitoring and agricultural applications to protect food safety (Eurekalerat, 2020, https://www.eurekalert.org/).

2.4.3 CROP PROTECTION

Crops are continuously threatened by biotic and abiotic elements, including pests, diseases, toxic agrochemicals, climate change, and so on, which form the basis of human health and the economy. Nanotechnology is a cutting-edge strategy used to safeguard crops and uphold long-term sustainability. It operates under the tenet of "minimal usage with maximum effect." Crop protection can be improved with the help of nanotechnology. There is great promise for reducing chemical pesticide doses, boosting crop productivity, and fostering sustainable development through the nanoformulation or encapsulation of insecticides, fungicides, herbicides, and bactericides using NMs (Kumar et al., 2019). Pesticides, fungicides, and other agrochemicals can be encapsulated in NAIPs to improve targeted delivery and minimize environmental effects (Ojha et al., 2018). The agricultural nanoapplications include nanopesticide formulations, nanosensors, and the development of crop varieties with nanoparticle-mediated resistance. Different polymer-based NMs (like nanospheres, nanocapsules, nanogels, and micelles), porous inorganic materials (porous hollow nanoparticles, silica nanoparticles), lipid-based NMs (solid lipid nanoparticles, liposomes), and clay-based NMs are used for controlled release to ensure stability, long-term effects, and solubility for crop protection. Nanoparticles made of magnesium oxide, zinc oxide, copper oxide, and silver have all demonstrated positive antibacterial action against phytopathogens (Shaikh, 2022; Medina-Pérez et al., 2019; Ojha et al., 2018). According to numerous studies (Liu et al., 2021; Kah and Hofmann, 2014), nanotype pesticides are more effective at eliminating pests and are less likely to have negative effects on people. For example, when permethrin- and spinosad-loaded chitosan NPs were applied to *Drosophila melanogaster*, they displayed greater efficacy and bioavailability even at lower doses than free spinosad and free permethrin, and the nanocomposites caused the least damage to the ecological environment and humans (Sharma et al., 2019). The nanoinsecticide particles shrink and become more centralized after being encapsulated by NPs, giving them stability and a gradual release capability. These characteristics make insecticides more effective and less hazardous to people. Additionally, the biological toxicity of highly concentrated NPs offers a special route for directly preventing the growth of pathogens, bacteria, and viruses. For instance, a number of nanoinsecticides benefit from the poisonous properties of metallic NPs. Compared to bulk Al_2O_3 treatment, aluminum oxide (Al_2O_3) nanoparticles showed excellent potential for eradicating *Sitophilus oryzae* in stored rice (Das et al., 2019). Moreover, nanocarriers also amplify the efficacy and benefits of sex pheromones. For instance, when methyl eugenol-loaded nanogels were applied to guava orchards, more trap catches were made than in the group receiving only methyl eugenol as a control (Hellmann et al., 2011; Bhagat et al., 2013).

The production of nanobactericides and nanofungicides by metallic NPs appears to have a particularly high potential (Elmer et al., 2018). For instance, nanobactericides made of silver nanoparticles (Ag NPs) and holy basil leaf extract showed higher inhibition of *Xanthomonas axonopodis* pv. punicae on pomegranate when compared to free-leaf extracts (Sherkhane et al., 2018). Copper nanoparticles (Cu NPs) in two respective concentrations (1500 or

300 ppm) were also found to limit the development of five bacteria in bacterial activity experiments, including *Pseudomonas savastanoi* pv. Savastanoi, *Erwinia amylovora*, *Dickeya dadantii*, *Agrobacterium tumefaciens*, and *Dickeya dadantii* (Varympopi et al., 2020). Additionally, nonmetallic NPs can control plant diseases, whereas metallic NPs have the ability to inhibit bacteria and fungi. For instance, compared to the untreated group, chitosan NPs efficiently suppressed *Xanthomonas campestris* infection in chilli peppers (Esyanti et al., 2020).

Metallic NPs also prevent the growth of phytopathogenic fungi. For instance, compared to untreated plants, nickel ferrite ($NiFe_2O_4$) and cobalt ferrite ($CoFe_2O_4$) nanoparticles reduced the incidence of *Fusarium* wilt. Moreover, these NPs had no adverse impacts on the growth of capsicum plants (Sharma et al., 2017). Additionally, the complexes of nanoparticles (NPs) and fungicides exhibit concentrated particle size distributions and substantial specific surface areas, which enhance antifungal activity and lengthen fungicide release periods (Elmer and White, 2018). According to numerous investigations, fungicides encapsulated by NPs were incredibly effective at suppressing fungus (Liu et al., 2021; Pariona et al., 2019; Maluin et al., 2019). In comparison to pure hexaconazole, chitosan-hexaconazole NPs crosslinked with tripolyphosphate (TPP) improved the suppression of *Ganoderma boninense* growth (Maluin et al., 2019). In a variety of ways, including by interacting with nucleic acids, inducing plant immune responses, and delivering RNAi systems, NPs have emerged as viable management tools to stop viral invasion. CNTs applied topically to tobacco effectively reduced tobacco mosaic virus (TMV) symptoms compared to a control group that wasn't treated. In plants treated with CNTs, there was a drop in the relative expression level of viral coat proteins. Salicylic acid and abscisic acid concentrations in plants treated with CNTs were much higher than those in the control group. These findings demonstrate that CNTs prevent TMV infection by impeding viral mobility and multiplication.

Moreover, phytoviruses with high genetic diversity and dynamic evolution also cause significant losses in crop productivity (Farooq et al., 2021). Nanoparticles (NPs) have become effective management tools to thwart viral invasion in a number of ways, including by interacting with nucleic acids, activating plant immune responses, and delivering RNAi systems. TMV symptoms were dramatically reduced by the topical application of CNTs to tobacco as compared to a control group that received no treatment. Viral coat proteins' relative expression level decreased in plants exposed to CNTs. The levels of salicylic acid and abscisic acid in plants exposed to CNTs were significantly greater than those in the control group. These results show that CNTs inhibit viral movement and proliferation, thereby protecting against TMV infection (Adeel et al., 2021). As per the findings of El-Dougdoug et al. (2018), metallic nanoparticles (NPs), such as Ag NPs, can interact with the coat proteins of ToMV and PVY (Potato virus Y) viruses and stimulate a plant's immune system to prevent infection. When tomatoes were sprayed with Ag NPs, phytoviruses ToMV and PVY infections were lessened relative to an untreated control. In the meantime, ToMV-infected tomatoes had considerably higher levels of total soluble protein (TSP) and polyphenol oxidase (PPO) activity compared to the control. The effects of nanoparticles (NPs) in different concentrations and formulations on insects, pests, pathogens, and weeds are elaborated in Table 2.4.

2.4.4 WATER MANAGEMENT

The foundation of terrestrial ecosystems and agriculture is soil, which also plays a role in the nitrogen cycle, water infiltration and storage, plant support, and microbial activity. Rapid soil degradation and desertification processes reduce agricultural production, have a detrimental effect on rural families' economies, and lead to hunger and poverty. In areas experiencing desertification and/or drought, the application of nanotechnology (nanoclays and hydrogels) helps to improve the moisture retention capacity of soils and hence increase agricultural production. Nanoclays can incorporate a lot of water into their structure because of their tiny particles and large contact areas. Hydrogels are affordable, biodegradable, have little impact on the environment, and have a high water retention capacity (Fontagro, 2021, https://www.fontagro.org/). Mounier et al. (2023) examined the effects of iron oxide nanoparticles (Fe_3O_4, i.e., magnetite) on a lead-polluted water-soil-plant system during alternating periods of water stress. The outcome demonstrated that plants treated with iron oxide nanoparticles (Fe_3O_4) displayed a better water stress response and leaf water status. This was because of the enhanced retention of water in the soils containing nanomagnetite particles (NPsMagn). Additionally, Al-Saeedi (2022) investigated how five rates of silica nanoparticle (Si-NP) treatments affected cucumber (*Cucumis sativus* L.) yield, soil-soluble Si4+, the water characteristics curve, and water use efficiency after harvest. All study parameters revealed increases in values as the rate of silicon nanoparticle production increased. Similarly, Elahe et al. (2023) examined how oxide nanoparticles affected soil tensile strength

TABLE 2.4 Effects of Different Nanoparticles (NPs) in Crop Protection Systems.

Nanoparticles (NPs)	Type	Concentration	Original Pesticide	Target	Reference
Zinc oxide (ZnO)	Insecticide	20 mg/L	*Aspergillus niger*-mediated ZnO nanoparticles	*Holotrichia* sp.	Shukla et al. (2020)
Nanogel	Insecticide	12 mg/mL	Methyl eugenol	*Bactrocera dorsalis*	Bhagat et al. (2013)
Chitosan	Insecticide	10 mg/L	Spinosad/permethrin	*Drosophila melanogaster*	Sharma et al. (2019)
Aluminium oxide (Al_2O_3)	Insecticide	2 g/kg	---	*Sitophilus oryzae* L.	Das et al. (2019)
Chitosan	Fungicide	10 µg/L	Hexaconazole	*Ganoderma boninense*	Maluin et al. (2019)
Cobalt ferrite ($CoFe_2O_4$) and Nickel ferrite ($NiFe_2O_4$)	Fungicide	500 mg/L	---	*Dematophora necatrix* and *Colletotrichum gloeosporioides*	Sharma et al. (2017)
Copper (CuNPs)	Fungicide	0.5 mg/mL	---	*Fusarium oxysporum* and *Fusarium solani*	Pariona et al. (2019)
Nanogel	Herbicide	15 mL/L	Savory essential oil	weeds	Taban et al. (2020)
Solid lipid	Herbicide	0.3 kg/ha	Atrazine and simazine	*Raphanus raphanistrum*	De Oliveira et al. (2015)
Polysaccharide	Herbicide	0.5 g/L	Metsulfuron methyl	*Chenopodium album*	Kumar et al. (2017)
Chitosan	Bactericide	1 mg/mL	Streptomycin sulfate	*Fusarium oxysporum* and *Colletotrichum gloeosporioides*	Sharma et al. (2017)
Copper (CuNPs)	Bactericide	240 mg/L	---	*Pseudomonas savastanoi* pv. Savastanoi and *Pectobacterium carotovorum*	Varympopi et al. (2020)

TABLE 2.4 *(Continued)*

Nanoparticles (NPs)	Type	Concentration	Original Pesticide	Target	Reference
Silver (AgNPs)	Bactericide	15 mM	Holy basil (leaf extracts)	*Agrobacterium tumefaciens*, *Erwinia amylovora*, *Xanthomonas axonopodis* pv. Punicae, *Dickeya dadantii*	Sherkhane et al. (2018)
Silver (AgNPs)	Antiviral	50 mg/L	---	Tomato mosaic virus (TMV), Potato virus Y (PVY)	El-Dougdoug et al. (2018)
Carbon nanotubes (CNTs)	Antiviral	200 mg/L	---	Tobacco mosaic virus (TMV)	Adeel et al. (2021)
Silver (AgNPs)	Bactericide and fungicide	100 μM (bacteria) 200 μM (Fungi)	*Fusarium oxysporum* mediated NPs	Klebsiella pneumoniae, Bacillus anthracis, Aspergillus. *alternata, Aspergillus flavus,* *Trichoderma sp.*	Abdel-Hadi et al. (2023)

and water retention curves. The findings revealed that, compared to the control (0.39) and 5% nano-magnesium oxide (0.19), the levels of 1% and 3% nano-iron oxide considerably boosted water retention to 0.64 and 0.63, respectively. This increase in water retention is likely attributable to the treated soils' higher adsorption surfaces.

In Agriculture 4.0, NAIPs aid in effective water management. Nanosensors can measure the quantity of soil moisture, enabling farmers to use accurate irrigation schedules. Additionally, NMs can improve soil's ability to retain water, lowering the amount of water needed and decreasing water loss through evaporation (Bueno et al., 2017). In the current scenario, watering is now automated using wireless sensors. Numerous automation techniques can help save a lot of water. In agriculture 4.0, common and efficient automation techniques are applied, including simple Arduino-based irrigation, ARM-based irrigation, GSM-based irrigation, IoT-based irrigation systems, PLC-based irrigation, deep learning-based irrigation, machine learning-based irrigation, and fuzzy logic-based irrigation systems (Selvaraj et al., 2022). Extremely precise and responsive, nano-sensors are delicate, accurate, and sensitive tools for recognizing and reacting to physical, chemical, and biological impulses. Accurate assessment of soil moisture and temperature is greatly aided by the use of nanosensors in agriculture and, notably, soil science. These devices are more affordable, simpler to use, and more user-friendly than other similar instruments because they come with wireless communication networks. Some of the nanosensors used in soil analyses include graphene, CNTs, graphite wireless sensors, and nanotechnology-based microelectromechanical systems (Mahdizadeh and Najafi, 2019). Using computational techniques, Ruszczak and Boguszewska-Makowska (2022) evaluated the recalibration and precise characterization of frequently employed smart soil-moisture sensors. The study demonstrated the potential for the Extra Trees algorithm to reduce the median moisture estimation error from 2.035% for the baseline model to 0.808%.

2.4.5 *IMPROVED CROP QUALITY AND SHELF LIFE*

Excellent levels of nutrients and bioactive substances are typically found in fruits and vegetables. The key issues for the food processing sectors are maintaining their shelf life and preventing decay, quality deterioration, and microbiological spoilage of the fresh produce. Fresh vegetables can be preserved using a variety of methods, including physical, chemical, and biological preservation. Due to their increased cost and negative side effects, these treatments could not, however, be fully sustained. Nanotechnology, which has recently emerged, offers a cutting-edge, inventive, and environmentally friendly way to preserve fresh vegetables. As a result, there is a rise in marketability, less food waste, and enhanced product quality. Nano-agricultural inputs can lengthen shelf life and improve crop quality. Because of their large surface-to-volume ratio, higher barrier property, and better antimicrobial spectrum, inorganic, organic, and combined engineered NMs (nano-emulsions, nano-particles, nano-composites, nano-tracers, nano-sensors, and nano-packaging) are extensively used in improving the shelf life of fresh produce (Sagar et al., 2022; Dikshit et al., 2021).

Nanotechnology is a revolutionary strategy for developing new edible coating components with NMs having a size of 1–100 nanometers for the postharvest management of fresh fruit (Parisi et al., 2015). One of the most effective methods for increasing the shelf life of fresh fruits is the use of nanotechnology, and the NMs used for edible coatings also have special traits, including antibacterial and antifungal qualities (Singh et al., 2020). Nanoparticles (NMs), or nanocomposite materials, are used in a variety of packaging applications due to their ability to suppress microbial development. The most commonly used antibacterial NPs are metallic and semi-conducting. Different metallic NPs, such as Cu with Ag, Au, and Pt, exhibit varying levels of effectiveness. ZnO, TiO_2, WO_3, and MgO are well-known antibacterial NPs in the semiconducting family. Other antimicrobial NPs include synthetic antimicrobial agents like ethylenediaminetetraacetic acid (EDTA), quaternary ammonium salts, benzoic, propionic, and sorbic acids, as well as natural antimicrobial agents such as thymol, nisin, carvacrol, and antibiotics. Natural antimicrobial agents also include enzymes like lysozyme and peroxidase (Dikshit et al., 2021; Kalita and Baruah, 2019; Assefa and Admassu, 2013). One of the hardest problems in agriculture is reducing post-harvest losses in fruits and vegetables. Meena et al. (2020) presented a nanonet of Cu-chitosan nanoparticles, which has the capacity to increase the shelf-life of stored tomatoes, in order to address this problem. Cu-chitosan NPs (0.01–0.04%) treatment considerably reduced physiological weight loss, respiration rate, and fruit hardness during storage. Similarly, when spraying tomato fruits with zinc oxide nanoparticles after harvest, Sharifan et al. (2021) looked at whether this improved the fruits' ability to keep fresh and their zinc content. The findings also provided insight into the Zn metal's thermal conductivity, which facilitated tomato dehydration and color maturation by speeding up the process. In order to prolong the shelf life of tomatoes, this study offers crucial knowledge regarding the useful application of ZnO-NPs and their function in mineralization. Moreover, vegetable oils are usually microencapsulated for use in postharvest preservation, and nano-facilitated "active" and "smart" food packaging offers many advantages over traditional methods thanks to better antibacterial coatings, barrier characteristics, and mechanical strength (Mihindukulasuriya and Lim, 2014). Due to their antifungal or antibacterial qualities, nanocomposites play an excellent role in food packaging and preservation. Ag NPs-infused nanocomposites have been used to extend the shelf life of several crop-derived foods (Jiang et al., 2021; Bhuyan et al., 2019; Metak, 2015).

2.4.6 ENVIRONMENTAL SUSTAINABILITY

Nanotechnology makes use of numerous fields of study and technologies, including chemistry, biology, physics, and engineering. Manufacturing NMs, nanotubes, nanocomposites, nanofilters, and nanoparticles are a few examples of how nanotechnology is used to create systems with important environmental applications (Beni and Jabbari, 2022; Fang and Guo, 2018). NMs usually exhibit higher biogeochemical and chemical reactivity compared to other macroscale materials because they have a high surface-to-volume ratio. They are especially suitable for performing such operations because of their improved qualities and effectiveness. The creation and use of more nanoscale goods for environmental remediation have increased during the past few years. For instance, at numerous hazardous waste sites damaged by chlorinated solvents or oil spills, NMs have been used to clean up contaminated soil and groundwater. The physicochemical, surface, and optical-electronic properties of manufactured nanoparticles provide solutions to problems that were previously challenging to solve using conventional techniques. The environmental remediation process uses a range of techniques, such as absorption, surface assimilation, synthetic processes, catalysis using light, and filtering, to remove environmental pollutants (Jiang et al., 2018; Guerra et al., 2018). Sustainable agricultural practises are helped by NAIPs. They lessen the environmental effects of conventional farming practises by increasing nutrient use efficiency, decreasing pesticide use, and lowering water requirements. Further boosting sustainability, nanotechnology can facilitate the creation of environmentally friendly and biodegradable inputs (Roy et al., 2021).

Numerous nanodevices and sophisticated materials are or can be employed to increase driving safety and environmental impact. Intervehicle and road-to-vehicle communication, real-time processors with AI chips, sensors, and radars will all contribute to the safe implementation of autonomous or intelligent driving. The environmental impact of driving will be lessened by high-power semiconductor devices with wide band gap materials, high-performance motors with rare earth magnets, next-generation lithium (Li)-ion batteries, fuel cells with green hydrogen, and high-strength and lightweight materials like carbon-fiber-reinforced polymers (CFRP) (Pokrajac et al., 2021). Nanotechnology-based biofuel cells have the capacity to transform biological energy directly into electrical energy. In place of the metal present in typical cells, these cells use microbes and enzymes. These cells use wastes like carbon dioxide and human wastewater, which makes them unique and desirable. On the other hand, some people think that using nanotechnology in and of itself could cause fresh environmental issues, such as the emergence of new hazardous compounds and associated biological dangers (Rodríguez-Félix et al., 2021; Hosseini et al., 2018). When pollution detection technology is more widely available and less expensive, there is a need for ecosystem monitoring, process control, and environmental decision-making. Precise sensors have the capacity to support both the environment and human health in a sustainable way, and they can identify contaminants at the molecular level. In essence, a sensor is a form of energy converter that can recognize specific characteristics or occurrences (mechanical, physical, chemical, etc.) in its environment and present them as an output signal, which is often an electrical or optical signal (Brahmkhatri et al., 2021; Rasheed et al., 2020). The incorporation of platinum nanoparticles in the structure of CNTs boosts the sensor's sensitivity to NO_2. Moreover, the use of different metals in the construction of multiwalled CNTs provides sensors with great capabilities for the selective detection of harmful gases. Additionally, NH_3 and H_2S gases can be detected when silver and copper are present (Sharafeldin et al., 2021). Similarly, one of the most precise and sophisticated sensors for the detection of heavy metals is based on NMs called quantum dots. Due to their distinctive physicochemical characteristics, high specific surface area, and high reactivity, quantum dot NMs can be used as sensors (Yadav and Chowdhury, 2021). The technology for employing quantum dot nanoparticles also uses some biomass. Due to their great environmental and biological compatibility, these nanobiosensors are used as appealing fluorescent nanosensors. To find Fe (III) in effluents, for instance, carbon quantum dots produced from green algal waste are utilized (Liu et al., 2020). The use of NAIPs in farming practises has tremendous potential for

Agriculture 4.0, but their widespread adoption and regulatory issues need to be carefully considered.

2.5 POTENTIAL RISKS ASSOCIATED WITH THE USE OF NANOMATERIALS IN AGRICULTURE

Although the use of NAIPs has enhanced crop yields and revolutionized the industry, there are many profoundly negative impacts on aquatic and soil ecosystems, along with the health of the farmers and the general public who consume the nano-agri-input cultivated food (Yadav et al., 2023; Pathak et al., 2020). Some of the key risks are listed in the following sections.

2.5.1 ENVIRONMENTAL IMPACT

The area of nanotechnology includes ground-breaking inventions and cutting-edge uses for NMs. However, less is known about the toxicity and adverse consequences of NPs on biological systems and the environment. Inhalation, skin absorption, and ingestion are the three most prevalent ways that NPs are exposed, and particle size, surface chemistry, shape, and surface area are all used to categorize the toxicity of NPs. Due to their small size, NMs have a high potential for environmental dispersion and unforeseen ecological effects. They could contaminate the land, water, and air, interacting in unanticipated ways with creatures and ecosystems. Moreover, aquarian systems may be exposed to particles from solid waste, waste-water effluents, direct discharges, or unintentional spills (Jose et al., 2022).

2.5.2 HUMAN HEALTH CONCERNS

The potential health concerns brought on by exposure to nanoparticles are not well understood. NMs may be ingested, inhaled, or come into touch with skin in numerous ways, exposing farm workers and customers to them. To fully comprehend the long-term health impacts of exposure to these elements, more research is required. The toxicity of NPs may differ based on the precise nanoscale features that depend on reactivity, retention period, and distribution in the living body. By taking into account how living cells and NPs interact, NPs toxicity can be assessed. To ensure lower risk and sustainable application of nanotechnology, novel NMs must be synthesized biologically to lessen health risks (Jose et al., 2022; Ray et al., 2009). Researchers have extensively investigated the potential toxicologic effects of MNPs on human health using both *in vitro* and *in vivo* methods (Couto and Almeida, 2022). Insoluble particles can also hang around the lung for a very long time (Gupta and Xie, 2018; Blank et al., 2007). The lung may become injured and experience biological reactions if particles stay there for an extended period of time. Because of their incredibly small size, inhaled ultrafine particles (UFPs) may also settle in the olfactory mucosa and subsequently move on to the central nervous system (CNS), where they could cause neurotoxicity (Asgharian and Prince, 2006; Mühlfeld et al., 2008). In fact, due to their size and makeup, some nanoparticles have been proven to result in organ destruction and oxidative stress, which permanently harm cells. In a study conducted by Magrez et al. (2006) to evaluate the hazardous impact of carbon-based nanoparticles on lung cancer cells, the researchers found evidence that these particles may produce size-dependent cytotoxicity.

2.5.3 DEVELOPMENT OF RESISTANT PESTS

Due to their special characteristics, NMs have the potential to be an essential part of the majority of industrial products, fertilizers, medicines, and medication research. There is currently insufficient knowledge regarding the dangers of nanoparticle toxicity and how it affects the environment, crop protection systems, and human health. NMs have the ability to cross biological membranes, gain access to organs, and swallow or inhale them to enter the bloodstream. This might cause genotoxicity and biological effects. Continuous use of pesticides or antimicrobials based on NMs may cause the emergence of resistant pest populations, which could present serious difficulties for pest management tactics (Yousef et al., 2023; Forest, 2022). The effectiveness of pesticides with controlled release on the intended pathogen can be increased using nanopesticides. However, it has been noted that the targeted disease only receives 0.1% of the nanopesticide. In contrast, the remaining 99.9% goes to the pathogen's environment, which causes a loss of biodiversity, water and soil contamination, and increases disease and pest resistance (de Albuquerque et al., 2020).

2.5.4 BIOACCUMULATION IN THE FOOD CHAIN

The transfer of NMs from plants to animals and humans could result in their accumulation in the food chain. Concerns are raised regarding the potential effects of this bioaccumulation on higher trophic levels. Nanoparticles are released into waste streams from the constantly growing number of consumer products that incorporate NMs (Dinda, 2018). Judy et al. (2011) examined the transmission of gold (Au) NPs using *Nicotiana tabacum* L. cv. *Xanthi* and *Manduca sexta* (tobacco hornworm) as model organisms for plant uptake and the potential for trophic transfer. The findings show that gold nanoparticles were transferred trophically and biomagnified from a primary producer to a primary consumer. This finding has significant ramifications for the dangers of nanotechnology, including the possibility of human exposure. Similarly, Kuehr et al. (2021) assessed the bioaccumulation of NMs using freshwater invertebrate species. A scientific analysis of engineered NMs (ENMs) that bioaccumulate in organisms and biomagnify in the food web may put tremendous strain on the environment. Moreover, in the aquatic food chain, Li et al. (2022) investigated the effects of nano-TiO_2 on bioaccumulation and biomagnification. Overall, the research supported the hypothesis that nano-TiO_2 was harmful to plankton and posed a serious environmental risk to aquatic ecosystems.

2.5.5 REGULATION AND MONITORING CHALLENGES

Although nanotechnology is a rapidly developing industry, current regulations may not be adequate to address the particular hazards posed by using nanoparticles in agriculture. Monitoring and evaluating the effects of NM use on the environment and human health may be difficult. Global population growth has increased food consumption, so businesses have already started to offer nano-based products such as nano-encapsulated nutrients and agrochemicals, antimicrobials, and food packaging to meet the demand. Applications for approval for the marketing of nano-based products must demonstrate that such new products may be used safely without damaging the customer or the environment. Many countries have been actively examining whether their legal systems are adequate to handle nanotechnologies. Overall, as this field of study keeps delivering new scientific interventions that need to be screened by regulatory bodies, the safety precautions and rules for nanotechnology-based agri-products need to be revised from time to time (Kumari et al., 2023; Raj et al., 2021). In this context, the DBT, Ministry of Science and Technology, Government of India, in 2020, released guidelines for the evaluation of nano-based agricultural inputs (NAIP) and food products (DBT India, 2020, https://dbtindia.gov.in/).

2.5.6 ETHICAL AND SOCIETAL CONCERNS

In terms of their overall influence on sustainable agriculture, food security, and possibly unequal access to cutting-edge technologies among different areas or socioeconomic groups, the use of NMs in agriculture presents broader ethical and societal problems. There is a need for direction in deciding on hazards, risks, and controls because the possible health impacts of occupational exposure to nanoparticles are not yet fully understood scientifically. Decision-makers, in particular employers, employees, investors, and health authorities, may find it valuable to understand the ethical problems involved (Van de Voorde and Jeswani, 2021; Schulte and Salamanca-Buentello, 2007). Additionally, a focus on small enterprises and the adoption of a global perspective may help society's discussions about the dangers of nanotechnologies at work. Moreover, it is also essential to conduct thorough risk analyses, create suitable legislation, and set responsible usage standards for NMs in agriculture in order to manage these risks. The safe and sustainable integration of nanotechnology in agriculture depends on open communication between stakeholders, including scientists, politicians, farmers, and the general public (Khan, 2015; Bainbridge, 2004).

2.6 CONCLUSION AND FUTURE PERSPECTIVES

Nanotechnology is the manipulation of materials and structures at the nanoscale. It has the potential to revolutionize a number of industries, including agriculture, by providing innovative solutions to problems relating to crop productivity, resource use, and environmental sustainability. Using nanotechnology in agriculture is frequently referred to as "nano-agriculture" and is the backbone of resilient agriculture or Agriculture 4.0. The main ingredient of nano-agriculture is NMs that are used to improve nutrient delivery, plant growth, production, and plant protection. Their use in agriculture has drawn more and more attention. Different NMs as NAIPs (NFs, nanopesticides, nanoherbicides, nanofungicides, and nanosensors) have been shown to have potential impacts on Agriculture 4.0. As a futuristic technology, Agriculture 4.0 is the integration of cutting-edge technologies into agricultural practises, including the IoT, artificial intelligence, big data, and automation. Agriculture 4.0 and nanotechnology have the potential to completely transform how we produce, run, and maintain agricultural systems. However, it is necessary to examine the utilization of NMs' risk. Through improved management and conservation inputs, PA based on nanotechnology (i.e., Agriculture 4.0) may boost crop productivity. In conclusion, Agriculture 4.0 and the convergence of nanotechnology offer hope for making agriculture a more productive, sustainable, and efficient industry. But in order to maximize the advantages while minimizing potential hazards, rigorous implementation, regulatory supervision, and ongoing research are essential. Moreover, there is a requirement for the harmonization of regulatory frameworks for handling potential health hazards, transboundary movements or trade, and environmental risks associated with NAIPs worldwide.

KEYWORDS

- **Agriculture 4.0**
- **crop protection**
- **environmental sustainability**
- **ethical and societal concerns**
- **nano-agri-input-products (NAIPs)**
- **nanosensors**
- **nanotechnology**
- **precision agriculture**

REFERENCES

Abdel-Hadi, A., Iqbal, D., Alharbi, R., Jahan, S., Darwish, O., Alshehri, B., Banawas, S., Palanisamy, M., Ismail, A., Aldosari, S., Alsaweed, M., Madkhali, Y., Kamal, M., & Fatima, F. (2023). Myco-synthesis of silver nanoparticles and their bioactive role against pathogenic microbes. *Biology*, *12*, 661. https://doi.org/10.3390/biology12050661

Adeel, M., Farooq, T., White, J. C., Hao, Y., He, Z., & Rui, Y. (2021). Carbon-based nanomaterials suppress tobacco mosaic virus (TMV) infection and induce resistance in *Nicotiana benthamiana*. *Journal of Hazardous Materials*, *404*, 124167.

Adhikari, T., Kundu, S., Meena, V., & Rao, A. S. (2014). Utilization of nano rock phosphate by maize (*Zea mays* L.) crop in a vertisol of Central India. *Journal of Agricultual Science of Technology*, *4*, 5A.

Agrawal, S., & Rathore, P. (2014). Nanotechnology pros and cons to agriculture: A review. *International Journal of Current Microbiology and Applied Science, 3*, 43–55.

Ahmar, S., Mahmood, T., Fiaz, S., Mora-Poblete, F., Shafique, M. S., Chattha, M. S., & Jung, K. H. (2021). Advantage of nanotechnology-based genome editing system and its application in crop improvement. *Frontiers in Plant Science*, *12*, 663849. DOI: 10.3389/fpls.2021.663849

Ali, S., Rizwan, M., Noureen, S., Anwar, S., Ali, B., Naveed, M., Abd_Allah, E. F., Alqarawi, A. A., & Ahmad, P. (2019). Combined use of biochar and zinc oxide nanoparticle foliar spray improved the plant growth and decreased the cadmium accumulation in rice (Oryza sativa L.) plant. *Environmental Science and Pollution Research*, *26*, 11288–11299.

Al-Saeedi, A. H. (2022). Enhancement of soil water characteristics curve (SWCC) and water use efficiency of cucumber (*Cucumis sativus* L.) in sandy soils by using silica nanoparticles. *Journal of King Saud University-Science*, *34*(4), 10192. https://doi.org/10.1016/j.jksus.2022.101926

An, C., Sun, C., Li, N., Huang, B., Jiang, J., Shem, Y., Wang, C., Zhao, X., Cui, B., Wang, C., Li, X., Zhan, S., Gao, F., Zeng, Z., Cui, H., & Wang, Y. (2022). Nanomaterials and nanotechnology for the delivery of agrochemicals: Strategies towards sustainable agriculture. *Journal of Nanobiotechnology*, *20*, 11. https://doi.org/10.1186/s12951-021-01214-7

Antonacci, A., Arduini, F., Moscone, D., Palleschi, G., & Scognamiglio, V. (2018). Nanostructured (Bio)sensors for smart agriculture. *TrAC Trends in Analytical Chemistry*, *98*, 95–103.

Asgharian, B., & Price, O. T. (2006). Airflow distribution in the human lung and its influence on particle deposition. *Inhalation Toxicology*, *18*(10), 795–801.

Ashfaq, A., Khursheed, N., Fatima, S., Anjum, Z., & Younis, K. (2022). Application of nanotechnology in food packaging: Pros and Cons. *Journal of Agriculture and Food Research*, *7*, 100270. https://doi.org/10.1016/j.jafr.2022.100270

Assefa, Z., & Admassu, S. (2013). Development and characterization of antimicrobial packaging films. Journal of Food Processing & Technology, 4, 1–9.

Bainbridge, W. S. (2004). Social and ethical implications of nanotechnology. In B. Bhushan (Eds.), *Springer Handbook of Nanotechnology*. Springer, Berlin. https://doi.org/10.1007/3-540-29838-X_38

Bandyopadhyay, S., Peralta-Videa, J. R., & Gardea-Torresdey, J. L. (2013). Advanced analytical techniques for the measurement of nanomaterials in food and agricultural samples: A review. *Environmental Engineering Science*, *30*, 118–125.

Beegum, S., & Das, S. (2022). Nanosensors in agriculture. In S. Ghosh, S. Thongmee & A. Kumar, (Eds.), *Agricultural nanobiotechnology: Biogenic nanoparticles, nanofertilizers and nanoscale biocontrol agents* (pp. 465–478). Woodhead Publishing Series in Food Science, Technology and Nutrition. https://doi.org/10.1016/B978-0-323-91908-1.00012-2

Beni, A. A., & Jabbari, H. (2022). Nanomaterials for environmental applications. *Results in Engineering*, *15*, 100467. https://doi.org/10.1016/j.rineng.2022.100467

Bhagat, D., Samanta, S. K., & Bhattacharya, S. (2013). Efficient management of fruit pests by pheromone nanogels. *Scientific Reports*, *3*, 1294.

Bhuyan, D., George, W. G., & Ratul, K. D. (2019). Prospects and application of nanobiotechnology in food preservation: Molecular perspectives. *Critical Reviews in Biotechnology*, *39*(6), 759–778.

Blank, F., Gehr, P., & Rothen-Rutishauser, B. (2007). *In vitro* human lung cell culture models to study the toxic potential of nanoparticles. In S. C. Sahu & D. A. Casciano (Eds.) *Nanotoxicity: From in vitro and in vivo models to health risks* (pp. 379–96). Wiley.

Bouis, H. (2017). An overview of the landscape and approach for biofortification in Africa. African Journal of Food, Agriculture, Nutrition and Development, *17*(2). 10.18697/ajfand.78.HarvestPlus01

Boverhof, D. R., Bramante, C. M., Butala, J. H., Clancy, S. F., Lafranconi, M., West, J., & Gordon, S. C. (2015). Comparative assessment of nanomaterial definitions and safety evaluation considerations. *Regulatory Toxicology and Pharmacology*, *73*(1), 137–150.

Brahmkhatri, V., Pandit, P., Rababaware, P., D'Souza, A., & Kurkuri, M. D. (2021). Recent progress in detection of chemical and biological toxins in Water using plasmonic nanosensors. *Trends in Environmental Analytical Chemistry*, *30*(2019), e00117. DOI: 10.1016/j.teac.2021.e00117

Brandt, K., & Barrangou, R. (2019). Applications of CRISPR technologies across the food supply chain. *Annual Review of Food Science and Technology*, *10*, 133–150.

Bratovcic, A. (2020). Biosynthesis of green silver nanoparticles and Its UV-Vis characterization. *International Journal of Innovative Science, Engineering and Technology*, *7*, 170–176.

Bratovcic, A., Hikal, W., Said-Al Ahl, H., Tkachenko, K., Baeshen, R., Sabra, A., & Sany, H. (2021). Nanopesticides and nanofertilizers and agricultural development: Scopes, advances and applications. *Open Journal of Ecology*, 11, 301–316. DOI: 10.4236/oje.2021.114022

Bratovcic, A., & Suljagic, S. (2019). Micro- and nano-encapsulation in the food industry. *Croatian Journal of Food Science and Technology*, *11*, 113–121. https://doi.org/10.17508/CJFST.2019.11.1.17

Bronson, K. (2019). Looking through a responsible innovation lens at uneven engagements with digital farming. *NJAS - Wageningen Journal of Life Sciences*, *90–91*, 100294. https://doi.org/10.1016/j.njas.2019.03.001

Bueno, C. C., Garcia, P. S., Steffens, C., Deda, D. K., & de Lima Leite, F. (2017). Nanosensors. In O. Da Oliveira Jr, F. L. G. Marystela, F. D. Leite, & A. L. Da Róz. *Nanoscience and its applications: Micro and nano technologies* (pp. 121–153). https://doi.org/10.1016/B978-0-323-49780-0.00005-3

Burrell, J., Brooke, T., & Beckwith, R. (2004). Sensor and actuator networks—Vineyard computing: Sensor networks in agricultural production. *IEEE Pervasive ComputIng*, *3*, 38–45.

Chen, R., Ratnikova, T. A., Stone, M. B., Lin, S., Lard, M., Huang, G., Hudson, J. A. S., & Ke, P. C. (2010). Differential uptake of carbon nanoparticles by plant and mammalian cells. Small, 6, 612–617.

Cicek, S., & Nadaroglu, H. (2015). The use of nanotechnology in the agriculture. *Advanced Nano Res*earch, *3*, 207.

Couto, C., & Almeida, A. (2022). Metallic nanoparticles in the food sector: A mini-review. Foods, *11*, 402. DOI: 10.3390/foods11030402

Das, S., Yadav, A., & Debnath, N. (2019). Entomotoxic efficacy of aluminium oxide, titanium dioxide and zinc oxide nanoparticles against *Sitophilus oryzae* (L.): A comparative analysis. *Journal of Stored Products Research*, *83*, 92–96.

Dasgupta, N., Ranjan, S., & Ramalingam, C. (2017). Applications of nanotechnology in agriculture and water quality management. *Environmental Chemistry Letters*, *15*, 591–605.

de Albuquerque, F. P., Preisler, A. C., Fraceto, L. F., Oliveira, H. C., & de Castro, V. L. S. S. (2020). Overview of nanopesticide environmental safety aspects and regulatory issues: The case of nanoatrazine. In L. F. Fraceto, V. L. S. S. de Castro, R. Grillo, D. Ávila, H. C. Oliveira & R. Lima (Eds.), *Nanopesticides*. Springer, Cham. https://doi.org/10.1007/978-3-030-44873-8_9

Del Prado-Audelo, M. L., García Kerdan, I., Escutia-Guadarrama, L., Reyna-González, J. M., Magaña, J. J., & Leyva-Gómez, G. (2021). Nanoremediation: Nanomaterials and nanotechnologies for environmental cleanup. *Frontiers in Environmental Science, 9*, 793765. DOI: 10.3389/fenvs.2021.793765

De Oliveira, J. L., Campos, E. V. R., da Silva, C. M. G., Pasquoto, T., Lima, R., & Fraceto, L. F. (2015). Solid lipid nanoparticles co-loaded with simazine and atrazine: Preparation, characterization, and evaluation of herbicidal activity. *Journal of Agricultural and Food Chemistry, 63*, 422–432.

Department of Biotechnology, India (DBT, India). (2020). Guidelines for evaluation of nano-based agri-input (NAIP) and food products in India. https://dbtindia.gov.in/sites/default/files/uploadfiles/Guidlines%20for%20Evaluation%20of%20Nano%20Based%20Agri%20Input%20and%20Food%20Products%20in%20India.pdf

DeRosa, M. C., Monreal, C., Schnitzer, M., Walsh, R., & et al. (2010) Nanotechnology in fertilizers. *Nature Nanotechnology, 5*, 91. https://doi.org/10.1038/nnano.2010.2

Dikshit, P. K., Kumar, J., Das, A. K., Sadhu, S., Sharma, S., Singh, S., Gupta, P. K., & Kim, B. S. (2021). Green synthesis of metallic nanoparticles: Applications and limitations. *Catalysts, 11*, 902. https://doi.org/10.3390/catal11080902

Dimkpa, C. O., Andrews, J., Fugice, J., Singh, U., Bindraban, P. S., Elmer, W. H., Gardea-Torresdey, J. L., & White, J. C. (2020). Facile coating of urea with low-dose ZnO nanoparticles promotes wheat performance and enhances Zn uptake under drought stress. *Frontiers in Plant Science, 11*, 168.

Dinda, S. C. (2018). Bioaccumulation and toxic profiling of nanostructured particles and materials. In A. C. Gomes & M. P. Sarria (Eds.), *Unravelling the safety profile of nanoscale particles and materials*. DOI: 10.5772/intechopen.74802, https://www.intechopen.com/chapters/59716

El-Dougdoug, N. K., Bondok, A. M., & El-Dougdoug, K. A. (2018). Evaluation of silver nanoparticles as antiviral agent against ToMV and PVY in tomato plants. *Middle East Journal of Applied Sci*ence, *8*, 100–111.

Elemike, E. E., Uzoh, I. M., Onwudiwe, D. C., & Babalola, O. O. (2019). The role of nanotechnology in the fortification of plant nutrients and improvement of crop production. *Applied Sciences, 9*(3), 499. https://doi.org/10.3390/app9030499

Elmer, W. H., Ma, C., & White, J. C. (2018). Nanoparticles for plant disease management. *Current Opinion in Environmental Science & Health, 6*, 66–70.

Elmer, W., & White, J. C. (2018). The future of nanotechnology in plant pathology. *Annual Review of Phytopathology, 56*, 111–133.

Errebhi, M., Rosen, C. J., & Birong, D. E. (1998). Calibration of a petiole sap nitrate test for irrigated "Russet Burbank" potato. *Communications in Soil Science and Plant Analysis, 29*, 23–35.

Esser, B., Schnorr, J. M., & Swager, T. M. (2012). Selective detection of ethylene gas using carbon nanotube-based devices: Utility in determination of fruit ripeness. *Angewandte Chemie International Edition, 51*, 5752–5756.

Esyanti, R. R., Farah, N., Bajra, B. D., Nofitasari, D., Martien, R., Sunardi, S., & Safitri, R. (2020). Comparative study of nano-chitosan and synthetic bactericide application on chili pepper (*Capsicum annuum* L.) infected by *Xanthomonas campestris*. *AGRIVITA Journal of Agricultural Sci*ence, *42*, 11.

Eurekalerat. (2020). SMART researchers develop plant nanobionic sensor to monitor arsenic levels in soil. https://www.eurekalert.org/news-releases/660399 (accessed August 5, 2023)

Farooq, T., Adeel, M., He, Z., Umar, M., Shakoor, N., da Silva, W., Elmer, W., White, J. C., & Rui, Y. (2021). Nanotechnology and plant viruses: An emerging disease management approach for resistant pathogens. *ACS Nano*, *15*, 6030–6037.

Firbank, L. G., Attwood, S., Eory, V., Gadanakis, Y., Lynch, J. M., Sonnino, R., & Takahashi, T. (2018). Grand challenges in sustainable intensification and ecosystem services. *Frontiers in Sustainable Food Systems, 2*, 7. DOI: 10.3389/fsufs.2018.00007

Fontagro. (2021). Nanotechnology in soil moisture management. https://www.fontagro.org/new/proyectos/nanotecnologiahumedales/en (accessed June 5, 2023)

Forest, V. (2022). Experimental and computational nanotoxicology—complementary approaches for Nanomaterial Hazard Assessment. *Nanomaterials*, *12*(8), 1346. https://doi.org/10.3390/nano12081346

Fraceto, L. F., Grillo, R., de Medeiros, G. A., Scognamiglio, V., Rea, G., & Bartolucci, C. (2016). Nanotechnology in agriculture: Which innovation potential does it have? *Frontiers in Environmnental Science*, *4*, 20.

Gao, Y., Li, D., Li, D., Xu, P., Mao, K., Zhang, Y., Qin, X., Tang, T., Wan, H., Li, J., Guo, M., & He, S. (2020). Efficacy of an adhesive nanopesticide on insect pests of rice in field trials. *Journal of Asia-Pacific Entomology*, *23*, 1222–1227. https://doi.org/10.1016/j.aspen.2020.08.013

Grieve, D., Duckett, D., Collison, M., Boyd, L., West, J., Yin, H., Arvin, F., & Pearson, S. (2019). The challenges posed by global broadacre crops in delivering smart agri-robotic solutions: A fundamental rethink is required. *Global Food Security*, *23*, 116–124. https://doi.org/10.1016/j.gfs.2019.04.011

Grillo, R., Abhilash, P. C., & Fraceto, L. F. (2016). Nanotechnology applied to bio-encapsulation of pesticides. *Journal of Nanoscience and Nanotechnology*, *16*, 1231–1234. https://doi.org/10.1166/jnn.2016.12332

Guerra, F., Attia, M., Whitehead, D., & Alexis, F. (2018). Nanotechnology for environmental remediation: Materials and applications. *Molecules*, *23* (7), 1760.

Guilger-Casagrande, M., Germano-Costa, T., Pasquoto-Stigliani, T., Fraceto, L. F., & Lima, R. (2019). *Biosynthesis of silver nanoparticles employing trichoderma harzianum with enzymatic stimulation for the control of sclerotinia sclerotiorum. Scientific Reports*, *9*, 14351. https://doi.org/10.1038/s41598-019-50871-0

Gupta, R., & Xie, H. (2018). Nanoparticles in daily life: Applications, toxicity and regulations. *Journal of Environmental Pathology Toxicology and Oncology*, *37*(3), 209–230. DOI: 10.1615/JEnvironPatholToxicolOncol.2018026009. PMID: 30317972; PMCID: PMC6192267.

Gutiérrez, M., Alegret, S., Cáceres, R., Casadesús, J., Marfà, O., & del Valle, M. (2007). Application of a potentiometric electronic tongue to fertigation strategy in greenhouse cultivation. *Computers and Electronics in Agriculture*, *57*, 12–22.

Hellmann, C., Greiner, A., & Wendorff, J. H. (2011). Design of pheromone releasing nanofibers for plant protection. *Polymers for Advanced Technologies*, *22*, 407–413.

Hermans, F. (2018). The potential contribution of transition theory to the analysis of bio-clusters and their role in the transition to a bioeconomy. *Biofuels, Bioproducts and Biorefining*, *12*, 265–276.

Hosseini, M., Abdelrazek, A. H., Sadri, R., Mallah, A. R., Kazi, S. N., Chew, B. T., Rozali, S., & Yusoff, N. (2018). Numerical study of turbulent heat transfer of nanofluids containing eco- friendly treated carbon nanotubes through a concentric annular heat exchanger. *International Journal of Heat and Mass Transfer, 127*, 403–412. 10.1016/j.ijheatmasstransfer.2018.08.040

Hussain, T. (2017). Nanocides: Smart delivery system in agriculture and horticultural crops. *Advances in Plants and Agriculture Research, 6*(6), 175. DOI: 10.15406/apar.2017.06.00233

Javid, M., Haleem, A., Singh, R. P., & Suman, R. (2022). Enhancing smart farming through the applications of Agriculture 4.0 technologies. *International Journal of Intelligent Networks, 3*, 150–164. https://doi.org/10.1016/j.ijin.2022.09.004

Jiang, M., Song, Y., Kanwar, M. K., Ahammed, G. J., Shao, S., & Zhou, J. (2021). Phytonanotechnology applications in modern agriculture. *Journal of Nanobiotechnology, 19*, 430. https://doi.org/10.1186/s12951-021-01176-w

Jiang, D., Zeng, G., & Huang, D. (2018). Remediation of contaminated soils by enhanced nanoscale zero valent iron. *Environmental Research, 163*, 217–227.

Jose, J. P. A., Younus, L. A., Reddy, K. B., Sana, S. S., Gangadhar, L., Hou, T., Chakravorty, A., & Bhardwaj, P. (2022). Environmental impact on toxicity of nanomaterials. In R. P. Singh, A. R. Rai, A. Abdala & R. G. Chaudhary (Eds), *Biogenic sustainable nanotechnology: Trends and progress, micro and nano technologies* (pp. 315–355). https://doi.org/10.1016/B978-0-323-88535-5.00011-1

Judy, J. D., Unrine, J. M., & Bertsch, P. M. (2011). Evidence for biomagnification of gold nanoparticles within a terrestrial food chain. *Environmental Science & Technology, 45*(2), 776–781.

Kah, M., & Hofmann, T. (2014). Nanopesticide research: Current trends and future priorities. *Environment International, 63*, 224–235.

Kalita, D., & Baruah, S. (2019). The impact of nanotechnology on food. In Do Nascimento, R., Ferreira, O. P., De Paula, A. J. & Neto, V. D. O. S. (Eds.), Nanomaterials Applications for Environmental Matrices (pp. 369–379). Elsevier.

Kamel, A., Abd-Elsalam, K. A., & Alghuthaymi, M. A. (2015). Nanobiofungicides: Is it the next-generation of fungicides? *Journal of Nanotechnology and Material Science, 2*, 38–40.

Kaushal, M., & Wani, S. P. (2017). Nanosensors: Frontiers in precision agriculture. In *Nanotechnology* (pp. 279–291).

Khan, A. (2015). Ethical and social implications of nanotechnology. *QScience Proceedings (Engineering Leaders Conference 2014)*, (p. 57). http://dx.doi.org/10.5339/qproc.2015.elc2014.57

Klerkx, L., & Rose, D. (2020). Dealing with the game-changing technologies of agriculture 4.0: How do we manage diversity and responsibility in food system transition pathways? *Global Food Security, 24*, 100347. https://doi.org/10.1016/j.gfs.2019.100347

Kosaka, P. M., Pini, V., Ruz, J. J., Da Silva, R. A., González, M. U., Ramos, D., Calleja, M., & Tamayo, J. (2014). Detection of cancer biomarkers in serum using a hybrid mechanical and optoplasmonic nanosensor. *Nature Nanotechnology, 9*, 1047–1053.

Kubota, A., Thompson, T. L., Doerge, T. A., & Godin, R. E. (1997). A petiole sap nitrate test for broccoli. *Journal of Plant Nutrltion, 20*, 669–682.

Kuehr, S., Kosfeld, V., & Schlechtriem, C. (2021). Bioaccumulation assessment of nanomaterials using freshwater invertebrate species. *Environmental Science Europe, 33*, 9. https://doi.org/10.1186/s12302-020-00442-2

Kumar, S., Bhanjana, G., Sharma, A., Dilbaghi, N., Sidhu, M., & Kim, K. H. (2017). Development of nanoformulation approaches for the control of weeds. *Science of the Total Environment, 586*, 1272–1278.

Kumar, S., Nehra, M., Dilbaghi, N., Marrazza, G., Hassan, A. A., & Kim, K. H. (2019). Nano-based smart pesticide formulations: Emerging opportunities for agriculture. *Journal of Controlled Release*, *294*, 131–153.

Kumari, S., & Khan, S. (2017). Synthesis and applications of nanofungicides: A next-generation fungicide. In R. Prasad (Eds.), *Fungal Nanotechnology*. Springer, Cham. https://doi.org/10.1007/978-3-319-68424-6_6

Kumari, R., Suman, K., Karmakar, S., Mishra, V., Lakra, S. G., Saurav, G. K., & Mahto, B. K. (2023). Regulation and safety measures for nanotechnology-based agri-products. *Frontiers in Genome Editing*, *5*, 1200987. DOI: 10.3389/fgeed.2023.1200987

Lakshmeesha, T. R., Murali, M., Ansari, M. A., Udayashankar, A. C., Alzohairy, M. A., Almatroudi A., Alomary, M. N., Asiri, S. M. M., Ashwini, B. S., Kalagatur, N. K., Nayak, C. S., & Niranjana, S. R. (2020). Biofabrication of zinc oxidenanoparticles from melia azedarach and its potential in controlling soybeanseed-borne phytopathogenic fungi. *Saudi Journal of Biological Sciences*, *27*, 1923–1930. https://doi.org/10.1016/j.sjbs.2020.06.013

Li, H., Merkl, P., Sommertune, J., Thersleff, T., & Sotiriou, G. A. (2022). SERS hotspot engineering by aerosol self-assembly of plasmonic Ag nanoaggregates with tunable interparticle distance. *Advanced Science*, 2201133.

Li, S., Simonian, A., & Chin, B. A. (2010). Sensors for agriculture and the food industry. *The Electrochemical Society Interface*, *19*, 41–46.

Li, M., Zhang, Y., Feng, S., Zhang, X., Xi, Y., & Xiang, X. (2022). Bioaccumulation and biomagnification effects of nano-TiO_2 in the aquatic food chain. *Ecotoxicology*, *31*(6), 1023–1034. DOI: 10.1007/s10646-022-02572-0. Epub 2022 Jul 13. PMID: 35831721.

Liu, C., Zhou, H., & Zhou, J. (2021). The applications of nanotechnology in crop production. *Molecules*, *26*, 7070. https://doi.org/10.3390/ molecules26237070

Liu, F., Zhu, S., Li, D., Chen, G., & Ho, S. H. (2020). Detecting ferric iron by microalgal residue-derived fluorescent nanosensor with an advanced kinetic model. *iScience*, *23*, 101174. https://doi.org/10.1016/j.isci.2020.101174

Lowry, G. V., Avellan, A., & Gilbertson, L. M. (2019). Opportunities and challenges for nanotechnology in the agri-tech revolution. *Nature Nanotechnology*, 14, 517–522. https://doi.org/10.1038/s41565-019-0461-7

Lu, H., Wang, J., Stoller, M., Wang, T., Bao, Y., & Hao, H. (2016). An overview of nanomaterials for water and wastewater treatment. *Advances in Materials Science and Engineering*, 4964828. http://dx.doi.org/10.1155/2016/4964828

Lu, C. M., Zhang, C. Y., Wen, J. Q., Wu, G. R., & Tao, M. X. (2002). Research on the effect of nanometer materials on germination and growth enhancement of *Glycine Max* and its mechanism. *Soybean Science*, *3*, 168–172.

Magrez, A., Kasas, S., Salicio, V., Pasquier, N., Seo, J. W., Celio, M., Catsicas, S., Schwaller, B., & Forró, L. (2006). Cellular toxicity of carbon-based nanomaterials. *Nano Letter*, *6*(6), 1121–1125.

Mahdizadeh, M., & Najafi, N. (2019). Application of nano-sensors in the determination of soil moisture and temperature. *Land Management Journal*, *6*(2), 169–178. https://doi.org/10.22092/lmj.2019.118336

Maluin, F. N., Hussein, M. Z., Yusof, N. A., Fakurazi, S., Idris, A. S., Zainol Hilmi, N. H., & Jeffery Daim, L. D. (2019). Preparation of chitosan– hexaconazole nanoparticles as fungicide nanodelivery system for combating ganoderma disease in oil palm. *Molecules*, *24*, 2498.

Maruyama, C. R., Guilger, M., Pascoli, M., Bileshy-José, N., Abhilash, P., Fraceto, L. F., & De Lima, R. (2016). Nanoparticles based on chitosan as carriers for the combined herbicides imazapic and imazapyr. *Scientific Reports, 6*, 1–15.

Medina-Pérez, G., Fernández-Luqueño, F., Campos-Montiel, R., Sánchez-López, K. B., Afanador-Barajas, L. N., & Prince, L. (2019). Nanotechnology in crop protection: Status and future trends. In O. Koul (Ed.), *Nano-biopesticides: Today and future perspectives* (pp. 17–45). Academic Press. https://doi.org/10.1016/B978-0-12-815829-6.00002-4

Meena, M., Pilania, S., Pal, A., Mandhania, S., Bhushan, B., Kumar, S., Gohari, G., & Saharan, V. (2020). Cu-chitosan nano-net improves keeping quality of tomato by modulating physio-biochemical responses. *Scientific Reports*, *10*, 21914. https://doi.org/10.1038/s41598-020-78924-9

Metak, M. M. (2015). Effects of nanocomposite-based nano-silver and nano-titanium dioxide on food packaging materials. *International Journal of Applied Science & Technol*ogy, *5*, 26–40.

Mihindukulasuriya, S. D. F., & Lim, L. T. (2014). Nanotechnology development in food packaging: A review. *Trends Food Science Technology*, *40*, 149–167.

Mounier, L., Pédrot, M., Bouhnik-Le-Coz, M., & Cabello-Hurtado, F. (2023). Impact of iron oxide nanoparticles on a lead-polluted water–soil–plant system under alternating periods of water stress. *Environmetal Science Advances*, 2, 767–779. https://doi.org/10.1039/d2va00283c

Muhammad, A. I. (2019). Nano-fertilizers for sustainable crop production under changing climate: A global perspective. InMirza, H. (Ed.), *Sustainable Crop Production* (Chapter 18). IntechOpen.

Mühlfeld, C., Gehr, P., & Rothen-Rutishauser, B. (2008). Translocation and cellular entering mechanisms of nanoparticles in the respiratory tract. *Swiss Medical Weekly, 138*(27-28), 387–391.

NFU. (2023). The future of food 2040. https://www.nfuonline.com/nfu-online/news/thefuture-of-food-2040 (accessed July 20, 2023)

Nuruzzaman, M., Rahman, M. M., Liu, Y., & Naidu, R. (2016). Nanoencapsulation, nano-guard for pesticides: A new window for safe application. *Journal of Agricultural and Food Chemistry*, *64*, 1447–1483. https://doi.org/10.1021/acs.jafc.5b05214

Ojeda-Barrios, D. L., Morales, I., Juárez-Maldonado, A., Sandoval-Rangel, A., Fuentes-Lara, L. O., & Benavides-Mendoza, A. (2020). Importance of nanofertilizers in fruit nutrition. In A. K. Srivastava & C. Hu (Eds.), *Fruit crops: Diagnosis and management of nutrient constraints* (pp. 497–508). Elsevier. https://doi.org/10.1016/B978-0-12-818732-6.00035-6

Ojha, S., Singh, D., Sett, A., Chetia, H., Kabiraj, D., & Bora, U. (2018). Nanotechnology in crop protection. In D. K. Tripathi, P. Ahmad, S. Sharma, D. K. Chauhan & N. K. Dubey (Eds.), *Nanomaterials in plants, algae, and microorganisms: Concepts and controversies* (Vol. 1, pp. 345–391). Academic Press. https://doi.org/10.1016/B978-0-12-811487-2.00016-5

Okey-Onyesolu, C. F., Hassanisaadi, M., Bilal, M., Mahmood, B., Rahdar, A., Iqbal, J., & Kyzas, G. Z. (2021). Nanomaterials as nanofertilizers and nanopesticides: An overview. *Chemistry Select*, *6*(33), 8645–8663.

Onyeaka, H., Passaretti, P., Miri, T., & Al-Sharify, Z. T. (2022). The safety of nanomaterials in food production and packaging. *Current Research in Food Science*, *5*, 763–774. https://doi.org/10.1016/j.crfs.2022.04.005

Pariona, N., Mtz-Enriquez, A. I., Sánchez-Rangel, D., Carrión, G., Paraguay-Delgado, F., & Rosas-Saito, G. (2019). Green-synthesized copper nanoparticles as a potential antifungal against plant pathogens. *RSC Advances*, *9*, 18835–18843.

Parisi, C., Vigani, M., & Rodriguez-Cerezo, E. (2015). Agricultural nanotechnologies: What are the current possibilities? *Nano Today*, **10**(2), 124–127.

Park, H. J., Kim, S. H., Kim, H. J., & Choi, S. H. (2006). A new composition of nanosized silica-silver for control of various plant diseases. *The Plant Pathology Journal*, *22*, 295–302.

Pathak, S. O., Pandey, B. K., Singh, S., Hasanain, M., Shukla, G., & Maurya, D. K. (2020). Nanotechnology in agriculture: Benefits, risks and challenges. *Food and Scientific Reports*, *1* (4), 89–91.

Patil-Patankar, V., & Sanghvi, G. (2022). Nanoherbicides: A sustainable option for field applications. In J. Samuel, A. Kumar & J. Singh (Eds.), *Relationship between microbes and the environment for sustainable ecosystem services (Vol. 1). Microbial Products for Sustainable Ecosystem Services* (pp. 335–355). Elsevier. https://doi.org/10.1016/B978-0-323-89938-3.00015-3

Paul, A., & Roychoudhury, A. (2021). Go green to protect plants: Repurposing the antimicrobial activity of biosynthesized silver nanoparticles to combat phytopathogens. *Nanotechnology for Environmental Engineering*, *6*, 10.

Peixoto, S., Henriques, I., & Loureiro, S. (2021). Long-term effects of $Cu(OH)_2$ nanopesticide exposure on soil microbial communities. *Environmental Pollution*, *269*, 116113. https://doi.org/10.1016/j.envpol.2020.116113

Periakaruppan, R., Palanimuthu, V., Abed, S. A., & Danaraj, J. (2023). New perception about the use of nanofungicides in sustainable agriculture practices. *Archives of Microbiology*, *205*, 4. https://doi.org/10.1007/s00203-022-03324-8

Pestovsky, Y. S., & Martinez-Antonio, A. (2017). The use of nanoparticles and nanoformulations in agriculture. *Journal of Nanoscience and Nanotechnology*, *12*, 8699–8730. https://doi.org/10.1166/jnn.2017.15041

Pimentel, D. (2009). Pesticides and pest control. In *Integrated pest management: Innovation-development process* (pp. 83–87). Springer: Berlin/Heidelberg, Germany.

Pokrajac, L., Abbas, A., Chrzanowski, W., Dias, G. M., Eggleton, B. J., Maguire, S., Maine, E., Malloy, T., Nathwani, J., Nazar, L., Sips, A., Sone, J., Berg, A., Weiss, P., & Mitra, S. (2021). Nanotechnology for a sustainable future: Addressing global challenges with the international network4Sustainable nanotechnology. *ACS Nano*, *15*(12), 18608–18623. https://doi.org/10.1021/acsnano.1c10919

Raj, S. N., Anooj, E. S., Rajendran, K., & Vallinayagam, S. (2021). A comprehensive review on regulatory invention of nano pesticides in agricultural nano formulation and food system. *Journal of Molecular Structure*, *1239*, 130517. https://doi.org/10.1016/j.molstruc.2021.130517

Rajiv, P., Chen, X., Li, H., Rehaman, S., Vanathi, P., Abd-Elsalam, K. A., & Li, X. (2020). Silica-based nanosystems: Their role in sustainable agriculture. In K. A. Abd-Elsalam (Ed.), *Multifunctional Hybrid Nanomaterials for Sustainable Agri-Food and Ecosystems: Micro and Nano Technologies* (pp. 437–459). Elsevier.

Raliya, R., Saharan, V., Dimkpa, C., & Biswas, P. (2018). Nanofertilizer for precision and sustainable agriculture: Current state and future perspectives. *Journal of Agricultural and Food Chemistry*, *66*, 6487–6503. https://doi.org/10.1021/acs.jafc.7b02178

Rasheed, T., Hassan, A. A., Kausar, F., Sher, F., Bilal, M., & Iqbal, H. M. N. (2020). Carbon nanotubesassisted analytical detection – sensing/delivery cues for environmental and biomedical monitoring. *TrAC Trends in Analytical Chemistry*, *132*, 116066.https://doi.org/10.1016/j.trac.2020.116066

Ray, P. C., Yu, H., & Fu, P. P. (2009). Toxicity and environmental risks of nanomaterials: Challenges and future needs. *Journal of Environmental Science Health. Part C Environmental Carcinogenesis & Ecotoxicology Reviews*, *27*(1), 1–35. DOI: 10.1080/10590500802708267

Rico, C. M., Majumdar, S., Duarte-Gardea, M., Peralta-Videa, J. R., & Gardea-Torresdey, J. L. (2011). Interaction of nanoparticles with edible plants and their possible implications in the food chain. *Journal of Agricultural and Food Chem*istry, 59, 3485–3498.

Rodríguez-Félix, F., López-Cota, A. G., Moreno-Vásquez, M. J., Graciano-Verdugo, A. Z., Quintero-Reyes, I. E., Del-Toro-Sánchez, C. L., & Tapia-Hernández, J. A. (2021). Sustainable-green synthesis of silver nanoparticles using safflower (*Carthamus tinctorius* L.) waste extract and its antibacterial activity. *Heliyon, 7*(4), e06923. https://doi.org/10.1016/j.heliyon.2021.e06923

Rose, D. C., & Chilvers, J. (2018). Agriculture 4.0: Broadening responsible innovation in an era of smart farming. *Frontiers in Sustainable Food Syst*ems, *2*, 87. DOI: 10.3389/fsufs.2018.00087

Roy, A., Sharma, A., Yadav, S., Jule, L. T., & Krishnaraj, R. (2021). Nanomaterials for remediation of environmental pollutants. *Bioinorganic Chemistry & Applications*, *2021*, 1764647. https://doi.org/10.1155/2021/1764647

Ruszczak, B., & Boguszewska-Mańkowska, D. (2022). Soil moisture a posteriori measurements enhancement using ensemble learning. *Sensors*, *22*, 4591. https://doi.org/10.3390/s22124591

Sabzevari, S., & Hofman, J. A. (2022). Worldwide review of currently used pesticides' monitoring in agricultural soils. *Science of the Total Environment*, *812*, 152344.

Sagadevan, S., & Periasamy, M. (2014). Recent trends in nanobiosensors and their applications—A review. *Reviews on Advanced Materials Science*, *36*, 62–69.

Sagar, N. A., Kumar, N., Choudhary, R., Bajpai, V. K., Cao, H., Shukla, S., & Pareek, S. (2022). Prospecting the role of nanotechnology in extending the shelf-life of fresh produce and in developing advanced packaging. *Food Packaging and Shelf Life*, *34*, 100955. https://doi.org/10.1016/j.fpsl.2022.100955

Samra, C., & Hayrunnisa, N. (2015). The use of nanotechnology in the agriculture. *Advances in Nano Research*, *3*(4), 207–223. https://doi.org/10.12989/ANR.2015.3.4.207

Samreen, T., Rasool, S., Kanwal, S., Riaz, S., Sidra-Tul-Muntaha, & & Nazir, M. Z. (2022). Role of nanotechnology in precision agriculture. *Environmental Sciences Proceedings, 23*(1), 17. https://doi.org/10.3390/environsciproc2022023017

Schnoor, B., Elhendawy, A., Joseph, S., Putman, M., Chacón-Cerdas, R., Flores-Mora, D., Bravo-Moraga, F., Bravo-Moraga, F., & Salvador-Morales, C. (2018). Engineering atrazine loaded poly (lactic-co-glycolic acid) nanoparticles to ameliorate environmental challenges. *Journal of Agricultural and Food ChemIstry*, *66*, 7889–7898.

Schulte, P. A., & Salamanca-Buentello, F. (2007). Ethical and scientific issues of nanotechnology in the workplace. *Environmental Health Perspectives*, *115*(1), 5–12. DOI: 10.1289/ehp.9456

Selvaraj, S. K., Kumar, R., Balamurugan, K., Joany, R. M., Dorothy, R., Tuan, A. N., & Rajendran, S. (2022). Wireless nanosensor network for irrigation control. In A. Denizli, T. A. Nguyen, S. Rajendran, G. Yasin & A. K. Nadda. *Nanosensors for smart agriculture: Micro and nano technologies* (pp. 463–478). https://doi.org/10.1016/B978-0-12-824554-5.00005-7

Senapaty, M. K., Ray, A., & Padhy, N. (2023). IoT-enabled soil nutrient analysis and crop recommendation model for precision agriculture. *Computers*, *12*, 61.

Shaikh, H. S. (2022). Nanotechnology in crop protection: A review. *Research Journal of Science and Technology*, *14*(3), 177–2. DOI: 10.52711/2349-2988.2022.00029

Shaikh, A., Meroliya, H., Dagade-Gadale, S., & Waghmode, S. (2021). Applications of nanotechnology in precision agriculture: A review. *Research & Reviews in Biotechnology & Biosciences*, *8*(1), 105–117. http://doi.org/10.5281/zenodo.5118402

Shang, Y., Hasan, M. K., Ahammed, G. J., Li, M., Yin, H., & Zhou, J. (2019). Applications of nanotechnology in plant growth and crop protection: A review. *Molecules*, *24*, 2558.

Sharafeldin, I., Garcia-Rios, S., Ahmed, N., Alvarado, M., Vilanova, X., & Allam, N. K. (2021). Metal-decorated carbon nanotubes-based sensor array for simultaneous detection of toxic gases. *Journal of Environmental Chemical Engineering*, *9*(1), 104534. https://doi.org/10.1016/j.jece.2020.104534

Sharifan, H., Noori, A., Bagheri, M., & Moore, M. (2021). Postharvest spraying of zinc oxide nanoparticles enhances shelf-life qualities and zinc concentration of tomato fruits. *Crop and Pasture Science*, *73*(2), 22–31. https://doi.org/10.1071/CP21191

Sharma, B., Lakra, U., Sharma, R., & Sharma, S. R. (2022). A comprehensive review on nanopesticides and nanofertilizers—A boon for agriculture. In M. Ghorbanpour & M. A. Shahid (Eds.), *Nano-enabled Agrochemicals in Agriculture*. Academic Press Publication. https://doi.org/10.1016/B978-0-323-91009-5.00026-4

Sharma, P., Sharma, A., Sharma, M., Bhalla, N., Estrela, P., Jain, A., Thakur, P., & Thakur, A. (2017). Nanomaterial fungicides: In vitro and in vivo antimycotic activity of cobalt and nickel nanoferrites on phytopathogenic fungi. *Global ChallEnges*, *1*, 1700041.

Sharma, K., Sharma, R., Shit, S., & Gupata, S. (2012). Nanotechnological application on diagnosis of a plant disease. *Internationl Conference on Advances Biol. Med. Sci.*

Sharma, A., Sood, K., Kaur, J., & Khatri, M. (2019). Agrochemical loaded biocompatible chitosan nanoparticles for insect pest management. *Biocatalysis and Agricultural Biotechnology*, *18*, 101079.

Shaw, D. S., & Honeychurch, K. C. (2022). Nanosensor applications in plant science. *Biosensors*, *12*, 675. https:// doi.org/10.3390/bios12090675

Sherkhane, A. S., Suryawanshi, H. H., Mundada, P. S., & Shinde, B. P. (2018). Control of bacterial blight disease of pomegranate using silver nanoparticles. *Journal of Nanomedicine & Nanotechnology*, *9*, 1–5.

Shukla, G., Gaurav, S. S., & Singh, A. (2020). Synthesis of mycogenic zinc oxidenanoparticles and preliminary determination of its efficacy as a Larvicide Againstwhite Grubs (*Holotrichia* sp.). *International Nano Letters*, *10*, 131–139.

Singh, K., Singh, A., & Rai, S. (2022). A study on nanomaterials for water purification. *Materialstoday: Proceedings*, *51*(1), 1157–1163.

Singh, V., Yadav, V. K., & Mishra, V. (2020). Nanotechnology: An application in biofuel production. In M. Srivastava, N. Srivastava, P. K. Mishra & V. K. Gupta (Eds.), *Nanomaterials in Biofuels Research* (pp. 143–160), Springer.

Smith, I. F. (2000). Micronutrient interventions: Options for Africa. Food and Nutrition Bulletin, 21, 532–537.

Solanki, P., Bhargava, A., Chhipa, H., Jain, N., & Panwar, J. (2015). Nano-fertilizers and their smart delivery system. In M. Rai, C. Ribeiro, L. Mattoso & N. Duran (Eds.), *Nanotechnologies in Food and Agriculture*. Springer, Cham. https://doi.org/10.1007/978-3-319-14024-7_4

Sookhtanlou, M., Allahyari, M. S., & Surujlal, J. (2022). Health risk of potato farmers exposed to overuse of chemical pesticides in Iran. *Safety and Health at Work, 13*, 23–31.
Sousa, G. F. M., Gomes, D. G., Campos, E. V. R., Oliveira, J. L., Fraceto, L. F., Stolf-Moreira, R., & Oliveira, H. C. (2018). Post-emergence herbicidal activity of nanoatrazine against susceptible weeds. *Frontiers in Environmental Science*, *6*, 12.
Squire, H. J., Tomatz, S., Voke, E., González-Grandío, E., & Landry, M. P. (2023). The emerging role of nanotechnology in plant genetic engineering. *Nature Reviews Bioengineering*, 1, 314–328. https://doi.org/10.1038/s44222-023-00037-5
Subramanian, K. S., & Rajkishore, S. K. (2018). Regulatory framework for nanomaterials in agri-food systems. In M. Rai & J. K. Biswas (Eds.), *Nanomaterials: Ecotoxicity, safety, and public perception* (pp. 319–342). Springer International Publishing, Cham. https://doi.org/10.1007/978-3-030-05144-0_16
Stephens, N., Sexton, A. E., Driessen, C. (2019). Making sense of making meat: Key moments in the first 20 years of tissue engineering muscle to make food. *Frontiers in Sustainable Food Systems*, *3*, 45. DOI: 10.3389/fsufs.2019.00045
Taban, A., Saharkhiz, M. J., & Khorram, M. (2020). Formulation and assessment of nano-encapsulated bioherbicides based on biopolymers and essential oil. *Industrial Crops and Products*, *149*, 112348.
Taha, N. A., Hamden, S., Bayoumi, Y. A., Elsakhawy, T., El-Ramady, H., &Solberg, S. Ø. (2023). Nanofungicides with selenium and silicon can boost the growth and yield of common bean (*Phaseolus vulgaris* L.) and control alternaria leaf spot disease. *Microorganisms, 11*(3), 728. https://doi.org/10.3390/microorganisms11030728
Takeshita, V., Carvalho, L. B., Galhardi, J. A., Munhoz-Garcia, G. V., Pimpinato, R. F., Oliveira, H. C., Tornisielo, V. L., & Fraceto, L. F. (2022). Development of a preemergent nanoherbicide: From efficiency evaluation to the assessment of environmental fate and risks to soil microorganisms. *ACS Nanoscience Au*, *2*(4), 307–323. https://doi.org/10.1021/acsnanoscienceau.1c00055
The Fertiliser Association of India. (2021). The fertiliser (inorganic, organic, or mixed) control order, 1985 (as amended up to 2021), along with the fertilizer (movement control) order, 1973, and essential commodities act, 1955 (p. 257). https://www.faidelhi.org/general/FCO-July-21-content.pdf
Thomas, E., Rathor, I., & Tarafdar, J. C. (2016). Bio-inspired synthesis of nitrogen nanoparticles and its application on pearl millet (*Pennisetum americanum* L) cv. HHB 67. *Journal of Bionanoscience*, *10*, 1–7. DOI: 10.1166/jbns.2016.1374
Usman, M., Farooq, M., Wakeel, A., Nawaz, A., Cheema, S. A., Ur Rehman, H., Ashraf I., & Sanaullah M. (2020). Nanotechnology in agriculture: Current status, challenges, and future opportunities. *Science of the Total Environment*, *721*, 137778. https://doi.org/10.1016/j.scitotenv.2020.137778
Van de Voorde, M., & Jeswani, G. (2021). *Ethics in nanotechnology: Social sciences and philosophical aspects*. De Gruyter. https://doi.org/10.1515/9783110719932
Varympopi, A., Dimopoulou, A., Theologidis, I., Karamanidou, T., Kerou, A. K., Vlachou, A., Karfaridis, D., Papafotis, D., Hatzinikolaou, D. G., Tsouknidas, A., & Skandalis., N. (2020). Bactericides based on copper nanoparticles restrain growth of important plant pathogens. *Pathogens*, *9*, 1024.
Velu, G., Ortiz-Monasterio, I., Cakmak, I., Hao, Y., & Singh, R. P. (2014). Biofortification strategies to increase grain zinc and iron concentrations in wheat. Journal of Cereal Science, 59, 365–372.

Watanabe, C., Naveed, N., & Neittanmäki, P. (2018). Digital solutions transform the forest-based bioeconomy into a digital platform industry - A suggestion for a disruptive business model in the digital economy. *Technolgy in Society*, *54*, 168–188.

Wegner, L. H. (2012). Using the multifunctional xylem probe for in situ studies of plant water and ion relations under saline conditions. *Methods in Molecular Biol*ogy, *913*, 35–66.

WHO. (2000). Nutrition for Health and Development. WHO: Geneva, Switzerland.

Yadav, P., & Chowdhury, P. (2021). Functionalized CdTe fluorescence nanosensor for the sensitive detection of water borne environmentally hazardous metal ions. *Optical Materials*, *111*, 110584. https://doi.org/10.1016/j.optmat.2020.110584

Yadav, A., Yadav, K., Ahmad, R., & Abd-Elsalam, K. A. (2023). Emerging frontiers in nanotechnology for precision agriculture: Advancements, hurdles and prospects. *Agrochemicals*, *2*, 220–256. https://doi.org/10.3390/ agrochemicals2020016

Yan, Y., Zhu, X., Yu, Y., Li, C., Zhang, Z., & Wang, F. (2022). Nanotechnology strategies for plant genetic engineering. *Advanced Materials*, *34*(7), e2106945. DOI: 10.1002/ adma.202106945. Epub 2022 Jan 6. PMID: 34699644.

Yang, T., Doherty, J., Guo, H., Zhao, B., Clark, J. M., Xing, B., Hou, R., & He, L. (2019). Real-time monitoring of pesticide translocation in tomato plants by surface-enhanced raman spectroscopy. *Analytical Chemistry*, *91*, 2093–2099.

Yousef, H. A., Fahmy, H. M., Arafa, F. N., Arafa, F. N., Allah, M. Y. A., Tawfik, Y. M., El-Halwany, K. K., El-Ashmanty, B. A., Al-anany, F. S., Mohamed, M. A., & Bassily, M. E. (2023). Nanotechnology in pest management: Advantages, applications, and challenges. *International Journal of Tropical Insect Science*, *43*(1). https://doi.org/10.1007/ s42690-023-01053-z

Zambon, I., Cecchini, M., Egidi, G., Saporito, M. G., & Colantoni, A. (2019). Revolution 4.0: Industry vs. agriculture in a future development for SMEs. *Processes, 7*(1), 36. https://doi. org/10.3390/pr7010036

CHAPTER 3

Recent Advances in Nanomaterials Synthesis for Agriculture and the Environment

SUNANDANA DAS[1], PRITHU BHATTACHARYYA[2], TAMAL DAS[2], JISHNU BANERJEE[2], KRISHNENDU ACHARYA[3], and JOY SARKAR[4]

[1]*Department of Botany, Scottish Church College, Kolkata, West Bengal, India*

[2]*Department of Botany, Ramakrishna Mission Vivekananda Centenary College, Rahara, Khardaha, West Bengal, India*

[3]*Molecular and Applied Mycology and Plant Pathology Laboratory, Centre of Advanced Study, Department of Botany, University of Calcutta, Kolkata, India*

[4]*Department of Botany, Dinabandhu Andrews College, Kolkata, India*

ABSTRACT

Agriculture has been the most instrumental steppingstone for human civilization since time immemorial. Without proper food, the human race would have perished long ago. Since agriculture is so important and science always comes up with a way to improve human lives, the improvement of agriculture has always been a key part of research and development. Agriculture has evolved, and major chemical compounds have proven to improve cultivation processes under various circumstances. In modern times, nanotechnology and the synthesis of nanoparticles have turned the tables and have been substantially implemented in agriculture. Nanoparticles provide a wide range of opportunities to bring about beneficial changes in agriculture. From improved growth to better sustainability, nanoparticles have provided a lot of features in

Nanobiotechnology for Agricultural Sciences: Nano-Agri-Input Products for Crop Production and Environmental Protection. J. C. Tarafdar, Deepak Kumar, Sanjay K. Singh, and Krishna Pal Singh (Eds.)

agriculture. In this chapter, methods for nanomaterial synthesis via top-down, bottom-up, and chemical methods, the importance of green nanotechnology in agriculture, applications of nanomaterials in the manufacturing of nanofertilizers and nano-pesticides, and types of nanoparticles, including metal nanoparticles, metal oxide nanoparticles, and others, have been included. Nanomaterials can be widely used as nanofertilizers and nano-pesticides with specific actions and have also been proven to increase efficiency in monitoring crop requirements as nano-sensors. Also, since toxicity and contamination are major by-products of agriculture due to the heavy use of chemical fertilizers, remediation of contaminated soil and water using nanomaterials has been implemented, along with environmental safety concerns regarding the use of nanomaterials and nanocomposites, with a nod toward the future perspective of the use of nanomaterials in the field of agriculture.

3.1 INTRODUCTION

Agriculture has always served as the foundation of developing countries. Agriculture contributes significantly to their economy and is a major source of employment. Several issues affect the agricultural industry, including reduced output, which is exacerbated by climate change, and the widespread use of fertilizers, which causes ecological imbalance and the deterioration of soil and water resources (Singh et al., 2022). With an ever-increasing human population comes the problem of ensuring food security. While many reasons contribute to food scarcity, pests and crop weeds are among the most frequent and have received substantial attention (Liliane and Charles, 2020).

Fertilizers are important because the proportion of nutrients present in the soil may not always fulfill the requirements for ideal plant growth and output, so they must be supplied to crops externally (Bowonder, 1979; Parayil, 1992). However, overuse of chemical fertilizers depletes soil micronutrients and microflora and elevates salinity, further lowering crop production and nutritional quality (Singh, 2000). Conventional fertilizers have well-acknowledged disadvantages such as reduced bioavailability and increased losses owing to leaching and runoff, which deteriorate soil and water resources. One of the major disadvantages of nutrient-delivering fertilizers is their inefficiency, which leads to increased waste and renders them unsustainable in the long term (Qureshi et al., 2018).

As agriculture confronts multiple challenges, the concept of precision farming is gaining traction as a means of improving current farming techniques by utilizing nano-products such as nanofertilizers, nano-pesticides, and nano-sensors to change the current scenario of agricultural practices, which is largely unsustainable (Duhan et al., 2017; Fayaz et al., 2021; Sangeetha et al., 2021; Singh et al., 2021). Furthermore, the use of nanomaterials (NMs) for advanced agricultural applications can be extremely useful in terms of increasing yield, reducing farm input costs, improving plant stress resistance, and improving crop nutrient quality, as shown in Figure 3.1 (Singh et al., 2021).

NMs have a great surface area-to-volume ratio due to their small size, which results in enhanced surface activity and reactivity. Because of their extremely small size (less than 100 nm), these materials can penetrate biological barriers and permeate into plant tissue via root or foliar treatment, offering potentially effective routes of nutrition and pesticide delivery (Singh et al., 2021). As a result of their unique tuneable features, NMs have been

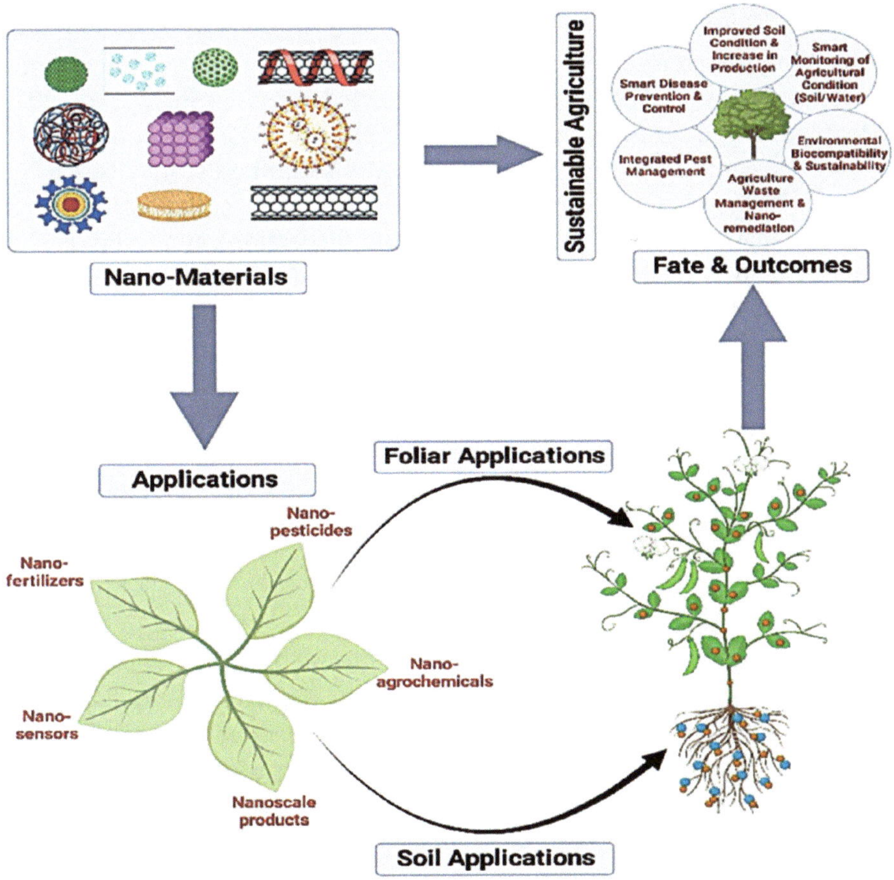

FIGURE 3.1 Applications of NMs for sustainable agriculture.

Source: Adapted from Singh et al. (2021).

extensively researched in a variety of sectors, including agriculture. NMs offer enormous potential for usage as nanofertilizers and nano-pesticides with targeted action and increased efficiency, as nano-sensors for successfully monitoring crop requirements, and as catalysts or adsorbers in soil remediation. In the current setting, when crop output and quality must be maximized while minimizing environmental effects, it is believed that NMs can considerably improve traditional agricultural practices and serve as an alternative tool for sustainable intensive agriculture (Singh et al., 2022).

3.2 NM SYNTHESIS METHODS

The method of synthesis is an essential aspect to consider when producing NM-based products since it has a significant impact on their physicochemical properties, which in turn control the impacts of their interaction with living cells. Nano-based products can have a focused impact by selecting an appropriate synthesis method based on their intended purpose (Singh et al., 2022). Many advanced synthesized metallic nanoparticles (Ag NPs, Cu/CuO NPs, TiO_2 NPs, carbon NPs, Fe NPs, Zn NPs, Mg NPs, Si NPs, K-NPs, Au-NPs, CeO_2-NPs, graphene-silver NPs, etc.), together with polymers (polymeric micelles, nanocrystals, nano sponges, nano dispersion), many lipids (liposomes, nano emulsions), lipid-polymer hybrid (nanocapsule, dendrimer), and carbon-based nanoparticles (carbon nanotube, graphene, fullerene) have already gained a lot of attention in modern agriculture because of their distinct electrical, optical, catalytic, and structural properties (Khanna et al., 2019; Sarkar et al., 2022). This section briefly discusses the many approaches utilized for synthesizing nanoparticles, which include physical, chemical, and biological processes, as shown in Figure 3.2. Bottom-up and top-down methodologies are used to classify these methods (Singh et al., 2022).

3.2.1 TOP-DOWN METHOD

The top-down methodology is a disruptive method that relies only on processes that crush or shred bulk materials into smaller particles in the nanometer range. The mechanical approach is the oldest and simplest top-down method, in which tungsten carbide, carbide steel, or iron balls are milled (de Carvalho et al., 2013; Saiphaneendra and Srivastava, 2017). The top-down approach also includes techniques such as lithography, laser ablation, sputtering, and electro-sputtering (Rafique et al., 2019; Singh et al., 2022; Tang et al., 2017; Verma et al., 2018a, 2018b).

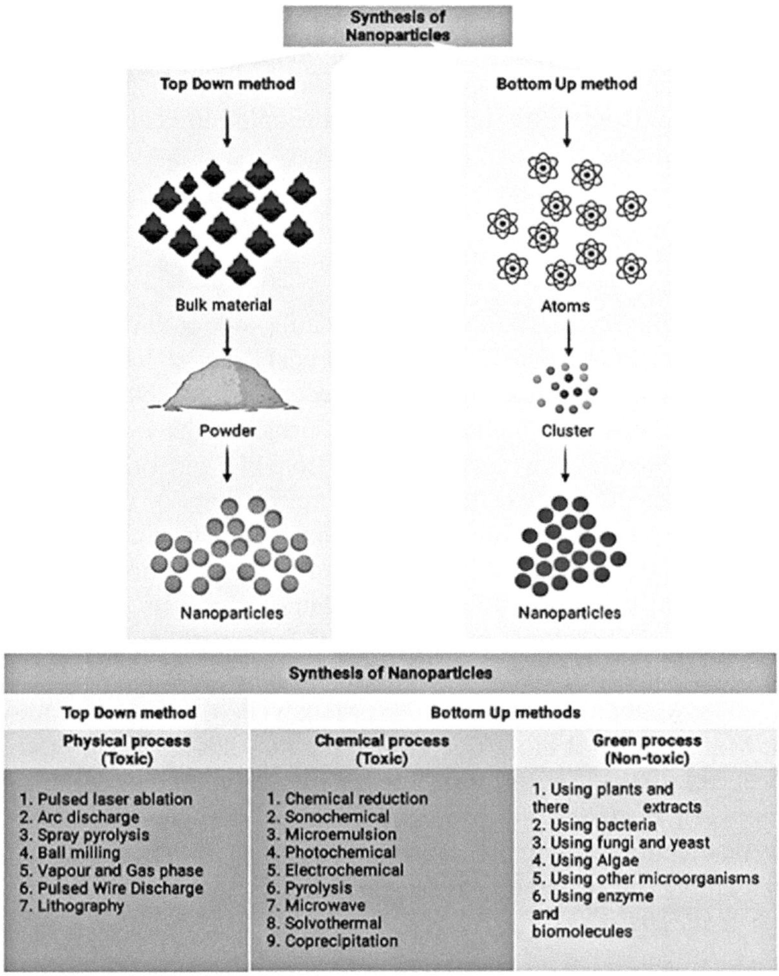

FIGURE 3.2 Methods of nanomaterial synthesis via bottom-up and top-down approaches.
Source: Adapted from Singh et al. (2022).

3.2.2 BOTTOM-UP METHOD

In this method, smaller components, such as atoms and molecules, are used as building blocks to create nanoparticles. This method essentially entails a chemical reaction that produces nuclei (atomic clusters), which are then used to generate nanoparticles with fewer flaws and a more homogenous chemical makeup (Singh et al., 2022). In general, the bottom-up methodology refers to the synthesis of nanoparticles through chemical reactions between atoms, ions, and molecules. Certain salts are used as starting materials to initiate chemical reactions among the involved atoms, ions, and molecules. These salts are combined with an appropriate chelating agent to generate a homogenous solution. Controlling the nature of the solution is also critical during the synthesis process (Singh et al., 2020).

3.2.3 CHEMICAL METHOD

Chemical methods provide the most versatile and easiest method for the synthesis of nanoparticles. With this approach, one is free to choose any number of characteristics, including precursors, reductants, surfactants, and concentrations. These characteristics influence the form, size, composition, and structure of nanoparticles. By making particular modifications to the synthesis conditions, the intended morphology of nanoparticles can be obtained (Ahari et al., 2018). Chemical methods include coprecipitation, thermal decomposition, solvothermal decomposition, sonochemical processes, self-assembly utilizing templates, pyrolysis, electrospinning, chemical vapor deposition, atomic or molecule condensation, and photochemical procedures (Cho and Kuo, 2016; Dulgerbaki et al., 2015; Maity et al., 2008; Maity and Agrawal, 2007; Manawi et al., 2018; Muradov et al., 2018; Muthukumar et al., 2018; Rycenga et al., 2009; Tricoli et al., 2016; Zhao et al., 2020). The electrochemical synthesis method entails a reaction between electrodes at different poles. Nanoparticles are formed at the electrode-electrolyte contact (Mazarío et al., 2016). Nanoparticles of various morphologies can be generated by altering the electrochemical conditions (Singh et al., 2022).

3.3 WHY GREEN NANOTECHNOLOGY FOR AGRICULTURE?

With the growing dependence on insecticides and pesticides and the concomitant awareness of the environmental impact of exposure to these compounds at low levels, there is a compelling need for agrotechnology to change toward greener technical systems to be employed in farming. After making substantial contributions, the utilization of nanotechnology in agriculture is now a proven solution for moving toward sustainability. However, many processes involved in nano-synthesis and application require high temperatures, pressures, acidity, and hazardous metals (Khan et al., 2011; Lou and Zeng, 2002; Santoro et al., 2013; Toshiki et al., 2003).

As a result, over the last two decades, there has been a significant shift in the approach to nano-synthesis from traditional methods to plant-based protocols, owing to numerous advantages such as a sustainable pathway, diverse plant metabolites, advanced extraction and separation strategies, the development of an exhaustive phytoconstituent database, and instrumental assistance provided by high-performance liquid chromatography-mass spectrometry, high-performance liquid chromatography-nuclear magnetic resonance, high-performance liquid chromatography-mass spectrometry, and centrifugal partition chromatography. Incredibly effective extraction methods, such as chemo-genomics and metabolomics in combination with ultrasound, enzymes, hydro-distillation, and supercritical fluid extraction, have enabled the manufacture of customized extracts that were not formerly feasible (Takla et al., 2018). Nanotechnology, on the other hand, can be employed in conjunction with these improvements to successfully establish a sustainable agricultural system, and a strong emphasis is being placed on plant-based nanotechnology formulations, which have a high potential to tackle significant farming problems (Malini et al., 2022).

3.4 EFFECTS OF SHAPE AND SIZE OF NMS

NMs are distinguished by their electrical, mechanical, and optical properties. The effects of quantum confinement and the local energy disorder of surface atoms impart different features to NMs (Kovalenko et al., 2015). NMs have substantial benefits over their bulk counterparts when it comes to chemical bioactivity, cellular/tissue/organ penetration capabilities, and significantly increased bioactivity, which can be utilized in applications. NMs' physico-chemical qualities are largely dependent on size and shape. This allows for the optimization of nanoscale qualities to develop various nanostructured materials with tailored properties to fulfill various requirements by changing the morphology of the nanoparticles (Hoshyar et al., 2016).

Nanoparticles come in a variety of shapes, including, but not limited to, nanowires, nanorods, nanoshells, nanocages, nanoprisms, cubic, spherical, nanobars, and flower-shaped particles. The variety of forms and sizes of nanoparticles is determined by numerous reaction parameters such as precursor quality, concentration, type of reactants, pH of the solution, heating rates, stirring settings, and reaction duration. Controlling the morphology of nanoparticles is critical, which is normally accomplished by altering the synthesis processes, reducing agents, and stabilizers used in nanoparticle formation (Varanda et al., 2019).

Nanoparticles with varying shapes and sizes have distinct colors and properties (Khan et al., 2019). Extensive research has been conducted to understand the relationship between size and the needed qualities of nanoparticles. For example, research on the antibacterial activity of various metal nanoparticles, such as silver colloids, has revealed that the smaller the particle size of the nanoparticles, the greater their antibacterial activity (Raza et al., 2016). Studies on the catalytic activity of nanoparticles have indicated a dependency on their size and shape, in addition to their structure and chemical-physical environment (Guo et al., 2018). The optical and magnetic properties of nanoparticles are also affected by their form and size (Moaied and Hong, 2019; Wu et al., 2016). The size and form of the nanoparticles influence their distribution rate and site-specificity in the delivery mechanism. Furthermore, shape-based preferential interaction with proteins is conceivable by employing nanoparticles (Zhang et al., 2016). The ideal size is determined by the intended action needed in the particular scenario. Extensive testing is needed when creating nanoparticles for uses involving interactions with living cells. Given that cellular absorption is influenced by the shape and size of the nanoparticles, it is essential to constantly prepare them to a precise size. The ability of nanoparticles to provide tailored action is therefore compromised by the fact that even little variations in particle size can produce a wide range of effects (Talamini et al., 2017). According to their size and shape, nanoparticles might be more or less harmful (Jeevanandam et al., 2018; Singh et al., 2022).

3.5 APPLICATIONS OF NMS

Precision farming is enabling nanotechnology to knock on the doors of the agricultural industry since it can be used effectively to suit the overall structure of sustainable agriculture, specifically by offering environmentally safe and resourceful technologies (Duhan et al., 2017; Fraceto et al., 2016; Raliya et al., 2018). There are now more prospects to use the foundations of nanotechnology in the agriculture industry thanks to the recent development of the technology in medications and pharmaceuticals (Robles-García et al., 2016; Vega-Vásquez et al., 2020). The development of novel commodities, technologies, and agricultural applications has been achieved through the convergence of nanotechnology and other scientific disciplines. The conservation of crop inputs, cattle, animal rearing, aquaculture, and fisheries is one of the many ways that nanotechnology has the potential to enhance the current agricultural system. By utilizing nanoscale products and continuous monitoring systems, agro-based nanotechnology's goal is to increase crop output while reducing the number of agricultural inputs used (Chhipa, 2019; Scott and Chen, 2013). Nanotechnology has the potential to increase crop quality and productivity at every level of the agricultural process. To maintain water quality and control nutrient inputs for sustainable agricultural production, nanotechnology is crucial (Mukhopadhyay, 2014). When NMs interact with plants, they may have a favorable impact on plant growth and soil quality. Plant morphology and physiology are enhanced by the liberation of NM-encapsulated micronutrients in the soil biota and organic compounds (Sabir et al., 2014; Saharan and Pal, 2016; Verma et al., 2018a, 2018b). Through the elimination of harmful compounds from the soil and subsequent improvement of the soil's natural condition, remediation of soil and water with the help of NMs will significantly aid in preserving sustainability. For the detection of heavy metals and environmental pollutants in soil media, NM-based optical and electrochemical sensors have been deployed (Das et al., 2015).

3.5.1 NANOFERTILIZERS

Agriculture currently confronts a wide range of issues, including nutrient deficiency, crop yield stagnation, reduced soil organic matter, limited water availability, nutrient deficiency in the soil, decreased land area owing to urbanization and land degradation, and labor shortages (Godfray et al., 2010). In addition, as the human population expanded to supply the surplus need for food, the "Green Revolution" was introduced in the 1950s and 1960s, bringing in the notion of widespread fertilizer use in agriculture. Fertilizers are nutrients that are applied to crops from outside sources, typically in the form of chemical compounds (Feregrino-Perez et al., 2018). Chemical fertilizers increase agricultural yield and are thus widely employed in a variety of applications (Liu and Lal, 2015). However, the leaching of minerals into the soil caused by the continual use of chemical fertilizers reduces soil fertility and increases the concentration of salts, which has a direct impact on soil fertility. It is critical to create nanofertilizers that can efficiently release nutrients in a gradual and regulated way to increase crop yield and quality (Feregrino-Perez et al., 2018). That's why the use of nanoscience and nanotechnology is growing rapidly, and new ways of producing unique and desired materials for crop production and management are constantly being presented. It is one of the most prominent concepts in the emerging science of precision agriculture, in which farmers use fertilizers and other inputs more efficiently. The uncontrolled population has resulted in the massive manufacturing of conventional fertilizers to enhance food production and crop protection, but this eventually reduces soil fertility and food quality. These chemical fertilizers not only contribute to the woes of the delicate ecology, but they also have an impact on human health if they are not employed. According to research presented by the United Nations Food and Agricultural Organization, worldwide demand for fertilizers is predicted to climb in the coming years. As a result, nanofertilizers are eco-friendly or smart fertilizers that can enhance fertilizer application rates while reducing nutrient loss, namely, phosphorus and nitrogen. It is critical to incorporate intelligent solutions for sustainable agriculture, such as nanotechnology. Nanofertilizers are environmentally friendly fertilizers or smart fertilizers with the potential to increase the application rates of fertilizers and reduce the loss of nutrients from them, mainly phosphorous and nitrogen (Singh et al., 2022).

When used at low concentrations, nanofertilizers have a good effect on plant growth parameters. Metallic nanoparticles, such as gold and silver nanoparticles, have been studied to increase the number of leaves, leaf area, plant height, chlorophyll content, and sugar content, resulting in higher agricultural output (Vuong et al., 2017). Zinc oxide (ZnO, 4500 mg/hectare), when applied to the brinjal plant combined with 25% (by amount) of the commonly used nitrogen phosphorus potassium (NPK) fertilizer, boosts production by 91% (Kale and Gawade, 2016). Field studies on rabi crops using 50% less normal NPK fertilizer and two sprays of nanonitrogen resulted in significantly improved crop output (Kumar et al., 2020). For formulations containing porous or laminated inorganic or carbonaceous NMs, adsorption nanofertilizers are preferable. Biochar nanocarriers formed by the carbonization of agricultural wastes or low-value biomass, in particular, have received a lot of interest owing to their low cost and exceptional physical adsorption characteristics (Feregrino-Perez et al., 2018). In biochar-based adsorption nanofertilizers, the kinetics of a range of nutrition elements such as N, P, K, Na, Mg, Ca, and Zn have been proven. Plant micronutrients include molybdenum (Mo), copper (Cu), iron (Fe), nickel (Ni), manganese (Mn), and zinc (Zn), and they are required in much lesser proportions than macronutrients for crop growth. Micronutrients used in composite formulations provide more nutrients while causing less environmental harm. In comparison to other traditional ions and salts, micronutrients packaged in NPs improve nutrient bioavailability and fertilizer usage efficiency (Dimkpa and Bindraban,

2016). A composite of three micronutrient NPs (ZnO, CuO, and B_2O_3) has been effectively developed for agronomic fortification in soybean plants to ameliorate drought stress (Dimkpa and Bindraban, 2018).

Various nano-fabricated fertilizer forms were created employing a variety of natural and synthetic materials, with the ultimate objective of achieving a controlled release rate and high absorption efficiency to fit crop development demands. We may divide nanofabrication fertilizers into three types: (i) nano-supported fertilizers, in which nanostructured materials are applied as additives to regulate the release of fertilizers; (ii) nanosized fertilizers, which refer to fertilizers made on the nanoscale; and (iii) nano-wrapped fertilizers, which apply NM wraps or coatings to contain regular-size fertilizers. Table 3.1 summarizes recently reported nano-fabricated fertilizers according to this classification (Shao et al., 2022).

3.5.1.1 MOLECULAR MECHANISM OF NANOFERTILIZERS

Since NMs are smaller than the pore size of the plant cell wall (up to 20 nm), they may be used as a nanofertilizer, which aids in nutrient absorption, as shown in Table 3.2. Nanofertilizers can be applied directly to the soil and/or leaves. As a result, there are two major techniques of application for nanofertilizer. The first is a foliar entry, followed by a root entry. The foliar entrance is provided by spraying the trough cuticle, stromata, hydathodes, lenticels, and wounds, whereas root access is provided via root tips, lateral roots, root hairs, and rhizodermis. Foliar applications of nanofertilizer are preferred or can be used when soil and weather conditions are poor, as shown in Figure 3.3.

Furthermore, this method of administration enhances direct nutrient uptake into the plant leaf system while decreasing fertilizer waste. As a result, foliar application of nanofertilizer results in increased NUE and a faster response to crop or plant development (Mahil and Baburai, 2019). Foliar entrance is aided by nanopores, the base of the trichome, and stromal openings in leaves, which enable NP absorption and transport inside the leaves. For example, it has been demonstrated that nanosized particles (43 nm) are deeply buried in the bulk of the leaf interior compartment in Faba beans (*Vicia faba*) and more (in *C. arabica* below 2.5 nm and *P. cerasus* below 100 nm), implying the effectiveness of nanofertilizer to enhance nutrient uptake (Eichert et al., 2008). It is recommended as a nanofertilizer for increased nutrient usage efficiency (NUE) due to its greater capacity to pass and supply nutrients through nanosized (50–60 nm) plasmodesmata holes, which normally transfer ions between cells (Iqbal et al., 2019). Furthermore, as compared to traditional

TABLE 3.1 Typical Nano-Fabricated Fertilizers and Their Mechanisms.

Category	Materials for Nanostructure	Fabrication Strategy/ Concrete Method	Nutrient-Release Mechanism	Reference
Nano-supported fertilizers	Calcium phosphate	Entrapment/doping nutrients into the nanocarrier formation system	Diffusion	Kottegoda et al., (2017); Shao et al. (2022)
	Liposome	Entrapment/solvent injection techniques or thin lipid-film hydration and extrusion methods	Integrity disruption caused by osmotic pressure	Farshchi et al. (2021); Karny et al., (2018) Farshchi et al. (2021); Karny et al. (2018)
	Carboxyl cellulose	Entrapment/chelation of carboxyl cellulose and Fe^{2+}	pH-Stimulated release, the nanostructure would be disintegrated in an acidic condition	Wang et al. (2016)
	Biochar	Entrapment/physical adsorption	Diffusion	Das et al. (2021); Khan et al. (2021)
	Inorganic porous materials: zinc aluminosilicate, zinc layered hydroxide-nitrate, zeolite, etc	Entrapment/physical adsorption	Diffusion	Jahangirian et al. (2020); Naseem et al. (2020)
	Porous halloysite nanotubes and chitosan	Entrapment/urea was physically adsorbed into porous halloysite nanotubes which were further coated with chitosan to block the nutrient	Glutathione produced by crops could break down chitosan	Wang et al. (2020)
	Chitosan and anionic compounds	Entrapment/electrostatic self-assembly	Diffusion and chitosan hydrolysis	Dhlamini et al. (2020); Ha et al. (2019); Kubavat et al. (2020)

TABLE 3.1 *(Continued)*

Category	Materials for Nanostructure	Fabrication Strategy/ Concrete Method	Nutrient-Release Mechanism	Reference
Nanosized fertilizers	Humic acids (HA) and other organic acids	Neutralization of Ca^{2+} and PO_4^{3-}, organic acids could be functionalized by dipping	Dissolving promoted by nanometerization and organic acid-functionalization	Samavini et al. (2018); Yoon et al. (2020)
	Humic acid and urea and thermoplastic starch	Mixing	Dissolving promoted by nanometerization and the soluble host matrixes of urea and starch	Giroto et al. (2015)
	Manganese zinc ferrite nanoparticle	Template-free microwave-assisted hydrothermal synthesis technique	Slow dissolving	Shebl et al. (2020)
	Leonardite potassium humate and $Fe_2(SO_4)_3$	Coprecipitation	Slow dissolving	Cieschi et al. (2019)
Nano-wrapped fertilizers	Nano-silica	Spraying standard-size urea with a combination of coated polymer and nano-silica	Reducing the porosity of coating through –OH cross-linking, thus extending the release longevity of the coated urea tablet	Liu et al. (2018)
	Sodium alginate-loaded hollow nano-silica	The polyurethane coating of normal-size urea was electrostatically adsorbed with hollow nano-silica that was loaded with sodium alginate	The sodium alginate would release to form a gel with Ca^{2+}, thus blocking the pores and cracks of the coating to regulate the release rate	Zhang et al. (2020)

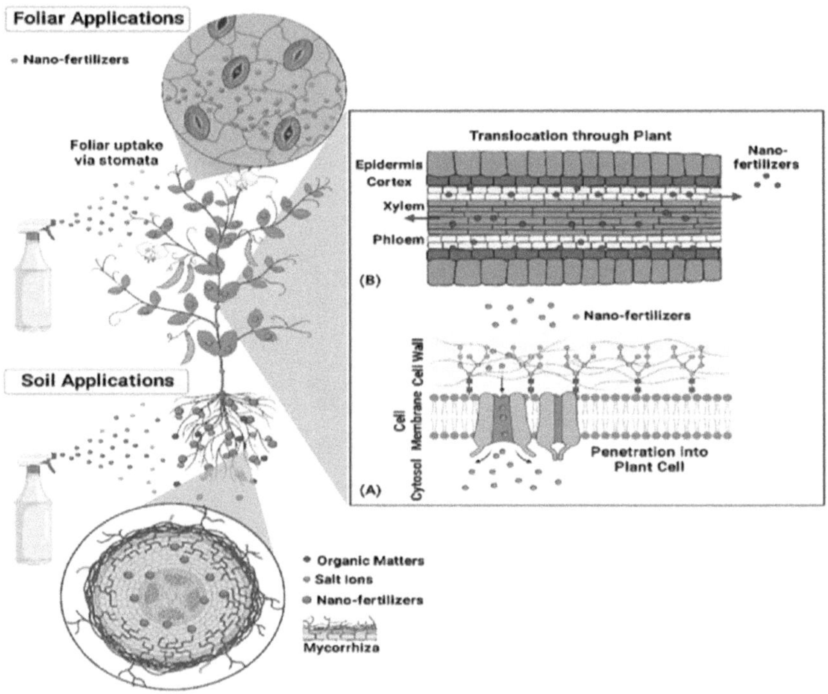

FIGURE 3.3 An overview of nanofertilizer application in agriculture.
NPs, nanoparticles; NFs, nanofertilizers.

Source: Adapted from Verma et al. (2022).

fertilizer or manure, plant root penetration exhibits substantial passage for nutrient absorption, such as being extremely permeable to nanofertilizer. The demonstration of fluorescently tagged monodispersed mesoporous silica NPs suggests that they accurately characterize the processes of NP uptake and translocation. NPs enter the root via symplastic and apoplastic routes and travel to the xylem tissue before reaching plant areas such as the leaves and stem (Sun et al., 2014). For example, ZnO and CeO_2 NPs are depicted by tracing travel routes from the ryegrass root cells to the vascular tissues of soybeans (*Glycine max*) (López-Moreno et al., 2010). Once inside the cell, these particles migrate apoplastically or symplastically from the plasmodesmata of one cell to the next until they reach the cytoplasm. They begin to spread to numerous cytoplasmic organelles and engage in diverse metabolic processes in cells in the cytoplasm. Furthermore, plants grown in hydroponic medium had greater NP absorption than plants cultivated in the sand (Solanki et al., 2015). As a result, nanofertilizers can transport and supply nutrients

TABLE 3.2 An Overview of Some Key Investigations on the Use of Various Agrochemical Nanoparticles and Their Potential to Enhance Agricultural Sustainability.

Nanomaterial as Fertilizer	Nutrient	Test Plant/Soil	Agronomic Findings/Application	Reference
Magnetite (Fe_2O_3) NPs	Micronutrient (Fe)	Coriander plant	• Growth promoter and antimicrobial substance	Fahad et al. (2022)
Manganese zinc ferrite NPs	Micronutrient (Mn, Zn)	Squash plant (*Cucurbita pepo* L)	• Vegetative growth, high mineral content • Improved squash plant yield	Shebl et al. (2020)
Nano zeolite	Nano-enhanced fertilizers (NPK)	Sage plant	• Improved vegetative growth • Increase in fresh and dry weight • Improved photosynthesis • Improved phenols, tannins, total flavonoids, oil constituents, macro, and micro elements	Mahmoud and Swaefy (2020)
Urea-loaded mesoporous nanofertilizers	Nano-enhanced fertilizers	*Oryza sativa*	• Regulated nutrient delivery • High yield • High nitrogen recovery efficiency (NRE)	Naseem et al. (2020)
Chitosan nanofertilizer	Micronutrient (Cu)	*Zea mays*	• Higher seedling vigor index • Increased chlorophyll contents	Sharma et al. (2020)
Zinc oxide nanoparticles (ZnO NPs)	Micronutrient (Zn)	Soybean (*Glycine max*)	• Improved crop yield • Better food quality	Yusefi-Tanha et al. (2020)
Zn and Cu nanoparticles (NPs)	Micronutrients (Zn and Cu)	Basil plant (*Ocimum basilicum*)	• Increased chlorophyll and carotenoid content in the leaves • Highest phenolic and flavonoid content	Abbasifar et al. (2020)
Copper nanoparticles	Micronutrient (Cu)	Sandy loam soil	• Improved soil nitrification kinetics • Slow-release of micronutrient fertilizer • Controls the bioavailability of Cu to soil bacteria	Vande Voort and Arai (2018)
ZnO nanoparticles	Micronutrient (Zn)	*Triticum aestivum*	• Enhances Zn uptake under drought stress • Grain nutrient accumulation	Dimkpa et al., (2020a, 2020b)

TABLE 3.2 *(Continued)*

Nanomaterial as Fertilizer	Nutrient	Test Plant/Soil	Agronomic Findings/Application	Reference
Urea-loaded mesoporous nanofertilizers	Nano-enhanced fertilizers	*Oryza sativa*	• High nitrogen recovery efficiency (NRE) • High yield • Controllable release of nutrients	Naseem et al. (2020)
Nano-Mg	Macronutrient (Mg)	*Vigna unguiculata*	• Increased yield • Increased stem Mg content	Delfani et al. (2014)
Calcium nanoparticles (CaNPs)	Macronutrient (Ca)	*Moringa oleifera*	• Enhanced nutritional quality of fruits	Azeez et al. (2020)
Calcium borate nanoparticles	Macronutrient (Ca) and Boron	Lettuce (*Lactuca sativa*) and zucchini (*Cucurbita pepo*)	• Improved plant yield and availability	Meier et al. (2020)
Hydroxyapatite NPs–urea	Macronutrient (N)	*Oryza sativa*	• Slow-release of nitrogen (35% less) • Increased plant NAE • Increased N and K content in leaves	Kottegoda et al. (2017)
Hydroxyapatite NPs–urea	Macronutrient (N, P)	Red-yellow oxisol soil	• Higher sustained phosphorus availability • Decreased phosphorus immobilization	Giroto et al. (2017)
Urea–hydroxyapatite–montmorillonite nanocomposites	Macronutrient (N)	*Oryza sativa*	• Slow and sustained release of urea into the soil • Higher crop yield	Madusanka et al. (2017)
Nano-chitosan	Macronutrient (NPK)	*Triticum aestivum*	• Increase in harvest and crop index • Increase in root and shoot length	Abdel-Aziz et al. (2016)
Calcium phosphate nanoparticles (CaP)	Macronutrient (NPK)	*Triticum durum*	• Enhanced the efficiency of fertilization • Precision agriculture • Increased efficiency of N uptake	Ramírez-Rodríguez et al. (2020)

from cell to cell via a narrower channel of plasmodesmata (50–60 nm). As a consequence, higher NUE and reduced nutrient loss during entrance to transport by nanofertilizer resulted in enhanced productivity (6–17%) and nutritional quality of field crops and plants (Iqbal et al., 2019).

3.5.2 NANOPESTICIDES

Since more than half of the world's population depends on plants and their products for a living, increasing plant production and maintaining plant health are major challenges. Pathogens, nutrient deficiencies, and soil and air pollution, for example, are all biotic and abiotic factors that can impair plant health by causing damage that leads to lower crop and fruit output. Plant illness is frequently caused by a diverse array of disease-causing organisms (or pests). As a result, preventing further plant withering is an essential imperative. Insects are important in agriculture because they destroy crops and overrun stored food, causing food quality to deteriorate and plant illnesses to spread. Pesticides are used to enhance agricultural productivity and efficiency by protecting plants from harmful elements such as plant diseases and insects (Jampílek and Kráľová, 2017). Herbicides, insecticides, bactericides, fungicides, and nematicides are popular pesticides used to target weeds, pests, plant diseases, nematodes, and microorganisms that destroy standing crops (Ecobichon, 2001). However, it has been shown that pesticide usage is harmful and poisonous to the environment; most of them are detrimental to human and animal health (Meena et al., 2020). Furthermore, the development of effective and safe pesticides is difficult and costly, but nanotechnology offers a fresh and better option (Sasson et al., 2007). Similar to those nanofertilizers, nanopesticides have garnered significant interest in replacing conventional pesticides due to their effectiveness against various pests and opportunities for targeted action, reducing the possibility of leakage to the environment, as shown in Table 3.3. Nano herbicides, nanoinsecticides, nano nematicides, and nano fungicides are the most commonly tested nanopesticides, as shown in Table 3.4.

Raj et al. in 2019, recently reported and elaborated on different elements of nanoagrochemicals, such as nanoencapsulation, the methodology involved in the synthesis, increased agricultural production, impact on horticulture, and contribution to crop conservation (Nisha Raj et al., 2021). One of the technology's prospective uses is the regulated and controlled release of pesticide chemicals, which considerably reduces waste pollution. Graphene oxide (GO) is cleverly exploited as a nanocarrier to revolutionize the delivery of pyrethroid insecticide against the phytophagous mite pest spider mite (Zhao et al., 2019). The newly created GO-pesticide nanocomposite has been shown to have a strong loading capacity and great release behavior both inside and outdoors. Its efficacy and potential for practical applications are demonstrated by its uniform distribution and excellent adsorption on the cuticle of spider mites and bean plant leaves. Additionally, engineered nanoinsecticides are helping the normal cotton crop as well as the insect-resistant transgenic (Bt cotton) crop, which was developed to battle the bollworm. Cu NPs can upregulate the expression of exogenous microbial proteins from *Bacillus thuringiensis* programmed through the Bt toxin in plant tissue to increase resistance against the bollworm at a low dosage (10 mg/L) (Zhao et al., 2020). TiO_2 and ZnO NPs, which are photochemically active, function as antibacterial agents when spot disease affects tomato plants. TiO_2 and ZnO NPs, which are photochemically active, produce excited electrons when they are exposed to light. These excited electrons then directly transfer to superoxide radicals in the presence of oxygen. NPs are designed to have antibacterial qualities and have a significant influence on agriculture as nanoinsecticides using photochemically active TiO_2. With regard to *X. perforans*, which causes spot disease in tomato plants, it exhibits significant photocatalytic activity as well as greater antibacterial capability (Álvarez et al., 2016).

TABLE 3.3 Types of Nanopesticides with Their Benefits.

Type	Relation to Nanomaterials	Benefits
Nanoemulsions	The active ingredient is located in nanoscale oil droplets that float in an aqueous solution	Improved distribution and increased uptake by an organism
Nanocapsule	The active ingredient is packed in a nanoscale solution	Increase uptake and thus increased effectiveness against pests
Nanogel	The active ingredient is distributed in a gel, which consists of nanoscale building blocks	Easier handling and decreased evaporation of active ingredient
Nanofibers	The active ingredient is incorporated in a nanoscaled fiber	Improved distribution, slow release of active ingredient
Liposomes	Liposomes are in nano-sized particles	Slow release of active ingredient

Source: Adapted from Sindhu et al. (2021).

TABLE 3.4 Various Investigated and Used Nanopesticides.

Nanomaterial	Nanoparticles Name	Concent Ration	Pest	Plant Disease	Advantage	Reference
Ag-based NPs	DNA-directed silver (Ag) nanoparticles grown on graphene oxide (GO)	16 mg/L	*Xanthomonas perforans*	Bacterial spot of tomato	Significantly decreased the activity of cultured *Xanthomonas perforans*	Ocsoy et al. (2013)
	Turnip leaf extracted green-synthesize Ag NPs	(~16 nm)	*Gloeophyllum abietinum, G. trabeum, Chaetomium globosum, and Phanerochaete sordida*	Several wood-degrading fungi	Showed broad-spectrum antifungal activity against several wood-degrading fungi	Narayanan and Park (2014)
	Artemisia absinthium extracted green-synthesize Ag NPs	10 mg/L	*Phytophthora*	*Phytophthora* infection	Suppressed *Phytophthora* infection on plants and improved plant survival	Ali et al. (2015)
Ti-based NPs	TiO_2 NPs (photochemically active)	500–800 mg/L	*Xanthomonas perforans*	The pathogen that induces spot disease in tomato plants	High photocatalytic activity and antibacterial potential to reduce bacterial spot severity	Paret et al. (2013)
Mg-based NPs	MgO NPs	0.1, 0.5, 0.7, or 1% (growth media: 50% vermiculite and 50% perlite)	*Ralstonia solanacearum*	Resistance of tomato plants to *Ralstonia solanacearum*	The pre-treatment of tomato roots with MgO NPs suspension (0.1, 0.5, 0.7, or 1%) significantly reduced the incidence of disease. The mechanism included the upregulation of salicylic acid-inducible PR1, jasmonic-acid-inducible LoxA, ethylene-inducible Osm, and systemic-resistance-related GluA	Imada et al. (2016)

TABLE 3.4 *(Continued)*

Nanomaterial	Nanoparticles Name	Concent Ration	Pest	Plant Disease	Advantage	Reference
	MgO NPs	200 or 250 mg/L	*Ralstonia solanacearum*	Induced tobacco bacterial wilt	Effectively suppressed *R. solanacearum*, which induced tobacco bacterial wilt. The proposed mechanism included physical injury of cell membranes and ROS accumulation	Imada et al. (2016)
Ce-based NPs	CeO_2 NP	50 and 250 mg/L by root and foliar path	*Fusarium wilt*	Mediated suppression of in tomato 250 mg/L	CeO_2 NPs significantly reduced disease severity, by 53% and 57%, respectively	Adisa et al. (2018)
Si-based NPs	Mesoporous silica nanoparticles (MSNs) with a chitosan coating (CTS-MSNs) or without	36 nm, 500 mg/L	*Fusarium wilt*	Suppress *Fusarium wilt* in soil-grown watermelon	Both MSNs and CTS-MSNs reduced disease severity, by 40% and 27%, respectively	Buchman et al. (2019)
Mn-based NPs	MnO NP	1 mg/L /40 nm	*Fusarium oxysporum*	*Fusarium oxysporum* f.sp. *lycopersici* Tomato disease	Reduced disease estimates [area under the disease progress curve (AUDPC)] by 28% in plants grown in soil	Buchman et al. (2019)

3.5.3 *COMMERCIALLY AVAILABLE NMS*

The spectrum of NMs includes everything from inorganic or organic, crystalline or amorphous particles that can be found as single particles, powders, aggregates, or dispersed in a matrix, through colloids, suspensions, emulsions, nanolayers, and films, all the way up to the class of fullerenes and their derivatives. Different types of NMs are used in commercial agricultural nanoformulations to make nanopesticides and nanofertilizers. The selected commercial NMs that are available on the market are listed in Table 3.5.

3.6 TYPES OF NANOPARTICLES IN AGRICULTURE

It is very essential in the current situation to meet the dietary needs of the constantly expanding world population. Nearly one-third of crops grown conventionally suffer damage, mostly as a result of pest infestations, microbiological assaults, natural disasters, poor soil properties, and a lack of available nutrients. To solve these problems, we urgently need more inventive technology (Mittal et al., 2020). According to recent scientific evidence, nanotechnology may have a favorable effect on the agri-food industry by reducing the negative effects of agricultural practices on the environment and human health and increasing food security and productivity (Fraceto et al., 2016). Atoms and molecules make up every element in the universe. It is simple to solve issues in all fields by adjusting or changing their nanosize. Similar to how it could solve many problems in the food and agriculture sectors. Pesticides, fertilizers, and biosensors are all effectively replaced by NMs. The use of nanotechnologies has made it easier to improve nutrition, deliver bioactive ingredients and micronutrients safely, and preserve food (Vijayakumar et al., 2022).

3.6.1 *METAL NANOPARTICLES*

Scientists have been fascinated by metallic nanoparticles for more than a century, and they are now widely used in engineering and biological sciences. Because of their enormous potential in nanotechnology, they are a focus of interest (Mody et al., 2010).

3.6.1.1 SYNTHESIS OF METAL NANOPARTICLES

In order to create metallic NPs, it is becoming more and more important to use natural alternatives, including plant and microbial enzymes, polysaccharides, vitamins, biodegradable polymers, and biological systems (Iravani, 2014). Depending on their properties, algal species may also be used in the production of metallic NPs. NM (NP) biosynthesis depends

TABLE 3.5 List of Selected Commercial Agricultural Nanoformulations.

Product Name	Manufacturer	Nanoparticle	Properties	Application	Reference
NanoZn™	Aqua-Yield Hub, USA	Zinc (nanocapsule)	• Plant growth regulation • Zinc delivery to crops • Nanofertilizer	Nanofertilizer	https://www.aquayield.com/
NANO FERTILIZERS	Silvertech Kimya Sanayi ve Ticaret Ltd., TURKEY	Zinc oxide titanium dioxide	• Plant growth acceleration • Crop yield enhancement • Photosynthesis Improvement • Soil salinization	Nanofertilizer	http://www.silvertech-kimya.com.tr/
NovaLand-F	Land Green & Technology Co, TAIWAN	Mn, Cu, Fe, Zn, Mo, N	• Preventing the growth of algae • Plant growth regulation • Plant growth acceleration • Increased resistance to plant diseases • Plant metabolism acceleration	Nanofertilizer	https://www.lgt.tw/
Groagro 4: Super Kalium Catalyst + T.E	Microwell Bio Solutions Sdn Bhd, MALAYSIA	Potassium	• Nutritional value improvement • Plant growth acceleration	Nanofertilizer	https://product.statnano.com/company/microwell-bio-solutions-sdn-bhd
NUBIOTEK® HYPER Fe+Mg	Bioteksa, Mexico	Iron (nanoparticle / nanopowder) Magnesium (nanocapsule)	• Plant growth regulation • Nutritional plant growth acceleration • Plant resistance enhancement	Nanofertilizer	https://bioteksa.com/
FÉRTIL CALMAG	HPL Agronegocios, Brazil	Magnesium (nanocapsule)	• Plant growth regulation • Absorption enhancement	Nanofertilizer	https://product.stat-nano.com/company/hpl-agronegocios
Nanovec TSS 80	Laboratórios Bio-Médicin, Brazil	Manganese (nanocapsule) Molybdenum (nanocapsule)	• Seed treatment	Nanofertilizer	https://product.stat-nano.com/company/laborat%C3%B3rios-bio-m%C3%A9dicin

TABLE 3.5 *(Continued)*

Product Name	Manufacturer	Nanoparticle	Properties	Application	Reference
FERTILE CALCIUM 25	HPL Agronegocios, Brazil	Calcium nanoparticle /nanopowder	• Nutritional absorption enhancement	Nanofertilizer	https://product.stat-nano.com/company/hpl-agronegocios
Lithocal	Litho Plant, Brazil	Calcium nanoparticle /nanopowder manganese nanoparticle /nanopowder	• Nutritional absorption enhancement	Nanofertilizer	https://lithoplant.com.br/
FOSVIT K30	Kimitec Group, Spain	Phosphorous (nanoparticle /nanopowder) Diameter: 2 nm Potassium (nanoparticle /nanopowder) Diameter: 2 nm	• Self-defence enhancement	Nanofertilizer	https://kimitec.com/en/
TAG NANO ZINC	Tropical Agrosystem India (P) Ltd., India	Zinc	• Plant growth regulation • Photosynthesis improvement • Root activity improvement	Nanofertilizer	https://www.tropicalagro.in/
Nano Zinc (Chelated)	Alert Biotech, India	Zinc (nanoparticle /nanopowder)	• pH stability • Zinc enrichment in crop produce	Nanofertilizer	https://alertbiotech.com/
Banner MAXX	Syngenta, Switzerland	Propiconazole concentrate emulsion	• Fungicidal activity	Nanopesticide	https://www.syngenta.com/en

heavily on biomolecules like proteins, carbohydrates, and numerous secondary metabolites produced by algae. Metal ions and extracellular polysaccharides are templated by algal membrane proteins, which minimize fluctuating metal ions and stabilize metal NPs. The size, shape, and layout of the NPs are controlled by the capping and stabilizing effects of flavonoids and terpenoids, which are useful in the production of metal nanoparticles. *Chlorella vulgaris* produces AgNPs, *Phaeodactylum tricornutum* produces CdS NPs, and *Stephanopyxis turris* is capable of producing Au NPs with ease. Further research in this field has been made possible by the screening of actinomycetes for their innate capacity to make NPs. An extreme actinomycete known as *Thermomonospora sp.* can extracellularly synthesize monodisperse, spherically shaped Au NPs having a typical size of 8 nm, as shown in Figure 3.4 (Jeevanandam et al., 2018; Khanna et al., 2021; Purohit et al., 2019; Siddiqi and Husen, 2016). In addition, metals can easily permeate and interact with the internal side chains of amino acids in the hollow interior cavity of the viral capsid that is present in virus particles. Thus, a natural nano-reactor that produces metal NPs of different sizes and shapes is provided by the virus' interior cavity (Purohit et al., 2019).

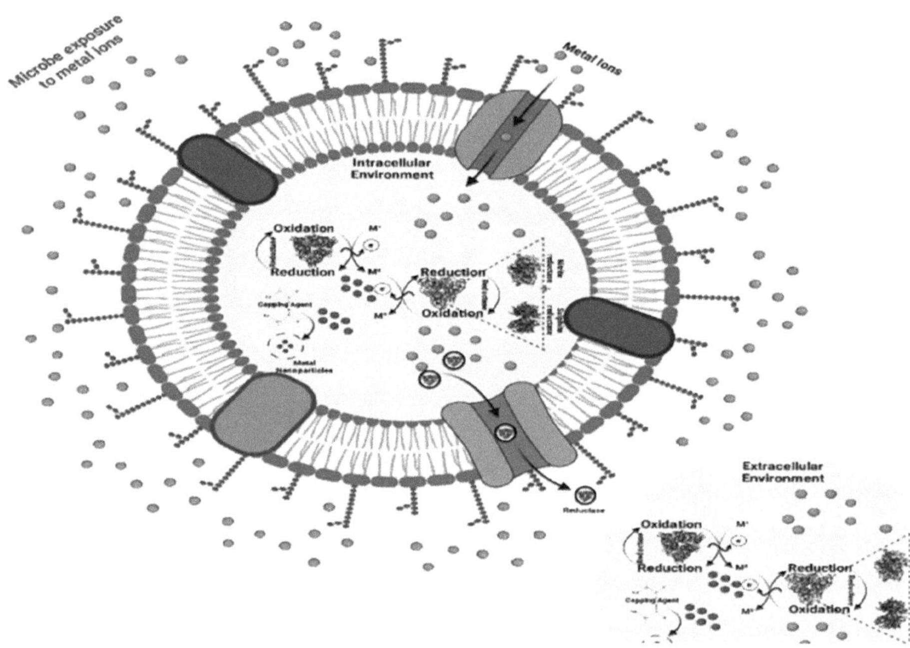

FIGURE 3.4 Nanoparticle biosynthesis mechanism in microbial cells against exposure to metal ions.

Source: Adapted from Khanna et al. (2021).

3.6.1.2 EFFECTS OF METAL NANOPARTICLES

Metals, metal oxides, and nanoparticles in bio-composites are examples of well-developed NMs (Sangeetha et al., 2017). Depending on the chemical characteristics, size, shape, surface charge, and dose of the particle, metal nanoparticles may have either beneficial or harmful effects on plants (Sarkar et al., 2022). For their assimilation, accumulation, translocation, and, more importantly, effects on the growth and development of crop plants, such as increased seed germination rates, root and shoot elongation, plant biomass, and plant yield under certain concentration levels, metal-based nanoparticles have been used (Chhipa, 2019; Nair et al., 2010; Sarkar et al., 2022). According to a meta-analysis by Guo et al. (2022), Ag NP treatments enhanced the final germination percentage (FGP) of agriculturally significant crop species by more than other NPs at a dosage of 50 mg/L. Additionally, Zn NPs elongated the roots (Guo et al., 2022).

Numerous cutting-edge synthesized metallic nanoparticles (Ag NPs, Cu/CuO NPs, TiO_2 NPs, carbon NPs, Fe NPs, Zn NPs, Mg NPs, Si NPs, K-NPs, Au-NPs, CeO_2-NPs, Graphene-silver NPs, etc.), as well as lipid (liposomes, nano emulsions), polymer (polymeric micelles, nano crystals, nano sponges, nano dispersion), lipid hybrid (nano-polymer), and carbon-based nanoparticles (carbon nanotube, graphene, fullerene), have already attracted significant attention in modern agriculture, as shown in Table 3.6 (Khanna et al., 2021; Luna et al., 2015; Sarkar et al., 2022). Numerous studies revealed that the harmful effects of metal and metal oxide nanoparticles on plants depended on the length of exposure (Sarkar et al., 2022).

E. coli, *S. typhimurium*, *S. aureus*, *S. enterica*, and *L. monocytogenes* have all been optically detected using metal NPs such as Au NPs, Ag NPs, and magnetic NPs (Singh et al., 2021). The antibacterial activity of several metal nanoparticles, including silver colloids and Ag NPs, has been found to increase with decreasing particle size (Raza et al., 2016; Shao et al., 2022). Considering these distinctive benefits, the majority of metal nanoparticles have only been produced at the laboratory scale, and more work is needed to adapt these systems for use in actual agricultural fields.

3.6.2 METAL OXIDE NANOPARTICLES

Making nanostructures or nanoarrays with unique features compared with those of bulk or single particle species is an aim in the growing field of nanotechnology (Rodriguez et al., 2000; Valden et al., 1998). Due to their small size and the great density of their corner or edge sites, oxide nanoparticles

TABLE 3.6 Diverse Metal and Metalloid Nanoparticles, Their Types and Methods of Manufacturing, How They are Used, and the Effects They Have on Various Plants.

NPs	Synthesis Method	Size	Concent Ration	Application	Types of Plants	Effects	Type of Stress	Reference
Ag NPs	Biological method (dates extract)	34–50 nm	25, 50, and 75, 100 mg/L	Soil or growth media	*Triticum aestivum*, *Oryza sativa*	Elevated root and shoot length, root counts, and plant morphological growth, such as leaf area and number, were all improved	Heat stress	Iqbal et al. (2019); Khattab et al. (2022); Mirzajani et al. (2013)
Cu-chitosan NPs	Procured	NA	0.01–0.16%	Bio fertilization of soil	*Zea mays*, *Hordeum vulgare*	Enhanced seed germination, improved root and shoot length, increased fresh and dry biomass, enhanced antioxidant activity, and higher anthocyanin and chlorophyll content all contribute to enhanced plant growth and grain weight	Drought stress	Choudhary et al. (2017); Kadri et al. (2022); Van Nguyen et al. (2022)
Au NPs	Various solution-based approaches	10–100 nm	50 and 70 DAS	Leaf tissues	*Brassica juncea*	Improved seed germination, cell water, and oxygen absorption, and seed capsule permeability	Oxidative stress	Arora et al. (2012); Ferrari et al. (2021); Yeh et al. (2012)
Si NPs	Solvent extraction method along with the Sol-gel method, nutrient solution	15–91 nm, 202 nm	150 and 300 mg/L	Soil application, spraying, or seed soaking	*Oryza sativa*, *Triticum aestivum*, *Solanum lycopersicum*	Significantly increased antioxidants to defend against UV-B radiation, increased nutrient uptake, improved carbon absorption, promoted plant growth, and generated oxidative stress	Salinity stress, UV-B stress	Khattab et al. (2022); Tripathi et al. (2017)

TABLE 3.6 *(Continued)*

NPs	Synthesis Method	Size	Concent Ration	Application	Types of Plants	Effects	Type of Stress	Reference
Fe NPs	Solvent extraction	50–100 nm	0, 25, 50, and 100 mg/kg	Potting soil	*Triticum aestivum*	Increased photosynthesis, increased Fe concentration, enhanced development and physiology, and decreased cadmium concentration	Cadmium and drought stress	Adrees et al. (2020); Farrell et al. (2003)
Mn NPs	Inert gas condensation, solvent extraction	NA	0.1, 0.5, and 1 mg/L	Nano priming	*Capsicum annuum*	Improved sustainable agriculture, plant salt stress control is important	Salinity stress	Buccolieri et al. (2017); Ward et al. (2006); Ye et al. (2020)

can display distinct physical and chemical characteristics. Any material's structural qualities, such as lattice symmetry and cell parameters, are likely to be significantly influenced by particle size, along with other significant groups of fundamental properties (Fernández-García and Rodriguez, 2011).

It was discovered that certain metal oxide nanoparticles have intriguing functions, as shown in Table 3.7. According to recent studies (Shao et al., 2022), the physiology of plants can be affected by the uptake, distribution, and impacts of metal oxide nanoparticles.

By detecting microbial infections in the environment adjacent to the plants, metal oxide NPs have also been investigated for their usefulness in controlling plant diseases, as shown in Table 3.8. For example, optical nano-biosensors for the diagnosis of aquatic and foodborne pathogens have been developed (Singh et al., 2021). There was also a discussion of other oxide NPs. Results from a study to assess the effects of nano-CeO_2 on *Spodoptera frugiperda*, an arthropod pest found abundantly in agricultural areas, revealed the toxicity of nano-CeO_2 for *S. frugiperda* under field trials (Castro et al., 2021). Nano-silica was also discovered to be successful in substituting chemical insecticides to safeguard stored goods (Zahran and Sayed, 2021).

3.6.3 OTHER TYPES OF NANOPARTICLES: THEIR PRODUCTION AND APPLICATIONS

Eco-friendly, biodegradable, economical, and biopolymer-based nanohybrid structures for promoting plant growth are becoming increasingly popular, as shown in Table 3.9. According to a recent study, applying chitosan nanoparticles (Ch NPs) to the leaves dramatically improved growth, yield, and mineral content (Fe, Zn, Mn, P, Ca, and Mg) when compared to controls (Shao et al., 2022). As a result of their biodegradability and biocompatibility, chitosan-based nanoparticles and nanoformulations have received substantial research (Singh et al., 2022). By cross-linking diuron disulfide linkages, chitosan nanoparticles were created for regulated administration based on glutathione content (Yu et al., 2015). Additionally, Ch NPs induced several defence-related enzymes (chitinase, beta-1,3 glucanase, chitosanase, protease inhibitors, peroxidase, and polyphenol oxidase) in the leaves of finger millet plants, and they are also involved in *Rhizopus sp.* and *Alternaria alternata*'s delayed mycelia growth and inhibition of spore germination, respectively (Chookhongkha et al., 2013; Saharan et al., 2013; Shao et al., 2022). Some sweet pepper genes that seem to be resistant to black mold disease may be induced by the use of salicylic acid nanoparticles (SA NPs). According to *in*

TABLE 3.7 Diverse Metal Oxide Nanoparticles, Their Types and Methods of Manufacturing, How They are Used, and the Effects They Have on Various Plants.

NPs	Synthesis Method	Size	Concentration	Application	Types of Plants	Effects	Type of Stress	Reference
ZnO	Procured	12–50 nm	20 mg/L; 0.2–1 µM; 0, 0.5, and 1 g/L; 10 mg/L; 100–500 mg/L	Soil application, hydroponics, foliar application	*Triticum aestivum*, *Nicotiana tabcum*, *Glycine max*, *Coffea arabica*, *Capsicum chinense*	Grain yield and biomass accumulation, fruit yield, and biomass accumulation, as well as improvements to plant developmental physiology, metabolites, enzymatic activities, and morphological characteristics	Drought	Davarpanah et al. (2016); García-López et al. (2018); Rossi et al. (2019)
TiO_2	Laser-ablation method/ procured	10–25 nm	0, 10, 100, and 500 mg/L	Leaf treatment	*Linum usitatissimum*, *Oryza sativa*	On rice grains grown in phosphorus-deficient soil, carotenoids, and chlorophyll content increased, MDA activity and levels dropped, and palmitic acid and glycerol levels increased	Drought	Aghdam et al. (2016); Zahra et al. (2017)
Fe_2O_3	Microwave-assisted hydrothermal method	27.3–34.62 nm	0, 10, 20, and 30 µM	Hoagland solution	*Mentha piperita*	Decreased amounts of proline and MDA, as well as decreased antioxidant enzyme activity	Salinity	Askary et al. (2017); Shebl et al. (2019)
Mn_3O_4	Microwave-assisted hydrothermal method	20–60 nm	20 ppm	Foliar application	Squash plant	Enhanced fruit vegetative development characteristics and increased yield	NA	Shebl et al. (2019)

TABLE 3.7 *(Continued)*

NPs	Synthesis Method	Size	Concentration	Application	Types of Plants	Effects	Type of Stress	Reference
CuO	Procured	NA	10–2000 mg/L	Seed soaking	*Hordeum vulgare*	Enhanced antioxidant activity, longer roots and shoots, fresher and drier biomass, and improved seed germination rates	NA	Kadri et al. (2022)
CeO	Microwave combustion (MCM) and conventional route sol-gel (CRSGM)	2.1 ± 1.4 nm	500 mg/L	Seed priming	*Gossypium hirsutum*	Reduced ROS levels, better root characteristics, and activation of mechanisms that control tolerance, such as ion homeostasis	Salinity stress	Mittal et al. (2020); Tamizhdurai et al. (2017)
Al_2O_3	Laser ablation	NA	2000 mg/L	Foliar application	*Zea mays*	Demonstrated harmful effects on root extension.	Oxidative stress	Ahmed et al., (2022); Li et al. (2020); Yang et al. (2017)
NiO_2	Chemical precipitation method using Triton X-100	54 nm	120 mg/Kg	Whole plant	*Triticum aestivum*	Decreased plant growth and photosynthetic capacity, but increased antioxidant content	Oxidative stress	Saleh et al. (2019)

TABLE 3.8 Various Forms of Metal Oxide That are Employed to Protect Crops.

NPs	Application	ConcentRation	Plant Species	Time Span	Effects	Reference
CuO NPs	Foliar application	150–340 µg/mL	*Lycopersicon esculentum*	11 days	*Phytophthora infestans*-caused late blight disease has been effectively managed	Giannousi et al. (2013); Hong et al. (2015)
MgO NPs	Drenching	7–10 µg/mL	*Solanum lycopersicum*	7 days	Prevented *Ralstonia solanacearum*, a pathogen, from spreading and causing bacterial wilt disease	Imada et al. (2016)
CeO_2 NPs	Hydroponics	10 mg/L	*Raphanus sativus*	21 days	Impacted radish growth and aggregated in the nutritional storage tissues of the root and shoot	Zhang et al. (2015)
TiO_2 and SiO_2 NPs	Foliar applications	20 and 30 mg/L	*Oryza sativa*	55 days	Increased development and decreased Cd toxicity by increasing antioxidant capacity and inhibiting Cd transfer	Rizwan et al. (2019)
ZnO and CuO NPs	Fruit spray	100 and 1000 µg/mL	*Prunus domestica*	4 days	Evaluated the effectiveness of NPs against several harmful plant pathogenic fungi and reduced symptoms of soil-borne illnesses and grey mold caused by *B. cinerea*	Malandrakis et al. (2019)

vitro research, SA NPs applied at 1.4 mM significantly inhibited the growth of *A. alternata* (Abdel-Rahman et al., 2021).

Additionally, biological and medicinal research employs graphene NMs. One important characteristic that distinguishes GO from other carbon NMs is how easily the surface may be modified. GO's photothermal action and drug transport capabilities allow it to be employed for simultaneous chemo- and photothermal treatment without the necessity of additional pharmaceutical carriers (Yusaf et al., 2022). The effects of pH, graphene, and photosynthetic factors on plant development, as well as the antioxidant defence system's enzymes on leaves and roots, were all studied, and significantly increased plant growth was found (Shao et al., 2022). With an eye on environmental remediation, materials such as silica, nonmagnetic/magnetic carbon nanotubes/nanorods, nano-clay/nano-membranes, MOFs, GO, and other NMs have been investigated in combination with carrageenan biopolymers (Evode et al., 2021).

TABLE 3.9 Additional Nanoparticles in Many Forms and Their Numerous Applications.

Nanoparticles	Applications	Reference
Graphene oxide	Improved watermelon ripeness and fruit sugar content; increased root length, leaf area, leaf number, and flower bud development in *Arabidopsis*.	Park et al. (2020)
Graphene nano sheets	Enhanced H_2O_2 level, sugar stimulation, and improved plant output and growth	Younes et al. (2019)
GO–gold nanostars	Improved Raman diffusion detection and drug delivery progressed surface	Wang et al. (2014)
Chitosan nanoparticles	Promoting the development of *Camellia sinensis* and *Cucumis sativus*, generating resistance to the disease known as powdery mildew, and initiating and enhancing immunological responses	Chandra et al. (2015); Jogaiah et al. (2020)
Chitosan-gold nanocomposite	Degradation of the soil and decreased biological nitrogen-fixing	Goyal and Sivanappan, (2017)

3.7 REMEDIATION OF CONTAMINATED SOIL AND WATER USING NMS

Global issues like pollution in the agro-ecosystem have potential solutions owing to nanotechnology. Due to harmful organic and inorganic pollutants, inappropriate handling of agricultural wastes, and other factors, soil and groundwater have been put at severe risk. The elimination of heavy metals, dyes, volatile organic compounds, polyaromatic hydrocarbons, biphenyls, and other organic contaminants from wastewater discharged from fields is made possible by nanotechnology (Dasgupta et al., 2017; Qadri et al., 2020).

Nano-remediation techniques have a range of benefits, including the ability to reduce clean-up times, do away with complicated methods for treating and disposing of contaminated soil, lower cleaning costs, and reduce the concentrations of some contaminants to almost undetectable levels (Dasgupta et al., 2020b; Fajardo et al., 2020). Metallic and metal oxide nanoparticles, carbon nanotubes, fullerenes, nanofibers, magnetic nanoparticles, nano-zero-valent iron quantum dots, and polymeric nanoparticles are the most often used NMs for agricultural water management (Qu et al., 2013). The removal of pollutants through adsorption, photocatalysis, reduction, and chemical oxidation is the most popular method of remediating contaminated water and soils (Guerra et al., 2018; Huang et al., 2019; Rodríguez-González et al., 2019).

NMs have been demonstrated to be effective in treating contaminated soils along with water remediation, as shown in Table 3.10. Through ion exchange/coordination interactions, the adsorption of harmful soil contaminants onto the surfaces of organic ligands, minerals, and NMs might alter their mobility and toxicity, as shown in Figure 3.5 (Hamid et al., 2019, 2020). The

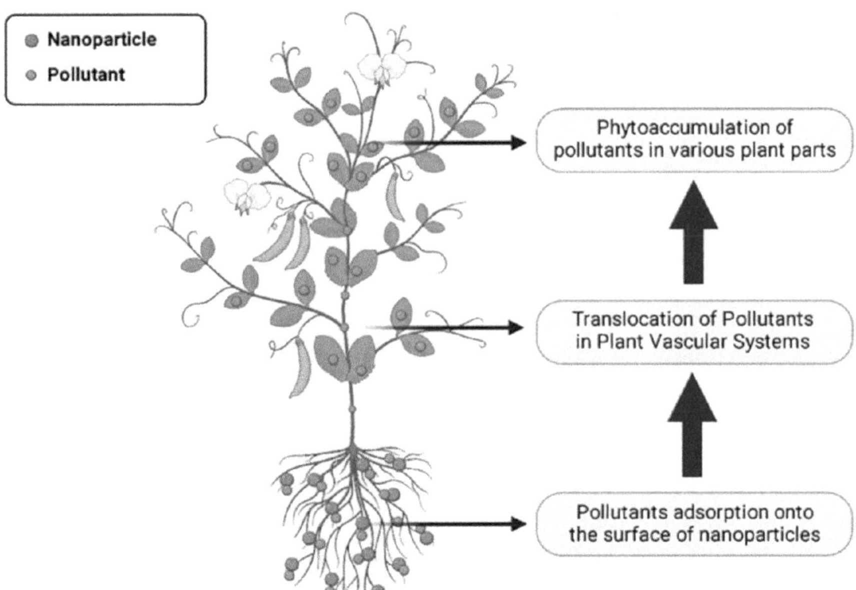

FIGURE 3.5 Soil remediation via phytoremediation by the use of nanoparticles.

Source: Adapted from Singh et al. (2022).

adsorbent NM's capacity to be recycled and reused is crucial for lowering remediation costs. In this context, super-paramagnetism, environmental friendliness, rapid separation after adsorption, and simplicity of synthesis are just a few benefits of magnetic iron oxide nanoparticles (Abd-Elsalam et al., 2019; Attia and Elsheery, 2020).

NMs have made significant strides in other areas as well, including the efficient biodegradation of organic pollutants in soil and water, including triclosan, lindane, polychlorinated biphenyls, dichlorodiphenyldichloroethylene, and polybrominated diphenyl ethers (Bebić et al., 2020; Nguyen et al., 2020; Singh et al., 2021).

TABLE 3.10 Examples of Soil Remediation Using Nano-Based Material.

Sr. No.	Nanomaterial	Primary Constituent	Contaminant in Soil	Reference
1.	Polyvinylpyrrolidone stabilized nano-zero-valent iron	Iron	Trichloroethylene	Tian et al. (2020)
2.	Nano-Fe/Cu	Iron and copper	Nitrate ions	Shubair et al. (2018)
3.	Multiwalled carbon nanotubes	Carbon or graphene	Chromium (VI)	Zhang et al. (2018)
4.	Modified carbon black nanoparticle	Carbon, hydrogen, and oxygen	Petroleum	Cheng et al. (2019)
5.	Single-walled carbon nanotubes	Carbon or graphene	DDT	Zhang et al. (2017)
6.	Goethite nanospheres	Ferrihydrite	Arsenic	Baragaño et al. (2020b)
7.	Nano-Fe_3O_4@C-COOH	Iron	Lead	Ma et al. (2020)
8.	Magnetite nanoparticles	Iron	Arsenic	Baragaño et al. (2020a)

3.8 TOXICITY AND ENVIRONMENTAL SAFETY CONCERNS

The potential harm that NMs, particularly bioactive nanoparticles, could cause to the environment and human health is well known and is being rigorously addressed. Recently, toxic contamination when nanoparticles are employed as an alternative to assisted phytoremediation has been analyzed (Perea Vélez et al., 2021). Additionally, the types, sizes, and pathways of transmission of nanoparticles as well as those who consume the treated plants as food are given consideration (Salama et al., 2021). For the benefit of researchers and decision-makers in the manufacturing and biomedical industries, the carcinogenicity of several nanoparticles, particularly metal nanoparticles, has also been examined (Liu and Kong, 2021). Ag has been included as a constituent of over 100 pesticides due to its antimicrobial qualities (Bergeson, 2010a). Since the destiny and lifecycle of NMs are still poorly recognized and generally unregulated, their properties may potentially pose a hazard to the environment. Due to the nanoparticles and the chemicals used in their transformation, there could be genotoxicity and phytotoxicity as a consequence (Bapat et al., 2022; Bergeson, 2010b).

According to Lu et al. (2010), citrate-coated colloidal Ag nanoparticles were not hazardous to humans in terms of genotoxicity (damage to DNA), cytotoxicity (damage to cells), or phototoxicity (toxicity caused by photo-degradation). However, quantitative information on the intake and composition of silver nanoparticles and their potential changes in soil was recently examined (Kulikova, 2021). It was observed that the only data currently available on the amount of Ag NPs in soil are simulation findings, and they ranged widely from 5.33×10^{-6} to 7.4 mg/kg at an annual input rate of 1.2×10^{-3} to 9.68 mg/kg. According to the analysis of the available theories of Ag NPs translocation from soil to plants, there is currently no risk of agricultural products being contaminated with Ag NPs. It was advised to evaluate the impacts of Ag NPs in the soil in long-term tests (over 90 days) at a nanoparticle concentration not exceeding 10 mg/kg after some data showed that the detrimental effects of Ag NPs on microorganisms were also time-dependent (Kulikova, 2021).

Like Ag NPs, the chemical activity of ZnO could have a variety of harmful effects on plants when used in large doses. These effects may differ depending on the type of plant used as well as the size and form of ZnO NPs (Thounaojam et al., 2021). Substantial research has been carried out to circumvent the antagonistic impact of ZnO NPs, with low doses and periods of exposure found to be advantageous. Furthermore, it was thought that ensuring NP stability could minimize the adverse effects of ZnO NPs in plants while also increasing their promoting efficacy (Thounaojam et al., 2021). Furthermore, the safety of TiO_2 NPs was investigated. Larue et al. (2012) discovered that TiO_2 NPs accumulated in wheat and rapeseed plantlets. However, no comprehensive investigation has yet been conducted to follow the metabolic pathway of TiO_2 NPs in crops (Shao et al., 2022).

3.9 CONCLUSION AND FUTURE PERSPECTIVE

Nanotechnology acts as a vital component, affecting many areas of the food sector. Although it has many potential uses, it is important to note that most of the information we currently know about it comes from laboratory-based studies. Understanding the practical application without knowing how it affects environmental toxicity is almost impossible. The sale of breakthrough nanoparticle-based agrochemicals is further restricted by a lack of regulations, regulatory authorities, and other equivalent entities. To advance and build futuristic studies based on identified knowledge shortages, it is necessary to carry out extensive trials. Generating a more workable strategy through detailed analysis of upcoming studies is required to ascertain the risks associated with the use of nanoparticle-based products in the agricultural sector. The potential of nanoparticles to bioaccumulate during field applications depends on the impact this has on nanotoxicity. Also, knowledge of tropic chain transfer in the ecosystem and determining the appropriate level of safety estimation need to be known first. Research on human and environmental toxins can help us comprehend the complicated interactions between the agroecosystem, nanoscale agrochemicals, exposure levels, and people. As of now, the use of NMs in agriculture-based industries is a double-edged sword. To overcome the cons, extensive research and practical usage must be put into play, or else it would just be shunted off to the "huge potential but no practical use" zone.

KEYWORDS

- **environmental safety concerns**
- **nanofertilizers**
- **nanoformulations**
- **nanomaterials**
- **nanopesticides**
- **sustainable agriculture**

REFERENCES

Abbasifar, A., Shahrabadi, F., & ValizadehKaji, B. (2020). Effects of green synthesized zinc and copper nano-fertilizers on the morphological and biochemical attributes of basil plant. *Journal of Plant Nutrition*, *43*(8), 1104–1118. https://doi.org/10.1080/01904167.2020.1724305

Abdel-Aziz, H. M. M., Hasaneen, M. N. A., & Omer, A. M. (2016). Nano chitosan-NPK fertilizer enhances the growth and productivity of wheat plants grown in sandy soil. *Spanish Journal of Agricultural Research, 14*(1), e0902. https://doi.org/10.5424/sjar/2016141-8205

Abdel-Rahman, F. A., Khafagi, E. Y., Soliman, M. S., Shoala, T., & Ahmed, Y. (2021). Preharvest application of salicylic acid induces some resistant genes of sweet pepper against black mold disease. *European Journal of Plant Pathology, 159*(4), 755–768. https://doi.org/10.1007/s10658-020-02199-z

Abd-Elsalam, K. A., Mohamed, M. A., & Prasad, R. (Eds.). (2019). *Magnetic Nanostructures.* Springer International Publishing. https://doi.org/10.1007/978-3-030-16439-3

Adisa, I. O., Reddy Pullagurala, V. L., Rawat, S., Hernandez-Viezcas, J. A., Dimkpa, C. O., Elmer, W. H., White, J. C., Peralta-Videa, J. R., & Gardea-Torresdey, J. L. (2018). Role of cerium compounds in fusarium wilt suppression and growth enhancement in tomato (Solanum lycopersicum). *Journal of Agricultural and Food Chemistry, 66*(24), 5959–5970. https://doi.org/10.1021/acs.jafc.8b01345

Adrees, M., Khan, Z. S., Ali, S., Hafeez, M., Khalid, S., Ur-Rehman, M. Z., Hussain, A., Hussain, K., Shahid Chatha, S. A., & Rizwan, M. (2020). Simultaneous mitigation of cadmium and drought stress in wheat by soil application of iron nanoparticles. *Chemosphere, 238*, 124681. https://doi.org/10.1016/j.chemosphere.2019.124681

Aghdam, M. T. B., Mohammadi, H., & Ghorbanpour, M. (2016). Effects of nanoparticulate anatase titanium dioxide on physiological and biochemical performance of Linum usitatissimum (Linaceae) under well-watered and drought stress conditions. *Brazilian Journal of Botany, 39*(1), 139–146. https://doi.org/10.1007/s40415-015-0227-x

Ahari, H., Karim, G., Anvar, A. A., Pooyamanesh, M., Sajadis, A., Mostaghim, A., & Heydari, S. (2018, August). Synthesis of the Silver Nanoparticle by Chemical Reduction Method and Preparation of Nanocomposite based on AgNPS, *Proceedings of the 4th World Congress on Mechanical, Chemical, and Material Engineering (MCM'18) Madrid, Spain.* https://doi.org/10.11159/iccpe18.125

Ahmed, B., Rizvi, A., Syed, A., Rajput, V. D., Elgorban, A. M., Al-Rejaie, S. S., Minkina, T., Khan, M. S., & Lee, J. (2022). Understanding the phytotoxic impact of Al3+, nano-size, and bulk Al2O3 on growth and physiology of maize (Zea mays L.) in aqueous and soil media. *Chemosphere, 300*, 134555. https://doi.org/10.1016/j.chemosphere.2022.134555

Ali, M., Kim, B., Belfield, K. D., Norman, D., Brennan, M., & Ali, G. S. (2015). Inhibition of Phytophthora parasitica and P. capsici by Silver Nanoparticles Synthesized Using Aqueous Extract of Artemisia absinthium. *Phytopathology, 105*(9), 1183–1190. https://doi.org/10.1094/PHYTO-01-15-0006-R

Álvarez, S. P., López, N. E. L., Lozano, J. M., Negrete, E. A. R., & Cervantes, M. E. S. (2016). Plant fungal disease management using nanobiotechnology as a tool. In *Advances and applications through fungal nanobiotechnology* (pp. 169–192). https://doi.org/10.1007/978-3-319-42990-8_8

Arora, S., Sharma, P., Kumar, S., Nayan, R., Khanna, P. K., & Zaidi, M. G. H. (2012). Gold-nanoparticle induced enhancement in growth and seed yield of Brassica juncea. *Plant Growth Regulation, 66*(3), 303–310. https://doi.org/10.1007/s10725-011-9649-z

Askary, M., Talebi, S. M., Amini, F., & Bangan, A. D. B. (2017). Effects of iron nanoparticles on Mentha piperita L. under salinity stress. *Biologija, 63*(1). https://doi.org/10.6001/biologija.v63i1.3476

Attia, T. M. S., & Elsheery, N. I. (2020). *Nanomaterials: Scope, Applications, and Challenges in Agriculture and Soil Reclamation* (pp. 1–39). https://doi.org/10.1007/978-3-030-33996-8_1

Azeez, L., Adejumo, A. L., Simiat, O. M., & Lateef, A. (2020). Influence of calcium nanoparticles (CaNPs) on nutritional qualities, radical scavenging attributes of Moringa oleifera and risk assessments on human health. *Journal of Food Measurement and Characterization*, *14*(4), 2185–2195. https://doi.org/10.1007/s11694-020-00465-6

Bapat, M. S., Singh, H., Shukla, S. K., Singh, P. P., Vo, D.-V. N., Yadav, A., Goyal, A., Sharma, A., & Kumar, D. (2022). Evaluating green silver nanoparticles as prospective biopesticides: An environmental standpoint. *Chemosphere*, *286*, 131761. https://doi.org/10.1016/j.chemosphere.2021.131761

Baragaño, D., Alonso, J., Gallego, J. R., Lobo, M. C., & Gil-Díaz, M. (2020a). Magnetite nanoparticles for the remediation of soils co-contaminated with As and PAHs. *Chemical Engineering Journal*, *399*, 125809. https://doi.org/10.1016/j.cej.2020.125809

Baragaño, D., Alonso, J., Gallego, J. R., Lobo, M. C., & Gil-Díaz, M. (2020b). Zero valent iron and goethite nanoparticles as new promising remediation techniques for As-polluted soils. *Chemosphere*, *238*, 124624. https://doi.org/10.1016/j.chemosphere.2019.124624

Bebić, J., Banjanac, K., Ćorović, M., Milivojević, A., Simović, M., Marinković, A., & Bezbradica, D. (2020). Immobilization of laccase from Myceliophthora thermophila on functionalized silica nanoparticles: Optimization and application in lindane degradation. *Chinese Journal of Chemical Engineering*, *28*(4), 1136–1144. https://doi.org/10.1016/j.cjche.2019.12.025

Bergeson, L. L. (2010a). Nanosilver pesticide products: What does the future hold? *Environmental Quality Management*, *19*(4), 73–82. https://doi.org/10.1002/tqem.20263

Bergeson, L. L. (2010b). Nanosilver: US EPA's pesticide office considers how best to proceed. *Environmental Quality Management*, *19*(3), 79–85. https://doi.org/10.1002/tqem.20255

Bowonder, B. (1979). Impact analysis of the green revolution in India. *Technological Forecasting and Social Change*, *15*(4), 297–313. https://doi.org/10.1016/0040-1625(79)90023-4

Buccolieri, A., Serra, A., Maruccio, G., Monteduro, A. G., Padmanabhan, S. K., Licciulli, A., Bonfrate, V., Salvatore, L., Manno, D., Calcagnile, L., & Giancane, G. (2017). Synthesis and characterization of mixed iron-manganese oxide nanoparticles and their application for efficient nickel ion removal from aqueous samples. *Journal of Analytical Methods in Chemistry*, *2017*, 1–9. https://doi.org/10.1155/2017/9476065

Buchman, J. T., Elmer, W. H., Ma, C., Landy, K. M., White, J. C., & Haynes, C. L. (2019). Chitosan-coated mesoporous silica nanoparticle treatment of Citrullus lanatus (Watermelon): Enhanced fungal disease suppression and modulated expression of stress-related genes. *ACS Sustainable Chemistry & Engineering*, *7*(24), 19649–19659. https://doi.org/10.1021/acssuschemeng.9b04800

Castro, B. M. M., Santos-Rasera, J. R., Alves, D. S., Marucci, R. C., Carvalho, G. A., & Carvalho, H. W. P. (2021). Ingestion and effects of cerium oxide nanoparticles on Spodoptera frugiperda (Lepidoptera: Noctuidae). *Environmental Pollution*, *279*, 116905. https://doi.org/10.1016/j.envpol.2021.116905

Chandra, S., Chakraborty, N., Dasgupta, A., Sarkar, J., Panda, K., & Acharya, K. (2015). Chitosan nanoparticles: A positive modulator of innate immune responses in plants. *Scientific Reports*, *5*(1), 15195. https://doi.org/10.1038/srep15195

Cheng, J., Sun, Z., Yu, Y., Li, X., & Li, T. (2019). Effects of modified carbon black nanoparticles on plant-microbe remediation of petroleum and heavy metal co-contaminated soils. *International Journal of Phytoremediation*, *21*(7), 634–642. https://doi.org/10.1080/15226514.2018.1556581

Chhipa, H. (2019). *Applications of nanotechnology in agriculture* (pp. 115–142). https://doi.org/10.1016/bs.mim.2019.01.002

Cho, Y. Y., & Kuo, C. (2016). Optical and electrical characterization of electrospun Al-doped zinc oxide nanofibers as transparent electrodes. *Journal of Materials Chemistry C*, *4*(32), 7649–7657. https://doi.org/10.1039/C6TC02586B

Chookhongkha, N., Sopondilok, T., & Photchanachai, S. (2013). Effect of chitosan and chitosan nanoparticles on fungal growth and chilli seed quality. *Acta Horticulturae*, *973*, 231–237. https://doi.org/10.17660/ActaHortic.2013.973.32

Choudhary, R. C., Kumaraswamy, R. V., Kumari, S., Pal, A., Raliya, R., Biswas, P., & Saharan, V. (2017). Synthesis, characterization, and application of chitosan nanomaterials loaded with zinc and copper for plant growth and protection. In *Nanotechnology* (pp. 227–247). Springer Singapore. https://doi.org/10.1007/978-981-10-4573-8_10

Cieschi, M. T., Polyakov, A. Y., Lebedev, V. A., Volkov, D. S., Pankratov, D. A., Veligzhanin, A. A., Perminova, I. V., & Lucena, J. J. (2019). Eco-friendly iron-humic nanofertilizers synthesis for the prevention of iron chlorosis in soybean (glycine max) grown in calcareous soil. *Frontiers in Plant Science*, *10*. https://doi.org/10.3389/fpls.2019.00413

Das, S. K., Ghosh, G. Kr., Mishra, V. Kr., Choudhury, B. U., Dutta, S. Kr., Hazarika, S., Kalita, H., Roy, A., Singh, N. U., Gopi, R., Devi, E. L., Mukherjee, I., Balusamy, A., Singh, M., Yadav, A., Kapoor, C., & Baruah, K. (2021). Utilizing dissimilar feedstocks derived biochar amendments to alter soil biological indicators in acidic soil of Northeast India. *Biomass Conversion and Biorefinery*, *13*, 10203–10214. https://doi.org/10.1007/s13399-021-01670-z

Das, S., Sen, B., & Debnath, N. (2015). Recent trends in nanomaterials applications in environmental monitoring and remediation. *Environmental Science and Pollution Research*, *22*(23), 18333–18344. https://doi.org/10.1007/s11356-015-5491-6

Dasgupta, N., Ranjan, S., & Lichtfouse, E. (Eds.). (2020a). *Environmental nanotechnology volume 3* (Vol. 27). Springer International Publishing. https://doi.org/10.1007/978-3-030-26672-1

Dasgupta, N., Ranjan, S., & Lichtfouse, E. (Eds.). (2020b). *Environmental nanotechnology volume 4* (Vol. 32). Springer International Publishing. https://doi.org/10.1007/978-3-030-26668-4

Dasgupta, N., Ranjan, S., & Ramalingam, C. (2017). Applications of nanotechnology in agriculture and water quality management. *Environmental Chemistry Letters*, *15*(4), 591–605. https://doi.org/10.1007/s10311-017-0648-9

Davarpanah, S., Tehranifar, A., Davarynejad, G., Abadía, J., & Khorasani, R. (2016). Effects of foliar applications of zinc and boron nano-fertilizers on pomegranate (*Punica granatum* cv. Ardestani) fruit yield and quality. *Scientia Horticulturae*, *210*, 57–64. https://doi.org/10.1016/j.scienta.2016.07.003

de Carvalho, J. F., de Medeiros, S. N., Morales, M. A., Dantas, A. L., & Carriço, A. S. (2013). Synthesis of magnetite nanoparticles by high energy ball milling. *Applied Surface Science*, *275*, 84–87. https://doi.org/10.1016/j.apsusc.2013.01.118

Delfani, M., Baradarn Firouzabadi, M., Farrokhi, N., & Makarian, H. (2014). Some physiological responses of black-eyed pea to iron and magnesium nanofertilizers. *Communications in Soil Science and Plant Analysis*, *45*(4), 530–540. https://doi.org/10.1080/00103624.2013.863911

Dhlamini, B., Paumo, H. K., Katata-Seru, L., & Kutu, F. R. (2020). Sulphate-supplemented NPK nanofertilizer and its effect on maize growth. *Materials Research Express*, *7*(9), 095011. https://doi.org/10.1088/2053-1591/abb69d

Dimkpa, C. O., Andrews, J., Fugice, J., Singh, U., Bindraban, P. S., Elmer, W. H., Gardea-Torresdey, J. L., & White, J. C. (2020a). Facile coating of urea with low-dose ZnO nanoparticles promotes wheat performance and enhances Zn uptake under drought stress. *Frontiers in Plant Science, 11*. https://doi.org/10.3389/fpls.2020.00168

Dimkpa, C. O., Andrews, J., Sanabria, J., Bindraban, P. S., Singh, U., Elmer, W. H., Gardea-Torresdey, J. L., & White, J. C. (2020b). Interactive effects of drought, organic fertilizer, and zinc oxide nanoscale and bulk particles on wheat performance and grain nutrient accumulation. *Science of The Total Environment, 722*, 137808. https://doi.org/10.1016/j.scitotenv.2020.137808

Dimkpa, C. O., & Bindraban, P. S. (2016). Fortification of micronutrients for efficient agronomic production: A review. *Agronomy for Sustainable Development, 36*(1), 7. https://doi.org/10.1007/s13593-015-0346-6

Dimkpa, C. O., & Bindraban, P. S. (2018). Nanofertilizers: New products for the industry? *Journal of Agricultural and Food Chemistry, 66*(26), 6462–6473. https://doi.org/10.1021/acs.jafc.7b02150

Duhan, J. S., Kumar, R., Kumar, N., Kaur, P., Nehra, K., & Duhan, S. (2017). Nanotechnology: The new perspective in precision agriculture. *Biotechnology Reports, 15*, 11–23. https://doi.org/10.1016/j.btre.2017.03.002

Dulgerbaki, C., Maslakci, N. N., Komur, A. I., & Oksuz, A. U. (2015). Electrochromic device based on electrospun WO 3 nanofibers. *Materials Research Bulletin, 72*, 70–76. https://doi.org/10.1016/j.materresbull.2015.07.024

Ecobichon, D. J. (2001). Pesticide uses in developing countries. *Toxicology, 160*(1–3), 27–33. https://doi.org/10.1016/S0300-483X(00)00452-2

Eichert, T., Kurtz, A., Steiner, U., & Goldbach, H. E. (2008). Size exclusion limits and lateral heterogeneity of the stomatal foliar uptake pathway for aqueous solutes and water-suspended nanoparticles. *Physiologia Plantarum, 134*(1), 151–160. https://doi.org/10.1111/j.1399-3054.2008.01135.x

Evode, N., Qamar, S. A., Bilal, M., Barceló, D., & Iqbal, H. M. N. (2021). Plastic waste and its management strategies for environmental sustainability. *Case Studies in Chemical and Environmental Engineering, 4*, 100142. https://doi.org/10.1016/j.cscee.2021.100142

Fahad, Balouch, A., Agheem, M. H., Memon, S. A., Baloch, A. R., Tunio, A., Abdullah, Pato, A. H., Jagirani, M. S., Panah, P., Gabole, A. A., & Qasim, S. (2022). Efficient mitigation of cadmium and lead toxicity in coriander plant utilizing magnetite (Fe3O4) nanofertilizer as growth regulator and antimicrobial agent. *International Journal of Environmental Analytical Chemistry, 102*(16), 3868–3879. https://doi.org/10.1080/03067319.2020.1776861

Fajardo, C., Sánchez-Fortún, S., Costa, G., Nande, M., Botías, P., García-Cantalejo, J., Mengs, G., & Martín, M. (2020). Evaluation of nanoremediation strategy in a Pb, Zn and Cd contaminated soil. *Science of The Total Environment, 706*, 136041. https://doi.org/10.1016/j.scitotenv.2019.136041

Farrell, D., Majetich, S. A., & Wilcoxon, J. P. (2003). Preparation and characterization of monodisperse Fe nanoparticles. *The Journal of Physical Chemistry B, 107*(40), 11022–11030. https://doi.org/10.1021/jp0351831

Farshchi, H. K., Azizi, M., Teymouri, M., Nikpoor, A. R., & Jaafari, M. R. (2021). Synthesis and characterization of nanoliposome containing Fe2+ element: A superior nano-fertilizer for ferrous iron delivery to sweet basil. *Scientia Horticulturae, 283*, 110110. https://doi.org/10.1016/j.scienta.2021.110110

Fayaz, M., Rabani, M. S., Wani, S. A., & Thoker, S. A. (2021). Nano-agriculture: A novel approach in agriculture. In *Microbiota and biofertilizers* (pp. 99–122). Springer International Publishing. https://doi.org/10.1007/978-3-030-48771-3_7

Feregrino-Perez, A. A., Magaña-López, E., Guzmán, C., & Esquivel, K. (2018). A general overview of the benefits and possible negative effects of the nanotechnology in horticulture. *Scientia Horticulturae*, *238*, 126–137. https://doi.org/10.1016/j.scienta.2018.03.060

Fernández-García, M., & Rodriguez, J. A. (2011). Metal oxide nanoparticles. In *Encyclopedia of inorganic and bioinorganic chemistry*. John Wiley & Sons, Ltd. https://doi.org/10.1002/9781119951438.eibc0331

Ferrari, E., Barbero, F., Busquets-Fité, M., Franz-Wachtel, M., Köhler, H.-R., Puntes, V., & Kemmerling, B. (2021). Growth-promoting gold nanoparticles decrease stress responses in arabidopsis seedlings. *Nanomaterials*, *11*(12), 3161. https://doi.org/10.3390/nano11123161

Fraceto, L. F., Grillo, R., de Medeiros, G. A., Scognamiglio, V., Rea, G., & Bartolucci, C. (2016). Nanotechnology in agriculture: Which innovation potential does it have? *Frontiers in Environmental Science*, *4*. https://doi.org/10.3389/fenvs.2016.00020

Giannousi, K., Avramidis, I., & Dendrinou-Samara, C. (2013). Synthesis, characterization and evaluation of copper-based nanoparticles as agrochemicals against Phytophthora infestans. *RSC Advances*, *3*(44), 21743. https://doi.org/10.1039/c3ra42118j

Giroto, A. S., Fidélis, S. C., & Ribeiro, C. (2015). Controlled release from hydroxyapatite nanoparticles incorporated into biodegradable, soluble host matrixes. *RSC Advances*, *5*(126), 104179–104186. https://doi.org/10.1039/C5RA17669G

Giroto, A. S., Guimarães, G. G. F., Foschini, M., & Ribeiro, C. (2017). Role of slow-release nanocomposite fertilizers on nitrogen and phosphate availability in soil. *Scientific Reports*, *7*(1), 46032. https://doi.org/10.1038/srep46032

Godfray, H. C. J., Beddington, J. R., Crute, I. R., Haddad, L., Lawrence, D., Muir, J. F., Pretty, J., Robinson, S., Thomas, S. M., & Toulmin, C. (2010). Food security: The challenge of feeding 9 billion people. *Science*, *327*(5967), 812–818. https://doi.org/10.1126/science.1185383

Goyal, M. R., & Sivanappan, R. K. (Eds.). (2017). *Engineering practices for agricultural production and water conservation*. Apple Academic Press. https://doi.org/10.1201/9781315365954

Guerra, F., Attia, M., Whitehead, D., & Alexis, F. (2018). Nanotechnology for environmental remediation: Materials and applications. *Molecules*, *23*(7), 1760. https://doi.org/10.3390/molecules23071760

Guo, K., Li, H., & Yu, Z. (2018). Size-dependent catalytic activity of monodispersed nickel nanoparticles for the hydrolytic dehydrogenation of ammonia borane. *ACS Applied Materials & Interfaces*, *10*(1), 517–525. https://doi.org/10.1021/acsami.7b14166

Guo, H., Liu, Y., Chen, J., Zhu, Y., & Zhang, Z. (2022). The effects of several metal nanoparticles on seed germination and seedling growth: A meta-analysis. *Coatings*, *12*(2), 183. https://doi.org/10.3390/coatings12020183

Ha, N. M. C., Nguyen, T. H., Wang, S.-L., & Nguyen, A. D. (2019). Preparation of NPK nanofertilizer based on chitosan nanoparticles and its effect on biophysical characteristics and growth of coffee in green house. *Research on Chemical Intermediates*, *45*(1), 51–63. https://doi.org/10.1007/s11164-018-3630-7

Hamid, Y., Tang, L., Hussain, B., Usman, M., Gurajala, H. K., Rashid, M. S., He, Z., & Yang, X. (2020). Efficiency of lime, biochar, Fe containing biochar and composite amendments for Cd and Pb immobilization in a co-contaminated alluvial soil. *Environmental Pollution, 257*, 113609. https://doi.org/10.1016/j.envpol.2019.113609

Hamid, Y., Tang, L., Sohail, M. I., Cao, X., Hussain, B., Aziz, M. Z., Usman, M., He, Z., & Yang, X. (2019). An explanation of soil amendments to reduce cadmium phytoavailability and transfer to food chain. *Science of The Total Environment, 660*, 80–96. https://doi.org/10.1016/j.scitotenv.2018.12.419

Hong, J., Rico, C. M., Zhao, L., Adeleye, A. S., Keller, A. A., Peralta-Videa, J. R., & Gardea-Torresdey, J. L. (2015). Toxic effects of copper-based nanoparticles or compounds to lettuce (*Lactuca sativa*) and alfalfa (Medicago sativa). *Environmental Science: Processes & Impacts, 17*(1), 177–185. https://doi.org/10.1039/C4EM00551A

Hoshyar, N., Gray, S., Han, H., & Bao, G. (2016). The effect of nanoparticle size on in vivo pharmacokinetics and cellular interaction. *Nanomedicine, 11*(6), 673–692. https://doi.org/10.2217/nnm.16.5

Huang, D., Wu, J., Wang, L., Liu, X., Meng, J., Tang, X., Tang, C., & Xu, J. (2019). Novel insight into adsorption and co-adsorption of heavy metal ions and an organic pollutant by magnetic graphene nanomaterials in water. *Chemical Engineering Journal, 358*, 1399–1409. https://doi.org/10.1016/j.cej.2018.10.138

Imada, K., Sakai, S., Kajihara, H., Tanaka, S., & Ito, S. (2016). Magnesium oxide nanoparticles induce systemic resistance in tomato against bacterial wilt disease. *Plant Pathology, 65*(4), 551–560. https://doi.org/10.1111/ppa.12443

Iqbal, M., Raja, N. I., Mashwani, Z.-U.-R., Hussain, M., Ejaz, M., & Yasmeen, F. (2019). Effect of silver nanoparticles on growth of wheat under heat stress. *Iranian Journal of Science and Technology, Transactions A: Science, 43*(2), 387–395. https://doi.org/10.1007/s40995-017-0417-4

Iravani, S. (2014). Bacteria in nanoparticle synthesis: Current status and future prospects. *International Scholarly Research Notices, 2014*, 1–18. https://doi.org/10.1155/2014/359316

García-López, I.J., Lira-Saldivar, R. H., Zavala-García, F., Olivares-Sáenz, E., Niño-Medina, G., Angélica Ruiz-Torres, N., Méndez-Argüello, B., & Díaz-Barriga, E. (2018). Effects of zinc oxide nanoparticles on growth and antioxidant enzymes of Capsicum chinense. *Toxicological & Environmental Chemistry, 100*(5–7), 560–572. https://doi.org/10.1080/02772248.2018.1550781

Jahangirian, H., Rafiee-Moghaddam, R., Jahangirian, N., Nikpey, B., Jahangirian, S., Bassous, N., Saleh, B., Kalantari, K., & Webster, T. J. (2020). Green synthesis of zeolite/Fe2O3 nanocomposites: Toxicity & cell proliferation assays and application as a smart iron nanofertilizer. *International Journal of Nanomedicine, 15*, 1005–1020. https://doi.org/10.2147/IJN.S231679

Jampílek, J., & Kráľová, K. (2017). Nanopesticides: Preparation, targeting, and controlled release. In *New pesticides and soil sensors* (pp. 81–127). Elsevier. https://doi.org/10.1016/B978-0-12-804299-1.00004-7

Jeevanandam, J., Barhoum, A., Chan, Y. S., Dufresne, A., & Danquah, M. K. (2018). Review on nanoparticles and nanostructured materials: History, sources, toxicity and regulations. *Beilstein Journal of Nanotechnology, 9*, 1050–1074. https://doi.org/10.3762/bjnano.9.98

Jogaiah, S., Satapute, P., De Britto, S., Konappa, N., & Udayashankar, A. C. (2020). Exogenous priming of chitosan induces upregulation of phytohormones and resistance against cucumber powdery mildew disease is correlated with localized biosynthesis of defense enzymes. *International Journal of Biological Macromolecules, 162*, 1825–1838. https://doi.org/10.1016/j.ijbiomac.2020.08.124

Kadri, O., Karmous, I., Kharbech, O., Arfaoui, H., & Chaoui, A. (2022). Cu and CuO nanoparticles affected the germination and the growth of barley (Hordeum vulgare L.) seedling. *Bulletin of Environmental Contamination and Toxicology*, *108*(3), 585–593. https://doi.org/10.1007/s00128-021-03425-y

Kale, P. A., & Gawade, S. N. (2016). Studies on nanoparticle induced nutrient use efficiency of fertilizer and crop productivity. *Green Chemistry & Technology Letters*, *2*(2), 88–92. https://doi.org/10.18510/gctl.2016.226

Karny, A., Zinger, A., Kajal, A., Shainsky-Roitman, J., & Schroeder, A. (2018). Therapeutic nanoparticles penetrate leaves and deliver nutrients to agricultural crops. *Scientific Reports*, *8*(1), 7589. https://doi.org/10.1038/s41598-018-25197-y

Khan, M. Z. H., Islam, M. R., Nahar, N., Al-Mamun, M. R., Khan, M. A. S., & Matin, M. A. (2021). Synthesis and characterization of nanozeolite based composite fertilizer for sustainable release and use efficiency of nutrients. *Heliyon*, *7*(1), e06091. https://doi.org/10.1016/j.heliyon.2021.e06091

Khan, N. A., Kang, I. J., Seok, H. Y., & Jhung, S. H. (2011). Facile synthesis of nano-sized metal-organic frameworks, chromium-benzenedicarboxylate, MIL-101. *Chemical Engineering Journal*, *166*(3), 1152–1157. https://doi.org/10.1016/j.cej.2010.11.098

Khan, I., Saeed, K., & Khan, I. (2019). Nanoparticles: Properties, applications and toxicities. *Arabian Journal of Chemistry*, *12*(7), 908–931. https://doi.org/10.1016/j.arabjc.2017.05.011

Khanna, P., Kaur, A., & Goyal, D. (2019). Algae-based metallic nanoparticles: Synthesis, characterization and applications. *Journal of Microbiological Methods*, *163*, 105656. https://doi.org/10.1016/j.mimet.2019.105656

Khanna, K., Kohli, S. K., Handa, N., Kaur, H., Ohri, P., Bhardwaj, R., Yousaf, B., Rinklebe, J., & Ahmad, P. (2021). Enthralling the impact of engineered nanoparticles on soil microbiome: A concentric approach towards environmental risks and cogitation. *Ecotoxicology and Environmental Safety*, *222*, 112459. https://doi.org/10.1016/j.ecoenv.2021.112459

Khattab, S., El Sherif, F., AlDayel, M., Yap, Y.-K., Meligy, A., & Ibrahim, H. I. M. (2022). Silicon dioxide and silver nanoparticles elicit antimicrobial secondary metabolites while enhancing growth and multiplication of Lavandula officinalis in-vitro plantlets. *Plant Cell, Tissue and Organ Culture (PCTOC)*, *149*(1–2), 411–421. https://doi.org/10.1007/s11240-021-02224-x

Kottegoda, N., Sandaruwan, C., Priyadarshana, G., Siriwardhana, A., Rathnayake, U. A., Berugoda Arachchige, D. M., Kumarasinghe, A. R., Dahanayake, D., Karunaratne, V., & Amaratunga, G. A. J. (2017). Urea-hydroxyapatite nanohybrids for slow release of nitrogen. *ACS Nano*, *11*(2), 1214–1221. https://doi.org/10.1021/acsnano.6b07781

Kovalenko, M. V., Manna, L., Cabot, A., Hens, Z., Talapin, D. V., Kagan, C. R., Klimov, V. I., Rogach, A. L., Reiss, P., Milliron, D. J., Guyot-Sionnnest, P., Konstantatos, G., Parak, W. J., Hyeon, T., Korgel, B. A., Murray, C. B., & Heiss, W. (2015). Prospects of nanoscience with nanocrystals. *ACS Nano*, *9*(2), 1012–1057. https://doi.org/10.1021/nn506223h

Kubavat, D., Trivedi, K., Vaghela, P., Prasad, K., Vijay Anand, G. K., Trivedi, H., Patidar, R., Chaudhari, J., Andhariya, B., & Ghosh, A. (2020). Characterization of a chitosan-based sustained release nanofertilizer formulation used as a soil conditioner while simultaneously improving biomass production of Zea mays L. *Land Degradation & Development*, *31*(17), 2734–2746. https://doi.org/10.1002/ldr.3629

Kulikova, N. A. (2021). Silver nanoparticles in soil: Input, transformation, and toxicity. *Eurasian Soil Science*, *54*(3), 352–365. https://doi.org/10.1134/S1064229321030091

Kumar, Y., Tiwari K. N., Nayak, R. K., Rai, A., Singh, S. P., Singh, A. N., Kumar, Y., Tomar, H., Singh, T., & Raliya, R. (2020). Nanofertilizers for increasing nutrient use efficiency, yield and economic returns in important winter season crops of Uttar Pradesh. *Indian Journal of Fertilisers*, 772–786.

Larue, C., Veronesi, G., Flank, A.-M., Surble, S., Herlin-Boime, N., & Carrière, M. (2012). Comparative uptake and impact of TiO2 nanoparticles in wheat and rapeseed. *Journal of Toxicology and Environmental Health, Part A*, *75*(13–15), 722–734. https://doi.org/10.1080/15287394.2012.689800

Li, Z., Wray, P. R., Su, M. P., Tu, Q., Andaraarachchi, H. P., Jeong, Y. J., Atwater, H. A., & Kortshagen, U. R. (2020). Aluminum oxide nanoparticle films deposited from a nonthermal plasma: Synthesis, characterization, and crystallization. *ACS Omega*, *5*(38), 24754–24761. https://doi.org/10.1021/acsomega.0c03353

Liliane, T. N., & Charles, M. S. (2020). Factors affecting yield of crops. In *Agronomy - Climate change and food security*. IntechOpen. https://doi.org/10.5772/intechopen.90672

Liu, L., & Kong, L. (2021). Research progress on the carcinogenicity of metal nanomaterials. *Journal of Applied Toxicology*, *41*(9), 1334–1344. https://doi.org/10.1002/jat.4145

Liu, R., & Lal, R. (2015). Potentials of engineered nanoparticles as fertilizers for increasing agronomic productions. *Science of The Total Environment*, *514*, 131–139. https://doi.org/10.1016/j.scitotenv.2015.01.104

Liu, L., Shen, T., Yang, Y., Gao, B., Li, Y. C., Xie, J., Tang, Y., Zhang, S., Wang, Z., & Chen, J. (2018). Bio-based large tablet controlled-release urea: Synthesis, characterization, and controlled-released mechanisms. *Journal of Agricultural and Food Chemistry*, *66*(43), 11265–11272. https://doi.org/10.1021/acs.jafc.8b04042

López-Moreno, M. L., de la Rosa, G., Hernández-Viezcas, J. Á., Castillo-Michel, H., Botez, C. E., Peralta-Videa, J. R., & Gardea-Torresdey, J. L. (2010). Evidence of the differential biotransformation and genotoxicity of ZnO and CeO2 nanoparticles on soybean (Glycine max) plants. *Environmental Science & Technology*, *44*(19), 7315–7320. https://doi.org/10.1021/es903891g

Lou, X. W., & Zeng, H. C. (2002). Hydrothermal synthesis of α-MoO $_3$ nanorods via acidification of ammonium heptamolybdate tetrahydrate. *Chemistry of Materials*, *14*(11), 4781–4789. https://doi.org/10.1021/cm0206237

Lu, W., Senapati, D., Wang, S., Tovmachenko, O., Singh, A. K., Yu, H., & Ray, P. C. (2010). Effect of surface coating on the toxicity of silver nanomaterials on human skin keratinocytes. *Chemical Physics Letters*, *487*(1–3), 92–96. https://doi.org/10.1016/j.cplett.2010.01.027

Luna, I. Z., Hilary, L. N., Chowdhury, A. M. S., Gafur, M. A., Khan, N., & Khan, R. A. (2015). Preparation and characterization of copper oxide nanoparticles synthesized via chemical precipitation method. *OALib*, *02*(03), 1–8. https://doi.org/10.4236/oalib.1101409

Ma, C., Liu, F.-Y., Wei, M.-B., Zhao, J.-H., & Zhang, H.-Z. (2020). Synthesis of novel core-shell magnetic Fe₃O₄@C nanoparticles with carboxyl function for use as an immobilisation agent to remediate lead-contaminated soils. *Polish Journal of Environmental Studies*, *29*(3), 2273–2283. https://doi.org/10.15244/pjoes/111232

Madusanka, N., Sandaruwan, C., Kottegoda, N., Sirisena, D., Munaweera, I., De Alwis, A., Karunaratne, V., & Amaratunga, G. A. J. (2017). Urea–hydroxyapatite-montmorillonite nanohybrid composites as slow-release nitrogen compositions. *Applied Clay Science*, *150*, 303–308. https://doi.org/10.1016/j.clay.2017.09.039

Mahil, E. I. T., & Baburai, N. A. K. (2019). Foliar application of nanofertilizers in agricultural crops -A review. *Journal of Farm Sciences*, *32*(3), 239–249.
Mahmoud, M., & Swaefy, H. (2020). Comparison between effect of commercial and nano Npk in presence of nano zeolite on sage plant yield and components under drought stress. *Zagazig Journal of Agricultural Research*, *47*(2), 435–457. https://doi.org/10.21608/zjar.2020.94486
Maity, D., & Agrawal, D. C. (2007). Synthesis of iron oxide nanoparticles under oxidizing environment and their stabilization in aqueous and non-aqueous media. *Journal of Magnetism and Magnetic Materials*, *308*(1), 46–55. https://doi.org/10.1016/j.jmmm.2006.05.001
Maity, D., Ding, J., & Xue, J.M. (2008). Synthesis of magnetite nanoparticles by thermal decomposition: Time, temperature, surfactants, and solvent effects. *Functional Materials Letters*, *01*(03), 189–193. https://doi.org/10.1142/S1793604708000381
Malandrakis, A. A., Kavroulakis, N., & Chrysikopoulos, C. V. (2019). Use of copper, silver and zinc nanoparticles against foliar and soil-borne plant pathogens. *Science of The Total Environment*, *670*, 292–299. https://doi.org/10.1016/j.scitotenv.2019.03.210
Malini, S., Raj, K., Madhumathy, S., El-Hady, K. M., Islam, S., & Dutta, M. (2022). Bioinspired advances in nanomaterials for sustainable agriculture. *Journal of Nanomaterials*, *2022*, 1–11. https://doi.org/10.1155/2022/8926133
Manawi, Y., Ihsanullah, Samara, A., Al-Ansari, T., & Atieh, M. (2018). A review of carbon nanomaterials' synthesis via the chemical vapor deposition (CVD) method. *Materials*, *11*(5), 822. https://doi.org/10.3390/ma11050822
Mazarío, E., Mayoral, A., Salas, E., Menéndez, N., Herrasti, P., & Sánchez-Marcos, J. (2016). Synthesis and characterization of manganese ferrite nanoparticles obtained by electrochemical/chemical method. *Materials & Design*, *111*, 646–650. https://doi.org/10.1016/j.matdes.2016.09.031
Meena, R., Kumar, S., Datta, R., Lal, R., Vijayakumar, V., Brtnicky, M., Sharma, M., Yadav, G., Jhariya, M., Jangir, C., Pathan, S., Dokulilova, T., Pecina, V., & Marfo, T. (2020). Impact of agrochemicals on soil microbiota and management: A review. *Land*, *9*(2), 34. https://doi.org/10.3390/land9020034
Meier, S., Moore, F., Morales, A., González, M.-E., Seguel, A., Meriño-Gergichevich, C., Rubilar, O., Cumming, J., Aponte, H., Alarcón, D., & Mejías, J. (2020). Synthesis of calcium borate nanoparticles and its use as a potential foliar fertilizer in lettuce (Lactuca sativa) and zucchini (Cucurbita pepo). *Plant Physiology and Biochemistry*, *151*, 673–680. https://doi.org/10.1016/j.plaphy.2020.04.025
Mirzajani, F., Askari, H., Hamzelou, S., Farzaneh, M., & Ghassempour, A. (2013). Effect of silver nanoparticles on Oryza sativa L. and its rhizosphere bacteria. *Ecotoxicology and Environmental Safety*, *88*, 48–54. https://doi.org/10.1016/j.ecoenv.2012.10.018
Mittal, D., Kaur, G., Singh, P., Yadav, K., & Ali, S. A. (2020). Nanoparticle-based sustainable agriculture and food science: Recent advances and future outlook. *Frontiers in Nanotechnology*, *2*. https://doi.org/10.3389/fnano.2020.579954
Moaied, M., & Hong, J. (2019). Size-dependent critical temperature and anomalous optical deispersion in ferromagnetic CrI3 nanotubes. *Nanomaterials*, *9*(2), 153. https://doi.org/10.3390/nano9020153
Mody, V., Siwale, R., Singh, A., & Mody, H. (2010). Introduction to metallic nanoparticles. *Journal of Pharmacy and Bioallied Sciences*, *2*(4), 282. https://doi.org/10.4103/0975-7406.72127

Mukhopadhyay, S. S. (2014). Nanotechnology in agriculture: Prospects and constraints. *Nanotechnology, Science and Applications*, *63*. https://doi.org/10.2147/NSA.S39409

Muradov, M. B., Balayeva, O. O., Azizov, A. A., Maharramov, A. M., Qahramanli, L. R., Eyvazova, G. M., & Aghamaliyev, Z. A. (2018). Synthesis and characterization of cobalt sulfide nanoparticles by sonochemical method. *Infrared Physics & Technology*, *89*, 255–262. https://doi.org/10.1016/j.infrared.2018.01.014

Muthukumar, K., Lakshmi, D. S., Acharya, S. D., Natarajan, S., Mukherjee, A., & Bajaj, H. C. (2018). Solvothermal synthesis of magnetic copper ferrite nano sheet and its antimicrobial studies. *Materials Chemistry and Physics*, *209*, 172–179. https://doi.org/10.1016/j.matchemphys.2018.02.004

Nair, R., Varghese, S. H., Nair, B. G., Maekawa, T., Yoshida, Y., & Kumar, D. S. (2010). Nanoparticulate material delivery to plants. *Plant Science*, *179*(3), 154–163. https://doi.org/10.1016/j.plantsci.2010.04.012

Narayanan, K. B., & Park, H. H. (2014). Antifungal activity of silver nanoparticles synthesized using turnip leaf extract (Brassica rapa L.) against wood rotting pathogens. *European Journal of Plant Pathology*, *140*(2), 185–192. https://doi.org/10.1007/s10658-014-0399-4

Naseem, F., Zhi, Y., Farrukh, M. A., Hussain, F., & Yin, Z. (2020). Mesoporous ZnAl2Si10O24 nanofertilizers enable high yield of Oryza sativa L. *Scientific Reports*, *10*(1), 10841. https://doi.org/10.1038/s41598-020-67611-4

Nguyen, T. H., Nguyen, T. T. L., Pham, T. D., & Le, T. S. (2020). Removal of lindane from aqueous solution using aluminum hydroxide nanoparticles with surface modification by anionic surfactant. *Polymers*, *12*(4), 960. https://doi.org/10.3390/polym12040960

Nisha Raj, S., Anooj, E. S., Rajendran, K., & Vallinayagam, S. (2021). A comprehensive review on regulatory invention of nano pesticides in Agricultural nano formulation and food system. *Journal of Molecular Structure*, *1239*, 130517. https://doi.org/10.1016/j.molstruc.2021.130517

Ocsoy, I., Paret, M. L., Ocsoy, M. A., Kunwar, S., Chen, T., You, M., & Tan, W. (2013). Nanotechnology in plant disease management: DNA-directed silver nanoparticles on graphene oxide as an antibacterial against Xanthomonas perforans. *ACS Nano*, *7*(10), 8972–8980. https://doi.org/10.1021/nn4034794

Parayil, G. (1992). The green revolution in India: A case study of technological change. *Technology and Culture*, *33*(4), 737. https://doi.org/10.2307/3106588

Paret, M. L., Vallad, G. E., Averett, D. R., Jones, J. B., & Olson, S. M. (2013). Photocatalysis: Effect of light-activated nanoscale formulations of TiO2 on xanthomonas perforans and control of bacterial spot of tomato. *Phytopathology*, *103*(3), 228–236. https://doi.org/10.1094/PHYTO-08-12-0183-R

Park, S., Choi, K. S., Kim, S., Gwon, Y., & Kim, J. (2020). Graphene oxide-assisted promotion of plant growth and stability. *Nanomaterials*, *10*(4), 758. https://doi.org/10.3390/nano10040758

Perea Vélez, Y. S., Carrillo-González, R., & González-Chávez, Ma. del C. A. (2021). Interaction of metal nanoparticles–plants–microorganisms in agriculture and soil remediation. *Journal of Nanoparticle Research*, *23*(9), 206. https://doi.org/10.1007/s11051-021-05269-3

Purohit, J., Chattopadhyay, A., & Singh, N. K. (2019). *Green synthesis of microbial nanoparticle: Approaches to application* (pp. 35–60). https://doi.org/10.1007/978-3-030-16534-5_3

Qadri, H., Bhat, R. A., Mehmood, M. A., & Dar, G. H. (Eds.). (2020). *Fresh water pollution dynamics and remediation*. Springer Singapore. https://doi.org/10.1007/978-981-13-8277-2

Qu, X., Alvarez, P. J. J., & Li, Q. (2013). Applications of nanotechnology in water and wastewater treatment. *Water Research*, *47*(12), 3931–3946. https://doi.org/10.1016/j.watres.2012.09.058

Qureshi, A., Singh, D. K., & Dwivedi, S. (2018). Nano-fertilizers: A novel way for enhancing nutrient use efficiency and crop productivity. *International Journal of Current Microbiology and Applied Sciences*, *7*(2), 3325–3335. https://doi.org/10.20546/ijcmas.2018.702.398

Rafique, M., Rafique, M. S., Kalsoom, U., Afzal, A., Butt, S. H., & Usman, A. (2019). Laser ablation synthesis of silver nanoparticles in water and dependence on laser nature. *Optical and Quantum Electronics*, *51*(6), 179. https://doi.org/10.1007/s11082-019-1902-0

Raliya, R., Saharan, V., Dimkpa, C., & Biswas, P. (2018). Nanofertilizer for precision and sustainable agriculture: Current state and future perspectives. *Journal of Agricultural and Food Chemistry*, *66*(26), 6487–6503. https://doi.org/10.1021/acs.jafc.7b02178

Ramírez-Rodríguez, G. B., Dal Sasso, G., Carmona, F. J., Miguel-Rojas, C., Pérez-de-Luque, A., Masciocchi, N., Guagliardi, A., & Delgado-López, J. M. (2020). Engineering biomimetic calcium phosphate nanoparticles: A green synthesis of slow-release multinutrient (NPK) nanofertilizers. *ACS Applied Bio Materials*, *3*(3), 1344–1353. https://doi.org/10.1021/acsabm.9b00937

Ramírez-Rodríguez, G. B., Miguel-Rojas, C., Montanha, G. S., Carmona, F. J., Dal Sasso, G., Sillero, J. C., Skov Pedersen, J., Masciocchi, N., Guagliardi, A., Pérez-de-Luque, A., & Delgado-López, J. M. (2020). Reducing nitrogen dosage in Triticum durum plants with urea-doped nanofertilizers. *Nanomaterials*, *10*(6), 1043. https://doi.org/10.3390/nano10061043

Raza, M., Kanwal, Z., Rauf, A., Sabri, A., Riaz, S., & Naseem, S. (2016). Size- and shape-dependent antibacterial studies of silver nanoparticles synthesized by wet chemical routes. *Nanomaterials*, *6*(4), 74. https://doi.org/10.3390/nano6040074

Rizwan, M., Ali, S., Ali, B., Adrees, M., Arshad, M., Hussain, A., Zia ur Rehman, M., & Waris, A. A. (2019). Zinc and iron oxide nanoparticles improved the plant growth and reduced the oxidative stress and cadmium concentration in wheat. *Chemosphere*, *214*, 269–277. https://doi.org/10.1016/j.chemosphere.2018.09.120

Robles-García, M. A., Rodríguez-Félix, F., Márquez-Ríos, E., Aguilar, J. A., Barrera-Rodríguez, A., Aguilar, J., Ruiz-Cruz, S., & Del-Toro-Sánchez, C. L. (2016). Applications of nanotechnology in the agriculture, food, and pharmaceuticals. *Journal of Nanoscience and Nanotechnology*, *16*(8), 8188–8207. https://doi.org/10.1166/jnn.2016.12925

Rodriguez, J. A., Jirsak, T., Pérez, M., Chaturvedi, S., Kuhn, M., González, L., & Maiti, A. (2000). Studies on the behavior of mixed-metal oxides and desulfurization: Reaction of H2S and SO2 with Cr2O3 (0001), MgO(100), and Cr x Mg 1- x O(100). *Journal of the American Chemical Society*, *122*(49), 12362–12370. https://doi.org/10.1021/ja003149j

Rodríguez-González, V., Terashima, C., & Fujishima, A. (2019). Applications of photocatalytic titanium dioxide-based nanomaterials in sustainable agriculture. *Journal of Photochemistry and Photobiology C: Photochemistry Reviews*, *40*, 49–67. https://doi.org/10.1016/j.jphotochemrev.2019.06.001

Rossi, L., Fedenia, L. N., Sharifan, H., Ma, X., & Lombardini, L. (2019). Effects of foliar application of zinc sulfate and zinc nanoparticles in coffee (Coffea arabica L.) plants. *Plant Physiology and Biochemistry*, *135*, 160–166. https://doi.org/10.1016/j.plaphy.2018.12.005

Rycenga, M., Camargo, P. H. C., & Xia, Y. (2009). Template-assisted self-assembly: A versatile approach to complex micro- and nanostructures. *Soft Matter*, *5*(6), 1129–1136. https://doi.org/10.1039/B811021B

Sabir, S., Arshad, M., & Chaudhari, S. K. (2014). Zinc oxide nanoparticles for revolutionizing agriculture: Synthesis and applications. *The Scientific World Journal, 2014*, 1–8. https://doi.org/10.1155/2014/925494

Saharan, V., Mehrotra, A., Khatik, R., Rawal, P., Sharma, S. S., & Pal, A. (2013). Synthesis of chitosan-based nanoparticles and their in vitro evaluation against phytopathogenic fungi. *International Journal of Biological Macromolecules, 62*, 677–683. https://doi.org/10.1016/j.ijbiomac.2013.10.012

Saharan, V., & Pal, A. (2016). *Chitosan based nanomaterials in plant growth and protection.* Springer India. https://doi.org/10.1007/978-81-322-3601-6

Saiphaneendra, B., & Srivastava, C. (2017). Synthesis of graphene–magnetite Nanoparticle composite using mechanical milling and electrochemical exfoliation. *JOM, 69*(7), 1143–1148. https://doi.org/10.1007/s11837-016-2189-2

Salama, D. M., Abd El-Aziz, M. E., Rizk, F. A., & Abd Elwahed, M. S. A. (2021). Applications of nanotechnology on vegetable crops. *Chemosphere, 266*, 129026. https://doi.org/10.1016/j.chemosphere.2020.129026

Saleh, A. M., Hassan, Y. M., Selim, S., & AbdElgawad, H. (2019). NiO-nanoparticles induce reduced phytotoxic hazards in wheat (Triticum aestivum L.) grown under future climate CO2. *Chemosphere, 220*, 1047–1057. https://doi.org/10.1016/j.chemosphere.2019.01.023

Samavini, R., Sandaruwan, C., De Silva, M., Priyadarshana, G., Kottegoda, N., & Karunaratne, V. (2018). Effect of citric acid surface modification on solubility of hydroxyapatite nanoparticles. *Journal of Agricultural and Food Chemistry, 66*(13), 3330–3337. https://doi.org/10.1021/acs.jafc.7b05544

Sangeetha, J., Hospet, R., Thangadurai, D., Adetunji, C. O., Islam, S., Pujari, N., & Al-Tawaha, A. R. M. S. (2021). Nanopesticides, nanoherbicides, and nanofertilizers: The greener aspects of agrochemical synthesis using nanotools and nanoprocesses toward sustainable agriculture. In *Handbook of nanomaterials and nanocomposites for energy and environmental applications* (pp. 1663–1677). Springer International Publishing. https://doi.org/10.1007/978-3-030-36268-3_44

Sangeetha, J., Thangadurai, D., Hospet, R., Harish, E. R., Purushotham, P., Mujeeb, M. A., Shrinivas, J., David, M., Mundaragi, A. C., Thimmappa, S. C., Arakera, S. B., & Prasad, R. (2017). Nanoagrotechnology for soil quality, crop performance and environmental management. In *Nanotechnology* (pp. 73–97). Springer Singapore. https://doi.org/10.1007/978-981-10-4573-8_5

Santoro, M., Gorelli, F. A., Bini, R., Haines, J., & van der Lee, A. (2013). High-pressure synthesis of a polyethylene/zeolite nano-composite material. *Nature Communications, 4*(1), 1557. https://doi.org/10.1038/ncomms2564

Sarkar, R., Rashid, M., Rahman, H., Kafi, A., Hosen, M. A., Rahman, M. I., & Khan, M. N. (2022). Recent advances in nanomaterials based sustainable agriculture: An overview. *Environmental Nanotechnology, Monitoring & Management, 18*, 100687. https://doi.org/10.1016/j.enmm.2022.100687

Sasson, Y., Levy-Ruso, G., Toledano, O., & Ishaaya, I. (2007). Nanosuspensions: Emerging novel agrochemical formulations. In *Insecticides design using advanced technologies* (pp. 1–39). Springer Berlin Heidelberg. https://doi.org/10.1007/978-3-540-46907-0_1

Scott, N., & Chen, H. (2013). Nanoscale science and engineering for agriculture and food systems. *Industrial Biotechnology, 9*(1), 17–18. https://doi.org/10.1089/ind.2013.1555

Shao, C., Zhao, H., & Wang, P. (2022). Recent development in functional nanomaterials for sustainable and smart agricultural chemical technologies. *Nano Convergence*, *9*(1), 11. https://doi.org/10.1186/s40580-022-00302-0

Sharma, G., Kumar, A., Devi, K. A., Prajapati, D., Bhagat, D., Pal, A., Raliya, R., Biswas, P., & Saharan, V. (2020). Chitosan nanofertilizer to foster source activity in maize. *International Journal of Biological Macromolecules*, *145*, 226–234. https://doi.org/10.1016/j.ijbiomac.2019.12.155

Shebl, A., Hassan, A. A., Salama, D. M., Abd El-Aziz, M. E., & Abd Elwahed, M. S. A. (2019). Green synthesis of nanofertilizers and their application as a foliar for Cucurbita pepo L. *Journal of Nanomaterials*, *2019*, 1–11. https://doi.org/10.1155/2019/3476347

Shebl, A., Hassan, A. A., Salama, D. M., Abd El-Aziz, M. E., & Abd Elwahed, M. S. A. (2020). Template-free microwave-assisted hydrothermal synthesis of manganese zinc ferrite as a nanofertilizer for squash plant (Cucurbita pepo L). *Heliyon*, *6*(3), e03596. https://doi.org/10.1016/j.heliyon.2020.e03596

Shubair, T., Eljamal, O., Khalil, A. M. E., & Matsunaga, N. (2018). Multilayer system of nanoscale zero valent iron and Nano-Fe/Cu particles for nitrate removal in porous media. *Separation and Purification Technology*, *193*, 242–254. https://doi.org/10.1016/j.seppur.2017.10.069

Siddiqi, K. S., & Husen, A. (2016). Green synthesis, characterization and uses of palladium/platinum nanoparticles. *Nanoscale Research Letters*, *11*(1), 482. https://doi.org/10.1186/s11671-016-1695-z

Sindhu, R. K., Chitkara, M., & Singh Sandhu, I. (2021). In R. K. Sindhu, M. Chitkara & I. S. Sandhu (Eds.), *Nanotechnology*. Jenny Stanford Publishing. https://doi.org/10.1201/9781003120261

Singh, R. B. (2000). Environmental consequences of agricultural development: a case study from the Green Revolution state of Haryana, India. *Agriculture, Ecosystems & Environment*, *82*(1–3), 97–103. https://doi.org/10.1016/S0167-8809(00)00219-X

Singh, R. P., Handa, R., & Manchanda, G. (2021). Nanoparticles in sustainable agriculture: An emerging opportunity. *Journal of Controlled Release*, *329*, 1234–1248. https://doi.org/10.1016/j.jconrel.2020.10.051

Singh, P. J., Kumar, M., Sharma, A., Pandey, G., Chae, K. H., & Lee, S. (2020). Bottom-up and top-down approaches for MgO. In *Sonochemical reactions*. IntechOpen. https://doi.org/10.5772/intechopen.91182

Singh, H., Sharma, A., Bhardwaj, S. K., Arya, S. K., Bhardwaj, N., & Khatri, M. (2021). Recent advances in the applications of nano-agrochemicals for sustainable agricultural development. *Environmental Science: Processes & Impacts*, *23*(2), 213–239. https://doi.org/10.1039/D0EM00404A

Singh, P. M., Tiwari, A., Maity, D., & Saha, S. (2022). Recent progress of nanomaterials in sustainable agricultural applications. *Journal of Materials Science*, *57*(24), 10836–10862. https://doi.org/10.1007/s10853-022-07259-9

Solanki, P., Bhargava, A., Chhipa, H., Jain, N., & Panwar, J. (2015). Nano-fertilizers and their smart delivery system. In *Nanotechnologies in food and agriculture* (pp. 81–101). Springer International Publishing. https://doi.org/10.1007/978-3-319-14024-7_4

Sun, D., Hussain, H. I., Yi, Z., Siegele, R., Cresswell, T., Kong, L., & Cahill, D. M. (2014). Uptake and cellular distribution, in four plant species, of fluorescently labeled mesoporous silica nanoparticles. *Plant Cell Reports*, *33*(8), 1389–1402. https://doi.org/10.1007/s00299-014-1624-5

Takla, S. S., Shawky, E., Hammoda, H. M., & Darwish, F. A. (2018). Green techniques in comparison to conventional ones in the extraction of Amaryllidaceae alkaloids: Best solvents selection and parameters optimization. *Journal of Chromatography A*, *1567*, 99–110. https://doi.org/10.1016/j.chroma.2018.07.009

Talamini, L., Violatto, M. B., Cai, Q., Monopoli, M. P., Kantner, K., Krpetić, Ž., Perez-Potti, A., Cookman, J., Garry, D., Silveira, C. P., Boselli, L., Pelaz, B., Serchi, T., Cambier, S., Gutleb, A. C., Feliu, N., Yan, Y., Salmona, M., Parak, W. J., … Bigini, P. (2017). Influence of size and shape on the anatomical distribution of endotoxin-free gold nanoparticles. *ACS Nano*, *11*(6), 5519–5529. https://doi.org/10.1021/acsnano.7b00497

Tamizhdurai, P., Sakthinathan, S., Chen, S.-M., Shanthi, K., Sivasanker, S., & Sangeetha, P. (2017). Environmentally friendly synthesis of CeO2 nanoparticles for the catalytic oxidation of benzyl alcohol to benzaldehyde and selective detection of nitrite. *Scientific Reports*, *7*(1), 46372. https://doi.org/10.1038/srep46372

Tang, N., Jiang, Y., Qu, H., & Duan, X. (2017). Conductive polymer nanowire gas sensor fabricated by nanoscale soft lithography. *Nanotechnology*, *28*(48), 485301. https://doi.org/10.1088/1361-6528/aa905b

Thounaojam, T. C., Meetei, T. T., Devi, Y. B., Panda, S. K., & Upadhyaya, H. (2021). Zinc oxide nanoparticles (ZnO-NPs): A promising nanoparticle in renovating plant science. *Acta Physiologiae Plantarum*, *43*(10), 136. https://doi.org/10.1007/s11738-021-03307-0

Tian, H., Liang, Y., Yang, D., & Sun, Y. (2020). Characteristics of PVP–stabilised NZVI and application to dechlorination of soil–sorbed TCE with ionic surfactant. *Chemosphere*, *239*, 124807. https://doi.org/10.1016/j.chemosphere.2019.124807

Toshiki, S., Hiromichi, Y., Takashi, S., Toshiya, O., Hisanori, S., and Shunji, B. (2003). New synthesis of high-quality double-walled carbon nanotubes by high-temperature pulsed arc discharge. *Nano Letters*, *3*(6), 769–773.

Tricoli, A., Nasiri, N., Chen, H., Wallerand, A. S., & Righettoni, M. (2016). Ultra-rapid synthesis of highly porous and robust hierarchical ZnO films for dye sensitized solar cells. *Solar Energy*, *136*, 553–559. https://doi.org/10.1016/j.solener.2016.07.024

Tripathi, D. K., Singh, S., Singh, V. P., Prasad, S. M., Dubey, N. K., & Chauhan, D. K. (2017). Silicon nanoparticles more effectively alleviated UV-B stress than silicon in wheat (Triticum aestivum) seedlings. *Plant Physiology and Biochemistry*, *110*, 70–81. https://doi.org/10.1016/j.plaphy.2016.06.026

Valden, M., Lai, X., & Goodman, D. W. (1998). Onset of catalytic activity of gold clusters on Titania with the appearance of nonmetallic properties. *Science*, *281*(5383), 1647–1650. https://doi.org/10.1126/science.281.5383.1647

Van Nguyen, D., Nguyen, H. M., Le, N. T., Nguyen, K. H., Nguyen, H. T., Le, H. M., Nguyen, A. T., Dinh, N. T. T., Hoang, S. A., & Van Ha, C. (2022). Copper nanoparticle application enhances plant growth and grain yield in maize under drought stress conditions. *Journal of Plant Growth Regulation*, *41*(1), 364–375. https://doi.org/10.1007/s00344-021-10301-w

Vande Voort, A., & Arai, Y. (2018). Macroscopic observation of soil nitrification kinetics impacted by copper nanoparticles: Implications for micronutrient nanofertilizer. *Nanomaterials*, *8*(11), 927. https://doi.org/10.3390/nano8110927

Varanda, L. C., Souza, C. G. S., Moraes, D. A., Neves, H. R., Souza Jr., J. B., Silva, M. F., Bini, R. A., Albers, R. F., Silva, T. L., Beck Jr., W. (2019). Size and shape-controlled nanomaterials based on modified polyol and thermal decomposition approaches. A brief review. *Anais Da Academia Brasileira de Ciências*, *91*(4). https://doi.org/10.1590/0001-3765201920181180

Vega-Vásquez, P., Mosier, N. S., & Irudayaraj, J. (2020). Nanoscale drug delivery systems: From medicine to agriculture. *Frontiers in Bioengineering and Biotechnology*, *8*. https://doi.org/10.3389/fbioe.2020.00079

Verma, S. K., Das, A. K., Patel, M. K., Shah, A., Kumar, V., & Gantait, S. (2018b). Engineered nanomaterials for plant growth and development: A perspective analysis. *Science of the Total Environment*, *630*, 1413–1435. https://doi.org/10.1016/j.scitotenv.2018.02.313

Verma, M., Kumar, V., & Katoch, A. (2018a). Synthesis of ZrO2 nanoparticles using reactive magnetron sputtering and their structural, morphological and thermal studies. *Materials Chemistry and Physics*, *212*, 268–273. https://doi.org/10.1016/j.matchemphys.2018.03.049

Verma, K. K., Song, X.-P., Joshi, A., Tian, D.-D., Rajput, V. D., Singh, M., Arora, J., Minkina, T., & Li, Y. R. (2022). Recent trends in nano-fertilizers for sustainable agriculture under climate change for global food security. *Nanomaterials*, *12*(1), 173. https://doi.org/10.3390/nano12010173

Vijayakumar, M. D., Surendhar, G. J., Natrayan, L., Patil, P. P., Ram, P. M. B., & Paramasivam, P. (2022). Evolution and recent scenario of nanotechnology in agriculture and food industries. *Journal of Nanomaterials*, *2022*, 1–17. https://doi.org/10.1155/2022/1280411

Vuong, L. D., Luan, N. D. T., Ngoc, D. D. H., Anh, P. T., & Bao, V.-V. Q. (2017). Green synthesis of silver nanoparticles from fresh leaf extract of Centella asiatica and their applications. *International Journal of Nanoscience*, *16*(01), 1650018. https://doi.org/10.1142/S0219581X16500186

Wang, C., He, Z., Liu, Y., Zhou, C., Jiao, J., Li, P., Sun, D., Lin, L., & Yang, Z. (2020). Chitosan-modified halloysite nanotubes as a controlled-release nanocarrier for nitrogen delivery. *Applied Clay Science*, *198*, 105802. https://doi.org/10.1016/j.clay.2020.105802

Wang, Y., Polavarapu, L., & Liz-Marzán, L. M. (2014). Reduced graphene oxide-supported gold nanostars for improved SERS sensing and drug delivery. *ACS Applied Materials & Interfaces*, *6*(24), 21798–21805. https://doi.org/10.1021/am501382y

Wang, M., Zhang, G., Zhou, L., Wang, D., Zhong, N., Cai, D., & Wu, Z. (2016). Fabrication of pH-controlled-release ferrous foliar fertilizer with high adhesion capacity based on nanobiomaterial. *ACS Sustainable Chemistry & Engineering*, *4*(12), 6800–6808. https://doi.org/10.1021/acssuschemeng.6b01761

Ward, M. B., Brydson, R., & Cochrane, R. F. (2006). Mn nanoparticles produced by inert gas condensation. *Journal of Physics: Conference Series*, *26*, 296–299. https://doi.org/10.1088/1742-6596/26/1/071

Wu, L., Mendoza-Garcia, A., Li, Q., & Sun, S. (2016). Organic phase syntheses of magnetic nanoparticles and their applications. *Chemical Reviews*, *116*(18), 10473–10512. https://doi.org/10.1021/acs.chemrev.5b00687

Yang, Y., Qin, Z., Zeng, W., Yang, T., Cao, Y., Mei, C., & Kuang, Y. (2017). Toxicity assessment of nanoparticles in various systems and organs. *Nanotechnology Reviews*, *6*(3), 279–289. https://doi.org/10.1515/ntrev-2016-0047

Ye, Y., Cota-Ruiz, K., Hernández-Viezcas, J. A., Valdés, C., Medina-Velo, I. A., Turley, R. S., Peralta-Videa, J. R., & Gardea-Torresdey, J. L. (2020). Manganese nanoparticles control salinity-modulated molecular responses in capsicum annuum L. through Priming: A sustainable approach for agriculture. *ACS Sustainable Chemistry & Engineering*, *8*(3), 1427–1436. https://doi.org/10.1021/acssuschemeng.9b05615

Yeh, Y.-C., Creran, B., & Rotello, V. M. (2012). Gold nanoparticles: Preparation, properties, and applications in bionanotechnology. *Nanoscale*, *4*(6), 1871–1880. https://doi.org/10.1039/C1NR11188D

Yoon, H. Y., Lee, J. G., Esposti, L. D., Iafisco, M., Kim, P. J., Shin, S. G., Jeon, J.-R., & Adamiano, A. (2020). Synergistic release of crop nutrients and stimulants from hydroxyapatite nanoparticles functionalized with humic substances: Toward a multifunctional nanofertilizer. *ACS Omega*, *5*(12), 6598–6610. https://doi.org/10.1021/acsomega.9b04354

Younes, N. A., Dawood, M. F. A., & Wardany, A. A. (2019). Biosafety assessment of graphene nanosheets on leaf ultrastructure, physiological and yield traits of Capsicum annuum L. and Solanum melongena L. *Chemosphere*, *228*, 318–327. https://doi.org/10.1016/j.chemosphere.2019.04.097

Yu, Z., Sun, X., Song, H., Wang, W., Ye, Z., Shi, L., & Ding, K. (2015). Glutathione-responsive carboxymethyl chitosan nanoparticles for controlled release of herbicides. *Materials Sciences and Applications*, *06*(06), 591–604. https://doi.org/10.4236/msa.2015.66062

Yusaf, T., Mahamude, A. S. F., Farhana, K., Harun, W. S. W., Kadirgama, K., Ramasamy, D., Kamarulzaman, M. K., Subramonian, S., Hall, S., & Dhahad, H. A. (2022). A comprehensive review on graphene nanoparticles: Preparation, properties, and applications. *Sustainability*, *14*(19), 12336. https://doi.org/10.3390/su141912336

Yusefi-Tanha, E., Fallah, S., Rostamnejadi, A., & Pokhrel, L. R. (2020). Zinc oxide nanoparticles (ZnONPs) as a novel nanofertilizer: Influence on seed yield and antioxidant defense system in soil grown soybean (Glycine max cv. Kowsar). *Science of The Total Environment*, *738*, 140240. https://doi.org/10.1016/j.scitotenv.2020.140240

Zahra, Z., Waseem, N., Zahra, R., Lee, H., Badshah, M. A., Mehmood, A., Choi, H.-K., & Arshad, M. (2017). Growth and metabolic responses of rice (Oryza sativa L.) cultivated in phosphorus-deficient soil amended with TiO2 nanoparticles. *Journal of Agricultural and Food Chemistry*, *65*(28), 5598–5606. https://doi.org/10.1021/acs.jafc.7b01843

Zahran, N. F., & Sayed, R. M. (2021). Protective effect of nanosilica on irradiated dates against saw toothed grain beetle, Oryzaephilus surinamensis (Coleoptera: Silvanidae) adults. *Journal of Stored Products Research*, *92*, 101799. https://doi.org/10.1016/j.jspr.2021.101799

Zhang, W., Ebbs, S. D., Musante, C., White, J. C., Gao, C., & Ma, X. (2015). Uptake and accumulation of bulk and nanosized cerium oxide particles and ionic cerium by radish (*Raphanus sativus* L.). *Journal of Agricultural and Food Chemistry*, *63*(2), 382–390. https://doi.org/10.1021/jf5052442

Zhang, J., Gong, J.-L., Zeng, G.-M., Yang, H.-C., & Zhang, P. (2017). Carbon nanotube amendment for treating dichlorodiphenyltrichloroethane and hexachlorocyclohexane remaining in Dong-ting Lake sediment — An implication for in-situ remediation. *Science of The Total Environment*, *579*, 283–291. https://doi.org/10.1016/j.scitotenv.2016.11.105

Zhang, X.-F., Shen, W., & Gurunathan, S. (2016). Silver nanoparticle-mediated cellular responses in various cell lines: An in vitro model. *International Journal of Molecular Sciences*, *17*(10), 1603. https://doi.org/10.3390/ijms17101603

Zhang, S., Yang, Y., Tong, Z., Gao, B., Gao, N., Shen, T., Wan, Y., Yu, Z., Liu, L., Ma, X., Guo, Y., Fugice, J., & Li, Y. C. (2020). Self-assembly of hydrophobic and self-healing bionanocomposite-coated controlled-release fertilizers. *ACS Applied Materials & Interfaces*, *12*(24), 27598–27606. https://doi.org/10.1021/acsami.0c06530

Zhang, Y., Yang, J., Zhong, L., & Liu, L. (2018). Effect of multi-wall carbon nanotubes on Cr(VI) reduction by citric acid: Implications for their use in soil remediation. *Environmental Science and Pollution Research*, *25*(24), 23791–23798. https://doi.org/10.1007/s11356-018-2438-8

Zhao, K., Hu, J., Ma, Y., Wu, T., Gao, Y., & Du, F. (2019). Topology-regulated pesticide retention on plant leaves through concave janus carriers. *ACS Sustainable Chemistry & Engineering*, *7*(15), 13148–13156. https://doi.org/10.1021/acssuschemeng.9b02319

Zhao, L., Lu, L., Wang, A., Zhang, H., Huang, M., Wu, H., Xing, B., Wang, Z., & Ji, R. (2020). Nano-biotechnology in agriculture: Use of nanomaterials to promote plant growth and stress tolerance. *Journal of Agricultural and Food Chemistry*, *68*(7), 1935–1947. https://doi.org/10.1021/acs.jafc.9b06615

CHAPTER 4

Recent Advances in Quality Assessment Tools for the Characterization of Nano-Agri-Input Products

NUSRAT IQBAL[1], SHRESHTA DUBEY[2], SAURABH DUBEY[2], and PRATIBHA SRIVASTAVA[3]

[1]Formulation Division, Institute of Pesticide Formulation Technology, Gurgaon, Haryana, India

[2]RS Formulation Development and Consultant, Bhiwadi, Rajasthan, India

[3]Bioprospecting Group, Agharkar Research Institute, Pune, Maharashtra, India

ABSTRACT

A growing area of interest is nano-agri-input products (NAIPs), which are composed of different formulations of nanomaterials, especially those created for agricultural uses. NAIPs have the potential to improve crop yield, minimize environmental pollution, and promote farming practices. To prevent any negative effects on crops, ecosystems, and systems supporting human health, it is crucial to ensure the quality and safety of the NAIPs. As a result, the fundamental requirement for NAIPs is adherence to guidelines for preparing them for quality assurance using cutting-edge methodologies for their characterization. Physicochemical parameters, such as dynamic light scattering, transmission electron microscopy (TEM), X-ray diffraction, and spectroscopic methods, provide information on the nanoparticle's size, shape, surface charge, and composition. These techniques help to understand the relationship between the physical properties of nanomaterials and their performance as NAIPs. This chapter provides an inclusive overview of

Nanobiotechnology for Agricultural Sciences: Nano-Agri-Input Products for Crop Production and Environmental Protection. J. C. Tarafdar, Deepak Kumar, Sanjay K. Singh, and Krishna Pal Singh (Eds.)

current approaches, essential considerations for the registration of any NAIP, and advanced NAIPs as formulations of pesticides and their application in agriculture.

4.1 INTRODUCTION

4.1.1 AGRICULTURE AND NANOTECHNOLOGY

Agriculture is the root cause of human development and contributes significantly to any country's economy. Soil conditions, minerals, nutrients, and water are essential parameters for crop production. The demand for food has exponentially increased with the growing population (Dijk et al., 2021). The involvement of science and technology can't be avoided in sustainable agricultural development (Pandey et al., 2022). The intervention of nanotechnology in agriculture has constructed a revolutionary area with the discovery of nanofertilizers, nanopesticides, nanofood, nanocomposites, nanocarriers, and nanoformulations, which showed the potential for improved agricultural output via increased plant productivity and enhanced crop protection but had its pros and cons (Awasthi et al., 2020). Therefore, well-to-do guidelines are issued by the Department of Biotechnology, Government of India, for nano-agri input products (NAIPs) and nano-agri products (NAPs) (DBT Guidelines, 2020). The guidelines are also in harmony with the international guidelines of REACH (REACH regulations, 2019), OECD (OECD, 2017), TSCA (TSCA, statuary definition, 2019), APVMA (APVMA guidelines, 2005), FDA/WHO (FDA guidelines, 2015), EFSA (EFSA Scientific Committee, 2011), and the principles of ICH (Quinlan et al., 2013).

With continuous research in science and technology, new nanomaterials (NMs) are introduced periodically. Efficacy assessment tools monitor the quality and agronomic potential of NAIPs. These include field trials, crop yield analyses, and nutrient uptake studies. By analyzing the impact of NAIPs on plant growth, yield, quality, and nutrient quantity, the efficacy of these products can be assured. There are no widely accepted final international guidelines for NAIPs and NAPs. The relationship between NMs' physio-chemical properties and other environmental conditions determines the effectiveness of their activity and the influence of their presence (Ahmed et al., 2023). Therefore, the novel multifunctional nanoproducts may undergo quality, safety, and efficacy assessments, which could benefit the economy and have a societal impact.

Agri-input products are used in nanoformulations, which can be used in agriculture or related fields for crop production, protection, control, harvesting, after-harvesting, and packaging. Agri-input products may be the new active ingredient (AI) or an existing AI with modified beneficial properties, dimensions, or phenomena linked with nanotechnology applications (Neme et al., 2012). They can be considered, but are not limited to, pest and disease prevention, regulation, and management, fertilizers, agrochemicals, biofertilizers, pheromone release, plant growth hormones, plant nutrients, and anti-transpirants, biostimulants, and genetically modified (GM) crops using NM as nanocarriers (Figure 4.1) for the betterment of crops (Fraceto et al., 2016). These products are categorized as NAIPs. In this chapter, we would like to emphasize the NAIPs for pest management and their quality assurance.

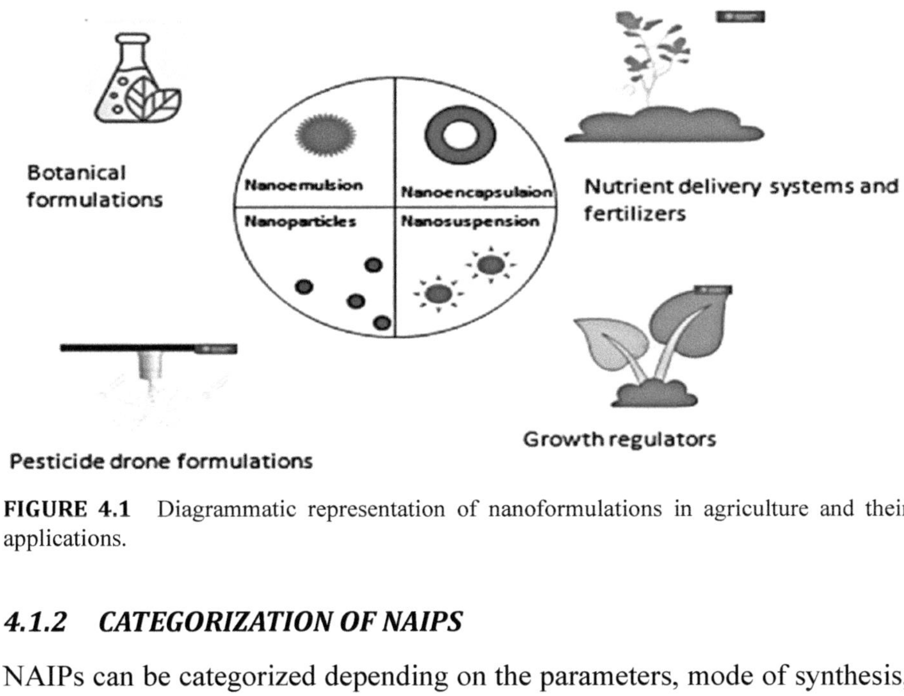

FIGURE 4.1 Diagrammatic representation of nanoformulations in agriculture and their applications.

4.1.2 CATEGORIZATION OF NAIPS

NAIPs can be categorized depending on the parameters, mode of synthesis, and functions of NMs, as well as the current products that incorporate synthetic and modified NMs (DBT Guidelines, 2020).

4.1.2.1 NANOFORMS VERSUS ACTIVE INGREDIENTS

Nanocarrier loaded with AI: Targeted agri-input NMs can be transported using a nanocarrier, which can be either a soft or hard substance. They have the potential for a gradual and monitored delivery of agricultural inputs.

AI transformed into nanoform: AIs can be converted into nanoforms, enhancing their potential for efficacy and stability.

4.1.2.2 BASED ON SYNTHESIS

NMs, which are synthesized using bioagents, are biologically synthesized NMs. The NMs synthesized with synthetic chemicals are called chemically synthesized NMs. Physically synthesized NMs are produced via methods including glow discharge, thermal and microwave-assisted ultrasonication, ball milling, laser ablation, and pulsed laser deposition.

4.1.2.3 BASED ON THE CHEMICAL NATURE OF NMS

Organic NMs: These are made up of organic substances, including carbohydrates, proteins, and lipids. Because of their low toxicity, their use in agriculture has been advocated.

Inorganic NMs: They are highly stable and simple to manufacture, like Ag, Fe, Au, Cu, and so on. NMs are a flexible option for agri-input products since they can be designed utilizing bottom-up methods and a variety of physio-chemical properties, including shape, size, and surface area. Composites are hybrid materials made at the nanometer scale by mixing polymers with inorganic solids (oxides).

4.1.2.4 BASED ON DEGRADABILITY

Biodegradable NMs: They are easily degradable by bio-agents and are used as nanocarriers.

Non-biodegradable NMs: Their use in NAIPS for controlled and slow-release of pesticides or fertilizer.

4.1.3 QUALITY ASSURANCE OF NAIPS

The quality of NAIPs can be assessed using a variety of factors. Any newly produced agri-products must adhere to the general standards outlined in the laws of CIB&RC, GCO, FSSAI, and BIS, regardless of whether they are based on nanotechnology (DBT Guidelines, 2020). However, based on different cases, an approach can be adopted for the evaluation of NAIPs for enhanced efficacy and safety due to the engrossment of multidisciplinary sciences and their intricate behavior. According to regulatory parameters, for the approval of NAIPs in the country, both AI and NMs should be registered.

Process control may be used if there is a unique process requirement for nanoformulations compared to traditional agri-input products. It is possible to indicate how NMs are provided, how waste is disposed of, and how these affect the environment.

Excipients are added to the formulations of NAIPs for the improvement and stability of the products. They inhibit agglomeration and aggregation and act as preservatives to restrict microbial contamination. Coupling reagents and surfactants alter the surface features of NMs. Excipients' information should be revealed.

Stability: NAIPs should be subject to the same storage requirements and stability standards as agriproducts. The OECD test recommendations (TG) for storage stability testing may be used. The products' shelf-life should be assessed under various storage conditions, including pH stability, photo-stability, ambient storage stability, low-temperature storage capability, and room-temperature storage stability.

According to OECD rules and provisions from the European Chemical Agency (ECA), the ecological toxicity test for effects on mammals should be undertaken as it is for traditional agricultural goods. As per the NAIP's features, in-silico, in-vitro, and in-vivo analyses should be conducted for the certifications. It is important to examine the safety of human health with regard to inhalation, genotoxicity, skin, and ocular contact.

For environmental safety, aquatic tests for growth susceptibility and freshwater algae, cyanobacteria, Daphnia acute immobilization (DAI), fish acute toxicity (FET), fish embryo acute toxicity, and terrestrial plant tests (seedling emergence and seedling growth) should be considered.

The registration dossier should contain information about the persistence of NAIPs, and residue analyses of used nanoAI, additives, and nanocarrier materials of NAIPs should be carried out during the life-cycle analysis of the product. The manufacturer must provide information about NM residue, the food or feed containing any nanosized degradable products, excipients, or surface coatings used on food contact material during product registration.

Registration of nanopesticides should be done in accordance with the current Insecticides Act, 1968, along with Rules, 1972, and maximum nano fertilizers and nano-micronutrients are necessary to register under the existing FCO, 1985, as per the defined data sets through the nodal agency of the Government of India and the Agriculture Department of State Governments.

4.1.4 CHARACTERIZATION OF NANOEMULSION

A variety of analytical techniques, including dynamic light scattering (DLS), small angle neutron scattering (SANS), viscosity, atomic force microscopy (AFM), zeta potential, and stability, can be used to characterize the physical characteristics of nanoemulsions, including nanosized droplets, high transparency, variable viscosity, and high stability (Mourdikoudis et al., 2018). DLS is one of the best techniques for estimating the size of a nanoemulsion droplet. The physical stability of nanoemulsions can be determined by measuring the size of the droplet as a function of time using DLS. DLS has also been reported for the assessment of the Ostwald ripening rate. Neutron wavelengths can analyze nanoscale structures, which makes the SANS method useful for generating bulk structures of nanoemulsions. Analyses such as viscosity, conductivity, and dielectric testing provide appropriate results for nanoemulsions at the macroscopic level. Nanoemulsions exhibit low viscosity with increasing water content and higher viscosity when the content of surfactant and co-surfactant is low. An alteration in viscosity helps determine the stability of nanoemulsion systems. To identify whether the developed nanoemulsions are o/w or w/o and to monitor the phase inversion phenomena, conductivity tests can be measured. Nanoemulsions' structural and dynamic properties can be assessed using dielectric measurements. AFM can be utilized to visualize the shape of the droplet nanoemulsion system. The stability of nanoemulsion systems can be monitored by different stability tests such as centrifugation assays, freeze-thaw cycles, heating-cooling tests, and ambient conditions. These tests determine the duration for which the nanoemulsion systems can sustain stability without exhibiting Ostwald ripening, coalescence, creaming, or sedimentation. It will be helpful to estimate the shelf-life of the prepared nanoemulsion.

Modern techniques like surface plasmon resonance (SPR) and autofluorescence can be used to analyze agrochemicals that have been nanoencapsulated. Among the several microscopic methods, electron microscopy (EM) is renowned for revealing the dimensions, form, and other elemental characteristics of NMs. One of the distinctive nanoscale imaging methods for the characterization of small-size (0.5 nm) NMs is transmission electron microscopy (TEM). Along with size, shape, and particle distribution, TEM in combination with an energy-dispersive X-ray (EDX) spectroscopy detector provides details on the elemental compositions of NMs. Inconsistencies in the polymer structure can be found using AFM. High-resolution pictures of biological material under a high vacuum and below room temperature are possible using cryo-TEM/cryo-SEM. Images of NMs aggregates in the root, epidermis, cortex, and xylem vessels can be obtained using confocal microscopy.

4.1.5 NANOFORMULATIONS FOR PEST MANAGEMENT

Sonkaria et al. (2012) and Pérez-de-Luque and Hermosn (2013) found that nanotechnology can be used to improve existing agricultural practices in order to ensure "sustainable intensification" of agricultural production, which will increase food production globally, enhance food quality, and lessen environmental losses. Agrochemical formulations containing nanoparticles (NPs) or encapsulated nanoscale materials with various coatings are suitable for improving the solubility and permeability or the bioavailability of an AI, lowering the dose of the AI and its dose-dependent toxicity, ensuring controlled release and targeted biodistribution of the AI, and reducing the environmental impact of the AI (Rai et al., 2015).

Nanoscale herbicides, insecticides, and fungicides can help lessen environmental contaminants, conserve biodiversity, and the spread of pests and illnesses in agriculture. Additionally, the application of NPs in the food industry can enhance food safety through disease detection, the development of intelligent food packaging systems, and the addition of nutraceuticals to food (Knauer and Bucheli, 2009). The significant negative aspects of NPs on the health of humans and the environment should be addressed in light of their growing utilization in many anthropogenic activities. Using nanotechnology in agriculture can considerably increase the efficacy of agricultural inputs, providing a key means of ensuring the long-term sustainability of agroecosystems. As a result, a thorough examination is needed to understand manufactured NMs, evaluate their mechanistic applicability, and determine their agroecological toxicity. This chapter aims to critically highlight the use of NPs in materials and pinpoint significant limitations in the application of nanotechnology to sustainable agriculture.

4.1.5.1 NANOTECHNOLOGY AND AGRICULTURAL SUSTAINABLE DEVELOPMENT

Nanopesticides, according to Kookana et al. (2014), are composed of extremely small particles of a pesticide's AI that have at least one size dimension in the 1–100 nm range. The 100 nm limit is predicated on the notion that there are distinguishing characteristics between nanoparticles and bulk material that often manifest at length scales lower than 100 nm, according to Hayles et al. (2017). In addition to the polymeric encapsulation of pesticides, nanoemulsions, and nanogels have also been proposed in various forms (Kah et al., 2014). Benefits of nanoparticle-based pesticide formulations include the capacity to delay the release of AIs, the ability to prevent pesticide degradation due to environmental factors, increased water solubility and stability of an AI, as well as increased insecticidal activity due to larger surface area and lower particle size (Balaji et al., 2015). In particular, it was anticipated that the mechanism of action of these nanopesticides against target pests would be improved in comparison to their bulk equivalents, as the small size of nanoparticles leads to more interaction with target pests and more excellent insecticidal activity may be seen (Gupta et al., 2023). Additionally, nanoparticles showed constant leaf cover and plant penetration, supporting their use in pest control and agriculture.

Nanotechnology can significantly improve production through nutrient control (Gruère, 2012), in addition to monitoring water quality and pesticides for the sustainable growth of agriculture (Prasad et al., 2017). Enormous features and applications of nanoparticles, is challenging to offer a comprehensive assessment of the health and environmental issues raised by them. Designed NPs may have characteristics such as chemical composition, shape, surface structure, surface charge, behavior, degree of particle aggregation (clumping), disaggregation, etc. that affect their toxicity in addition to their size (Zook et al., 2010). Because of this, the toxicity of different-sized or differently-shaped NMs with the same chemical makeup might vary. The application of nanotechnology research to the agriculture industry has evolved into an important part of environmentally friendly growth.

4.1.5.2 PEST MANAGEMENT WITH NANO-DELIVERY SYSTEMS

Nanoformulations are cutting-edge methods for creating water-insoluble pesticides with improved bioactivity, reduced toxicity, and lower environmental pollution. Modern techniques for producing water-insoluble pesticides with enhanced bioactivity, decreased toxicity, and decreased environmental pollution include nanoformulations. These formulations serve as nano-delivery vehicles for various agrochemicals used in agricultural practices and act as carriers for crop protection.

4.1.6 NANOEMULSIONS

Nanoemulsions are the mixing of two incompatible liquids stabilized by a suitable surfactant with an average droplet size of 500 nm (Figure 4.2) (Singh et al., 2017). Nanoemulsion is considered to be made up of extremely tiny oil/water emulsion nanoscale droplets with diameters less than 100 nm, according to Gutierrez et al. (2008) and Anton et al. (2008). The nanoemulsion is intended to enhance pest control and expand agrochemical delivery technology by enhancing pesticide solubility, bioavailability, stability, and wettability, claim Chen et al. (2017). For the protection of crops and the management of pests, pesticide nanoemulsions offer high potential. For managing agricultural pests and disease-carrying vectors, recent research has focused on nanoemulsions containing synthetic and biological pesticides.

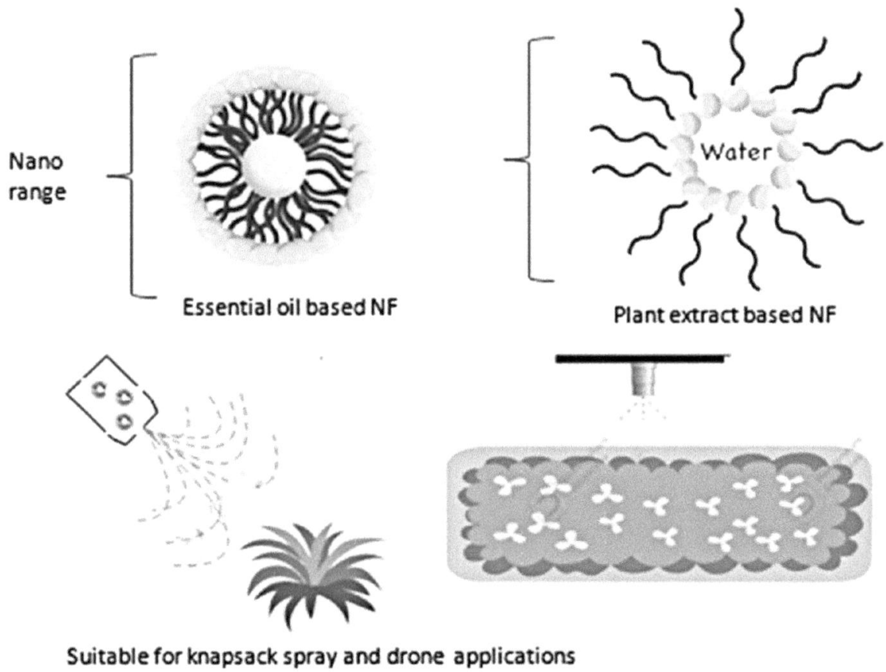

FIGURE 4.2 Diagrammatic representation of nanoemulsion and its application mechanism.

4.1.6.1 NANOEMULSIONS AND THEIR PREPARATION TECHNIQUES

Oil-in-water (O/W) or water-in-oil (W/O) are two opposing liquids that are dispersed in a biphasic manner and stabilized by an amphiphilic surfactant (Figure 4.3). From a thermodynamic perspective, smaller NEs are very tiny emulsion droplets, often oil droplets in water, with diameters in the order of 100 nm (typically 500 nm). They have advantageous properties such as a large surface area per unit volume, strong stability, an optically clear appearance, and flexible rheology (Sabry and Ragaei, 2018; Gupta et al., 2016; Anton and Vandamme, 2011).

The high-energy techniques used to create NEs include membrane emulsification, ultrasonic homogenization or sonication, microfluidization, high-amplitude ultrasonic technique, and jet disperser, and sonication or ultrasonic homogenization. By putting it under disruptive hydrodynamic stress, such as cavitations, turbulent shear, or laminar, the high-energy approaches are intended to provide the energy needed for emulsification. According to Anton et al. (2018), low-energy transitional emulsification techniques to create NEs are based on nonionic surfactant phase transitions (PEGylated surfactants) associated with a rapid shift in terms of how soluble they are in water and oil, respectively. According to Komaiko and McClements (2015), spontaneous emulsification is a low-energy process that just requires the addition of an organic phase (oil plus surfactant) to an aqueous phase. The size of the resulting droplets is influenced by the ratio of the surfactant to the oil, the kind of surfactant, the position of the surfactant, and the type of oil.

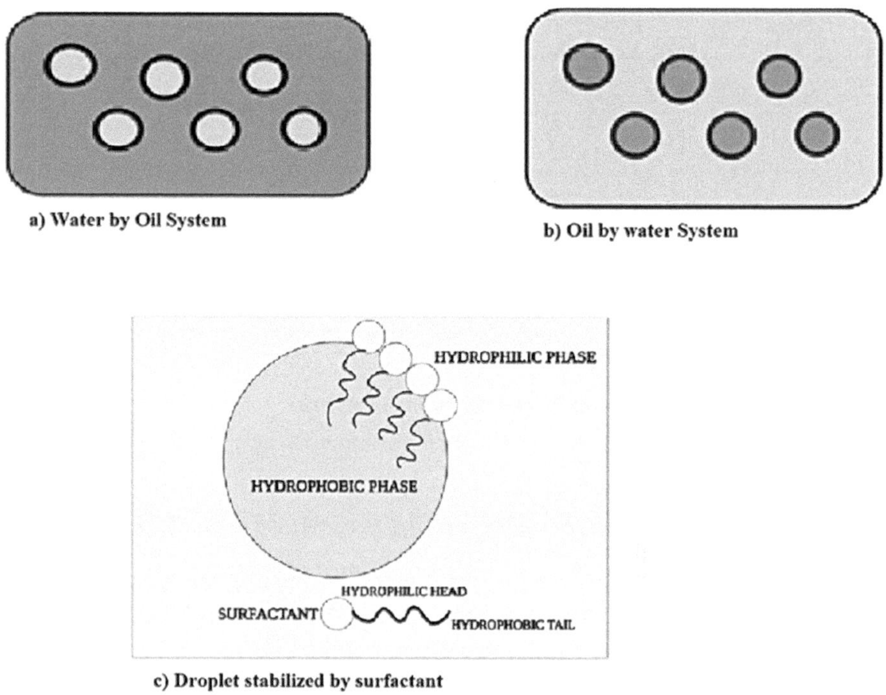

FIGURE 4.3 Diagrammatic image of nanoemulsion: (a) water-by-oil system; (b) oil-by-water system; (c) droplet stabilized by surfactant.

In comparison to typical emulsions, nanoemulsions have substantially higher values for the elastic modulus, Laplace pressure, surface area-to-volume ratio, and droplet size. Energy is a big issue for nanoemulsions. Therefore, it needs specialized equipment that can produce high levels of shear stress, such as an ultrasonic generator or a high-pressure homogenizer (Gupta et al., 2016). The "low-energy" approach for creating nanoemulsions described by Tadros et al. (2004) uses two liquid phases. The two phases come into touch with one another when one is a homogenous liquid made up of a lipophilic phase and a hydrophilic surfactant, as well as perhaps a solvent, polymer, or medication. The hydrophilic species present in the oily phase are then quickly dissolved into the aqueous phase, causing the oil to demix into nano-droplets that are then stabilized by amphiphiles. Thus, it appears that this approach is the simplest and requires no expensive or specialized equipment.

4.1.6.2 APPLICATIONS OF NANOEMULSIONS IN AGRICULTURE

Numerous substances with biological origin are employed to make bio-based NEs, such as encapsulating active components (such as essential oils (EOs)) and plant oils for the NE oil phase (such as sunflower oil, rapeseed oil, palm oil, and so on) (Abdou et al., 2018; Raviadaran et al., 2020; Kaci et al., 2018), biosurfactants (such as phospholipids, lactoferrin/alginate, Quillaja saponin, lecithin, whey protein, lactoferrin, gum arabic, and pectin), emulsifiers (Zhao et al., 2018; Liu et al., 2017; McClements and Gumus, 2016; Bai and McClements, 2016; Verma et al., 2016; Ozturk et al., 2014), targeting ligands like folate that are present on the NE's surface (Ganta et al., 2016), it is used to improve bioavailability, lessen negative side effects, minimize non-specific absorption, and enable precise targeting of certain target cells. Targeting technology often makes use of nanocarrier functionalization, which can be a surface modification (Attia et al., 2017) and/or ligand grafting (Geng et al., 2016).

A nanoemulsion of eucalyptus EO was proposed by Totan et al. (2020) as a defense against the main rice pests, *Triboliumcastaneum* (Herbst) and *Sitophilus oryzae* (L.). The nanoemulsions of fenpropathrin had excellent wetting, adhesion, and stretching capabilities with a narrow contact angle and great adherence to the target leaves (Wei et al., 2023). They also showed stability and exceptional insecticidal effects against *Helicoverpa armigera.* EO nanoemulsion from *Carlina acaulis* L. (Asteraceae) shows significant insecticidal efficacy and has no impact on aquatic or terrestrial species. The phytoparasitic nematode *Meloidogyne incognita* is controlled by a nanoemulsion of *C. acaulis* EO (Ntalli et al., 2023). According to Heydari et al. (2020), the nanoemulsion of *M. piperita* is a potential organic insecticide for controlling cotton aphids. A nanoemulsion with high physical stability that contains abamectin was discovered to make *Plutellaxylostella* more sensitive (Feng et al., 2020).

The phoxim nanoemulsion was shown to be more effective than commercial formulations, and literature also suggests improved insecticidal efficacy at lower dosages (Li et al., 2022). Due to their quick adsorption on the surfaces of plant leaves, low surface tension, and long shelf life, oil-in-water (O/W) nanoemulsions of bifenthrin in biodiesel have demonstrated high bio-efficacy against cabbage maggots (Yan et al., 2019). Table 4.1 represents recent nanoemulsions prepared for their use in agriculture.

4.1.7 NANOENCAPSULATION

By encapsulating the pesticide within another material, frequently a polymer, whose size is maintained in the nano-range, nanocapsules may be constructed to release the pesticide under certain environmental circumstances. By encapsulating the bioactive chemicals, nanoscale carriers or nanoencapsulation materials can improve the bioavailability of active substances and decrease toxicity, enhancing the timed release of active substances and enabling the exact targeting of active substances. Crop protection chemicals now use nanoencapsulation technology to either achieve these goals or be nanosized, which lowers the quantity sprayed (Figure 4.4). The results of nanoencapsulation in fulfilling the objectives of a good pesticide formulation have so far been the subject of various studies (Table 4.2). Although the objectives of nanoencapsulation are connected, the following are the principal objectives:

A nanoencapsulated pesticide formulation has enhanced solubility, specificity, and stability against diverse environmental conditions, including pH and temperature (Madhavi et al., 2020). It also has controlled release capabilities. Acetamiprid can be made more poisonous and effective against insects by being nanoencapsulated with sodium alginate and PEG (Ebadollahi et al., 2022). Since low concentrations of pesticides slowly release over time, making them less dangerous for species other than their intended targets and extending their shelf life, nanoencapsulation of pesticides is a growing area for pest management techniques. Atrazine was nanoencapsulated in polycapro-lactone, and it was shown to be less toxic to non-target soil invertebrates than free atrazine (Gomes et al., 2019). Due to the widespread use of polymeric nanocapsules, research into the most efficient way to create them has increased.

TABLE 4.1 Details of Recent Nanoemulsion Formulations in Agriculture.

Formulation	Pesticide	Bioactivity	Target Pest	References
Thymol NE *Quillaja* saponin as a natural surfactant	Biopesticide	Lower disease incidence (3.3–29%) enhancements of plant growth.	*Xanthomonas axonopodis* pv. *glycine*,	Kumari et al. (2018)
Glyphosate isopropylamine	Herbicide	Potentiation of the biological activity	*Asystasia gangetica*, *Diodiao cymifolia*, and *Paspalum conjugatum*.	Lim et al. (2013)
Neem oil tebuconazole NE	Pesticide	100% of mortality after 2 days	*Sitophilus oryzae* (L.) and *Tribolium castaneum* (Herbst)	Marziyeh et al. (2017)
Manilkara subsericea	Pesticide	Mortality	*Dysdercus peruvianus*	Fernandes et al. (2014)
Neem and citronella oil	Antifungal activity	The respective LC_{50} for *R. solani* and *S. rolfsii* were 13.67 and 25.64 mg/L and 14.71 and 20.88 mg/L, accordingly.	*Rhizoctonia solani* and *Scletotium rolfsii*	Ali et al. (2017)
Clove oil and lemongrass oil	Antifungal activity	Preventing the growth of mycelium (MIC 4000 mg/L)	*Fusarium oxysporum* f. sp. *lycopersici*	Hashem et al. (2018)
Pimpinella anisum L. EO	Insecticidal	81.33 ± 0.08% of adult mortality, and when the progeny were exposed to 10% NE, there was a drop of 70.85%, and the LC50 value after 72 hours was 9.84%.	*Tribolium castaneum*	Abd-Elsalam et al. (2015)
Eugenol NE	Antifungal activity	Reduction in the size and number of conidiospores and reduction in hyphae pigmentation	*Fusarium oxysporum* f. sp. *vasinfectum*	Nuruzzaman et al. (2016)

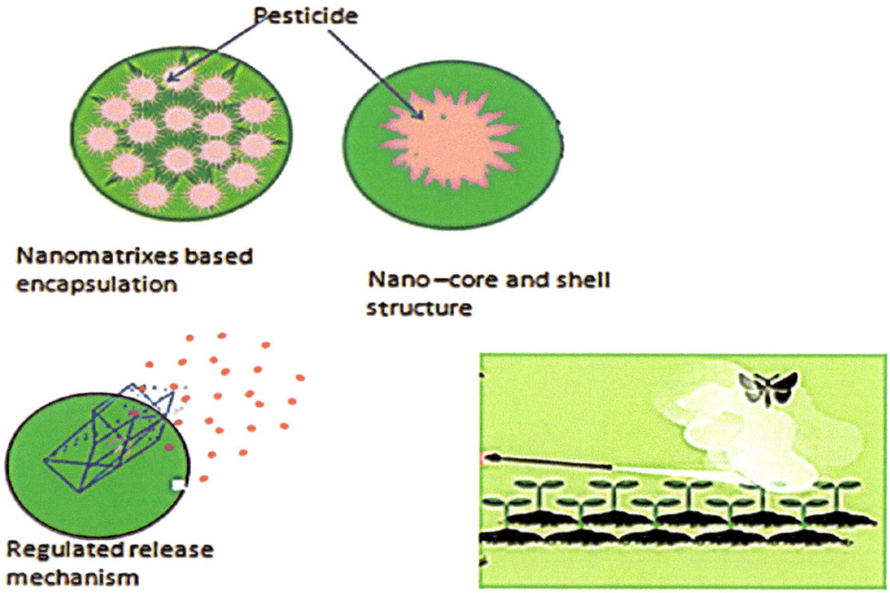

FIGURE 4.4 Diagrammatic representation of nanoencapsulation.

Researchers now have a variety of options for creating nanocapsules through the synthesis of various polymers and their inherent characteristics. The most commonly used techniques are layer-by-layer deposition, double emulsification, emulsification-coacervation, solvent evaporation, and nanoprecipitation. But in addition to the above-mentioned methods, many additional ones, such as melt dispersion, emulsion, polymerization, interfacial polymerization, interfacial deposition method, solvent displacement technique, emulsion-evaporation, and so on, were also identified in the literature. However, several ways of creating nanocapsules have been reported in other places.

4.1.7.1 APPLICATION OF NANOENCAPSULATED AGROCHEMICALS IN PRACTISE

Nanoencapsulated pesticide formulations are capable of lowering pesticide dosages and reducing human exposure to them, which is beneficial for the environment and crop protection. However, the use of nanoencapsulation materials for pesticide administration was hampered by ignorance of the production mechanism and a lack of a cost-benefit analysis.

Nanoencapsulation is the technology that appears to offer the most promise for practical application right now. Crop protection in agriculture is one of the most significant uses of nanoencapsulated goods.

Chemicals like glyphosate, imidazolinenones, and sulfonylureas, for instance, can be enclosed in polymeric shells. Germination promoters may also be put into the soil inside nanocapsules to protect them against degradation. A formulation of a pesticide that is nanoencapsulated offers better solubility, specificity, permeability, stability, and controlled release properties. Compared to traditional pesticides, they are more able to withstand severe environmental processes such as leaching, evaporation, photolytic hydrolysis, and microbial destruction because of their design. They have built-in switches to regulate the release and subsequent availability of pesticides so that they can be utilized efficiently for a specific amount of time.

The use of bifenthrin on a broad scale in agriculture may be safer because of the nanoencapsulation (Blewett et al., 2019). Over a period of days, cedarwood (*Cedrus deodara*) EO contained in pectin nanocapsules outperformed free EO in its effectiveness against the malaria vector *Anopheles culicifacies* (Kala et al., 2020). In Table 4.2, a list of the recently developed nanoncapsulated formulations is mentioned.

TABLE 4.2 Recent Nanoencapsulation Formulations in Agriculture.

Pesticides	Mode of Action	Targeted Pests	References
Bifenthrin (BFT) and λ-cyhalothrin (LAM) with the synergists piperonyl butoxide (PBO) and diethyl maleate (DEM)	Improve control efficacy	*C. includens*	Boff et al. (2022)
Eucalyptus globulus **and** *Zataria multiflora* **essential oils**	Contact insecticidal activity	*Ephestia kuehniella*	Emamjomeh et al. (2023)
Zanthoxylum rhoifolium	Reductions of 95% in the overall number of nymphs and eggs	*Bemisia tabaci*	Christofoli et al. (2015)
Azadirachtin	Enhanced UV stability	**Fall Armyworm**	Bae et al. (2022)
Essential oils of *Zataria multiflora* **and** *Rosmarinus officinalis*	LC_{50} values 103.53 and 112.64 μL/L air,	*Tribolium confusum*	Ahsaei et al. (2020)

Silva-Pereira et al. (2015) described the greater efficiency and less genotoxicity of nanosized polymerized atrazine (483.1 nm) by using epsilon-caprolactone as a monomer. Grillo et al. (2014) studied the controlled release properties of 3,4-dichlorophenoxy acetate of zinc-aluminium in different aqueous mediums (phosphate > carbonate > sulfate> chloride), with a release value in the range of 8.9–18.7. Nanoencapsulated herbicides can provide a useful means of reducing adverse impacts on humans as well as the environment.

4.1.8 NANOSUSPENSION

Improvements in solubility, wettability, and insecticidal activity are the key outcomes of nanosuspension with a uniform particle size below 100 nm due to its huge specific surface area, quick-dissolving rate, and great penetrability. The nanosuspension also lowered the amount of surfactant needed and reduced the requirement for organic solvents. Therefore, it is reasonable to anticipate that lambda-cyhalothrin nanosuspensions (LCNSs) will be a viable pesticide formulation for protecting plants and the environment (Chin et al., 2011).

By dissolving in organic solvents or employing emulsions, poorly soluble substances can be formed into organic solutions. With the exception of organic solutions and emulsions, large quantities of chemicals and organic solvents are frequently needed to improve the solubility necessary for certain formulations, which may result in unintended side effects. It would be preferable to prepare any poorly soluble medication using a universal formulation method. Using micronization to create a suspension concentration (SC), a traditional formulation method for such poorly soluble pharmaceuticals, involves milling a coarse drug powder with water and additives to an ultra-fine powder with a mean particle size generally in the range of 13 μm (Figure 4.5).

The solubility of many modern pesticides is quite low, and micronization does not result in bioavailability that is adequately high. Making these medications into nanocrystals, which are generally between 200 and 500 nm in size, is the next stage in the process of going from micrometer- to nanometer-sized particles. The two primary techniques for creating nanocrystals are bottom-up (controlled precipitation/crystallization), top-down technologies, and nanonizing (reducing large drug particle sizes to smaller ones by mechanical grinding down). When using the bottom-up method, the medication is dissolved in an organic solvent and treatment with an anti-solvent for precipitation while a stabilizer is present. As a result, homogeneous particles of smaller sizes are created. Besides, it can lead to the creation of amorphous drug nanoparticles, which have a high rate of dissolution and saturation solubility. While the top-down approach uses a variety of methods, including media milling, high-pressure homogenization, and microfluidization, to reduce particle size, For the purpose of producing more homogeneous nanoparticles, combination procedures that combine a pre-treatment with a subsequent size-reduction phase are also frequently used.

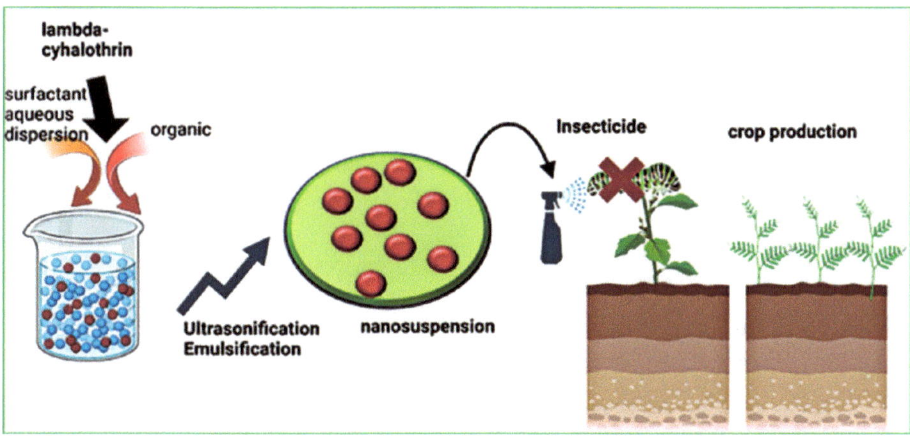

FIGURE 4.5 Diagrammatic representation of nanosuspension for crop improvement.

4.1.8.1 APPLICATION OF NANOSUSPENSION AGROCHEMICALS IN PRACTICE

Hazafaet al. (2022) have prepared nanosuspension of *Chrysanthem umcoronarium* and *Azadirachta indica* for pesticidal use against the *Rhyzopertha dominica* (lesser grain borer) and *Tribolium castaneum* (red flour beetle) using response surface methodology with particle sizes of 120.1 and 170.1 nm, respectively. An abamectin nanosuspension was developed as a pesticide against diamondback moth by Cui et al. (2020) with a particle size up to 90 nm using the wet-milling method. Spinetoram nanosuspension was prepared by milling techniques NS1 and NS2 with various sizes of ZrO_2 beads, which resulted in a size reduction of 2299 nm to 100 nm. The generated nanosuspension showed increased attachment and penetration of spinetoram into the leaf surface of *Brassicas*. The activity of nano-pyridayl suspension has enhanced threefold the pesticidal activity of *Helicoverpa armigera* compared to commercial formulations. By using the melt emulsification method, the LCNS was prepared by Wang et al. (2021), having a particle size of 12 nm (Table 4.3).

TABLE 4.3 Nanosuspensions in Agriculture.

Pesticides	Pesticide Class	Targeted Pests	References
Chrysanthemum coronarium and *Azadirachta indica*	Insecticides and fungicide	*Psidium guajava, Aedes aegypti, C. uinquefasciatus, Myzus persicae, Plutella xylostella, Spodoptera litura, Tribolium castaneum,* and *Helicoverpa armigera.*	Hazafa et al. (2022)
Avermectin	Insecticide	*Plutella xylostella*	Cui et al. (2020)
Abamectin	Insecticide	*Plutella xylostella*	Cui et al. (2019)
Spinetoram	Insecticide	*Plutella xylostella*	Rahwanudin et al. (2022)
Pyridalyl	Insecticide	*Helicoverpa armigera*	Saini et al. (2012)
Lambda-Cyhalothrin	Insecticide	*Mustard aphid*	Wang et al. (2021)

4.1.9 NANOPARTICLES IN PEST MANAGEMENT

Recently, there has been a lot of scientific interest in the development of novel insecticides using nanoparticles created by diverse synthesis processes. However, detailed information on their method of action against pests is lacking. Several studies have been carried out to investigate their hazardous potential against a variety of insect pests (Athanassiou et al., 2018; Rai et al., 2014). Since nanopesticides are relatively new substances and have not been thoroughly explored, there have only been a few studies looking at the toxicodynamics or toxicokinetics of NM-based pesticides against storage grain pests. In general, toxicokinetics refers to the movement and transformations an insecticide goes through inside an organism, including distribution, absorption, metabolism, and discharge, whereas toxicodynamics refers to the effects of molecules on the physiological, biochemical, and molecular levels as well as the mechanisms by which they act (Alzogaray et al., 2017). Moreover, several studies have examined the mode of action of nanopesticides based on alumina, silica, graphene oxide, and silver nanoparticles against insects (Benelli et al., 2018). Silver nanoparticles influence antioxidants and detoxifying enzymes, which decrease oxidative stress, acetylcholinesterase activity, and cell death. By up- or down-regulating important insect genes, they are also known to decrease gonadotropin release and protein synthesis, resulting in developmental harm and reproductive failure (Nair and Choi, 2011). By attaching to P and S in nucleic acids and proteins, metallic nanoparticles decrease membrane permeability, causing denaturation of organelles and enzymes and ultimately cell death. In addition to having an impact on growth and reproduction, nanoparticles of gold also function as trypsin inhibitors (Patil et al., 2016; Small et al., 2016). When waxes and lipids are physically absorbed by the insect cuticle after silicon dioxide and aluminum oxide nanopesticides attach to the insect, the result is dehydration of the insect (Arumugam et al., 2016). The *T. confusum*'s cuticle was punctured and broken by the nanozeolite formulation after adhering to the insect's body, which finally caused the bug to become dehydrated and die (Ibrahim et al., 2019). Nanostructured alumina's toxicity and method of action against *S. oryzae* were reported by Stadler et al. (2017). According to the investigations, charged nanostructured alumina adhered to the cuticle of the beetle due to triboelectric forces, which caused the insect to become dehydrated by sorbing away its wax covering owing to surface area events. Table 4.4 represents the exploration of nanoparticles for the development of NAIPs.

TABLE 4.4 Recent Developments of Nanoparticles in Pest Management.

Nano-Particles	Pesticides	Target Pests	References
Chitosan nanoparticles	Spinosad and permethrin	*Drosophila melanogaster*	Sharma et al. (2019)
Chitosan/Gum Arabic nanoparticles	Geraniol	Whitefly (*Bemisiatabaci*)	De Oliveira et al. (2018)
Zein nanoparticles	Eugenol and cinnamaldehyde	Spider mite (*Tetranychus urticae*) and soybean looper (*Chrysodeixis includes*)	de Oliveira et al. (2019)
Metal nanoparticles	Ag and Ag-Zn	Aphid	Rouhani et al. (2012)
Chitosan nanoparticles	Peppermint essential oil	Stored-grain pest	Rajkumar et al. (2020)
Silica nanoparticles	Indoxacarb	*Plutella xylostella*	Bilal et al. (2020)
Silica nanoparticles	Buprofezin	*Nilaparvata lugens*	Yang et al. (2021)
Metal nanoparticle	Nickel-oxide	Pulse beetle	Rahman et al. (2021)
Pungamoil-based Metal nanoparticle	Ag and Au	*Pericallia ricini*	Sahayaraj et al. (2016)
Copper nanoparticles	Copper	*Tribolium castaneum*	El-Saadony et al. (2020)

4.1.10 FUTURE ASPECTS OF NANO-TECHNOLOGY IN PEST MANAGEMENT

The research on nanotechnology in agriculture is anticipated to increase its precision and sustainability, given the enormous advances made in the field of medicine. Although nanotechnology in agriculture is still in its infancy, there is a lot of room for the development of precise nanoagrochemicals with regard to their containment in appropriate carrier materials, fabrication techniques, mode of application, and, most importantly, the assessment of the safety and toxicological effects of the nanoproducts on the plant-body and the end agricultural products.

ACKNOWLEDGMENTS

Biorender software is used for image preparation. The images are made using information from literature but with slight modifications.

KEYWORDS

- **Nano-agri-input-products (NAIPs)**
- **nanoformulations**
- **nanomaterials**
- **pest management**
- **quality assessment tools**
- **sustainable agriculture**

REFERENCES

Abd-Elsalam, K. A., & Khokhlov, A. R. (2015). Eugenol oil nanoemulsion: antifungal activity against *Fusarium oxysporum* f. sp. *vasinfectum* and phytotoxicity on cottonseeds. *Applied Nanoscience,* 5, 255–265.

Abdou, E. S., Galhoum, G. F., & Mohamed E. N. (2018). Curcumin loaded nanoemulsions/pectin coatings for refrigerated chicken fillets. *Food Hydrocolloids, 83*, 445–453.

Ahmed, A., He, P., He, P., Wu, Y., He, Y. & Munir, S. (2023). Environmental effect of agriculture-related manufactured nano-objects on soil microbial communities. *Environment International, 173*, 107819.

Ahsaei, S. M., Rodríguez-Rojo, S., Salgado, M., Cocero, M. J., Talebi-Jahromi, K., & Amoabediny, G. (2020). Insecticidal activity of spray dried microencapsulated essential oils of *Rosmarinusofficinalis* and *Zatariamultiflora* against *Triboliumconfusum*. *Crop Protection, 128*, 104996.

Ali, E. O. M., Shakil, N. A., Rana, V. N., Sarkar, D. J., Majumder, S., Kaushik, P., Braj Singh, B. B., & Kumar, J. (2017). Antifungal activity of nano emulsions of neem and citronella oils against phytopathogenic fungi, *Rhizoctoniasolani* and *Sclerotiumrolfsii*. *Industrial Crops and Products, 108*, 379–387.

Alzogaray, R. A., & Zerba, E. N. (2017). *Rhodniusprolixus* intoxicated. *Journal of Insect Physiology, 97*, 93–113.

Anton, N., Akram, S., Thierry, & Vandamme, F. (2018). Transitional nanoemulsification methods. In S. M. Jafari, D. J. McClements (Eds.), *Nanoemulsions* (pp. 77–100). Academic Press.

Anton, N., Benoit, J. P., & Saulnier, P. (2008). Design and production of nanoparticles formulated from nano-emulsion templates: A review. *Journal of Controlled Release, 128*, 185–199.

Anton, N., & Vandamme, T. F. (2011). Nano-emulsions and micro-emulsions: clarifications of the critical differences. *Pharmaceutical Research, 28*, 978–985.

APVMA Guidelines (2005). Guidelines for the generation of storage stability data of agricultural chemical products, Kingston ACT Australia 2604, December 2005.

Arumugam, G., Velayutham, V., Shanmugavel, S., & Sundaram, J. (2016). Efficacy of nanostructured silica as a stored pulse protector against the infestation of bruchid beetle, *Callosobruchus maculatus* (Coleoptera: Bruchidae). *Applied Nanoscience, 6*, 445–450.

Athanassiou, C. G., Kavallieratos, N. G., Benelli, G., Losic, D., Usha Rani, P., & Desneux, N. (2018). Nanoparticles for pest control: current status and future perspectives. *Journal of Pest Science, 91*, 1–15.

Attia, M. F., Dieng, S. M., Collot, M., Klymchenko, A. S., Bouillot, C., Serra, C. A., Schmutz, M., Er-Rafik, M., Vandamme, T. F., & Anton, N. (2017). Functionalizing nanoemulsions with carboxylates: Impact on the biodistribution and pharmacokinetics in mice. *Macromolecular Bioscience, 17*, 1600471.

Awasthi, G., Singh, T., Tiwari, Y., Awasthi, A., Tripathi, R. D., Srivastava S., Vajpayee, P., Kumar, A., & Awasthi, K. K. (2020). A review on nanotechnological intervention on plant growth and production. *Materials Today: Proceedings*. 31, 685-693.

Bae, M., Lewis, A., Liu, S., Arcot, Y., Lin, Y. L., Bernal, J. S., Cisneros-Zevallos, L., & Akbulut, M. (2022). Novel biopesticides based on nanoencapsulation of azadirachtin with whey protein to control fall armyworm. *Journal of Agricultural and Food Chemistry, 70*, 7900–7910.

Bai, L., & McClements, D. J. (2016). Development of microfluidization methods for efficient production of concentrated nanoemulsions: comparison of single- and dual-channel microfluidizers. *Journal of Colloid and Interface Science, 466*, 206–212.

Balaji, A. P. B., Mishra, P., Suresh Kumar, R. S., Mukherjee, A., & Chandrasekaran, N. (2015). Nanoformulation of poly (ethylene glycol) polymerized organic insect repellent by PIT emulsification method and its application for Japanese encephalitis vector control. *Colloids and Surfaces B: Biointerfaces, 128*, 370–378.

Benelli, G., Maggi, F., Pavela, R., Murugan, K., Govindarajan, M., Vaseeharan, B., Petrelli, R., Cappellacci, L., Kumar, S., Hofer, A., & Youssefi, M. R. (2018). Mosquito control

with green nanopesticides: towards the One Health approach? A review of non-target effects. *Environmental Science and Pollution Research, 25*, 10184–10206.
Bilal, M., Xu, C., Cao, L., Zhao, P., Cao, C., Li, F., & Huang, Q. (2020). Indoxacarb-loaded fluorescent mesoporous silica nanoparticles for effective control of *Plutellaxylostella* L. with decreased detoxification enzymes activities. *Pest Management Science, 76*, 3749–3758.
Blewett, T. A., Qi, A. A., Zhang, Y., Weinrauch, A. M., Blair, S. D., Folkerts, E. J., Sheddy, C., Nilsson, D., & Goss, G. G. (2019). Toxicity of nanoencapsulated bifenthrin to rainbow trout (*Oncorhynchus mykiss*)'. *Environmental Science: Nano, 6*, 2777–2785.
Boff, S., Conrad, T., Raizer, J., Wehrhahn, M., Bayer, M., Friedel, A., Theodorou, P., Schmitt, T., & Lupi, D. (2022). Low toxicity crop fungicide (fenbuconazole) impacts reproductive male quality signals leading to a reduction of mating success in a wild solitary bee. *Journal of Applied Ecology, 59*, 1596–1607.
Chen, F., Liang, L., Zhang, Z., Deng., Decker, E. A., & McClements, D. J. (2017). Inhibition of lipid oxidation in nanoemulsions and filled microgels fortified with omega-3 fatty acids using casein as a natural antioxidant. *Food Hydrocolloids, 63*, 240–248.
Chin, C. P., Wu, H. S., & Wang, S. S. (2011). New approach to pesticide delivery using nanosuspensions: Research and applications. *Industrial & Engineering Chemistry Research, 50*, 7637–7643.
Christofoli, M., Costa, E. C. C., Bicalho, K. U., Domingues, V. D. C., Márcio Fernandes Peixoto, M. F., Alves., Araújo, W. L., & Cazal, C. D. L. (2015). Insecticidal effect of nanoencapsulated essential oils from *Zanthoxylumrhoifolium* (Rutaceae) in Bemisiatabaci populations. *Industrial Crops and Products, 70*, 301–308.
Cui, B., Gao, F., Zeng, Z., Wang, C., Wang, Y., Sun, C., Zhao, X., Guo, L., Shen, Y., Liu, G., & Cui, H. (2020). Construction and characterization of avermectin B_2 solid nanodispersion. *Scientific Reports, 10*, 996.
Cui, B., Lv, Y., Gao, F., Wang, C., Zeng, Z., Wang, Y., Sun, C., Zhao, X., Shen, Y., Liu G., & Cui, H. (2019). Improving abamectin bioavailability via nanosuspension constructed by wet milling technique. *Pest Management Science, 75*, 2756–2764.
DBT Guidelines (2020). Guidelines for the evaluation of nano-based agri-input and food products in India, Department of Biotechnology, Govt. of India.
de Oliveira, E., Cecchin, D., Miyagaki, D. C., de Moura, A. L. C., Disarz, A., Souza, M. A., & Farina, A. P. (2019). Effect of different protocols of eugenol removal on the bond strength between the fibre post and root dentin. *Australian Endodontic Journal, 45*, 177–183.
de Oliveira, J. L., Campos, E. V. R., Pereira, A. E. S., Nunes, L. E. S., da Silva, C. C. L. Pasquoto, T., Lima, R., Smaniotto, G., Polanczyk, R. L., & Fraceto, L. F. (2018). Geraniol encapsulated in chitosan/gum arabic nanoparticles: a promising system for pest management in sustainable agriculture. *Journal of Agricultural and Food Chemistry, 66*, 5325–5334.
Dijk, M. V., Morley, T., Rau, M. L., & Saghai, Y. (2021). A meta-analysis of projected global food demand and the population at risk of hunger for the period of 2010–2050. *Nature Food, 2*, 494–501.
Ebadollahi, A., Valizadeh, B., Panahandeh, S., Mirhosseini, H., Zolfaghari, M., & Changbunjong, T. (2022). Nanoencapsulation of Acetamiprid by sodium alginate and polyethylene glycol enhanced its insecticidal efficiency. *Nanomaterials, 12*, 2971.
EFSA Scientific Committee (2011). Scientific opinion on guidance on the risk assessment of the application of nanoscience and nanotechnologies in the food and feed chain. *EFSA Journal, 9*(5), 2140.

El-Saadony, M. T., Abd El-Hack, M. E., Taha, A. E., Fouda, M. M. G., Ajarem, J. S., N. Maodaa, S., Allam, A. A., & Elshaer, N. (2020). Ecofriendly Synthesis and Insecticidal Application of Copper Nanoparticles against the Storage Pest *Triboliumcastaneum*. *Nanomaterials, 10*, 587.

Emamjomeh, L., Imani, S., Jahromi K. T., & Moharramipour, S. (2023). Nanoencapsulation enhances the contact toxicity of *Eucalyptus globulus* Labill and *Zataria multiflora* Boiss essential oils against the third instar larvae of *Ephestiakuehniella* (Lepidoptera: Pyralidae), *International Journal of Pest Management, 69*, 207–214.

FDA (2015). Guidance for industry: Use of nanomaterials in food for animals. https://www.fda.gov/downloads/AnimalVeterinary/GuidanceComplianceEnforcement/GuidanceforIndustry/UCM401508.pdf (accessed October 21, 2019).

Feng, X., Tjia, J. Y. Y., Zhou, Y., Liu, Q., Fu, C., & Yang, H. (2020). Effects of tocopherol nanoemulsion addition on fish sausage properties and fatty acid oxidation. *LWT, 118*, 108737.

Fernandes, C. P., de Almeida, F. B., Silveira, A. N., Gonzalez, M. S., Mello, C. B., Feder, D., Apolinário, R., Santos, M. G., Carvalho, J. C. T., Tietbohl, L. A. C., & Rocha, L. (2014). Development of an insecticidal nanoemulsion with *Manilkara subsericea* (Sapotaceae) extract. *Journal of Nanobiotechnology, 12*, 22.

Fraceto, L. F., Grillo, R., de Medeiros, G. A., Scognamigilo, V., Rea, G., Bartolucci, C., Ria, G., & Bartolucci, C. (2016). Nanotechnology in agriculture: which innovation potential does it have? *Frontiers in Environmental Science, 4*, 186737.

Ganta, S., Singh, A., Rawal, Y., Cacaccio, J., Patel, N. R., Kulkarni, P., Ferris, C. F., Amiji, M. M., & Coleman, T. P. (2016). Formulation development of a novel targeted theranosticnanoemulsion of docetaxel to overcome multidrug resistance in ovarian cancer. *Drug Delivery, 23*, 968–980.

Geng, T. Y., Xu, C., & Xu, G. H. (2016). Albumin conjugated lipid nanoemulsion for site specific delivery of rapamycin at inflammatory site of spinal cord injury. *International Journal of Clinical and Experimental Medicine, 9*, 21028–21037.

Gomes, S. I. L., Scott-Fordsmand, J. J., Campos, E. V. R., Grillo, R., Fraceto, L. F., & Amorim, M. J. B. (2019). On the safety of nanoformulations to non-target soil invertebrates – an atrazine case study. *Environmental Science: Nano, 6*, 1950–1958.

Grillo, R., Anderson E. S., Pereira, Nishisaka, C. S., de Lima, R., Oehlke, K., Greiner, R., & Fraceto, L. F. (2014). Chitosan/tripolyphosphate nanoparticles loaded with paraquat herbicide: An environmentally safer alternative for weed control, *Journal of Hazardous Materials, 278*, 163–171.

Gruère, G. P. (2012). Implications of nanotechnology growth in food and agriculture in OECD countries. *Food Policy, 37*, 191–198.

Gupta, A., Eral, H. B., Hatton, T. A., & Doyle, P. S. (2016). Nanoemulsions: formation, properties and applications. *Soft Matter, 12*, 2826–2841.

Gupta, A., Eral H. B., Alan Hatton, T., & Doyle, P. S. (2016). Nanoemulsions: formation, properties and applications.*Soft Matter, 12,* 2826–2841.

Gupta, R., Saxena, T., Mehra, N., Arora, R., & Sahgal, A. (2023). Nanopesticides: Promising Future in Sustainable Pest Management. *Journal of Advanced Applied Scientific Research, 5,* 22–37.

Gutierrez, J. M., Gonzalez, C., Maestro, A., Sole, I., Pey, C. M., & Nolla, J. (2008). Nano-emulsions: New applications and optimization of their preparation. *Current Opinion in Colloid & Interface Science, 13*, 245–251.

Hashem, A. S., Awadalla, S. S., Zayed, G. M., Maggi, F., & Benelli, G. (2018). *Pimpinella anisum* essential oil nanoemulsions against *Triboliumcastaneum*—insecticidal activity and mode of action. *Environmental Science and Pollution Research, 25*, 18802–18812.

Hayles, J., Johnson, L., Worthley, C., & Losic, D. (2017). Nanopesticides: A review of current research and perspectives. *New Pesticides and Soil Sensors, 2017*, 193–225.

Hazafa, A., Jahan, N., Zia, M. A., Rahman, K. U., Sagheer, M., & Naeem, M. (2022). Evaluation and optimization of nanosuspensions of *Chrysanthemumcoronarium* and *Azadirachtaindica* using Response Surface Methodology for pest management. *Chemosphere, 292*, 133411.

Heydari, M., Amirjani, A., Bagheri, M., Sharifian, I., & Sabahi Q. (2020). Eco-friendly pesticide based on peppermint essential oil nanoemulsion: preparation, physiochemical properties, and its aphicidal activity against cotton aphids. *Environmental Science and Pollution Research International, 27*, 6667–6679.

Ibrahim, S. S., & Salem, N. Y. (2019). Insecticidal Efficacy of Nano Zeolite against *Triboliumconfusum* (Col., Tenebrionidae) and *Callosobruchus maculatus* (Col., Bruchidae). *Bullentin of the National Research Centre, 43*, 1–8.

Kaci, M., Belhaffef, A., Meziane, S., Dostert, G., Menu, P., Velot, E., Desobry, S., & Arab-Tehrany, E. (2018). Nanoemulsions and topical creams for the safe and effective delivery of lipophilic antioxidant coenzyme Q10. *Colloids Surface B: Biointerfaces, 167*, 165–175.

Kah, M., Machinski, P., Koerner, P., Tiede, K., Grillo, R., Fraceto, L. F., & Hofmann, T. (2014). Analysing the fate of nanopesticides in soil and the applicability of regulatory protocols using a polymer-based nanoformulation of atrazine. *Environmental Science and Pollution Research, 21*, 11699–11707.

Kala, S., Sogan, N., Naik, S. N., Agarwal, A., & Kumar, J. (2020). Impregnation of pectin-cedarwood essential oil nanocapsules onto mini cotton bag improves larvicidal performances. *Scientific Reports, 10*, 14107.

Knauer, K., & Bucheli, T. D. (2009). Nano-materials: research needs in agriculture. *Revue Suisse d Agriculture, 41*, 341–345.

Komaiko, J., & McClements, D. J. (2015). Low-energy formation of edible nanoemulsions by spontaneous emulsification: factors influencing particle size. *Journal of Food Engineering, 146*, 122–128.

Kookana, R. S., Boxall, A. B., Reeves, P. T., Ashauer, R., Beulke, S., Chaudhry, Q., Cornelis, G., Fernandes, T. F., Gan, J., Kah, M., & Lynch, I. (2014). Nanopesticides: Guiding principles for regulatory evaluation of environmental risks. *Journal of Agricultural and Food Chemistry, 62*, 4227–4240.

Kumari, S., Kumaraswamy, R. V., Choudhary, R. C., Sharma, S. S., Pal, A., Raliya, R., Biswas, P., & Saharan, V. (2018). Thymol nanoemulsion exhibits potential antibacterial activity against bacterial pustule disease and growth promotory effect on soybean. *Scientific Reports, 8*, 6650.

Li, K; Zhao, Y., Yang, J., & Gu, J. (2022). Nanoemulsion-directed growth of MOFs with versatile architectures for the heterogeneous regeneration of coenzymes. *Nature Communications, 13*, 1879.

Lim, C. J., Basri, M., Omar, D., Abdul Rahman, M. B., Salleh, A. B., & Abdul Rahman, R. N. Z. (2013). Green nanoemulsion-laden glyphosate isopropylamine formulation in suppressing creeping foxglove (*A. gangetica*), slender buttonweed (*D. ocimifolia*) and buffalo grass (*P. conjugatum*). *Pest Managemnet Science, 69*, 104–111.

Liu, Y., Yu, X. M., Sun, R. J., & Pan, X. L. (2017). Folate-functionalized lipid nanoemulsion to deliver chemo-radiotherapeutics together for the effective treatment of nasopharyngeal carcinoma. *AAPS PharmSciTech, 18*, 1374–1381.

Madhavi, V., Vijaya A., Reddy, B., Madhavi, G., &Reddy, N. B. (2020). Nanoencapsulation of pesticides: sustainable perspective in agriculture. *AIP Conference Proceedings* 4 November 2280, 040029.

Marziyeh, C., Dzolkhifli, O., Mahiran, B., & Norhayu,A. (2017). Effect of nanoemulsion formulation of neem oil against red beetle weevil and flour beetle adult. *Journal of Pest Science, 42*, 158–165.

McClements, D. J., & Gumus, C. E. (2016). Natural emulsifiers – Biosurfactants, phospholipids, biopolymers, and colloidal particles: molecular and physicochemical basis of functional performance. *Advances in Colloid and Interface Science, 234*, 3–26.

Mourdikoudis, S., Pallares, R. M., & Thanh, N. T. K. (2018). Characterization techniques for nanoparticles: comparison and complementary upon studying nanoparticle properties. *Nanoscale, 10*, 12821–12954.

Nair, P. M. G., & Choi, J. (2011). Identification, characterization and expression profiles of *Chironomus riparius* glutathione S-transferase (GST) genes in response to cadmium and silver nanoparticles exposure. *Aquatic Toxicology 101*, 550–560.

Neme, K., Nafady, A., Uddin S., & Tola, Y. B. (2012). Application of nanotechnology in agriculture, post-harvest loss reduction and food processing: food security implication and challenges. *Heliyon, 7*, e08539.

Ntalli, N., Zochios, G., Nikolaou, P., Winkiel, M., Petrelli, R., Bonacucina, G., Perinelli, D. R., Spinozzi, E., Maggi, F., & Benelli, G. (2023).*Carlinaacaulis* essential oil nanoemulsion for managing Meloidogyne incognita. *Industrial Crops and Products, 193*, 116180.

Nuruzzaman, M., Rahman, M. M., Liu, Y., & Naidu, R. (2016). Nanoencapsulation, nano-guard for pesticides: a new window for safe application. *Journal of Agricultural and Food Chemistry, 64*, 1447–1483.

OECD (2017). Test No. 318: Dispersion stability of nanomaterials in simulated environmental media, OECD guidelines for the testing of chemicals, Section 3, OECD Publishing, Paris.

Ozturk, B., Argin, S., Ozilgen, M., & McClements, D. J. (2014). Formation and stabilization of nanoemulsion-based vitamin E delivery systems using natural surfactants: Quillaja saponin and lecithin. *Journal of Food Engineering, 142*, 57–63.

Pandey, C., Sethi, P. K., Kumari, S., Vishwakarma, J, & Tande, V. (2022). Smart Agriculture: Technological Advancement on Agriculture-Asystematical review. In *Deep learning in sustainable agriculture*. Elsevier.

Patil, C. D., Borase, H., Suryawanshi, R. K., & Patil, S. V. (2016). Trypsin inactivation by latex fabricated gold nanoparticles: A new strategy towards insect control. *Enzyme and Microbial Technology, 92*, 18–25.

Pérez-de-Luque. A., & Hermosín, M. C. (2013). Nanotechnology and its use in agriculture. In D. Bagchi, M. Bagchi, H. Moriyama, F. Shahidi (Eds.), *Bio-nanotechnology: A Revolution in Food, Biomedical and Health Sciences* (pp. 299–405). Wiley-Blackwell.

Prasad, R., Kumar, V., & Prasad, K. S. (2017). Nanotechnology in sustainable agriculture: Present concerns and future aspects. *African Journal of Biotechnology, 13*, 705–713.

Quinlan, M., Stroup, W., Schwenke, J., & Christopher (2013). Evaluating the performance of the ICH guidelines for shelf-life estimation. *Journal of Biopharmaceutical Statistics, 23*, 881–896.

Rahman, M. R., Parvin, A., Khan, M. J. H., War A. R., Lingaraju, K., Prasad, R., Das, S., Hussain, B., & Bhattacharyya, A. (2021). Efficacy of the green synthesized nickel-oxide nanoparticles against pulse beetle, *Callosobruchus maculatus* (F.) in black gram (*Vigna mungo* L.). *International Journal of Pest Management, 67*, 306–314.

Rahwanudin, I., Susanto, A., Panatarani, C., Zainuddin, A., & Setiawati, W. (2022). Study of spinetoram nano suspension for environmentally friendly control of *Plutellaxylostella* L. (Lepidoptera: Plutellidae). *Biocatalysis and Agricultural Biotechnology, 44*, 102456.

Rai, M., Ribeiro, C., Mattoso, L., & Duran N. (2015). Nanotechnologies in Food and Agriculture. Heidelberg: Springer.

Rai, A. B., Halder, J., & Kodandaram, M. H. (2014). Emerging insect pest problems in vegetable crops and their management in India: An appraisal. *Pest Management in Horticultural Ecosystems, 20*, 113–122.

Rajkumar, V., Gunasekaran, C., Dharmaraj, J. Chinnaraj, P., Paul, C. A., & Kanithachristy, I. (2020). Structural characterization of chitosan nanoparticle loaded with Piper nigrum essential oil for biological efficacy against the stored grain pest control. *Pesticide Biochemistry and Physiology, 166*, 104566.

Raviadaran, R., Ng, M. H., Manickam, S., & Chandran, D. (2020). Ultrasound-assisted production of palm oil-based isotonic W/O/W multiple nanoemulsion encapsulating both hydrophobic tocotrienols and hydrophilic caffeic acid with enhanced stability using oil-based Sucragel. *Ultrasonics Sonochemistry, 64*, 104995.

REACH Regulation (2019). Registration, Evaluation, Authorisation and Restriction of Chemicals Regulation (EC) No.1907/2006. https://osha.europa.eu/en/legislation/directives/regulation-ec-no-1907-2006-of-the European-parliament-and-of-the-council (accessed October 21, 2019).

Rouhani, M., Samih, M. A., & Kalantari, S. (2012). Insecticide effect of silver and zinc nanoparticles against *Aphisnerii* Boyer of fonscolombe (*Hemiptera*: Aphididae). *Chilean Journal of Agricultural Research, 72*, 590–594.

Sabry, A. H., & Ragaei, M. (2018). Nanotechnology and their applications in insect's pest control. In K. A. Abd-Elsalam, R. Prasad (Eds.), *Nanobiotechnology applications in plant protection, nanotechnology in the life sciences* (pp. 1–28). Springer International Publishing.

Sahayaraj, K., Madasamy M., & Radhika, S. A. (2016). Insecticidal activity of bio-silver and gold nanoparticles against *Pericalliaricini* Fab. (Lepidaptera: Archidae, Bionanoparticles for pest management. *Journal of Biopesticides, 9*, 63–72.

Saini, P., & Gopal, M. (2012). Bioefficacy of Pyridalyl Nanosuspension and its Commercial Formulation against *Helicoverpaarmigera* (Lepidoptera: Noctuidae). *The Journal of Rural and Agricultural Research, 21*, 80–83.

Sharma, A., Sharma, N. K., Srivastava, A., Kataria, A., Dubey, S., Sharma, S., & Kundu, B. (2018). Clove and lemongrass oil based non-ionic nanoemulsion for suppressing the growth of plant pathogenic *Fusariumoxysporum*f.sp. lycopersici. *Industrial Crops and Products, 123*, 353–362.

Sharma, A., Sood, K., Kaur, J., & Khatri, M. (2019). Agrochemical loaded biocompatible chitosan nanoparticles for insect pest management. *Biocatalysis and Agricultural Biotechnology, 18*, 101079.

Silva-Pereira, M. C., Teixeira, J. A., Pereira-Júnior, Ricardo, V. A., & Stefani, R. (2015). Chitosan/corn starch blend films with extract from *Brassica oleraceae* (red cabbage) as a visual indicator of fish deterioration. *LWT-Food Science and Technology, 61*, 258–262.

Singh, Y., Meher, J. G., Raval, K., Khan, F. A., Chaurasia, M., Jain, N. K., & Chourasia, M. K. (2017). Nanoemulsion: Concepts, development and applications in drug delivery. *Journal of Controlled Release, 252*, 28–49.

Small, T., Zapater, M. A. O., Gallello,G., Ribera, A., Romero, F. M., Torreblanca, A., & Garcera, M. D. (2016). Gold-nanoparticles ingestion disrupts reproduction and development in the German cockroach. *Science of the Total Environment, 565*, 882–888.

Sonkaria, S., Ahn, S. H., & Khare, V. (2012). Nanotechnology and its impact on food and nutrition. *Recent Patents on Food, Nutrition & Agriculture, 4*, 8–18.

Stadler, T., López, G., Guillermo, P., Gitto, J. G, & Butler, M. (2017). Nanostructured alumina: biocidal properties and mechanism of action of a novel insecticide powder. *Bulletin of Insectology, 70*, 17–25.

Tadros, T., Izquierdo, P., Esquena, J., & Solans, C. (2004). Formation and stability of nano-emulsions. *Advances in Colloid and Interface Science, 108–109*, 303–318.

Totan, A., Barik, N., Patil, N. B., Govindharaj, G. P. P., Gadratagi, B. G., Mahendiran, A., Mukherjee, A. K., Rath, & P. C. (2020). Nanoemulsion of eucalyptus oil: An alternative to synthetic pesticides against two major storage insects (*Sitophilusoryzae* (L.) and *Triboliumcastaneum* (Herbst)) of rice. *Industrial Crops and Products, 143*, 111849.

TSCA§3(9),15U.S.C.§2602(9). EPA's regulatory definition of a "new chemical substance" tracks the statutory definition. https://www.epa.gov/tsca-inventory (accessed October 21, 2019).

Verma, P., Meher, J. G., Asthana, S., Pawar, V. K., Chaurasia, M., & Chourasia, M. K. (2016). Perspectives of nanoemulsion assisted oral delivery of docetaxel for improved chemotherapy of cancer. *Drug Delivery, 23*, 479–488.

Wang, Y., Song, S., Chu, X., Feng, W., Li, J., Huang, X., Zhou, N., & Shen, J. (2021). A new temperature-responsive controlled-release pesticide formulation – poly(N-isopropylacrylamide) modified graphene oxide as the nanocarrier for lambda-cyhalothrin delivery and their application in pesticide transportation. *Colloids and Surfaces A-Physicochemical and Engineering Aspects, 612*, 125987.

Wei, N., Hou, C., Liu, Z., Liang, Q., Lv, Z., Meng, X., & Feng, J. (2023). Preparation of fenpropathrinnanoemulsion for eco-friendly management of *Helicoverpaarmigera:* improved insecticidal activity and biocompatibility. *Colloids and Surfaces A-Physicochemical and Engineering Aspects,* 656, 130442.

Yan, H., Bao, C., Chen, X., Yu, C., Kong, D., Shi, J., & Lin, Q. (2019). Preparation of biodiesel oil-in-water nanoemulsions by mixed surfactants for bifenthrin formulation. *RSC Advances, 9*, 11649–11658.

Yang, J., Feng, J., He, K., Chen, Z., Chen, W., Cao, H., & Yuan, S. (2021). Preparation of thermosensitive buprofezin-loaded mesoporous silica nanoparticles by the sol–gel method and their application in pest control. *Pest Management Science, 77*, 4627–4637.

Zhao, C. H., Shen, X., Guo, & M. R. (2018). Stability of lutein encapsulated whey protein nano-emulsion during storage. *PLOS One, 13*, e0192511.

Zook, J. M., Macusspie, R. I., Locascio, L., Halter, M. D., & Elliot J. (2010). Stable nanoparticle aggregates/agglomerates of different sizes and the effect of their size on hemolytic cytotoxicity. *Nanotoxicology, 5*, 517–530.

CHAPTER 5

Global Regulatory Framework for Nano-Agri-Input Products for Commercialization

ARKADEB MUKHOPADHYAY

Division of Agricultural Chemicals, ICAR-Indian Agricultural Research Institute, New Delhi, India

ABSTRACT

Several pieces of research on nanotechnology applications in the agriculture sector have been carried out in the past two decades to achieve food security as well as agricultural sustainability. Nano-enabled input products have gained much attention as compared to conventional input products due to their higher bioefficacy, lower dose, higher input use efficiency, sustained release to the target site, and generation of fewer waste products. Therefore, the importance of nanotechnology in agriculture is imperative for crop production as well as for crop protection. Nano-Agri-Input Products (NAIPs) may include active ingredients in the nano-range or nanocarrier for sustained and/or targeted delivery of pesticides and fertilizers. Nano-enabled products have exerted several toxicological effects, such as phytotoxicity, genotoxicity, and ecotoxicity, after application. Different guidelines are available for the evaluation of safety aspects of nano-enabled agricultural inputs, including the safety assessment of nontarget organisms and the environment. To date, several countries have developed regulatory guidelines for NAIPs for large-scale application in agriculture. However, the lack of a uniform legislative framework across the globe and even the nondevelopment of guidelines for the regulation of nano-enabled input products for agricultural use in many countries are

Nanobiotechnology for Agricultural Sciences: Nano-Agri-Input Products for Crop Production and Environmental Protection. J. C. Tarafdar, Deepak Kumar, Sanjay K. Singh, and Krishna Pal Singh (Eds.)

major bottlenecks in the large-scale acceptance and commercialization of these nano-agri-input products. Therefore, a spotlight in the form of this chapter has been shone upon the present regulatory frameworks and safety assessment guidelines of different countries for Nano-Agri-Input products, along with an address to the present limitations and future opportunities hidden underneath them. The present chapter is expected to help policymakers prepare and/or fine-tune the legislative frameworks for the development, commercialization, and application of nano-based agri-input products with greater societal acceptance and environmental sustainability in the near future.

5.1 INTRODUCTION

The total agricultural production will need to rise by 60% over 2005 levels to accommodate the predicted world population of 9.7 billion people in 2050 (Kah et al., 2018). From a crop production and protection perspective, agrochemicals are an inescapable part of agriculture. Agrochemical use is obligatory in crop protection since diverse pests cause crop losses of around 15–20% each year (Rathee and Dalal, 2018). Fertilizers are also considered an indispensable input for crop production to supply the crop nutrition demand (Jaga and Patel, 2012). However, through drift hazards, run-off, or leaching, a large amount of the applied agri-inputs is lost into the soil, air, or water bodies (Ghormade et al., 2011). The indiscriminate use of these agri-inputs may also lead to health risks, residual issues, environmental degradation, loss of biodiversity, and pesticide resistance in plants (Sarkar et al., 2021). The development of formulations with controlled or gradual release properties is crucial to addressing these issues. These kinds of formulations can provide targeted as well as sustained delivery in a way that is both ecologically responsible and practical from an economic standpoint. A more environmentally friendly approach for creating intelligent agri-input delivery systems is nanotechnology (Sekhon, 2014; Sarkar et al., 2021). Pest control, nutrient supply, preparation of controlled release formulations, detection of agrochemicals, and other applications of nanotechnology are already in use (Elizabath et al., 2019; Majumdar and Keller, 2021; Pramanik et al., 2020). The utilization of nanomaterials as a delivery tool for agri-inputs has several advantages, including enhanced systemic properties due to advanced mobility, decreased toxicity of pesticide residues due to the reduction of chemical solvents generally used in conventional crop protection chemical formulations, and improvements in input use efficiency due to a larger surface area (Kundu et al., 2021). Most of the nanopesticides and nanofertilizers that have been presented thus far reformulate registered active ingredients to outperform the existing AIs and mitigate the primary shortcomings of the present conventional agrochemical formulations (Gogos et al., 2012).

These nano-agri-input products (NAIPs) cannot be solely judged as benefits, though. According to recent findings, the environment is being loaded with nanoparticles (NPs) that are toxic to its inhabitants and different environmental compartments (Kolesniko et al., 2019). The ecosystems nearby experienced a rapid spread and accumulation of these nano-sized components. These materials' extremely minute structure, reactivity, and mobility may make them bioavailable and produce diverse harmful effects (Sengul and Asmatulu, 2020). The majority of the toxicities associated with nanoparticles have a long-term nature (Paul, 2021). There are many criteria available for evaluating the safety of nano-enabled agricultural inputs, including safety assessments toward nontarget organisms and the environment (Hoseini-Alfatemi et al., 2020; Tkachenko et al., 2021). The "European Union Scientific Committees and Agencies," the "International Organisation for Standardisation" (ISO), the "Organisation for Economic Cooperation and Development" (OECD), and the "United States Food and Drug Administration" (US FDA) are some of the expert organizations involved in this regard. To date, several nations have created regulatory standards for NAIPs for use on a broad scale in agriculture. However, a major barrier to the widespread adoption and commercialization of these nano-agri-input products is the lack of societal acceptance, the absence of a unified legal framework around the world, and even the nonestablishment of guidelines for the regulation of nano-enabled input products for agricultural use in many countries. Therefore, the current regulatory frameworks and safety evaluation criteria of various nations for nano-agri-input products have been highlighted in this chapter, along with an analysis of their current drawbacks and untapped potential in the future.

5.2 DEFINITION OF NANOMATERIALS AND NANO-AGRI-INPUT PRODUCTS

Particle size is the major criterion for defining a nanomaterial (NM) as per the European Union (EU). The EU guidelines describe any material to be regarded as NM "if 50% or more of the constituent particles have a particle size distribution of one or more dimensions in the range of 1–100 nm." In some specific cases, the threshold limit can be lowered to 1–50%. Moreover, any material having a volume-specific surface area greater than the nominal value of 60 m^2 cm^{-3} should also be considered an NM (Rauscher et al., 2017). According to ISO, "any material having an external dimension or internal structure/ surface structure in the nanoscale (approximately 1–100 nm) is considered a nanomaterial" (ISO, 2015). In other embodiments, the NMs have been described as particles having at least one dimension in the range of 1–100 nm. However, these embodiments consider particles having a size range up to 1000 nm as NMs if they possess nanoscale properties (Subramanian and Rajkishore, 2018).

"Agri-input products in the nanoform of finished formulations and active ingredients of new material or an already approved material with altered beneficial properties, dimensions, or phenomena associated with the application of nanotechnology that are intended for use in agriculture and related sectors such as crop management, production, protection, harvesting, post-harvesting, and packaging" are termed as nano-agri-input products. The uses may include, but are not limited to, biostimulants for agricultural benefits, antitranspirants, agrochemical delivery, and plant growth regulators (Government of India, 2020).

5.3 ENVIRONMENTAL FATE OF NAIPS

Environmental fate and persistence behavior are important factors to take into account when evaluating the risk of nano-enabled agri-inputs. For various environmental compartments, currently a multitude of fate and transport models are available, including FOCUS models like PEARL, MACRO, PRZM, and so on. The schematic representation of the environmental fate of NAIPs is shown in Figure 5.1. Different physicochemical and environmental factors influence the environmental fate and persistency behavior of NAIPs. The matrix pH, salinity, ionic strength, presence of organic matter, and microbes are some of the considerable factors in the aqueous compartment. For the soil compartment, different factors such as porosity of the soil matrix, soil type, mineral composition, water flow, pH, flora and faunal community, soil temperature, organic matter, and electrolytes are considered (Grillo et al., 2021).

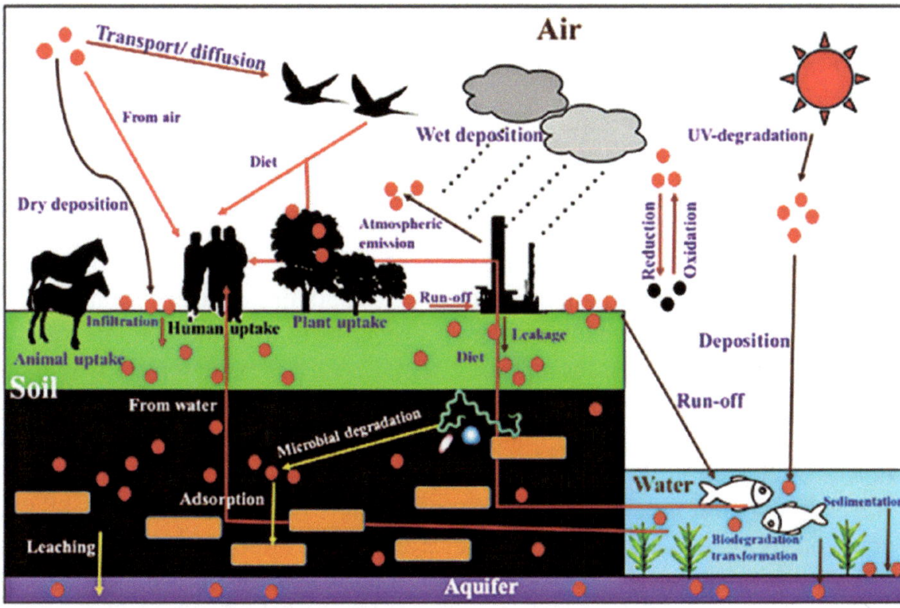

FIGURE 5.1 Environmental fate of NAIPs.

A change in the particle properties may also be due to the interaction of NMs with their surrounding environmental compartments and the sorption of different moieties onto the surface of the nanoparticle. The surface-binding properties of natural organic materials, particularly humic acids, are significant in this regard. It has also been demonstrated that humic acid adsorption affects the mobility behavior of negatively charged colloids through porous media (Franchi and O'Melia, 2003). As a result, mineral soils are more likely to have mobile nanoparticles than soils with high levels of organic matter. Nanoformulations of pesticides appear to have little impact on soil degradation, whereas they have a more noticeable impact on photodegradation (Kah et al., 2018). Nanomaterials' fate and persistency behavior in the environment can still be greatly influenced by the biotransformation that occurs in the environment as a result of interaction with biofilms, biological organisms, and plants. With respect to the biokinetics of nano-agri-input-based formulations, the possible translocation of delivery tools present in the nano range across the gastrointestinal tract is also conceivable.

5.4 DIFFERENT TYPES OF TOXICITY RELATED TO NAIPS

5.4.1 PHYTOTOXICITY

The extent of the phytotoxicity varied depending on various properties of nanomaterials, such as composition, size of nanoparticles, reactivity, concentration, plant species, and crop growth stages. In one study, smaller-sized CuO nanoparticles having diameters of 25 nm were shown to be more phytotoxic to soybeans than dissolved Cu^{2+} and bigger-sized CuO nanoparticles having diameters of 50 and 250 nm (Yusefi-Tanha et al., 2020). NPs with positive charges are more harmful than those with negative charges. Nano-fertilizers comprising metal oxide nanomaterials that build up in various plant bodies not only may demonstrate abnormal physiology in plants but also possess serious human health risks when consumed (Paul et al., 2020). In soils with higher levels of nano-cerium oxide, soybeans were almost completely unable to fix nitrogen (Priester et al., 2013). Higher concentrations of copper nanoparticles have been shown to have harmful effects on wheat and green gram (Lee et al., 2008), as well as lower biomass and root length in *Cucurbita pepo* (Stampoulis et al., 2009), in contrast to increased fungicidal capabilities and improved photosynthesis. A number of hydroponically grown plants have been found to be affected by myriad manufactured nanomaterials, which raises questions about their potential long-term consequences for the food supply (Rico et al., 2011).

AgNPs are the most commonly used nanoparticle in the agriculture sector due to their fungicidal, plant growth stimulator, and fruit-ripening activities. However, silver ions (Ag^+) that have persistency, bioaccumulation, and extremely poisonous behavior toward different organisms can be released by AgNPs (Yan and Chen, 2019). In one experiment by Geisler-Lee et al. (2012), AgNPs' phytotoxicity was found to be both size and concentration-dependent. AgNPs significantly reduced the total carotenoids and chlorophyll levels in rice seedlings after seven days of treatment (Nair and Chung, 2014). In another study, AgNPs were found to be accumulated in mustard seedlings and severely inhibit photosynthesis (Vishwakarma et al., 2017). With the bioaccumulation of Ag NPs in exo-polymeric components of algae, it has been observed that nanotoxicity in algae increases, posing hazards to the aquatic food chain (Zhou et al., 2016). Another study revealed that the smaller roots of *Arabidopsis thaliana* plants resulted in a large concentration of Ag NPs in their root cells (Geisler-Lee et al., 2012).

Several factors, including the size and concentration of NPs, their surface coating, test crop species, and growth stages, all play a significant role in the phytotoxicity of ZnO NPs (Shen et al., 2022). A recent study revealed that *Vicia faba* seed germination and seedling growth were improved by ZnO NPs at lower concentrations, but phytotoxicity was induced by concentrations in the higher range (Youssef and Elamawi, 2020). Another key component for ZnO NPs' phytotoxic action was detected to be the pH of the soil matrix (García-Gómez et al., 2018).

Because of their lower position in the trophic level, the plants act as a bridge for the movement of nanoparticles and their further accumulation in consumers belonging to the high trophic levels via the food chain. Mattiello et al. (2015) observed the phytotoxic as well as genotoxic effects of titanium oxide (TiO_2) and cerium oxide (CeO_2) nanoparticles on barley in a dose-dependent manner. In a study by Lian et al. (2020), intriguingly, the TiO_2-NP treatment increased some nutrient mineral uptakes, including N, P, and K, while reducing chlorophyll and plant growth in maize. Similar findings were observed by Teszlák et al. (2018), where TiO_2 NPs demonstrated metabolic inhibition of photosynthesis by dramatically reducing net CO_2 uptake and increasing stomatal conductance in grapevine. In a study on maize and nanoparticle interaction by Ullah et al. (2020), the PbS NPs exerted phytotoxic effects on root elongation as well as seed germination.

5.4.2 MAMMALIAN TOXICITY

Nanomaterials contain unique biological characteristics that are dual in nature. First, at the cellular level, the NPs are taken up by numerous pathways and might alter cellular activities. Second, at the whole-body level, the selection of NPs, their biodistribution, and accumulation on the target organs may cause them to act differently from conventional materials (Wu and Tang, 2018). Skin contact, ingestion, inhalation, and injection are the most common ways that NPs are exposed to living organisms.

Nano-sized agricultural inputs may improperly penetrate the skin and then become poisonous as a result of improper handling. According to a study by Inkielewicz-Stepniak et al. (2014), Ag NPs and F together caused cellular damage in human fibroblasts by penetrating the cell membrane and accumulating in the mitochondria. Ickrath et al. (2017) reported that ZnO NPs can accumulate inside human stem cells over time, causing cytotoxicity and genotoxicity. The CuO NPs can cause genotoxicity in human pulmonary epithelial cells due to oxidative stress (Ahamed et al., 2010). By causing cell death and reducing cellular adhesiveness, single-walled carbon nanotubes (SWCNTs) reduced the proliferation of kidney cells in cell culture. After injection, these materials also caused lung irritation. The *in vitro* administration of SWCNTs to bronchial epithelial cells and keratinocytes increases oxidative stress indicators. Multiwalled carbon nanotubes (MWCNTs) remain in the deep lung after inhalation, where these materials might cause inflammatory and fibrotic reactions (Agrawal and Rathore, 2014).

In a comparative study by Hong et al. (2013), it was discovered that mice were more harmful to intraperitoneal injections than to gastrointestinal gavage administration. Furthermore, the test animals given intraperitoneal exposure to ZnO NPs showed elevated levels of the liver enzymes glutamic pyruvic transaminase and alkaline phosphatase, suggesting liver damage. Due to their small size, NPs made of metal or metal oxides were easier to accumulate in the brain. Ze et al. (2013) showed that intranasal treatment of TiO_2 NPs for 90 days continuously produces considerable spongiocyte growth and hemorrhage in the mouse brain. Nanoparticles used for agricultural purposes may also have cytotoxic or genotoxic effects. For instance, at every level of DNA damage assessed in both *in vitro* and *in vivo* experiments, AgNPs have been found to exert genotoxic effects (Rodriguez-Garraus et al., 2020).

5.5 SAFETY ASSESSMENT STRATEGIES FOR NAIPS

The four steps of the risk assessment—hazard classification, dose-response analysis, exposure assessment, and risk characterization—are widely acknowledged and adopted from UN publications. The primary step in the risk evaluation and management of nanomaterials produced is hazard identification. To quantitatively assess the hazard potential, two methods—weight of evidence (WoE) and Bayesian Networks (BN)—are particularly relevant. A statistical framework called the WoE is built around expert elicitation. On the other hand, the potential danger of metals and their oxide-based nanomaterials like TiO_2 and ZnO can be predicted using the probabilistic machine learning approach known as BN (Subramanian and Rajkishore, 2018). The association between risks and a substance's concentration in a particular environment is quantitatively determined by the dose-response assessment. To draw conclusions about the type and scope of hazards, it wisely combined information from all the steps of risk assessment. As a result, risk characterization entails estimating the likelihood that unfavorable health impacts would materialize, assessing uncertainty, and summarizing the risk data.

Any product registration must provide comprehensive details on the *in vitro* and *in vivo* toxicity of numerous nontarget organisms. To date, it has been documented that nanomaterials employed in crop production and protection have a number of toxicity problems, including phytotoxicity, genotoxicity, mammalian toxicity, and toxicity to nontarget organisms (Najahi-Missaoui et al., 2020). Therefore, studying genotoxicity is one of the most crucial factors in this domain for registration purposes. NMs have the capacity to alter genetic material, which could result in mutations and ultimately different kinds of malignancies (Elespuru et al., 2018). Studies on the genotoxicity of NAIPs may be carried out using the existing OECD guidelines. Need-based cytotoxicity tests, or toxicity to cells, should also be carried out in order to register any nano-agri-input product. These studies must be undertaken using different existing techniques such as ATP Cell Titer-Glo, cell impedance, trypan blue, live and dead cell counts, and colony-forming efficiency (Saifi et al., 2018). The different protocols for testing the toxicity parameters of nanoparticles as suggested by the EFSA Scientific Committee in 2011 and guidelines for the evaluation of NAIPs by the Government of India in 2020 have been enlisted in Table 5.1.

TABLE 5.1 Toxicity Assessment Methods of NAIPs.

Test Parameter	OECD Test Method Number
Genotoxicity (*in vitro*)	476
	487
Genotoxicity (*in vivo*)	474
	488
	489
Dermal toxicity	427 (*in vivo*)
	428 (*in vitro*)
Inhalation exposure toxicity	403
Eye irritation	437
Chronic toxicity	453
Development and reproduction toxicity	414
	415
	416
Repeated-dose 90-day oral toxicity (*in vivo*)	408
Soils and sediments	222
Soil microbes	216
	217

TABLE 5.1 *(Continued)*

Test Parameter	OECD Test Method Number
Water	201
	202
	211
	Other test methods
Cytotoxicity	LDH release, MTT, XTT, Trypan blue, Alamar blue
Digestion studies (*in vitro*) using digestion models	Dynamic *in-vitro* human stomach
	Dynamic gastrointestinal digestion system
	Human gastric simulator

According to the regulations of the relevant regulatory bodies, comparative efficacy studies of conventional versus nano-based agri-inputs should also be conducted. Data on bio-efficacy, including bio-effectiveness, phytotoxicity, input residues, and persistence behavior in different environmental compartments, must be produced. Additionally, it is necessary to evaluate the effect of these nano-based input products on natural enemies, subsequent crops (in the case of nanoherbicide), soil microbes, honey bees, and earthworms (Chakraborty et al., 2021). The registration dossier should also address the method used to determine the residues of all formulants, any degradation products, the quantity of residues, the expected risks associated with the created residue, and bioaccumulation behavior.

5.6 REGULATION OF NAIPS IN DIFFERENT COUNTRIES

5.6.1 EUROPEAN UNION (EU)

In the European Union, crop protection products are regulated by Regulation (EC) No. 1107/2009. All of the products before market release should get prior authorization. Therefore, two levels of authorization are generally practised. The first one is a national-level evaluation and issue of a licence by EU member states, whereas the second one is the inspection of the active chemicals present in the products by the "European Food Safety Authority" (EFSA). The chemical residue levels in foods are governed by Regulation (EC) No. 396/2005.

The "Registration, Evaluation, Authorization, and Restriction of Chemicals" (REACH) evaluates the potential risk of any chemical, including those in the nano-range, toward the health of workers and different environmental compartments. To accomplish this, all chemical manufacturers and importers should undertake the safety assessment guidelines as mentioned in REACH

Regulation (EC) No. 1907/2006. Any additional test for understanding the material's properties should be conducted in this regard. Moreover, approval of any active ingredient having pesticidal properties does not necessarily include the same for its nano-ranged analogue unless it is specifically mentioned. Any pesticide synthesized in nanoform requires separate approval after satisfactory safety assessment results. Safety assessments should also be carried out for the formulation auxiliaries of the developed nanopesticide formulations taken together (Kookana et al., 2014).

5.6.2 UNITED STATES OF AMERICA (USA)

Under the "Federal Food, Drug, and Cosmetic Act," the "Food and Drug Administration" (FDA) in the USA is in charge of ensuring the safety of food as well as feed additives and food contact materials that are put on the market. The FDA does not categorically categorize nanotechnology products as being beneficial or harmful on their terms. Instead, the FDA regulates nanotechnology products in accordance with the precise legal requirements relevant to each type of product falling under the Agency's purview and in accordance with a science-based, product-focused regulatory strategy (FDA, 2020).

According to the "Federal Insecticide, Fungicide, and Rodenticide Act" (FIFRA), the "US Environmental Protection Agency" is in charge of regulating pesticides in the USA. The US-EPA is investigating a number of NMs, including carbon nanotubes, titanium dioxide, nano-silver, iron, cerium dioxide, and micronized copper, to create effective evaluation methods for engineered nanomaterials. As part of a complete environmental assessment (CEA) methodology that combines the risk assessment paradigm with a product life-cycle approach, it has released case studies that concentrate on specific examples. Pesticides containing nanosilver and/or other nanometals or metal oxides have been taken into consideration by the Scientific Advisory Panel of the FIFRA (FIFRA-SAP, 2010). The Panel reasoned that the present approaches may not be sufficient to address or forecast the environmental fate and consequences and that the potential environmental and human health concerns of such nanomaterials may differ from those posed by traditional pesticides.

5.6.3 CANADA

Nanoproducts are subject to regulation in Canada under the current Food and Drugs Act. Different agencies, including "Health Canada," "Environment and Climate Change Canada," and "Agriculture and Agri-Food Canada," are presently working on the safe use of nanomaterials in different sectors, including agriculture. The basic aim of these agencies is to develop a regulatory framework for risk mitigation of nano-enabled products, with a focus on the product life cycle and exposure effects. The Canadian regulatory framework follows the OECD guidelines for the safety evaluation of nanomaterials (Allan et al., 2021). Under the "US-Canada Regulatory Cooperation Council Nanotechnology Initiative," regulatory organizations in both countries tried to harmonize their approaches to the regulation of NMs.

5.6.4 NON-EUROPEAN COUNTRIES

In 2007, a "Russian Corporation of Nanotechnologies" was founded, and several federal programmes have defined the Russian strategy for the utilization of nanotechnology in the industrial sector. In the case of titanium dioxide, single-walled carbon nanotubes, and silver, safe reference levels for reducing the chances of harmful effects on the human body were developed through several federal programmes and incorporated into sanitary regulations (Amenta et al., 2015).

In Switzerland, protocols and laws already in place for dealing with conventional chemicals safeguard the safety of NMs. NMs must be documented along with particular details on their composition, particle size, shape, surface area, coatings, aggregation state, and functionalization in crop protection chemical products that are submitted for registration (TERI Web Desk, 2020).

5.6.5 AUSTRALIA

In October 2014, the "Australian Pesticides and Veterinary Medicines Authority" (APVMA) released a document titled, "Regulatory Considerations for Nanopesticides and Veterinary Nanomedicines" that discusses various definitions, risk guidelines, and regulatory frameworks (APVMA, 2014). The report included the potential and future outlooks of these nanoproducts in terms of efficacy and safety. In Australia, nano-enabled veterinary and agricultural chemical products are assessed in terms of their different safety aspects on a case-by-case basis during import, manufacture, and supply. The legislation, "Agricultural and Veterinary Chemicals Code Act, 1994," includes efficacy, safety, and trade criteria for agrochemical products. The nano-agri-input products also come under the purview of this regulation for getting APVMA consent.

5.6.6 *ASIAN COUNTRIES*

In India, the "Guidelines for Evaluation of Nano-Based Agri-Input and Food Products in India" were released by the Government of India in July 2020 to enhance the product quality and safety of nano-enabled agri-input products for widespread acceptance by end users (Government of India, 2020). These guidelines suggested the registration of nanofertilizers and nanopesticides as per the existing "Fertiliser Control Order, 1985" and "Insecticides Act, 1968," respectively. However, additional characterization experiments using Fourier-transformed infrared spectroscopy, dynamic light-scattering techniques, transmission electron spectroscopy, etc., stability studies, and different toxicological tests should be carried out for these nano-agri-input products.

In Taiwan, to certify nanoproducts, the "Taiwan Nanotechnology Industry Development Association" devised the nanoMark system (Mohammadi and Jafari, 2021).

In China, to control nano-agri-inputs in accordance with the Chinese food safety law, China has established the "Commission on Nanotechnology Standardisation" and the "National Centre for Nanoscience and Technology" (Wang and Wu, 2017).

In South Korea, the "Nanosafety and Ethics Strategic Plan (2017–2021 for Sustainable Nanotechnology for National Development)" has been designed to ensure health and environmental safety and security (Anonymous, 2018).

5.7 GLOBAL COMMERCIALIZATION STATUS OF NAIPS

It is widely known that nano-formulations perform far better than their bulk counterparts (Bratovcic et al., 2021). Nano-agri-input products significantly reduce the chemical load and aid in reducing environmental pollution due to their increased effectiveness and ability to precisely control the quantities of nutrient and pesticide components released (Sarkar et al., 2021). Globally, the market for nano-agri-input products is rising exponentially. Numerous countries are investing in the development of nanotechnology-based agricultural input products. The commercialized NAIPs have been listed in Table 5.2. The main obstacle to the widespread commercialization of NAIPs is still seen as the lack of worldwide standardized testing techniques for toxicity and other parameters, uniform legal frameworks for international trade, and large-scale societal acceptance.

TABLE 5.2 Commercialized Nano-Agri-Input Products.*

Trade Name	Nutrient/Pesticidal Active Ingredient/Mode of action	Formulation Composition Related Information	Industry	Country
		Nanofertilizer		
TAG Nano NPK	N, P, K	Proteino-lacto-gluconate chelated	Tropical Agrosystem India (P) Ltd	India
TAG Nano Potash	K			
TAG Nano Phos	P			
TAG Nano Zinc	Zn			
TAG Nano Cal	Ca			
Nano urea	N	–	IFFCO	
Nanomax NPK	NPK	Proteino-lacto-gluconate chelated	JU Agri Sciences Pvt. Ltd.	
Nanomax-Cal	Ca			
Nanomax-Potash	K			
Nanomax-Zinc	Zn			
Nano-Ag Answer®	NPK	Blends of sea kelp, microorganisms, and mineral electrolytes	Urth Agriculture	USA
Nano Calcium	Ca	–	AC International Network Co. Ltd.	Germany
		Nanoinsecticide and fungicide		
CruiserMaxx	Mefenoxam, Thiamethoxam, Fludioxonil	–	Syngenta	USA
		Nanoinsecticide		
NANO TRACER	Inhibition of synthesis as well as release of moulting hormones from prothoracic glands	Combination of herbal and organic extracts	Annadata Organic	India
NANO WIPER	-Do-			
NANO LASER	-Do-			

TABLE 5.2 *(Continued)*

Trade Name	Nutrient/Pesticidal Active Ingredient/Mode of action	Formulation Composition Related Information	Industry	Country
		Nanofungicide		
Knockout Nano	Nano copper	–	Zeal Biologicals	India
NANOCU	Nano copper	–	Bio Nano Technology	Egypt
Nano-antimicrobial		–		
AGRO 2400	Nano silver	Nano silver 2400 ppm, citric acid, wetting polymers, trisodium citrate, distilled water	Silvertech Kimya San. Tic. Ltd. Şti.	Turkey
AGRO 2475	Nano copper and silver	Nano silver 2400 ppm, citric acid, Nano copper 7500 ppm, wetting polymers, trisodium citrate, distilled water		
AGRO 2490	Nano chitosan, copper, and silver	Nano chitosan 1500 ppm, Nano silver 2400 ppm, citric acid, Nano copper 7500 ppm, wetting polymers, trisodium citrate, distilled water		

*Information obtained from industries' websites.

5.8 PRESENT LIMITATIONS IN EXISTING REGULATORY FRAMEWORKS FOR COMMERCIALIZATION OF NAIPS

There are many limitations in the existing regulatory frameworks for the commercialization of NAIPs, which are as follows:

- The effective long-term method to assure the safety related to nano-agri-input formulations should not be a case-by-case risk assessment approach for each NM, as is presently advised and practised in some nations, including the EU. This would consume too many resources and stifle innovation.
- In addition to industry, various governmental, nongovernmental, and international organizations are also engaged in several activities to determine the factors that determine NM risks and implement more effective strategies, such as grouping NMs and comparing hazard or exposure statistics. Different guidelines are available for the safety assessment of NMs in the food and feed sectors, along with nano-enabled plant protection products (Amenta et al., 2015). The guidelines for nanofertilizers, nano-biostimulants, or other nano-agri-input products are still scarce or in an early stage of development.
- It is also advised to take risk management steps to reduce exposure to NMs for agricultural as well as environmental sustainability in the long term. However, the exposure limitation may not be the expected outcome in some circumstances, such as when the nano-form should enable enhanced availability of nutrients or pesticidal active ingredients toward the target.
- Harmonized strategies addressing the marketing of NAIPs and their safe use would be advantageous, provided nanotechnology-based agri-input products join the worldwide trade. Therefore, the standards and guidelines addressing suitable test procedures for risk assessment can be harmonized globally and re-evaluated as technology advances.
- Most importantly, along with the globally harmonized legislative frameworks, the societal acceptance and improvement of people's perspectives toward the use of nanomaterials in agriculture should be considered one of the most important factors for the large-scale commercialization of nano-agri-input products in the near future.

5.9 CONCLUSION AND FUTURE PROSPECTS

Recent developments in nanotechnology have opened the door for a variety of novel methods, materials, and goods across various agro-industries, and they have also begun to open the door for advancements in the agri-food industry. Particularly recently, many research activities have been concentrated on the creation of nano-agri-input formulations to improve input use efficacy and hence lower the level of agrochemical usage in crop production as well as environmental protection as a step toward environmental sustainability. The present chapter investigated the state-of-the-art in regulatory and toxicological aspects of nano-agri-inputs to enable their safe and sustainable development and usage. It also revealed several gaps that must be addressed in future research experiments. In this context, the NAIPS's environmental fate and toxicological consequences should also be addressed in depth in future research studies, as they may differ from those of their conventional counterparts and have different impacts and environmental implications. Each component must be taken into account both individually and collectively as a nano-agri-input entity in the risk analysis of the developed nano-enabled products. This viewpoint also suggests that, for the reduction of trade barriers and the spread of nano-agriculture, the development of global regulatory homogeneity should be the top priority for national and international policymakers.

KEYWORDS

- **Nanotechnology**
- **nano-agri-input products**
- **phytotoxicity**
- **guidelines**
- **pesticides**
- **fertilizers**
- **regulation**

REFERENCES

Agrawal, S., & Rathore, P. (2014). Nanotechnology pros and cons to agriculture: a review. *International Journal of Current Microbiology and Applied Sciences, 3*(3), 43–55.

Ahamed, M., Siddiqui, M. A., Akhtar, M. J., Ahmad, I., Pant, A. B., & Alhadlaq, H. A. (2010). Genotoxic potential of copper oxide nanoparticles in human lung epithelial cells. *Biochemical and Biophysical Research Communications, 396*, 578–583.

Allan, J., Belz, S., Hoeveler, A., Hugas, M., Okuda, H., Patri, A., *et al.* (2021). Regulatory landscape of nanotechnology and nanoplastics from a global perspective. *Regulatory Toxicology and Pharmacology, 122*, 104885.

Amenta, V., Aschberger, K., Arena, M., Bouwmeester, H., Moniz, F. B., Brandhoff, P. *et al.* (2015). Regulatory aspects of nanotechnology in the agri/feed/food sector in EU and non-EU countries. *Regulatory Toxicology and Pharmacology, 73*(1), 463–476.

Anonymous (2018). (https://www.nanotec.or.th/en/wp-content/uploads/2018/09/Nanosafety-Soft-file.pdf (accessed July 3, 2022).

APVMA (2014). https://apvma.gov.au/sites/default/files/docs/report-draft-regulatory-considerations-nanopesticides-veterinary-nanomedicines.pdf (accessed August 24, 2022).

Bratovcic, A., Hikal, W. M., Said-Al Ahl, H. A., Tkachenko, K. G., Baeshen, R. S., Sabra, A. S., & Sany, H. (2021). Nanopesticides and nanofertilizers and agricultural development: Scopes, advances and applications. *Open Journal of Ecology, 11*(4), 301–316.

Chakraborty, R., Mukhopadhyay, A., Sarkar, S., & Kundu, A. (2021). An insight into data requiremnts for registration of nanopesticides and nanofertilizers in India. *Agriculture & Food: e-Newsletter, 3*, 453–455.

EFSA Scientific Committee (2011). Scientific opinion on guidance on the risk assessment of the application of nanoscience and nanotechnologies in the food and feed chain. *EFSA Journal, 9*(5), 2140.

Elespuru, R., Pfuhler, S., Aardema, M. J., Chen, T., Doak, S. H., Doherty, A., *et al.* (2018). Genotoxicity assessment of nanomaterials: recommendations on best practices, assays, and methods. *Toxicological Sciences, 164*(2), 391–416.

Elizabath, A., Babychan, M., Mathew, A. M., & Syriac, G. M. (2019). Application of nanotechnology in agriculture. *International Journal of Pure & Applied Bioscience, 7*(2), 131–139.

FDA (2020). https://www.fda.gov/media/140395/download (accessed July 23, 2022).

FIFRA-SAP (2010). Evaluation of the hazard and exposure associated with nanosilver and other nanometal pesticide products. http://www.epa.gov/scipoly/sap/meetings/2009/november/110309ameetingminutes.pdf (accessed on June 5, 2022).

Franchi, A., and C. R. O'Melia (2003). Effects of natural organic matter and solution chemistry on the deposition and re-entrainment of colloids in porous media. *Environmental Science & Technology, 37*(6), 1122–1129.

García-Gómez, C., Obrador, A., González, D., Babín, M., & Fernández, M. D. (2018). Comparative study of the phytotoxicity of ZnO nanoparticles and Zn accumulation in nine crops grown in a calcareous soil and an acidic soil. *Science of the Total Environment, 644*, 770–780.

Geisler-Lee, J., Wang, Q., Yao, Y., Zhang, W., Geisler, M., Li, K., *et al.* (2012). Phytotoxicity, accumulation and transport of silver nanoparticles by *Arabidopsis thaliana*. *Nanotoxicology, 7*(3), 323–337.

Ghormade, V., Deshpande, M. V., & Paknikar, K. M. (2011). Perspectives for nano-biotechnology enabled protection and nutrition of plants. *Biotechnology Advances, 29*(6), 792–803.

Gogos, A., Knauer, K., & Bucheli, T. D. (2012). Nanomaterials in plant protection and fertilization: current state, foreseen applications, and research priorities. *Journal of Agricultural and Food Chemistry, 60*(39), 9781–9792.

Government of India (2020). Guidelines for evaluation of nano-based agri-input and food products in India. https://static.pib.gov.in/WriteReadData/userfiles/NanoAgri_15.6.2020.pdf (accessed June 21, 2021).

Grillo, R., Fraceto, L. F., Amorim, M. J., Scott-Fordsmand, J. J., Schoonjans, R., & Chaudhry, Q. (2021). Ecotoxicological and regulatory aspects of environmental sustainability of nanopesticides. *Journal of Hazardous Materials, 404*, 124148.

Hong, T.-K., Tripathy, N., Son, H.-J., Ha, K.-T., Jeong, H.-S., & Hahn, Y.-B. (2013). A comprehensive in vitro and in vivo study of ZnO nanoparticles toxicity. *Journal of Materials Chemistry B, 1*(23), 2985–2992.

Hoseini-Alfatemi, S. M., Fallah, F., Armin, S., Hafizi, M., Karimi, A., & Kalanaky, S. (2020). Evaluation of blood and liver cytotoxicity and apoptosis-necrosis induced by nanochelating based silver nanoparticles in mouse model. *Iranian Journal of Pharmaceutical Research: IJPR, 19*(2), 207.

Ickrath, P., Wagner, M., Scherzad, A., Gehrke, T., Burghartz, M., Hagen, R., Radeloff, K., Kleinsasser, N., & Hackenberg, S. (2017). Time-dependent toxic and genotoxic effects of zinc oxide nanoparticles after long-term and repetitive exposure to human mesenchymal stem cells. *International Journal of Environmental Research and Public Health, 14*, 1590.

Inkielewicz-Stepniak, I., Santos-Martinez, M. J., Medina, C., & Radomski, M. W. (2014). Pharmacological and toxicological effects of co-exposure of human gingival fibroblasts to silver nanoparticles and sodium fluoride. *International Journal of Nanomedicine, 9*, 1677.

ISO (2015). ISO/TS 80004-1:2015 nanotechnologies—vocabulary—Part 1: core terms ISO/TS 80004-1:2015 nanotechnologies—vocabulary—Part 1: core terms. Geneva, Switzerland.

Kah, M., Kookana, R. S., Gogos, A., & Bucheli, T. D. (2018). A critical evaluation of nanopesticides and nanofertilizers against their conventional analogues. *Nature Nanotechnology, 13*(8), 677–684.

Kolesnikov, S. I., Timoshenko, A. N., Kazeev, K. S., Akimenko, Y. V., & Myasnikova, M. A. (2019). Ecotoxicity of copper, nickel, and zinc nanoparticles assessment on the basis of biological indicators of chernozems. *Eurasian Soil Science, 52*(8), 982–987.

Kookana, R. S., Boxall, A. B., Reeves, P. T., Ashauer, R., Beulke, S., Chaudhry, Q., *et al.* (2014). Nanopesticides: guiding principles for regulatory evaluation of environmental risks. *Journal of Agricultural and Food Chemistry, 62*(19), 4227–4240.

Kundu, A., Mukhopadhyay, A., & Sarkar, S. (2021). Role of nanocomposites as herbicide delivery tools in crop protection. *Agriculture & Food: e-Newsletter, 3*, 499–501.

Lee, W. M., An, Y. J., Yoon, H., & Kweon, H. S. (2008). Toxicity and bioavailability of copper nanoparticles to the terrestrial plants mung bean (*Phaseolus radiatus*) and wheat (*Triticum aestivum*): plant agar test for water-insoluble nanoparticles. *Environmental Toxicology and Chemistry: An International Journal, 27*, 1915–1921.

Lian, J., Zhao, L., Wu, J., Xiong, H., Bao, Y., Zeb, A., *et al.* (2020). Foliar spray of TiO2 nanoparticles prevails over root application in reducing Cd accumulation and mitigating Cd-induced phytotoxicity in maize (*Zea mays* L.). *Chemosphere, 239*, 124794.

Majumdar, S., & Keller, A. A. (2021). Omics to address the opportunities and challenges of nanotechnology in agriculture. *Critical Reviews in Environmental Science and Technology, 51*(22), 2595–2636.

Mattiello, A., Filippi, A., Poščić, F., Musetti, R., Salvatici, M. C., Giordano, C., *et al.* (2015). Evidence of phytotoxicity and genotoxicity in Hordeum vulgare L. exposed to CeO_2 and TiO_2 nanoparticles. *Frontiers in Plant Science, 6*, 1043.

Mohammadi, Z., & Jafari, S. M. (2021). Regulatory principles on food nanoparticles legislated by Asian and Oceanian countries. In *Safety and Regulatory Issues of Nanoencapsulated Food Ingredients* (pp. 201-238). Academic Press.

Nair, P. M. G., & Chung, I. M. (2014). Physiological and molecular level effects of silver nanoparticles exposure in rice (*Oryza sativa* L.) seedlings. *Chemosphere, 112*, 105–113.

Najahi-Missaoui, W., Arnold, R. D., & Cummings, B. S. (2020). Safe nanoparticles: Are we there yet? *International Journal of Molecular Sciences, 22*(1), 385.

Paul, S. (2021). Fate and Toxicities of Nanoparticles in Agriculture. *Agriculture & Food: e-Newsletter, 3*, 547–549.

Paul, S., Das, A., & Mukhopadhyay, A. (2020). Constraints of nanotechnology application in agriculture. *Agriculture & Food: e-Newsletter, 2*, 467–468.

Pramanik, P., Krishnan, P., Maity, A., Mridha, N., Mukherjee, A., & Rai, V. (2020). Application of nanotechnology in agriculture. In *Environmental Nanotechnology* (Vol. 4, pp. 317–348). Springer.

Priester, J. H., Ge, Y., Mielke, R. E., Horst, A. M., Moritz, S. C., Espinosa, K., Gelb, J., *et al.* (2013). Soybean susceptibility to manufactured nanomaterials with evidence for food quality and soil fertility interruption. *Proceedings of the National Academy of Sciences of the United States of America, 7*, 2451–2456.

Rathee, M., & Dalal, P. (2018). Emerging insect pests in Indian agriculture. *Indian Journal of Entomology, 80*(2), 267–281.

Rauscher, H., Rasmussen, K., & Sokull-Klüttgen, B. (2017). Regulatory aspects of nanomaterials in the EU. *Chemie Ingenieur Technik, 89*(3), 224–231.

Regulatory considerations for nanopesticides and veterinary nanomedicines (2014). Vol. 1. Australia: Australian Pesticides and Veterinary Medicines Authority. Retrieved from: https://apvma.gov.au/sites/default/files/docs/report-draft-regulatory-considerations-nanopesticides-veterinary-nanomedicines.pdf (accessed July 11, 2022).

Rico, C. M., Majumdar, S., Duarte-Gardea, M., Peralta-Videa, J. R., & Gardea-Torresdey, J. L. (2011). Interaction of nanoparticles with edible plants and their possible implications in the food chain. *Journal of Agricultural and Food Chemistry, 59*(8), 3485–3498.

Rodriguez-Garraus, A., Azqueta, A., Vettorazzi, A., & Lopez de Cerain, A. (2020). Genotoxicity of silver nanoparticles. *Nanomaterials, 10*(2), 251.

Saifi, M. A., Khan, W., & Godugu, C. (2018). Cytotoxicity of nanomaterials: Using nanotoxicology to address the safety concerns of nanoparticles. *Pharmaceutical Nanotechnology, 6*(1), 3–16.

Sarkar, N., Chaudhary, S., & Kaushik, M. (2021). Nano-fertilizers and nano-pesticides as promoters of plant growth in agriculture. In plant-microbes-engineered nano-particles (PM-ENPs) nexus in agro-ecosystems (pp. 153–163). Springer.

Sarkar, N., Chaudhary, S., & Kaushik, M. (2021). Nano-fertilizers and nano-pesticides as promoters of plant growth in agriculture. In *Plant-Microbes-Engineered Nano-particles (PM-ENPs) Nexus in Agro-Ecosystems* (pp. 153–163). Springer.

Sarkar, S., Kundu, A., Chakraborty, R., & Mukhopadhyay, A. (2021). A review on nanocomposites and their role in insecticide delivery. *Journal of Entomology and Zoology Studies, 9*(1), 1985–1988.

Sekhon, B. S. (2014). Nanotechnology in agri-food production: an overview. *Nanotechnology Science and Applications, 7*, 31.
Sengul, A. B., & Asmatulu, E. (2020). Toxicity of metal and metal oxide nanoparticles: a review. *Environmental Chemistry Letters, 18*(5), 1659–1683.
Shen, M., Liu, W., Zeb, A., Lian, J., Wu, J., & Lin, M. (2022). Bioaccumulation and phytotoxicity of ZnO nanoparticles in soil-grown *Brassica chinensis* L. and potential risks. *Journal of Environmental Management, 306*, 114454.
Stampoulis, D., Sinha, S. K., & White, J. C. (2009). Assay-dependent phytotoxicity of nanoparticles to plants. *Environmental Science and Technology, 43*, 9473–9479.
Subramanian, K. S., & Rajkishore, S. K. (2018). Regulatory framework for nanomaterials in agri-food systems. In *Nanomaterials: ecotoxicity, safety, and public perception* (pp. 319–342). Springer.
TERI Web Desk (2020). Guidelines for evaluation of nano-based agri-input and food products in India. https://www.teriin.org/policy-brief/guidelines-evaluation-nano-based-agri-input-and-food-products-india (accessed May 15, 2022).
Teszlák, P., Kocsis, M., Scarpellini, A., Jakab, G., & Kőrösi, L. (2018). Foliar exposure of grapevine (Vitis vinifera L.) to TiO_2 nanoparticles under field conditions: Photosynthetic response and flavonol profile. *Photosynthetica, 56*(4), 1378–1386.
Tkachenko, A., Onishchenko, A., Butov, D., & Tkachenko, M. (2021). Flow cytometry in nanotoxicology. *Inter Collegas, 8*(4), 278–289.
Ullah, H., Li, X., Peng, L., Cai, Y., & Mielke, H. W. (2020). In vivo phytotoxicity, uptake, and translocation of PbS nanoparticles in maize (*Zea mays* L.) plants. *Science of The Total Environment, 737*, 139558.
Vishwakarma, K., Upadhyay, N., Singh, J., Liu, S., Singh, V. P., Prasad, S. M. et al. (2017). Differential phytotoxic impact of plant mediated silver nanoparticles (AgNPs) and silver nitrate (AgNO3) on *Brassica* sp. *Frontiers in Plant Science, 8*, 1501.
Wang, L., & Wu, S. (2017). The National Center for Nanoscience and Technology, China (NCNST): an international innovation engine for nano research. *National Science Review, 4*(3), 500–509.
Wu, T., & Tang, M. (2018). Review of the effects of manufactured nanoparticles on mammalian target organs. *Journal of Applied Toxicology, 38*(1), 25–40.
Yan, A., & Chen, Z. (2019). Impacts of silver nanoparticles on plants: a focus on the phytotoxicity and underlying mechanism. *International Journal of Molecular Sciences, 20*(5), 1003.
Youssef, M. S., & Elamawi, R. M. (2020). Evaluation of phytotoxicity, cytotoxicity, and genotoxicity of ZnO nanoparticles in *Vicia faba*. *Environmental Science and Pollution Research, 27*(16), 18972–18984.
Yusefi-Tanha, E., Fallah, S., Rostamnejadi, A., & Pokhrel, L. R. (2020). Particle size and concentration dependent toxicity of copper oxide nanoparticles (CuO NPs) on seed yield and antioxidant defense system in soil grown soybean (*Glycine max* cv. Kowsar). *Science of The Total Environment,* 715, 136994.
Ze, Y., Zheng, L., Zhao, X., Gui, S., Sang, X., Su, J. et al. (2013). Molecular mechanism of titanium dioxide nanoparticles-induced oxidative injury in the brain of mice. *Chemosphere,* 92(9), 1183–1189.
Zhou, K., Hu, Y., Zhang, L., Yang, K., & Lin, D. (2016). The role of exopolymeric substances in the bioaccumulation and toxicity of Ag nanoparticles to algae. *Scientific Reports, 6*(1), 1–11.

PART II

Green Synthesis of Nanomaterials and Their Impact on Soil Fertility and Biotic/Abiotic Stress Management

CHAPTER 6

Role of Fungi in Biogenic Synthesis of Nanofertilizers and Nanopesticides

SOMANJANA KHATUA[1] and SANDIPTA GHOSH[2]

[1]*Department of Botany, Faculty of Science, University of Allahabad, Prayagraj, India*

[2]*School of Biological and Environmental Sciences, Shoolini University of Biotechnology and Management Sciences, Solan, India*

ABSTRACT

The increasing demand for sustainable agriculture has led to the exploration of eco-friendly alternatives to conventional fertilizers and pesticides. Among the promising candidates are nanotechnology-based solutions, which offer enhanced efficiency and reduced environmental impacts. Fungi have emerged as a vital player in the biogenic synthesis of these nanoformulations owing to their unique biochemical and metabolic capabilities. This review highlights the essential role of fungi in the green synthesis of nanofertilizers and nanopesticides, elucidating underlying mechanisms. Fungi such as *Aspergillus*, *Candida*, *Penicillium*, *Fusariam*, *Beauveri*, *Rhizopus*, and *Trichoderma* sp. possess diverse enzymes and biomolecules that may act as reducing and stabilizing agents in the synthesis process, where silver and zinc nanoforms are showing great hope. The controlled release of nutrients and active compounds from these nanofertilizers and nanopesticides allows better crop uptake and sustained pest control, thereby minimizing wastage and improving overall agricultural productivity. Further research is imperative to fully harness the potential of fungal-mediated nanoparticle synthesis in sustainable agriculture, making it increasingly viable for practical application.

Nanobiotechnology for Agricultural Sciences: Nano-Agri-Input Products for Crop Production and Environmental Protection. J. C. Tarafdar, Deepak Kumar, Sanjay K. Singh, and Krishna Pal Singh (Eds.)

6.1 INTRODUCTION

In November 2022, the global population exceeded 8 billion individuals, representing a growth of 1 billion people since 2010 (UN, 2023). Projections indicate that it is expected to reach 8.6 billion by 2030, 9.8 billion by 2050, and 11.2 billion by 2100, with several million "undernourished" (UN, 2017). In response to the challenges posed by food scarcity due to population growth, diverse agricultural alternatives, including the utilization of chemical or synthetic fertilizers, pesticides, and insecticides, have been adopted to achieve high crop yields in the shortest possible time and safeguard them from insect and pest infestations during and after harvest (Daniel et al., 2022). The excessive use of large amounts of chemical fertilizers, however, leads to environmental pollution and health risks. This in turn causes deterioration of the natural properties and fertility of the soil and the accumulation of heavy metals in plant tissues, thereby compromising the nutritional value of fruits and other crops (Nadarajan and Sukumaran, 2021). Besides, only 20–50% of the applied fertilizers are used efficiently; the other 50–80% are lost through leaching, emissions, or incorporation into the soil by microorganisms in the long term, generating ecological problems and economic losses (Avila Quezada et al., 2022).

On the other hand, agricultural industries are facing a major issue related to the loss of crop yield due to several abiotic and biotic factors. These determining issues pose a significant risk to farms, leading to crop yield reduction when they are not correctly monitored and managed (Raza et al., 2019). These aspects can be grouped into three categories, which include technological, biological, and environmental factors (Pouthika and Madhumitha, 2023). Among the biological influences, insect pests are causing a major decline in agricultural yield. Various microorganisms, including viruses, bacteria, and fungi, are responsible for the majority of plant diseases, while both soil-borne and above-ground insect pests significantly impact crop production. It is reported that insect pests are causing a huge loss (18–20%) in crop production throughout the world (Ofuya et al., 2023). Several commercially available chemical pesticides are generally used to reduce the occurrence of pests in fields. But the major problems with their applications are improper delivery of agrochemicals and management, while only 0.1% of agrochemicals used in crop protection are pest-specific and the rest are lost, entering the environment. Indeed, direct exposure to pesticides or handling foodstuffs containing pesticide residues can lead to a range of health issues, such as cancer, diabetes mellitus, respiratory disorders, neurological disorders, reproductive syndromes, and oxidative stress (Rani et al., 2021). Thus, the safe application of conventional agrochemicals is a major concern (Abdollahdokht et al., 2022).

As a result, environmentalists have shifted their focus toward organic farming, a method that emphasizes the cultivation of unpolluted crops using biofertilizers and biopesticides. These natural alternatives ensure optimal nutrient availability for plants while effectively managing pathogens and pests (Nayak and Solanki, 2021; Umesha et al., 2018). Recently, green nanotechnology has emerged as a promising field with the potential to revolutionize agricultural sciences through the development of pesticides and fertilizers. The utilization of nanopesticides and nanofertilizers offers several advantages, including the ability to predict reduced environmental pollution and controlled as well as delayed delivery. Furthermore, these nanoscale chemicals facilitate easy absorption and exhibit higher efficiency at lower doses, making them more eco-friendly options (Sharma et al., 2022). In this context, fungi hold great promise and significance in the production of biogenic nanoparticles (NPs) with diverse sizes, shapes, and properties. Their exceptional bioavailability, biological activity, and elevated metal tolerance make fungi ideal candidates for the synthesis of biogenic NPs (Sonawane et al., 2022). This review provides a brief overview of the increasing significance of fungal-mediated NP synthesis and its potential applications in agriculture, addressing the future implications of this emerging technology.

6.2 BIOFERTILIZERS

It is anticipated that worldwide food demand might climb by about 70% in 2050 due to the increasing population (Verma et al., 2022b). The demand for agricultural crops is thus projected to surge by 60% by the year 2030. In the pursuit of self-sufficiency, countries have extensively relied on the use of chemical fertilizers to boost crop yields (Allouzi et al., 2022). In a study, Yousaf et al. (2017) reported significant 19–41% and 61–76% increases in rice and rapeseed yields through the combined application of NPK fertilizers. Nonetheless, the extensive use of these chemical fertilizers has led to severe environmental pollution, resulting in reduced water-holding capacity and pH alterations in the soil, along with a decline in the soil microorganism population and overall fertility. In addition to being absorbed by plants, hazardous substances also accumulate in groundwater, leading to detrimental effects on the soil. Consequently, it becomes imperative to shift our focus toward the creation of secure and eco-friendly approaches to achieve sustainable crop production (Allouzi et al., 2022). For this reason, bio-fertilizers are acknowledged as indispensable components of sustainable agriculture, providing enduring advantages for enhancing soil fertility (Mitter et al., 2021). Bio-based organic fertilizers consist of beneficial microorganisms, encompassing plant, animal, and living or dormant microbial cells. These microorganisms play a pivotal role in augmenting nutrient availability, improving nutrient bio-accessibility, and fostering plant growth (Chaudhary et al., 2022). So far, arbuscular mycorrhizal fungi, plant growth-promoting rhizobacteria, and nitrogen-fixing rhizobia are mostly being used for inoculant production (Raimi et al., 2021). The market potential for bio-fertilizers has thus become substantial, and it is projected to grow at a rate of 14.2% to reach USD 3124.5 million by the end of 2024 (Joshi and Gauraha, 2022). In this context, nanofertilizers have garnered worldwide attention in the agricultural sector, captivating the interest of agricultural engineers, environmentalists, and soil scientists due to their manifold benefits (Kumar et al., 2023).

6.3 NANOFERTILIZERS

Nanofertilizers are emerging as a superior alternative to chemical fertilizers, primarily due to their enhanced resource utilization efficiency, reduced pollution, minimal fertilizer consumption, and controlled slow release of agrochemicals. Furthermore, they contribute to increased crop production and yield while minimizing disruption to soil structure. Additionally, nanofertilizers play a significant role in promoting nutritional quality, extending shelf life, conserving water, and enhancing plant resistance against pests and disease pathogens (Kumari and Singh 2020). Nanofertilizers, with a diameter of <100 nm, can disperse as either powder or liquid, effectively enhancing nutrient uptake by plants and promoting plant fertilization (El-Saadony et al., 2021). Exposure of plants to nanofertilizers results in various morphological and physiological changes, including alterations in germination frequency, shoot-root length, biomass, chlorophyll fluorescence yield, and photosynthetic efficiency (Verma et al., 2022a). According to Prasad et al. (2012), application of ZnONPs (25 nm) promoted seed germination and seedling vigor in peanuts, leading to early flowering and increased levels of leaf chlorophyll. Similarly, use of silver NPs (AgNPs) on wheat at different concentrations and durations (10–30 min) demonstrated positive effects on various physiological characteristics of the plants, such as increased shoot and root lengths, leaf surface area, and content of chlorophyll, carbohydrates, and proteins (Salama, 2012).

Nanofertilizers can be produced through various methods, including physical and chemical processes as well as green (biological) synthesis.

Among these, the green approach is preferred due to its simplicity, cleanliness, sustainability, and cost-effectiveness. Moreover, the use of green synthesis provides greater biocompatibility for nanoparticles (NPs) applications (Urnukhsaikhan et al., 2021; Guilger-Casagrande and de Lima, 2019).

6.4 MYCOSYNTHESIS OF NANOPARTICLES (NPS)

Utilizing eukaryotic organisms, such as bacteria (e.g., Actinomycetes), fungi, and yeasts, or their metabolic byproducts as reducing and stabilizing agents for NPs biosynthesis is a growing trend (Guilger-Casagrande and de Lima, 2019; Singh et al., 2016). Among these organisms, fungi are the preferred agents for biogenic NPs synthesis due to their abundant secretion of enzymes like reductases, ease of cultivation and growth in laboratories, manageable handling, and high metal tolerance (Chauhan et al., 2023; Sonawane et al., 2022; Yassin et al., 2021; Yadav et al., 2015). Implementation of fungal cultures is even more advantageous than that of bacterial systems due to their ability to yield good biomass production, diverse enzyme secretion, and the absence of additional steps for filtrate extraction. Compared to NPs synthesis using plants, fungal mycelial mass demonstrates greater resistance to agitation and pressure, making it more suitable for large-scale syntheses (Guilger-Casagrande and de Lima, 2019). Furthermore, by adjusting culture conditions such as time, temperature, pH, and biomass quantity, it is possible to manipulate the fungal metabolism to obtain NPs with desired characteristics, such as specific size and morphology (Guilger-Casagrande and de Lima, 2019). Consequently, fungal-synthesized NPs offer several benefits, including ease of scaling up, economic feasibility, and rapid coverage of significant surface areas through appropriate mycelia development (Michael et al., 2022).

Fungi possess a natural inclination to generate extracellular metabolites that serve as agents for their own survival when faced with toxic environmental stresses, such as metallic ions, predators, and fluctuations in temperature. In the process of biosynthesizing metal NPs using fungi, the mycelia are exposed to a salt solution. As a result, the toxic metal ions undergo conversion into nontoxic metallic solid NPs, facilitated by the catalytic effects of the extracellular enzymes and metabolites of fungi (Khan et al., 2018). Currently, two common methods are employed for synthesizing NPs from fungi: extracellular (outside or on the surface of fungal cells) and intracellular (within the fungal cell) methods (Chauhan et al., 2023). In the intracellular synthesis approach, a metal precursor is introduced into the mycelial culture and internalized within the biomass. Subsequently, the NPs need to be extracted after synthesis, involving chemical treatment, centrifugation, and filtration to disrupt the biomass and release the NPs (Guilger-Casagrande and de Lima, 2019). Intracellular synthesis yields NPs of smaller size, but the downstream processing becomes more challenging (Subashini and Bhuvaneswari, 2018). Conversely, in extracellular synthesis, the metal precursor is introduced into an aqueous filtrate containing solely fungal biomolecules. This results in the formation of free NPs in the dispersion (Figure 6.1). UV-visible spectroscopy can be used to observe color changes in the fungal filtrate due to variations in the optical properties of the material (Chauhan et al., 2023). The extracellular method is widely preferred as it is cost-effective, does not require additional procedures to release the NPs from cells, and is suitable for easier downstream processing and biomass handling. Nevertheless, the dispersion of the NPs needs to undergo purification to remove fungal residues and impurities. Purification can be achieved through methods such as simple filtration, membrane filtration, gel filtration, dialysis, and ultracentrifugation (Guilger-Casagrande and de Lima, 2019).

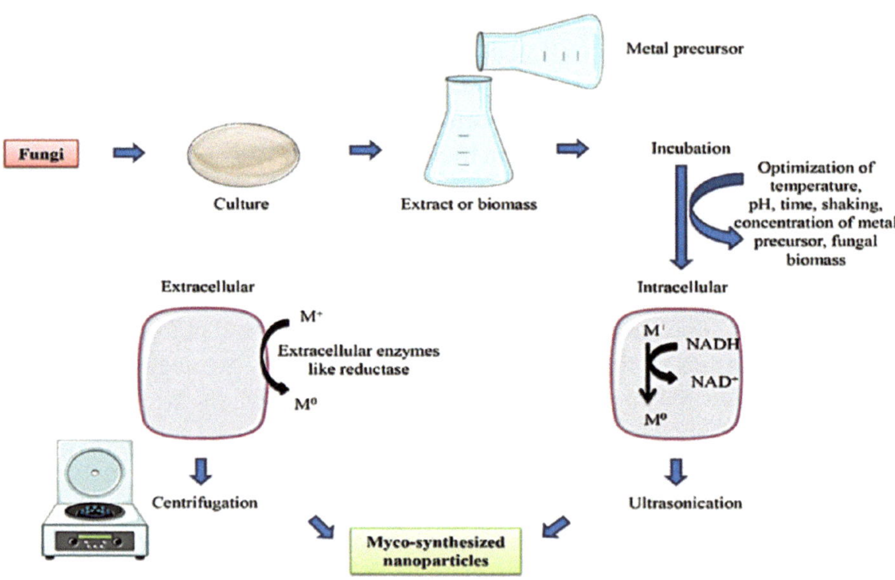

FIGURE 6.1 Steps in synthesis of fungi-mediated nanoparticles (NPs).

The earliest report of using fungi for NPs biosynthesis dates back to a letter in Nature in 1989, which described the production of CdSe NPs by *Candida albicans* (Xue et al., 2016). Since then, an increasing number of fungi have been found to possess the ability to synthesize NPs, including species like *Trichoderma asperellum*, *Cladosporium cladosporioides*, *Fusarium* spp., and *Aspergillus* spp. (Wang et al., 2021). Filamentous fungi and yeasts have recently attracted significant research interest for their capacity to synthesize large quantities of metal NPs at a low cost, their high regenerative ability, selective metal uptake, and survival under diverse environmental conditions of pH, ionic strength, and temperature (Beltrán Pineda et al., 2022). Numerous research groups have successfully reported the myco-synthesis of silver, gold, platinum, and titanium NPs in various sizes and shapes with different electrical and magnetic properties (Chauhan et al., 2023; Khan et al., 2018; Velusamy et al., 2016; Dhillon et al., 2012). Additionally, various oxides, such as $BaTiO_3$, Bi_2O_3, $CoFe_2O_4$, Co_3O_4, Fe_2O_3, Fe_3O_4, NiO, TiO_2, ZnO, and ZrO_2, have also been synthesized (Šebesta et al., 2022). Among all the elements studied, the biological synthesis of silver NPs in fungi has been the most extensively explored, primarily due to their wide applicability in various fields (Table 6.1).

6.4.1 MYCOSYNTHESIS OF SILVER NANOPARTICLES

The research papers on nanosilver mycosynthesis are numerous and continuously increasing, with several reviews published in recent years that delve into the production of silver NPs using fungal cultures (Beltrán Pineda et al., 2022; Guilger-Casagrande and Lima, 2019; Khan et al., 2018; Zhao et al., 2018). Up to now, scientists have explored 120 species of fungi from diverse taxa, including Ascomycota, Mucoromycota, and Basidiomycota, for their capability to produce nano-silver (Loshchinina et al., 2022). Filamentous fungi have been found to produce silver NPs with favorable morphological characteristics and stability (Wang et al., 2021). Basidiomycetes, in particular, have garnered significant interest as promising microorganisms for NPs fabrication. Among the Basidiomycetes studied for mycosynthesis, many belong to edible and medicinal mushrooms, which are often cultivated in artificial culture. These fungi generate a wide range of biologically active molecules that can serve as capping and stabilizing agents for NPs (Loshchinina et al., 2022). Moreover, researchers have investigated endophytic fungi living in symbiotic association with plants for their potential to synthesize silver NPs. For instance, the endophytic fungi *Penicillum* sp. isolated from curcuma leaves have shown exceptional capability as silver NPs producers (Subashini and Bhuvaneswari 2018).

Studies have indicated that there are two main routes for fungal-mediated silver NPs synthesis: (a) using cell-free broth or extract and (b) using dried or wet mycelia (Velhal et al., 2016). In the extracellular synthesis of NPs, enzymes present in the fungal filtrate act as reducing agents, converting silver ions to elemental silver (Ag^0) at a nanometric scale. As a result of the reaction, the color of the filtrate changes from pale yellow to brown, and UV-visible spectroscopy can be utilized to observe surface plasmon resonance bands, indicating alterations in the material's optical properties. The absorbance wavelengths of these bands typically fall within the range of 400–450 nm, and an absorbance peak at a longer wavelength suggests the presence of larger NPs. The size of the NPs is influenced by various synthesis conditions, such as the fungus species, temperature, pH, dispersion medium, and presence of capping on the NPs. Additionally, the color of the NPs dispersion directly correlates with the surface plasmon resonance (Guilger-Casagrande and de Lima, 2019). Numerous studies have demonstrated that myco-synthesized silver NPs possess various beneficial properties, including antibacterial, antifungal, anticancer, antioxidant, larvicidal, and other activities (Table 1). These diverse properties offer promising prospects for the application of mycogenic silver NPs in various fields.

6.4.2 MYCOSYNTHESIS OF GOLD NANOPARTICLES

Gold NPs have some unique medicinal implications, which include diagnostic medication, targeted drug delivery, and other biomedical applications (Dykman and Khlebtsov, 2011). However, only a few pieces of information on the process of preparation of Au-NP from fungal extracts and its fabrication using green synthetic methods can be found in research articles and reviews (Zhang et al., 2011; Sheikhloo and Salouti, 2011). To date, only a limited number of fungal species, less than 30, have been explored for their ability to synthesize gold NPs (AuNPs) (Kitching et al., 2015). The initial demonstration of using eukaryotic organisms for NPs synthesis was achieved with *Verticillium* sp. for the production of gold NPs (Mukherjee et al., 2001). Several fungi, including *Fusarium oxysporum*, *Aspergillus niger,* and *Candida albicans,* have been found to synthesize extracellular gold NPs. Additionally, *Pleurotus florida* has been reported to synthesize catalytically active AuNPs, and the glucan content of the mushroom was suggested to be responsible for stabilizing the synthesized NPs (Subashini and Bhuvaneswari, 2018).

In the biosynthetic process, the main precursor of AuNPs is $HAuCl^4$, which dissociates into Au^{3+} ions. Fungi commonly exhibit extracellular AuNP formation, where Au^{3+} ions are trapped and reduced by proteins present in the cell wall. On the other hand, intracellular AuNP formation occurs when Au^{3+} ions diffuse through the cell membrane and are reduced by cytosolic redox mediators (Kitching et al., 2015). The reduction reaction is facilitated by reductases, and the appearance of a purple color indicates the formation of gold NPs in the reaction mixture. The characteristic pink-purple color of gold solutions arises from the excitation of surface plasmon vibrations in the NPs, providing a convenient spectroscopic signature of their formation (Mukherjee et al., 2002).

6.4.3 MYCOSYNTHESIS OF OTHER NANOPARTICLES

Fungi have demonstrated the capability to synthesize copper NPs as well. *Penicillium vaksmanii*, *Penicillium aurantiogriseum*, and *Penicillium citrinum*, isolated from soil, have been utilized for the synthesis of copper NPs. The morphology of these NPs was influenced by factors such as monodispersity, pH, and concentration (Honary et al., 2012). Elemental selenium NPs were successfully synthesized by the white-rot fungus *Phanerochaete chrysosporium* (Subashini and Bhuvaneswari, 2018). Additionally, researchers employed *Alternaria alternata* to produce iron NPs, which were found to have a cubic shape with a size of 9 nm (Mohamed et al., 2015).

6.5 FUNGI-MEDIATED NANOFERTILIZER SYNTHESIS AND APPLICATION

Various NPs have been utilized in agricultural research to create nanofertilizers and growth enhancers for plants, but there have been limited studies on myco-synthesized NPs applied to plants. In one study, researchers used *Aspergillus fumigatus* to prepare ZnO NPs, which were then applied to *Vigna radiata* (mung bean). The application of these NPs led to significant increases in stem height, root length, root area, and nodule formation compared to the control group, with growth improvements of 32.8, 29.9, 53.3, and 58.9%, respectively. Additionally, the myco-synthesized ZnO NPs positively influenced the activity of various enzymes in the rhizosphere soil, including acid phosphatase, phytase, alkaline phosphatase, and dehydrogenase. This resulted in 10.8% increase in the mobilization of phosphorous. Moreover,

TABLE 6.1 Characters and Bioactivity of Fungi-Derived Nanoparticles (NPs).

NPs Type	Source	Synthesis Location	Synthesis Conditions	Shape	Bioactivity	References
Ag	*Alternaria* sp.	Extracellular	25°C, 1 mM $AgNO_3$, shaking	Spherical, 3–10 nm	Antifungal	Win et al. (2020)
	Arthroderma fulvum	Extracellular	55°C, pH 10, 12 h, 1.5 mM $AgNO_3$	Spherical, 20.56 nm	Antifungal	Xue et al. (2016)
	Aspergillus clavatus	Extracellular	1 mM $AgNO_3$	Polydisperse spherical or hexagonal, 10–25 nm	Antimicrobial	Verma et al. (2010)
	Aspergillus fumigatus	Extracellular	25°C, 1 mM $AgNO_3$, shaking	Spherical, 5–25 nm	Not done	Bhainsa and D'souza, (2006)
	Aspergillus fumigatus	Extracellular	25°C, pH 6, 1 mM $AgNO_3$, biomass 7 g, fungal culture age 7 days	Spherical, 322.8 nm	Antibacterial	Shahzad et al. (2019)
	Aspergillus japonicus	Extracellular	30◦C, 1 min, 0.8 mol/L $AgNO_3$, 1.5 mol/L NaOH	Spherical, 3.8–9.1 nm	Antibacterial	Li et al. (2021)
	Aspergillus niger	Extracellular	25°C, 72 h	Spherical, 20–70 nm	Pesticide	Soni and Prakash (2013)
	Aspergillus oryzae	Extracellular	1–8 mM $AgNO_3$	Spherical, 7–27 nm	Antibacterial	Phanjom and Ahmed (2017)
	Aspergillus sydowii	Extracellular	50°C, pH 8.0 and substrate concentration 1.5 mM	Spherical or close to spherical, 12 nm	Antifungal, antiproliferative	Wang et al. (2021)
	Aspergillus terreus	Extracellular	28°C, 24 h, 10 mmol/L $AgNO_3$	Polydispersed spherical, 1–20 nm	Antimicrobial	Li et al. (2012)
	Aspergillus terrerus	Intracellular	45°C, 0.7 g/100 ml fungal biomass, 6 mM $AgNO_3$	2 nm	Antibacterial	Velhal et al. (2016)
	Bjerkandera sp	Extra- and intracellular	144 h, 1 mM $AgNO_3$	Spherical, 70–90 nm	Not done	Osorio-Echavarría et al. (2021)
	Cladosporium cladosporioides	Extracellular	27°C, 78 h	Polydisperse small sphere to large spheres, 10–100 nm	Not done	Balaji et al. (2009)

TABLE 6.1 *(Continued)*

NPs Type	Source	Synthesis Location	Synthesis Conditions	Shape	Bioactivity	References
	Colletotrichum incarnatum	Extracellular	1 mM $AgNO_3$	Aggregated, spherical and crystalline, 5–25 nm	Antithrombin, cytotoxic	Chandankere et al. (2020)
	Colletotrichum sp.	Extracellular	50–80°C, alkaline pH, 1 mM $AgNO_3$	Myriad shapes, 5–60 nm	Antibacterial	Azmath et al. (2016)
	Coriolus versicolor	Extra- and intracellular	Alkaline	Spherical, 10 nm	Not done	Sanghi and Verma (2009)
	Duddingtonia flagans	Extracellular	60°C, pH 10	Spherical, 30–409 nm	Not done	Costa Silva et al. (2017)
	Epicoccum nigrum	Extracellular	55°C, alkaline pH, 1 mM $AgNO_3$	Spherical to nearly spherical, 1–22 nm	Antifungal	Qian et al. (2013)
	Fusarium oxysporum	Extracellular	pH 9–11, 40–60°C, 190.7 Lux	Spherical, 10–20 nm	Not done	Birla et al. (2013)
	Fusarium oxysporum	Extracellular	28°C, 72 h, with agitation	Spherical, 3.4–26.8 nm	Antifungal	Ishida et al. (2014)
	Fusarium oxysporum	Extracellular	28°C	Spherical, 24 nm	Antibacterial	Hamedi et al. (2017)
	Fusarium oxysporum	Intracellular	5 mM $AgNO_3$, 4.96 g/L biomass	Nearly spherical, 25–50 nm	Not done	Korbekandi et al. (2013)
	Fusarium oxysporum	Extracellular	50°C, pH 6, 11 g biomass, 10–2 M $AgNO_3$	Spherical, 5–13 nm	Antibacterial, antitumor	Husseiny et al. (2015)
	Fusarium oxysporum	Extracellular	1 mM $AgNO_3$	Globular, 50–100 nm	Antimicrobial	Abdel-Hadi et al. (2023)
	Guignardia mangiferae	Extracellular	30°C, pH 7, 10 g biomass, 1 mM $AgNO_3$	Spherical, 5–30 nm	Cytotoxic	Balakumaran et al. (2015)
	Isaria fumosorosea	Extracellular	30°C, pH 8.5, 150 rpm	Spherical, 51.31–111.02 nm	Pesticide	Banu and Balasubramanian (2014a)

TABLE 6.1 *(Continued)*

NPs Type	Source	Synthesis Location	Synthesis Conditions	Shape	Bioactivity	References
	Neurospora crassa	Extracellular	10–3 M of $AgNO_3$	Spherical, 11 nm	Not done	Castro-Longoria et al. (2011)
	Penicillium oxalicum	Extracellular	24 h, 24°C, 1 mM $AgNO_3$	Spherical, 13–23 nm	Antimicrobial, antioxidant, anticancer	Gupta et al. (2022)
	Penicillium oxalicum	Extracellular	pH 12	Spherical, 36.53 nm	Catalytic reduction of methylene blue	Du et al. (2015)
	Penicillium oxalicum	Extracellular	60°C, pH 7.0 and 1.5 mM $AgNO_3$	10–40 nm	Antimicrobial	Rose et al. (2019)
	Penicillium purpurogenum	Extracellular	pH 8, 1 mM $AgNO_3$	Spherical, 8–10 nm	Antimicrobial	Nayak et al. (2011)
	Penicillium verrucosum	Extracellular	25°C, 10 mmol/L $AgNO_3$	Polydisperse, 10–12 nm	Antifungal	Yassin et al. (2021)
	Phaenerochaete chrysosporium	Extracellular	37°C, 1 mM $AgNO_3$, 200 rpm, dark	Pyramidal, 50–200 nm	Not done	Vigneshwaran et al. (2006)
	Phoma glomerata	Extracellular	25°C, 1 mM $AgNO_3$, 120 rpm	19–65 nm	Not done	Gade et al. (2014)
	Pleurotus eryngii	Extracellular	1 mM $AgNO_3$	18.45 nm	Antimicrobial	Acay and Baran (2019)
	Rhizoctonia solani	Extracellular	Potato glucose broth	Spherical, 2–22 nm	Not done	Ashrafi et al. (2013)
	Rhizopus stolonifer	Extracellular	40°C, 10 mM $AgNO_3$	Spherical, 2.86 nm	Not done	AbdelRahim et al. (2017)
	Sclerotinia sclerotiorum	Extracellular	80°C, pH 11, 10 g biomass, 2 mM $AgNO_3$	Spherical, 10–15 nm	Antibacterial	Saxena et al. (2016)
	Trametes trogii	Extra- and intracellular	72 h, pH 13, 5 mM $AgNO_3$, with shaking	Aggregated, capped, spherical	Not done	Kobashigawa et al. (2019)
	Trichoderma harzianum	Extracellular	4 h, 40°C, 1 mM to 9 mM $AgNO_3$	Spherical, 51.10 nm	Antibacterial	Ahluwalia et al. (2014)

TABLE 6.1 *(Continued)*

NPs Type	Source	Synthesis Location	Synthesis Conditions	Shape	Bioactivity	References
	Trichoderma longibrachiatum	Extracellular	5, 10, 15, and 20 g fungal biomass, without shaking, 72 h	Spherical, 10 nm	Antifungal	Elamawi et al. (2018)
	Trichoderma viride	Extracellular	27°C	Spherical, 2–4 nm	Not done	Fayaz et al. (2009)
	Yeast strains HX-YS and LPP-12Y	Extracellular	10 mM $AgNO_3$	Spherical, 50–100 nm	Antitumor, antibacterial	Liu et al. (2021)
	Yeast	Extracellular	pH 10	Spherical, 13.8 nm	Antibacterial, cytotoxicity	Shu et al. (2020)
Au	*Colletotrichum* sp.	Extracellular	10^{-3} M $HAuCl_4$	Decahedral and icosahedral, 20–40 nm	Not done	Shankar et al. (2003)
	Fusarium oxysporum	Extracellular	72 h	Spherical and triangular, 20–40 nm	Not done	Mukherjee et al. (2002)
	Neurospora crassa	Extracellular	10^{-3} M of $HAuCl_4$	Spherical, 32 nm	Not done	Castro-Longoria et al. (2011)
	Phoma sp.	Extracellular	10^{-3} M of $HAuCl_4$	Spherical, 10–100 nm	Antimicrobial	Soltani Nejad et al. (2022)
	Trichoderma hamatum	Extracellular	38°C, pH 7, 0.5 mM gold chloride	Spherical, pentagonal, and hexagonal, 5–30 nm	Antimicrobial	Abdel-Kareem and Zohri (2018)
CdS	*Schizosaccharomyces pombe*	Extracellular	–	Hexagonal lattice, 1–1.5 nm	Not done	Kowshik et al. (2002)
Magnetite	*Fusarium oxysporum* and *Verticillium* sp.	Extracellular	27°C, 24 h	20–50 nm	Not done	Bharde et al. (2006)
Selenium	*Fusarium semitectum*	Extracellular	28°C, 3 days, 0.2 g of sodium selenite	92.33 nm	Antimicrobial, antioxidant, anticancer	Abbas and Baker (2020)

TABLE 6.1 *(Continued)*

NPs Type	Source	Synthesis Location	Synthesis Conditions	Shape	Bioactivity	References
Silica and Titanium	*Fusarium oxysporum*	Extracellular	27°C, 10–3M K_2SiF_6 (pH 3.1), K_2TiF_6 (pH 3.5), 200 rpm	5–15 nm	Not done	Bansal et al. (2005)
Te	*Aspergillus welwitschiae*	Extracellular	30°C, 48 h, 4 mmol K2TeO3	Oval to spherical, 60.8 nm	Antimicrobial	Abo Elsoud et al. (2018)
TiO_2	*Sachharomyces cerevisae*	Extracellular	60°C, 10–20 min, 0.025 (M) $TiO{\cdot}(OH)_2$	Spherical, 8–35 nm	Not done	Jha et al. (2009)
Zirconium	*Penicillium notatum*	Extracellular	27°C, 24 h, pH 3.6, 10–3 M K_2ZrF_6, 200 rpm	7.3 nm	Not done	Bansal et al. (2004)
ZnO	*Aspergillus fumigatus*	Extracellular	0.1 mM of salt solution zinc nitrate	Nearly spherical, 22.4 nm	Enhanced phosphorus mobilization	Raliya et al. (2016)
	Rhizoctonia bataticola	Extracellular	62 h	15–25 nm	Enhanced crop production	Tarafdar et al. (2014)
	Trichoderma harzianum	Extracellular	72 h, 150 rpm	Spherical, rod and hexagonal, 8–23 nm	Antifungal	Zaki et al. (2021)
	Xylaria acuta	Extracellular	1 g Zn $(NO_3)_2.6H_2O$, 400–700°C	34–55 nm	Antimicrobial, anticancer	Sumanth et al. (2020)

the application of these NPs enhanced leaf protein and chlorophyll contents in the plants (Raliya et al., 2016). Another study by Zaki et al. (2021) demonstrated that ZnO NPs synthesized using a cell-free aqueous filtrate of *Trichoderma harzianum* exhibited strong antifungal properties against soil-borne pathogens affecting cotton plants. Similarly, ZnO NPs were synthesized using strains of *Trichoderma* sp. (*T. harzianum*, *T. reesei*, and *T. reesei*) extracted from soil. These NPs showed potent antibacterial activity against *Xanthomonas oryzae* pv. *oryzae*, the bacterium responsible for causing Bacterial Leaf Blight in rice. Interestingly, when *Trichoderma* spp. was cocultured, the antibacterial results were even more effective compared to the monocultures. Furthermore, Fouda and Sofy (2022) synthesized ZnO NPs using cell-free extracts of *Penicillium chrysogenum* and compared their activity with chemically synthesized NPs and their ionic equivalent, $ZnSO_4$. When these myco-synthesized NPs were applied to plants at a concentration of 100 mg/l through foliar application, they showed promising results, leading to increased biosynthesis of photosynthetic pigments and improved yield. Additionally, the NPs enhanced the soluble sugar, protein, phenol, and oil content of canola seeds. Finally, Tarafdar et al. (2014) developed zinc nanofertilizers and applied them to pearl millet (*Pennisetum glaucum*). The results indicated significant enhancements in crop yield, shoot-root growth, chlorophyll content, and total soluble leaf protein. Moreover, the application of these nanofertilizers led to increased activity of alkaline phosphatase, dehydrogenase, acid phosphatase, and phytase enzymes in the rhizosphere, promoting the mobilization of native phosphorous for plant uptake. The findings also suggested that the NPs might enhance the population and activity of beneficial microorganisms in the rhizosphere, further aiding nutrient mobilization for plants.

6.6 NANOPESTICIDES

A growing concern regarding food waste and losses greatly reinforces the application of nanopesticides. Food waste and loss caused at various points in the chain process are gradually increasing the demand for the use of nanopesticides among crop growers and suppliers in order to reduce food waste. Additionally, global consumers prefer the consumption of fresh fruits and vegetables (Massaglia et al., 2019). Thus, growing demand also meets the urgency of supply for seasoned and unseasoned goods throughout the world. As a result, agro-industries are focused on exporting fresh vegetables and fruits, thereby boosting the need for the application of nanopesticides in the packing of these fresh items to retain product freshness. Further, during the COVID-19 epidemic, the food and agriculture industries were mostly hampered, but due to the health benefits associated with fruit and vegetable intake, there was an increase in population intake throughout the pandemic (Aday and Aday, 2020). In this context, nanopesticides are used in fruit and vegetable packing to maintain the freshness of these biodegradable foods, improve the performance of original pesticides, target transportation, and control the release of pesticides (Huang et al., 2018). Indeed, nanopesticides exhibit several advantages over traditional pesticides, as mentioned below:

1. Nanopesticides provide sustained release kinetics with high stability, efficient solubility, and enhanced permeability.
2. They help promote pest control effectiveness for a longer period of time.
3. They prevent degradation of the encapsulated molecules under adverse environmental conditions.
4. Application of nanopesticides leads to an improvement in crop quality and quantity as well.
5. Due to their smaller size, they are more uniformly distributed on crop leaves.
6. NPs with antimicrobial properties can improve the shelf life and freshness of packaged food.

It is well established that nanopesticides could significantly reduce or even destroy some specific organisms, and their utilization rate also greatly improved agricultural yield compared with conventional pesticides. But still, a huge proportion of nanopesticides cannot act on targeted organisms (Sun et al., 2019). Additionally, most of the microemulsions used in nanopesticides are polar in nature, which also has high acute toxicity (Pratap and Bhowmick, 2008). These polar solvents can easily enter the environment, are not biodegradable, and could widely cause acute poisoning in aquatic biota and even the death of animals (Chaud et al., 2021).

6.6.1 RECENT MARKET DRIVERS OF NANOPESTICIDES

Pesticides have been outlawed in several countries across the world due to their wide range of damaging impacts on people's livelihoods. Therefore, nations have devised several safety guidelines to address the issue and commercialize nanopesticides with better effects at a low chemical dose

(Kumari et al., 2023). The worldwide nanopesticide business is expanding gradually due to its widespread application in pest control as well as packaging. Reports stated Europe is on the leading edge of nanopesticide usage, accounting for more than 34% of yearly revenue creation in 2021. Asia-Pacific is projected as the second-highest market, and in the coming future, they are believed to register robust growth in the nanopesticide market. Besides market values, some other factors are also propelling the nanopesticide market in these regions, including demand for food grains, extensive exports of fresh foods and vegetables, and others (Rajna et al., 2019).

6.6.2 NANOPESTICIDES FROM PLANT ORIGIN

Nanopesticides, being a great substitute for conventional pesticides, offer a lot of potential in their application. Nanopesticides have a high surface area, allowing them to be target-specific, and their high degree of solubility further improves bioavailability at low dosages. These are some of the contrastive characteristics of nanopesticides over chemical pesticides that help to diminish the global harmful impact of pesticides (Shekhar et al., 2021). Generally, the procedure of NPs synthesis involves the addition of strong and weak chemical reducing agents like sodium borohydride, alcohols, and others, which are toxic and flammable in nature and also not eco-friendly. These facts prompted the researchers to explore eco-friendly options that could be target-specific and able to destroy pests that affect pre- and postharvest production. For these reasons, several plants and microorganisms have been used for the biogenic production of NPs. Green synthesis of NPs was considered a safe, sustainable, affordable, eco-friendly, and nontoxic substitute for chemical methods. In recent times, nanotechnology in plant disease management has increasingly gained attention, as several plants are already reported to be used as antiseptic agents against fungal-borne diseases of plants. "Diyarex Gold," a nano-biofungicide, is one such plant-based nanopesticide dominating the industry (Bhattacharyya et al., 2016). Green nanoformulations are the latest trend, being less toxic than chemical pesticides. Production of plant-based nanomaterials is beneficial because of their affordability, stability, and eco-friendliness (Pal et al., 2022). Thereby, various extracts from different plant parts, such as the stem, leaf, root, seed, and fruit pericarp, were utilized to synthesize NPs (Table 6.2).

TABLE 6.2 Pesticidal Activity of Some Plant-Based Nano-Biomaterials.

Nano-Biomaterials	Raw Materials	Parts Used	Target Organisms	References
Gold NPs	*Jasminum nervosum*	Leaf extract	*Culex quinquefasciatus*	Lallawmawma et al. (2015)
Silver NPs	*Berberis tinctoria*	Leaf extract	*Aedes albopictus, Toxorhynchites splendens* and *Mesocyclops thermocyclopoides*	Kumar et al. (2016)
Silver NPs	*Hedychium coronarium*	Rhizome extract	*Aedes aegypti, Megacyclops formosanus*	Kalimuthu et al. (2017)
Silver NPs	*Suaeda maritima*	Leaf extract	*Aedes aegypti* and *Spodoptera litura*	Suresh et al. (2018)
Silver NPs	*Nerium olender*	Leaf extract	*Leucinodes orbonalis* and *Anopheles stephensi*	Gogate et al. (2018)
Silver NPs	*Leonotis nepetifolia*	Leaf extract	*Spodoptera litura, Helicoverpa armigera, Aedes aegypti, Culex quinquefasciatus*	Manimegalai et al. (2020)
Nickel-palladium NPs	*Cocus nucifera*	Coir dust	*Callasobruchus maculates, Aedes ageypti*	Elango et al. (2016)
Selenium NPs	*Clausena dentata*	Leaf extract	*Aedes aegypti, Aedes stephensi, and Culex quinquefasciatus*	Sowndarya et al. (2017)
Silica NPs	Rice	Rice husk	Cotton aphids and mealybugs	Pavitra et al. (2018)
Zinc oxide NPs	*Pongamia pinnata*	Leaf extract	*Callosobruchus maculatus*	Malaikozhundan and Vinodhini (2018)
Nickel oxide NPs	*Rauvolfia serpentine*	Leaf extract	*Callosobruchus maculatus*	Rahman et al. (2020)
Iron oxide NPs	*Trigonella foenum-graecum*	Leaf extract	*Tuta absoluta*	Ramkumar et al. (2020)
Iron oxide NPs	*Cymbopogon citratus*	Leaf extract	*Caenorhabditis elegans*	Patiño-Ruiz et al. (2020)
Silver NPs	*Aristolochia indica*	Leaf extract	*Helicoverpa armigera*	Siva and Kumar (2015)
Nanoformulations	*Azadirachta indica*	Bark gum extract	*Helicoverpa armigera* and *Spodoptera litura*	Kamaraj et al. (2018)
Nanocapsule	*Azadirachta indica*	Neem seed kernel and neem oil	*Plutella xylostella*	Formin et al. (2013)
Nanoformulations	*Citrus sinensis*	Citrus peel and essential oil	*Tuta absoluta*	Compolo et al. (2017)

TABLE 6.2 *(Continued)*

Nano-Biomaterials	Raw Materials	Parts Used	Target Organisms	References
Titanium oxide NPs	*Moringa oleifera*	Leaf etract	Red palm weevil, *Rhynchophorus ferrugineus*	Al-Barly et al. (2015)
Nanoemulsion of essential oil	*Pimpinella anisum*, lemongrass and peppermint	Seeds	Oat aphid, *Rhopalosiphum padi*	Pascual-Villalobos et al. (2017)
Nanospheres of essential oil	*Zanthoxylum rhoifolium*	Seeds	Cotton whitefy, *Bemisia tabaci*	Christofoli et al. (2015)
Silver NPs	*Euphorbia prostrata*	Leaf extract	Rice weevil, *Sitophilus oryzae*	Zahir et al. (2012)
Silver and lead NPs	*Avicennia marina*	Leaf extract	*Sitophilus oryzae*	Sankar et al. (2015)
NPs loaded with essential oil	*Allium sativum*	Several parts of plant and essential oil	Red four beetle, *Tribolium castaneum*	Yang et al. (2009)
Silver NPs	*Camelina sativa*	Leaf extract	*Oryzaephilus surinamensis* and *Sitophilus granarius*	Rehman et al. (2021)
Copper NPs	*Lantana camara*	Leaf extract	*Anopheline mosquitoes*	Hassanain et al. (2019)
Copper nanocapsules	*Phaseolus vulgaris*	-	*Tetranychus urticae*	Dorri et al. (2018)
Titanium oxide NPs	*Desmostachya bipinnata*	Leaf extract	*Aedes aegypti* and *Spodoptera litura*	Shyam-Sundar et al. (2023)
Zinc oxide NPs	*Ficus racemosa*	Leaf extract	*Aedes albopictus*	Ragavendran et al. (2023)
Silver nanoperticles	*Euphorbia prostrata*	Leaf extract	*Sitophilus oryzae*	Zahir et al. (2012)
Silver NPs	*Nerium oleander*	Leaf extract	*Tribolium castaneum* and *Callosobruchus maculatus*	Jafer and Annon (2018)
Nickel NPs	*Cocos nucifera*	Leaf extract	*Callasobruchus maculates*	Elango et al. (2016)
NPs	*Azadirachta indica*	Leaf extract	*Plutella xylostella*	Forim et al. (2013)

6.6.3 NANOPESTICIDES FROM FUNGAL ORIGIN

In addition to plants, various fungal species have also shown remarkable potential to be used in the biogenic synthesis of NPs with different pesticidal characteristics (Guilger-Casagrande and Lima, 2019). Specifically, the basidiomycetous group of fungi, that is, mushrooms, demonstrate vast variability in their morphology, habitat, and cultivation. Several bioactive compounds have already been isolated and purified from mushrooms, such as beta-glucan, proteins, phenols, minerals, and complex terpenes (Khatua and Acharya, 2021). Mushroom-derived NPs possess some unique qualities, like extended shelf life, biocompatibility, high stability, and water solubility. According to Owaid (2020), NPs prepared using spent mushroom substrate from Oyster mushrooms were reported for the first time in 2007 to have remarkable medical potential toward several pathogenic microorganisms. Therefore, mushrooms seem to be a very potential candidate for biogenic synthesis of eco-friendly, safe, and stable nanomaterials. Currently, research on myco-NPs has been in focus, and several articles have reported the potential of NPs synthesis from wild and edible species as well. However, it has been speculated that mushrooms are a rich source of riboflavin, which further facilitates the reduction of metal ions into NPs (Bhat et al., 2013). An interesting study has reported that during the recovery of NPs (silver NPs, selenium NPs, gold NPs, and silicon NPs), fungal phenol oxidases and Mn-peroxidases are two important enzymes needed, and the study was done on four popular edible mushrooms, including *Ganoderma lucidum*, *Pleurotus ostreatus*, *Grifola frondosa,* and *Lentinula edodes* (Vetchinkina et al., 2017).

6.6.4 MODES OF ACTION OF FUNGI-DERIVED NANOPESTICIDES

The potential of biogenic nanopesticides synthesized from unicellular as well as multicellular fungi has been considered a sustainable and cost-effective technique for the pest management of agricultural crops. The material composition of nanofamilies includes metals, salts, metal oxides, organic nanomaterials, and nonmetal elements. These components also have biological relevance as antibacterials, antivirals, and insecticides. The vectors and insect pests cause great harm to agricultural crops, which results in public health concerns and a collapse of agricultural yield (Tudi et al., 2021). Recently, a huge number of studies have delineated the insecticidal importance of NPs toward various insect pests to highlight their potential to be developed as excellent insecticides, specifically nature-based NPs.

6.6.4.1 SILVER NANOPARTICLES AS NANOPESTICIDES

Research on NPs application for crop protection has significantly gained consideration over the last few years and motivates researchers to explore the effects of nanopesticides (Kah and Hofmann, 2014). In this context, metallic NPs of natural origin have proven useful against several plant pests, which will give them the supremacy to be selected as new pesticide formulations. However, one of the vital issues in the application of NPs is their toxicity, but silver has proven beneficial, particularly for its nontoxicity toward humans (Salleh et al., 2020). The antimicrobial efficacy of AgNPs has already been proven because of their larger surface area-to-volume ratio, which further increases their contact and penetration abilities inside the target organisms (Bruna et al., 2021).

Sundaravadivelan and Padmanabhan (2014) reported the pesticidal activity of AgNPs synthesized using *T. harzianum* against the different developmental stages of the dengue vector *Aedes aegypti*. They noticed that these biogenic AgNPs tested at the highest concentration (0.25%) promoted the death of I–IV instars and pupa of *Aedes aegypti.* They hypothesized that AgNPs produced from the filtrates of *T. harzianum* contain extracellular enzymes and proteins that play a prime role in the disruption of the cell membrane, which further leads to loss of cell permeability and cell death. However, the exact mechanism behind the larvicidal property of this AgNPs is not clear. Their study has shown that AgNPs could be utilized for the integrated mosquito control program.

Banu and Balasubramanian (2014b) reported the mosquitocidal activity of biogenic AgNPs synthesized using *Isaria fumosorosea* against *Culex quinquefasciatus* and *A. aegypti*. It was observed that the AgNPs were more effective against I–IV instars of *A. aegypti* than *Culex quinquefasciatus*. Moreover, the reason behind the lethality has not been studied yet. Similarly, some other literature also reported the larvicidal activity of AgNPs biosynthesized using fungi such as *Beauveria bassiana*, *Cochliobolus lunatus,* and others (Salunkhe et al., 2011; Banu and Balasubramanian, 2014b). Besides that, a nanoconjugate made up of silver NPs and pesticidal metabolites extracted from the fungal strain *Nomuraea rileyi* was prepared by adapting the *in vitro* green science method, which revealed amorphous and monodispersive NPs. Further, the pesticidal activity of the nanoconjugate was determined against *Spodoptera litura*, a lepidopteron pest, using different *in vitro* assays. The nanoconjugate was found to reduce different developmental parameters of *S. litura,* such as larval period, pupal period, adult emergence, and adult longevity. Further, different life stages of this lepidopteron pest were found to be susceptible to the NPs. Further investigation is required on the field trial of this nanoconjugate to be formulated as a biogenic nanopesticide against *Spodoptera litura* (Namasivayam and Bharani, 2021).

6.6.4.2 GOLD NANOPARTICLES AS NANOPESTICIDES

The reports of pesticidal activity of fungi-mediated gold nanopesticides are limited. One such gold NPs biosynthesis reported from a fungus, *Chrysosporium tropicum,* has shown potential insecticidal activity (Soni and Prakash, 2012a). It is reported that the fungus has its own components that are intrinsically toxic to larvae; thus, a strong cumulative insecticidal activity against *A. aegypti* was noticed. The authors also studied a comparative overview of lethality between gold and silver NPs produced from the same fungus. It was observed that at a concentration of 0.77 ppm (mg/l), the Au-NPs have been tested as a larvicide against the I instar larvae of *A. aegypti*, showing 100% mortality after 24 h. On the other hand, the gold NPs have promoted mortality at three times higher concentrations than the silver ones. Therefore, the result suggested that *C. tropicum* can be used as an environmentally safer approach for the development of gold NPs for mosquito control. Further, the same group of researchers also studied the larvicidal effect of gold NPs prepared using *A. niger* against the larvae of 3 mosquito species: *Anopheles stephensi*, *C. quinquefasciatus*, and *A. aegypti* (Soni and Prakash, 2012b). It was found that the mortality rates of larval instars (I–IV) of *C. quinquefasciatus* after the treatment of Au-NPs were greater than those of *A. aegypti* and *A. stephensi*. The effect was very prominent for *C. quinquefasciatus* just after 48 h, whereas the mortality rate was recorded only after 72 h for *A. aegypti* and *A. stephensi*. Thus, these NPs could lead to a new nanopesticide for mosquito control.

6.6.4.3 COPPER NANOPARTICLES AS NANOPESTICIDES

Copper NPs and their conjugated complexes are reported to be used as antimicrobial, antiviral, and wound-healing agents (Sharmin et al., 2021). Further, it demonstrated a promising role as a larvicidal antifeedant, as investigated by Rahman et al. (2022) using the polyphagous pest *Spodoptera frugiperdai*. With respect to fungal origin, Kalaimurugan et al. (2019) conducted an investigation on the biosynthesizing of copper NPs from whole cell biomass of *Fusarium proliferatum*. The NPs efficiently inhibited the growth of *A. aegypti*, *C. quinquefasciatus*, and *A. stephensi* with LC_{90} of 2130.99, 894.90, and 173.92 μg/ml, respectively. Thus, they suggested that it could be used as a potential mosquitocidal agent. In this context, another group of researchers prepared

Cu-NPs using the aqueous extract of the fungus *Metarhizium robertsii* and determined its pesticidal efficacy against *A. aegypti*, *C. quinquefasciatus*, and *A.* stephensi (Vivekanandhan et al., 2021). All the strains were found to be susceptible to these NPs, thus indicating their potential to be formulated as a new generation of biogenic nanopesticides. However, the LC_{90} of the Cu-NPs against the IV instar larvae of *A. aegypti*, *C. quinquefasciatus*, and *A. stephensi* was recorded as 65.14, 14.99, and 23.71 μg/ml respectively, which was much less than the previous finding, demonstrating its superiority over the previous one. However, related in-depth studies are limited so far regarding its hazardous impact on humans and ecosystems. An interesting report concerning the novel application of gelatin–copper bio-NPs against the cotton crop pest *Earias insulana* was published (Hala and Eman, 2021). The gelatin–copper NPs were synthesized using filtrates of *Aspergillus wentii*, *Aspergillus mottae,* and *Penecillium expansum.* It was noticed that the mortality rates of the larva treated with gelatin–copper NPs were higher than those of Cu-NPs produced using the same fungi. The reason behind these significant effects of NPs on larval and adult stages of the pest might be due to the penetration of the NPs, thus causing damage to the exoskeleton. Consequently, they elucidated the extraordinary insecticidal efficacy of these biofabricated NPs to be employed as biocontrol agents against spiny bollworm.

6.6.4.4 *OTHER METAL-BASED NPS AS NANOPESTICIDES*

Magnesium oxide (MgO) NPs are generally used in pharmaceutical preparations with diverse medicinal applications and are also used in wastewater treatment (Abinaya et al., 2021). However, little information is available regarding the pesticidal activity of MgO-NPs of fungal origin. *Penicillium chrysogenum*-synthesized MgO-NPs were shown to produce inhibitory effects on the larvae and pupae of *A. stephensi.* The LC_{50} value is found in the range of 12.4–29.8 ppm depending on its different growth stages (Fouda et al., 2021). Recently, Kumaravel et al. (2021) claimed bimetallic NPs can be used as nanopesticides. They focused on the biosynthesis of nanoformulations using fungal metabolites from *Metarhizium anisopliae* and investigated their pesticidal activity against *Spodoptera frugiperda.* They prepared bimetallic zinc oxide and titanium dioxide NPs, which were found to be spherical in shape and crystalline in nature. According to them, upon treatment with 100 μg/ml for 48 h, pupicidal and antifeedent activities were prominently observed, which resulted from the toxic manifestation of the NPs within the insect body. Moreover, MgO nanostructures cause distortion in the body axes of larvae, whereas ZnO-exposed larvae exhibit hyperpigmentation and structural abnormalities in the respiratory area (Chakrabarti and Patra, 2020). Cadmium oxide (CdO) NPs were synthesized using the culture filtrates of *Nigrospora oryzae,* which demonstrated insecticidal activity against III-instar larvae of *A. aegypti*, *C. quinquefasciatus*, and *A. stephensi* (Gowri et al., 2018). Therefore, it would appear that metallic NPs have outstanding mosquitocidal properties and control other insect pests as well, particularly larval and pupal populations. However, only a few reports have been published on the targeted approach to myco-synthesized NPs in the adult stage of the stored product pests. Recently, a group of researchers made an attempt to compare the pesticide efficacy of six types of NPs (silver, titanium dioxide, selenium, copper oxide, silicon dioxide, and zinc oxide) prepared using *Fusarium solani* (Helmy et al., 2023). Eggs and larvae of stored bean pests such as *Callosobruchus chinensis* and *Callosobruchus maculate* were used as target organisms. A species-dependent susceptibility to the NPs was observed, and eggs were found to be more susceptible than larvae. Among the NPs, selenium and titanium dioxide NPs decreased the hatchability of *C. chinensis* eggs by 23% and 18%, respectively, whereas in *C. maculatus*, titanium dioxide NPs treatment reduced the larva-to-adult survival rate by 11%, which further leads to a reduction in egg-to-adult survival. Thus, the results recommended that direct application of biosynthesized selenium and titanium dioxide NPs onto the eggs of the stored pest would be the most effective strategy to control the pests.

Researchers have developed a chitosan nanocomposite comprising chitosan and insecticidal metabolites of the fungus *Nomuraea rileyi,* and it was found to be highly stable with uniformly dispersed particles and high entrapment efficiency (Namasivayam et al., 2018). Pesticidal activity of the nanocomposite was measured by observing mortality rates of larval instars of *Spodoptera litura.* Further, the nanocomposite demonstrated remarkable potential against all the larval stages, showing high mortality, drastic midgut reduction, and depletion of hemolymph macromolecules. However, a biocompatibility assay revealed it to be nontoxic and not able to induce hemolysis in zebra fish. Thus, the finding suggested that this fabricated nanocomposite of the fungus is a suitable candidate for pest control without poisoning nontarget organisms.

6.7 ROLE OF ENTOMOPATHOGENIC FUNGI AS NANOPESTICIDES

An entomopathogenic fungus is a microorganism, capable of infecting, parasitizing, and killing of insect, and also during their life cycle they use insect as their temporary host. Thus, studying and manipulating their behavior as pest control agent would be beneficial for integrated pest management of agricultural crops. These microorganisms also participate in regulation of insect populations naturally within the environment and recycling organic matter (Liu et al., 2023). They represent a great phylogenetic diversity and are mostly distributed in phyla like Microsporidia, Oomycota, Chytridiomycota, Entomophtoromycota, Basidiomycota, and Ascomycota and mostly found abundantly in soils of terrestrial ecosystem rich in organic matter and humus. Generally, they are used as a substitute of traditional chemical pesticides in agricultural sectors because of their parasitic nature. The first successful use of *B. bassiana*, an entomopathogenic fungus, as a biocontrol agent to eradicate *Cylas formicarius* causing disease in sweet potato was executed in 1956 in China (Bihal et al., 2023).

6.7.1 CRITICAL POINTS TO BE ADDRESSED

In order to find eco-friendly, cost-effective, and potential pesticides to eradicate the plant pest, some major parameters need to be considered. This includes the preparation technique, which must be easy and simple, available at affordable rates, and target-specific with broad-spectrum application. Further, it must be nontoxic to humans and the environment, not accumulate in the food chain, and work efficiently at low dosages. Thus, there is a great deal of concern regarding the applicability of nanobiopesticides, which needs to be resolved through critical research. Further, the interaction of nanopesticides between soil and its organic matter is another crucial parameter that needs to be scientifically addressed. The fungi with pesticidal properties may play a prime role in nanopesticide development, thereby acknowledging their importance and influencing the future formation of NPs. In the context of metallic nanopesticides, the issue of whether the insecticidal activity of Au and AgNPs mainly comes from the components of their raw materials or their specific nanoscale properties is still pending. The utilization of biocompatible substances as raw materials in the mentioned studies contributes to the insecticidal activity of the NPs, which could be attributed to metal ions or physicochemical stimuli. Future work should focus on verifying the insecticidal efficacy of these biogenic NPs in a mimicked wild environment and optimizing the dosage to exhibit favorable effects against harmful insects while simultaneously being nontoxic to nontarget organisms. Hotspots for future research may incorporate the precise orchestration of form, size, and surface chemistry to enhance the insecticidal potency of myco-NPs.

6.8 CONCLUSION AND FUTURE PERSPECTIVES

Overall, the involvement of fungi in the biogenic synthesis of nanofertilizers and nanopesticides offers a promising pathway for sustainable agriculture. Its future potential is both dynamic and exhilarating, with several key perspectives emphasizing the significance of ongoing research and development in this domain.

1. Future research should focus on exploring and identifying novel fungal strains with enhanced capabilities for NPs synthesis. By understanding the genetic and metabolic basis of nanoparticle production in different fungi, we can tailor their properties to specific agricultural needs, such as nutrient release profiles or pest-specific targeting.
2. Developing innovative formulations that optimize stability, solubility, and interaction with plants or pests will be crucial for maximizing their effectiveness.
3. Research should also prioritize comprehensive studies to ensure the ecological safety of these nanoproducts.
4. Global collaboration among scientists, researchers, and policymakers will foster knowledge exchange, enabling rapid progress in the field. In conclusion, the future of fungal-mediated biogenic synthesis of nanofertilizers and nanopesticides holds immense potential for sustainable agriculture. By harnessing the unique abilities of fungi and integrating them with nanotechnology, we can pave the way for environmentally friendly, efficient, and targeted solutions to enhance crop productivity and pest management while minimizing the ecological footprint of agriculture.

KEYWORDS

- **Biogenic synthesis**
- **crop yield**
- **mode of action**
- **mosquitocidal**
- **nanoparticles**
- **plant growth**
- **silver**

REFERENCES

Abbas, H., & Baker, D. A. (2020). Biological evaluation of selenium nanoparticles biosynthesized by *Fusarium semitectum* as antimicrobial and anticancer agents. *Egyptian Journal of Chemistry*, *63*(4), 18–19.

Abdel-Hadi, A., Iqbal, D., Alharbi, R., Jahan, S., Darwish, O., Alshehri, B., Banawas, S., Palanisamy, M., Ismail, A., Aldosari, S., Alsaweed, M., Madkhali, Y., Kamal, M., & Fatima, F. (2023). Myco-synthesis of silver nanoparticles and their bioactive role against pathogenic microbes. *Biology*, *12*(5), 661.

Abdel-Kareem, M. M., & Zohri, A. A. (2018). Extracellular mycosynthesis of gold nanoparticles using *Trichoderma hamatum*: optimization, characterization and antimicrobial activity. *Letters in Applied Microbiology*, *67*(5), 465–475.

AbdelRahim, K., Mahmoud, S. Y., Ali, A. M., Almaary, K. S., Mustafa, A. E. Z. M. A., & Husseiny, S. M. (2017). Extracellular biosynthesis of silver nanoparticles using *Rhizopus stolonifer*. *Saudi Journal of Biological Sciences*, *24*(1), 208–216.

Abdollahdokht, D., Gao, Y., Faramarz, S., Poustforoosh, A., Abbasi, M., Asadikaram, G., & Nematollahi, M. H. (2022). Conventional agrochemicals towards nano-iopesticides: An overview on recent advances. *Chemical and Biological Technologies in Agriculture*, *9*(13).

Abinaya, S., Kavitha, H. P., Prakash, M., & Muthukrishnaraj, A. (2021). Green synthesis of magnesium oxide nanoparticles and its applications: A review. *Sustainable Chemistry and Pharmacy, 19*(7), 100368.

Abo Elsoud, M. M., Al-Hagar, O. E. A., Abdelkhalek, E. S., & Sidkey, N. M. (2018). Synthesis and investigations on tellurium myconanoparticles. *Biotechnology Reports (Amsterdam, Netherlands), 18*, e00247.

Acay, H., & Baran, M. F. (2019). Biosynthesis and characterization of silver nanoparticles using king oyster (*Pleurotus eryngii*) extract: Effect on some microorganisms. *Applied Ecology and Environmental Research*, *17*(4), 9205–9214.

Aday, S., & Aday, M. S. (2020). Impact of COVID-19 on the food supply chain. *Food Quality and Safety*, *4*(4), 167–180.

Ahluwalia, V., Kumar, J., Sisodia, R., Shakil, N. A., & Walia, S. (2014). Green synthesis of silver nanoparticles by *Trichoderma harzianum* and their bio-efficacy evaluation against *Staphylococcus aureus* and *Klebsiella pneumonia.* Industrial Crops *and* Products, *55*, 202–206.

Al-Barly, A., & Hamza, R. Z. (2015). Larvicidal, anti-oxidant activities and perturbation of transminases activities of titanium dioxide nanoparticles synthesized using *Moringa oleifera* leaves extract against the red palm weevil (*Rhynchophorus ferrugineus*). *Environmental Science, Biology, Chemistry*, *2*, 49–54.

Allouzi, M. M. A., Allouzi, S. M. A., Keng, Z. X., Supramaniam, C. V., Singh, A., & Chong, S. (2022). Liquid biofertilizers as a sustainable solution for agriculture. *Heliyon*, *8*(12), e12609.

Ashrafi, S. J., Rastegar, M. F., Ashrafi, M., Yazdian, F., Pourrahim, R., & Suresh, A. K. (2013). Influence of external factors on the production and morphology of biogenic silver nanocrystallites. *Journal of Nanoscience and Nanotechnology*, *13*(3), 2295–2301.

Avila-Quezada, G., Ingle, A., Golińska, P., & Rai, M. (2022). Strategic applications of nano-fertilizers for sustainable agriculture: Benefits and bottlenecks. *Nanotechnology Reviews*, *11*(1), 2123–2140.

Azmath, P., Baker, S., Rakshith, D., & Satish, S. (2016). Mycosynthesis of silver nanoparticles bearing antibacterial activity. *Saudi Pharmaceutical Journal, 24*(2), 140–146.

Balaji, D. S., Basavaraja, S., Deshpande, R., Mahesh, D. B., Prabhakar, B. K., & Venkataraman, A. (2009). Extracellular biosynthesis of functionalized silver nanoparticles by strains of *Cladosporium cladosporioides* fungus. *Colloids Surfaces. B, Biointerfaces*, *68*(1), 88–92.

Balakumaran, M. D., Ramachandran, R., & Kalaicheilvan, P. T. (2015). Exploitation of endophytic fungus, *Guignardia mangiferae* for extracellular synthesis of silver nanoparticles and their *in vitro* biological activities. *Microbiological Research, 178*, 9–17.

Bansal, V., Rautaray, D., Ahmad, A., & Sastry, M. (2004). Biosynthesis of zirconia nanoparticles using the fungus *Fusarium oxysporum*. *Journal of Materials Chemistry, 14*(22), 3303–3305.

Bansal, V., Rautaray, D., Bharde, A., Ahire, K., Sanyal, A., Ahmad, A., & Sastry, M. (2005). Fungus-mediated biosynthesis of silica and titania particles. *Journal of Materials Chemistry, 15*(26), 2583–2589.

Banu, A. N., & Balasubramanian, C. (2014a). Optimization and synthesis of silver nanoparticles using *Isaria fumosorosea* against human vector mosquitoes. *Parasitology Research, 113*(10), 3843–3851.

Banu, A. N., & Balasubramanian, C. (2014b). Myco-synthesis of silver nanoparticles using *Beauveria bassiana* against dengue vector, *Aedes aegypti* (Diptera: Culicidae). *Parasitology Research, 113*(8), 2869–2877.

Beltrán Pineda, M. E., Lizarazo Forero, L. M., & Sierra Y. C. A. (2022). Mycosynthesis of silver nanoparticles: A review. *Biometals: An International Journal on the Role of Metal Ions in Biology, Biochemistry, and Medicine*, *36*(4), 745–776. https://doi.org/10.1007/s10534-022-00479-1.

Bhainsa, K. C., & D'souza, S. F. (2006). Extracellular biosynthesis of silver nanoparticles using the fungus *Aspergillus fumigatus*. *Colloids and Surfaces B: Biointerfaces*, *47*(2), 160–164.

Bharde, A., Rautaray, D., Bansal, V., Ahmad, A., Sarkar, I., Yusuf, S. M., Sanyal, M., & Sastry, M. (2006). Extracellular biosynthesis of magnetite using fungi. *Small*, *2*(1), 135–141.

Bhat, R., Sharanabasava, V. G., Deshpande, R., Shetti, U., Sanjeev, G., & Venkataraman, A. (2013). Photo-bio-synthesis of irregular shaped functionalized gold nanoparticles using edible mushroom *Pleurotus florida* and its anticancer evaluation. *Journal of Photochemistry and Photobiology B: Biology*, *125*, 63–69.

Bhattacharyya, A., Duraisamy, P., Govindarajan, M., Buhroo, A. A., & Prasad, R. (2016). Nano-Biofungicides: Emerging trend in insect pest control. In R. Prasad (Eds.), *Advances and Applications Through Fungal Nanobiotechnology* (pp 307–319). Springer, Cham.

Bihal, R., Al-Khayri, J. M., Banu, A. N., Kudesia, N., Ahmed, F. K., Sarkar, R., Arora, A., & Abd-Elsalam, K. A. (2023). Entomopathogenic fungi: An eco-friendly synthesis of sustainable nanoparticles and their nanopesticide properties. *Microorganisms*, *11*(6), 1617.

Birla, S. S., Gaikwad, S. C., Gade, A. K., & Rai, M. K. (2013). Rapid synthesis of silver nanoparticles from *Fusarium oxysporum* by optimizing physicocultural conditions. *The Scientific World Journal*, *2013*, 796018.

Bruna, T., Maldonado-Bravo, F., Jara, P., & Caro, N. (2021). Silver nanoparticles and their antibacterial applications. *International Journal of Molecular Sciences*, *22*(13), 7202.

Castro-Longoria, E., Vilchis-Nestor, A. R., & Avalos-Borja, M. (2011). Biosynthesis of silver, gold and bimetallic nanoparticles using the filamentous fungus *Neurospora crassa*. *Colloids and Surfaces B: Biointerfaces, 83*(1), 42–48.

Chakrabarti, A., & Patra, P. (2020). Relative larvicidal property of common oxide nanostructures against *Culex quinquefasciatus*. *IET Nanobiotechnology*, *14*(5), 389–395.

Chandankere, R., Chelliah, J., Subban, K., Shanadrahalli, V. C., Parvez, A., Zabed, H. M., Sharma, Y. C., & Qi, X. (2020). Pleiotropic functions and biological potentials of silver nanoparticles synthesized by an endophytic fungus. *Frontiers in Bioengineering and Biotechnology*, *8*, 95.

Chaud, M., Souto, E. B., Zielinska, A., Severino, P., Batain, F., Oliveira-Junior, J., & Alves, T. (2021). Nanopesticides in agriculture: Benefits and challenge in agricultural productivity, toxicological risks to human health and environment. *Toxics*, *9*(6), 131.

Chaudhary, P., Singh, S., Chaudhary, A., Sharma, A., & Kumar, G. (2022). Overview of biofertilizers in crop production and stress management for sustainable agriculture. *Frontiers in Plant Science*, *13*, 930340.

Chauhan, A., Anand, J., Parkash, V., & Rai, N. (2023). Biogenic synthesis: A sustainable approach for nanoparticles synthesis mediatedby fungi. *Inorganic and Nano-Metal Chemistry*, *53*(5), 460–473.

Christofoli, M., Costa, E. C. C., Bicalho, K. U., de Cassia, D. V., Peixoto, M. F., Alves, C. C. F., & de Melo, C.C. (2015). Insecticidal effect of nanoencapsulated essential oils from *Zanthoxylum rhoifolium* (Rutaceae) in Bemisia tabaci populations. *Industrial Crops and Products, 70*, 301–308.

Compolo, O., Cherif, A., Ricapero, M., Siscaro, G., Grissa-Lebdi, K., Russo, A., Cucci, L. M., Pierto, D. D., Satriano, C., Demeu, X. N., Blondi, A., Zappala, Z., & Palmeri, V. (2017). Citrus peel essential oil nanoformulations to control the tomato borer, *Tuta absoluta*: chemical properties and biological activity. *Scientific Reports, 7*(1), 13036.

Costa Silva, L. P., Oliveira, J. P., Keijok, W. J., Silva, A. R., Aguiar, A. R., Guimarães, M. C. C., Ferraz, C. M., Araújo, J. V., Tobias, F. L., & Braga, F. R. (2017). Extracellular biosynthesis of silver nanoparticles using the cell-free filtrate of nematophagus fungus *Duddingtonia flagans*. *International Journal of Nanomedicine*, *12*, 6373–6381.

Daniel, A. I., Fadaka, A. O., Gokul, A., Bakare, O. O., Aina, O., Fisher, S., Burt, A. F., Mavumengwana, V., Keyster, M., & Klein, A. (2022). Biofertilizer: The future of food security and food safety. *Microorganisms*, *10*(6), 1220.

Dhillon, G. S., Brar, S. K., Kaur, S., & Verma, M. (2012). Green approach for nanoparticle biosynthesis by fungi: current trends and applications. *Critical Reviews in Biotechnology*, *32*(1), 49–73.

Dorri, H. R., Khaghani, S., Moghadam, A., Ghanbari, D., & Bihamta, M. R. (2018). The effect of copper nano-capsules on the control of two spotted spider mite (Tetranychusurticae). *Journal of Nanostructures*, *8*(3), 316–324.

Du, L., Xu, Q., Huang, M., Xian, L., & Feng, J. -X. (2015). Synthesis of small silver nanoparticles under light radiation by fungus *Penicillium oxalicum* and its application for the catalytic reduction of methylene blue. *Materials Chemistry and Physics, 160*, 40–47.

Dykman, L. A., & Khlebtsov, N. G. (2011). Gold nanoparticles in biology and medicine: Recent advances and prospects. *Acta Naturae*, *3*(2), 34–55.

Elamawi, R. M., Al-Harbi, R. E. & Hendi, A. A. (2018). Biosynthesis and characterization of silver nanoparticles using *Trichoderma longibrachiatum* and their effect on phytopathogenic fungi. *Egyptian Journal of Biological Pest Control*, *28*(28).

El-Saadony, M. T., ALmoshadak, A. S., Shafi, M. E., Albaqami, N. M., Saad, A. M., El-Tahan, A. M., Desoky, E. M., Elnahal, A. S. M., Almakas, A., Abd El-Mageed, T. A., Taha, A. E., Elrys, A. S., & Helmy, A. M. (2021). Vital roles of sustainable nano-fertilizers in improving plant quality and quantity-an updated review. *Saudi Journal of Biological Sciences, 28*(12), 7349–7359.

Fayaz, A. M., Balaji, K., Kalaichelvan, P. T., & Venkatesan, R. (2009). Fungal based synthesis of silver nanoparticles—an effect of temperature on the size of particles. *Colloids and Surfure. B, Biointerfaces, 74*(1), 123–126.

Forim, M. R., Costa, E. S., Da Silva, M. F. D. G. F., Fernandes, J. B., Mondego, J. M., & Boiça, J. A. L. (2013). Development of a new method to prepare nano-/microparticles loaded with extracts of *Azadirachta indica*, their characterization and use in controlling *Plutella xylostella. Journal of Agricultural and Food Chemistry*, *61*(38), 9131–9139.

Fouda, A., Awad, M. A., Eid, A. M., Saied, E., Barghoth, M. G., Hamza, M. F., Awad, M. F., Abdelbary, S., & Hassan, S. E. (2021). An eco-friendly approach to the control of pathogenic microbes and *Anopheles stephensi* malarial vector using magnesium oxide nanoparticles (Mg-NPs) fabricated by *Penicillium chrysogenum. International Journal of Molecular Sciences*, *22*(10), 5096.

Fouda, H., & Sofy, M. (2022). Effect of biological synthesis of nanoparticles from *Penicillium chrysogenum* as well as traditional salt and chemical nanoparticles of zinc on canola plant oil productivity and metabolic activity. *Egyptian Journal of Chemistry, 65*(3), 507–516.

Gade, A., Gaikwad, S., Duran, N., & Rai, M. (2014). Green synthesis of silver nanoparticles by *Phoma glomerata. Micron, 59*, 52–59.

Gowri, S., Gopinath, K., & Arumugam, A. (2018). Experimental and computational assessment of mycosynthesized CdO nanoparticles towards biomedical applications. *Journal of Photochemistry and Photobiology B: Biology*, *180,* 166–174.

Guilger-Casagrande, M., & de Lima, R. (2019). Synthesis of silver nanoparticles mediated by fungi: A review. *Frontiers in Bioengineering and Biotechnology, 7*, 287.

Gupta, P., Rai, N., Verma, A., Saikia, D., Singh, S. P., Kumar, R., Singh, S. K., Kumar, D., & Gautam, V. (2022). Green-based approach to synthesize silver nanoparticles using the fungal endophyte *Penicillium oxalicum* and their antimicrobial, antioxidant, and *in vitro* anticancer potential. *ACS Omega*, *7*(50), 46653–46673.

Hamedi, S., Ghaseminezhad, M., Shokrollahzadeh, S., & Shojaosadati, S. A. (2017). Controlled biosynthesis of silver nanoparticles using nitrate reductase enzyme induction of filamentous fungus and their antibacterial evaluation. *Artificial Cells, Nanomedicine, and Biotechnology*, *45*(8), 1588–1596.

Hassanain, N. A. E. H., Shehata, Z. A., Mokhtar, M. M., Shaapan, M. R., Hassanain, M. A. E. H., & Zaky, S. (2019). Comparison between insecticidal activity of *Lantana camara* extract and its synthesized nanoparticles against anopheline mosquitoes. *Pakistan Journal of Biological Sciences, 22*(*7*), 327–334.

Helmy, E. A. M., San, P. P., Zhang, Y. Z., Adarkwah, C., & Tuda, M. (2023) Entomotoxic efficacy of fungus-synthesized nanoparticles against immature stages of stored bean pests. *Scientific Reports, 13*(8508).

Honary, S., Barabadi, H., Gharaei-Fathabad, E., & Naghibi, F. (2012). Green synthesis of copper oxide nanoparticles using *Penicillium aurantiogriseum*, *Penicillium citrinum* and *Penicillium waksmanii. Digest Journal of Nanomaterials and Biostructures*, *7*(3), 999–1005.

Huang, B., Chen, F., Shen, Y., Qian, K., Wang, Y., Sun, C., Zhao, X., Cui, B., Gao, F., Zeng, Z., & Cui, H. (2018). Advances in targeted pesticides with environmentally responsive controlled release by nanotechnology. *Nanomaterials*, *8*(2), 102.

Husseiny, S. M., Salah, T. A., & Anter, H. A. (2015). Biosynthesis of size-controlled silver nanoparticles by *Fusarium oxysporum*, their antibacterial and antitumoral activities. *Beni-Suef University Journal of Basic and Applied Sciences*, *4*(3), 225–231.

Ishida, K., Cipriano, T. F., Rocha, G. M., Weissmüller, G., Gomes, F., Miranda, K., & Rozental, S. (2014). Silver nanoparticle production by the fungus *Fusarium oxysporum*: nanoparticle characterisation and analysis of antifungal activity against pathogenic yeasts. *Memórias do Instituto Oswaldo Cruz, 109*(2), 220–228.

Jafer, F. S., & Annon, M. R. (2018). Larvicidal effect of pure and green-synthesized silver nanoparticles against *Tribolium castaneum* (Herb.) and *Callosobruchus maculatus* (Fab.). *Journal of Global Pharma Technology, 10*(3), 448–454.

Jha, A. K., Prasad, K., & Kulkarni, A. R. (2009). Synthesis of TiO_2 nanoparticles using microorganisms. *Colloids and Surfaces B: Biointerfaces*, *71*(2), 226–229.

Joshi, S. K., & Gauraha, A. K. (2022). Global biofertilizer market: Emerging trends and opportunities. In R. Soni, D. C. Suyal, A. N. Yadav, & R. Goel (Eds.), *Trends of applied microbiology for sustainable economy: Developments in applied microbiology and biotechnology* (pp. 689–697). Academic Press.

Kah, M., & Hofmann, T. (2014). Nanopesticide research: Current trends and future priorities. *Environment International, 63*, 224–235.

Kalaimurugan, D., Sivasankar, P., Lavanya, K., Shivakumar, M. S., & Venkatesan, S. (2019). Antibacterial and larvicidal activity of *Fusarium proliferatum* (YNS2) whole cell biomass mediated copper nanoparticles. *Journal of cluster science, 30*(4), 1071–1080.

Kamaraj, C., Gandhi, P. R., Elango, G., Karthi, S., Chung, I. M., & Rajakumar, G. (2018). Novel and environmentally friendly approach: impact of Neem (*Azadirachta indica*) Gum Nano Formulation (NGNF) on *Helicoverpa armigera* (Hub.) and *Spodoptera litura* (Fab.). *International Journal of Biological Macromolecules, 107*(Pt A), 59–69.

Khan, A. U., Malik, N., Khan, M., Cho, C. H., & Khan, M. M. (2018). Fungi-assisted silver nanoparticle synthesis and their applications. *Bioprocess and Biosystems Engineering, 41*(1), 1–20.

Khatua, S., & Acharya, K. (2021). Exploration of macrofungal wealth of West Bengal in 21st Century. *Journal of Mycopathological research, 59*(3), 207–224.

Kitching, M., Ramani, M., & Marsili, E. (2015). Fungal biosynthesis of gold nanoparticles: mechanism and scale up. *Microbial Biotechnology, 8*(6), 904–917.

Kobashigawa, J. M., Robles, C. A., Martínez Ricci, M. L., & Carmarán, C. C. (2019). Influence of strong bases on the synthesis of silver nanoparticles (AgNPs) using the ligninolytic fungi *Trametes trogii. Saudi Journal of Biological Science*, *26*(7), 1331–1337.

Korbekandi, H., Ashari, Z., Iravani, S., & Abbasi, S. (2013). Optimization of biological synthesis of silver nanoparticles using *Fusarium oxysporum. Iranian Journal of Pharmaceutical Research, 12*(3), 289–298.

Kowshik, M., Deshmukh, N., Vogel, W., Urban, J., Kulkarni, S. K., & Paknikar, K. M. (2002). Microbial synthesis of semiconductor CdS nanoparticles, their characterization, and their use in the fabrication of an ideal diode. *Biotechnology and Bioengineering, 78*(5), 583–588.

Kumar, N., Samota, S. R., Venkatesh, K., & Tripathi S. C. (2023). Global trends in use of nano-fertilizers for crop production: Advantages and constraints-A review. *Soil and Tillage Research, 228*, 105645.

Kumaravel, J., Lalitha, K., Murugan, A., & Shivakumar, M. S. (2021). Mycosynthesis of bimetallic zinc oxide and titanium dioxide nanoparticles for control of *Spodoptera frugiperda. Pesticide Biochemistry and Physiology*, *178*, 104910.

Kumari, R., & Singh, D. P. (2020). Nano-biofertilizer: An emerging eco-friendly approach for sustainable agriculture. *Proceedings of the National Academy of Sciences*, *90*(4), 733–741.

Kumari, R., Suman, K., Karmakar, S., Mishra, V., Lakra, S. G., Saurav, G. K., & Mahto, B. K. (2023). Regulation and safety measures for nanotechnology-based agri-products. *Frontiers in Genome*, *5*, 1200987.

Lallawmawma, H., Sathishkumar, G., Sarathbabu, S., Ghatak, S., Sivaramakrishnan, S., Gurusubramanian, G., & Kumar, N. S. (2015). Synthesis of silver and gold nanoparticles using *Jasminum nervosum* leaf extract and its larvicidal activity against filarial and arboviral vector *Culex quinquefasciatus* Say (Diptera: Culicidae). *Environmental Science and Pollution Research International,* 22(22), 17753–17768.

Li, G., He, D., Qian, Y., Guan, B., Gao, S., Cui, Y., Yokoyama, K., & Wang, L. (2012). Fungus-mediated green synthesis of silver nanoparticles using *Aspergillus terreus*. *International Journal of Molecular Science, 13*(1), 466–476.

Li, P. J., Pan, J. J., Tao, L. J., Li, X., Su, D. L., Shan, Y., & Li, H. Y. (2021). Green synthesis of silver nanoparticles by extracellular extracts from *Aspergillus japonicas* PJ01. *Molecules*, *26*(15), 4479.

Liu, D., Smagghe, G., & Liu, T. X. (2023). Interactions between entomopathogenic fungi and insects and prospects with glycans. *Journal of Fungi*, *9*(5), 575.

Liu, X., Chen, J. L., Yang, W. Y. Qian, Y. C., Pan, J. Y., Zhu, C. N., Liu, L., Ou, W. B., Zhao, H. X., & Zhang, D. P. (2021). Biosynthesis of silver nanoparticles with antimicrobial and anticancer properties using two novel yeasts. *Scientific Reports, 11*(1), 15795.

Loshchinina, E. A., Vetchinkina, E. P., & Kupryashina, M. A. (2022). Diversity of biogenic nanoparticles obtained by the fungi-mediated synthesis: A review. *Biomimetics (Basel), 8*(1), 1.

Malaikozhundan, B., & Vinodhini, J. (2018). Nanopesticidal effects of *Pongamia pinnata* leaf extract coated zinc oxide nanoparticle against the Pulse beetle, *Callosobruchus maculatus*. *Materials Today Commununications*, *14,* 106–115.

Massaglia, S., Borra, D., Peano, C., Sottile, F., & Merlino, V. M. (2019). Consumer preference heterogeneity evaluation in fruit and vegetable purchasing decisions using the best–worst approach. *Foods*, *8*(7), 266.

Michael, A., Singh, A., Roy, A., & Islam, M. R. (2022). Fungal- and algal-derived synthesis of various nanoparticles and their applications. *Bioinorganic Chemistry and Applications*, *2022*, 3142674.

Mitter, E. K., Tosi, M., Obregón, D., Dunfield, K. E., & Germida, J. J. (2021). Rethinking crop nutrition in times of modern microbiology: Innovative biofertilizer technologies. *Frontiers in Sustainable Food System*, *5*, 606815.

Mohamed, Y. M., Azzam, A. M., Amin, B. H., & Safwat, N. A. (2015). Mycosynthesis of iron nanoparticles by *Alternaria alternata* and its antibacterial activity. *African Journal of Biotechnology*, *14*(14), 1234–1241.

Mukherjee, P., Ahmad, A., Mandal, D., Senapati, S., Sainkar, S. R., Khan, M. I., Parishcha, R., Ajaykumar, P., Alam, M., Kumar, R., & Sastry, M. (2001). Fungus-mediated synthesis of silver nanoparticles and their immobilization in the mycelial matrix: A novel biological approach to nanoparticle synthesis. *Nano Letters*, *1*(10), 515–551.

Mukherjee, P., Senapati, S., Mandal, D., Ahmad, A., Khan, M. I., Kumar, R., & Sastry, M. (2002). Extracellular synthesis of gold nanoparticles by the fungus *Fusarium oxysporum*. *ChemBioChem*: *a European Journal of Chemical Biology*, *3*(5), 461–463.

Nadarajan, S., & Sukumaran, S. (2021). Chemistry and toxicology behind chemical fertilizers. *Controlled release fertilizers for sustainable agriculture* (pp 195–229). Elsevier Newtork.

Namasivayam, S. K. R., & Bharani, R. S. A. (2021). Biocompatible silver nanoparticles-loaded fungal metabolites nanoconjugate (AgNp–FM) preparation for the noteworthy pesticidal activity. *National Academy Science Letters, 44*(6), 511–517.

Namasivayam, S. K. R., Bharani, R. S. A., & Karunamoorthy, K. (2018). Insecticidal fungal metabolites fabricated chitosan nanocomposite (IM-CNC) preparation for the enhanced larvicidal activity - An effective strategy for green pesticide against economic important insect pests. *International Journal of Biological Macromolecules, 120*(Pt A), 921–944.

Nayak, P., & Solanki, H. (2021). Pesticides and indian agriculture- a review. *International Journal of Research – GRANTHAALAYAH, 9*(5), 250–263.

Nayak, R. R., Pradhan, N., Behera, D., Pradhan, K. M., Mishra, S., Sukla, L. B., & Kanta, M. B. (2011). Green synthesis of silver nanoparticle by *Penicillium purpurogenum* NPMF: The process and optimization. *Journal of Nanoparticle Research, 13*(8), 3129–3137.

Ofuya, T. I., Okunlola, A. I., & Mbata, G. N. (2023). A review of insect pest management in vegetable crop production in Nigeria. *Insects, 14*(2), 111.

Osorio-Echavarría, J., Osorio-Echavarría, J., Ossa-Orozco, C. P., & Gómez-Vanegas, N. A. (2021). Synthesis of silver nanoparticles using white-rot fungus Anamorphous *Bjerkandera* sp. R1: influence of silver nitrate concentration and fungus growth time. *Scientific Reports, 11*(1), 3842.

Owaid, M. N. (2020). Biomedical applications of nanoparticles synthesized from mushrooms. In J. Patra, L. Fraceto, G. Das, & E. Campos (Eds.), *Green nanoparticles* (pp. 289–303). Springer: Cham, Switzerland.

Pal, K., Chakroborty, S., & Nath, N. (2022). Limitations of nanomaterials insights in green chemistry sustainable route: Review on novel applications. *Green Processing and Synthesis, 11(1)*, 951-964.

Pascual-Villalobos, M. J., Canio-Tejero, M., Vallejo, R., Guirao, P., Rodriez-Rojo, S., & Cocero, M. J. (2017). Use of nanoemulsions of plant essential oils on aphid repellents. *Industrial Crops and Products, 110,* 45–57.

Patiño-Ruiz, D., Sánchez-Botero, L., Tejeda-Benitez, L., Hinestroza, J., & Herrera, A. (2020). Green synthesis of iron oxide nanoparticles using *Cymbopogon citratus* extract and sodium carbonate salt: Nanotoxicological considerations for potential environmental applications. *Environmental Nanotechnology, Monitoring and Management,* 14, 100377.

Pavitra, G., Sushila, N., Sreenivas, A. G., Ashok, J., & Sharanagouda, H. (2018). Biosynthesis of green silica nanoparticles and its effect on cotton aphid, *Aphis gossypii* glover and mealy bug, *Phenacoccus solenopsis* tinsley. *International Journal of Current Microbiology and Applied Sciences, 7*(10), 1450–1460.

Phanjom, P., & Ahmed, G. (2017). Effect of different physicochemical conditions on the synthesis of silver nanoparticles using fungal cell filtrate of *Aspergillus oryzae* (MTCC No. 1846) and their antibacterial effects. *Advances in Natural Sciences: Nanoscience and Nanotechnology, 8*(4), 1–13.

Pouthika, K., & Madhumitha, G. (2023). A review on plant-derived nanomaterials: an effective and innovative insect-resistant strategy for alternate pesticide development. *International journal of Environmental Science and Technology, 21*(4).

Prasad, T. N. V. K. V., Sudhakar, P., Sreenivasulu, Y., Latha, P., Munaswamy, V., Reddy, R. K. T. S., Sreeprasad, P. R., & Sajanlal, P. T. (2012). Effect of Nanoscale zinc oxide particles on the germination, growth and yield of Peanut. *Journal of Plant Nutrition, 35*(6), 905–927.

Pratap, A. P., & Bhowmick, D. N. (2008). Pesticides as microemulsion formulations. *Journal of Dispersion Science and Technology*, *29*(9), 1325–1330.

Qian, Y., Yu, H., He, D., Yang, H., Wang, W., Wan, X., & Wang, L. (2013). Biosynthesis of silver nanoparticles by the endophytic fungus *Epicoccum nigrum* and their activity against pathogenic fungi. *Bioprocess and Biosystems Engineering, 36*(11), 1613–1619.

Ragavendran, C., Kamaraj, C., Jothimani, K., Priyadharsan, A., Kumar, D. A., Natarajan, D., & Malafaia, G. (2023). Eco-friendly approach for ZnO nanoparticles synthesis and evaluation of its possible antimicrobial, larvicidal and photocatalytic applications. *Sustainable Materials and Technologies, 36*, e00597.

Rahman, A., Pittarate, S., Perumal, V., Rajula, J., Thungrabeab, M., Mekchay, S., & Krutmuang, P. (2022). Larvicidal and antifeedant effects of copper nano-pesticides against *Spodoptera frugiperda* (J. E. Smith) and its immunological response. *Insects*, *13*(11), 1030.

Rahman, M. A., Parvin, A., Khan, M. S. H., Lingaraju, K., Prasad, R., Das, S., Hussain, B., War, A. R., & Bhattacharyya, A. (2020). Efficacy of the green synthesized nickel-oxide nanoparticles against pulse beetle, *Callosobruchus maculatus* (F.) in black gram (*Vigna mungo* L.). *International Journal of Pest Management, 67(4),* 306–314.

Raimi, A., Roopnarain, A., & Adeleke, R. (2021). Biofertilizer production in Africa: Current status, factors impeding adoption and strategies for success. *Scientific African, 11*, e00694.

Rajna, S., Paschapur, A. U., & Raghavendra, K. V. (2019). Nanopesticides: Its scope and utility in pest management. *Indian Farmer*, *6(1)*, 17–21.

Raliya R., Tarafdar J.C., & Biswas P. (2016). Enhancing the mobilization of native phosphorus in the mung bean rhizosphere using ZnO nanoparticles synthesized by soil fungi. *Journal of Agricultural and Food Chemistry, 64*(16), 3111–3118.

Ramkumar, G., Asokan, R., Ramya, S., & Gayathri, G. (2020). Characterization of *Trigonella foenum-graecum* derived iron nanoparticles and its potential pesticidal activity against *Tutaa bsoluta (*Lepidoptera). *Journal of Cluster Science,* 8, 47–82.

Rani, L., Thapa, K., Kanojia, N., Sharma, N., Singh, S., Grewal, A. S., Srivastav, A. L., & Kaushal, J. (2021). An extensive review on the consequences of chemical pesticides on human health and environment. *Journal of Cleaner Production, 283*, 124657.

Raza, A., Razzaq, A., Mehmood, S. S., Zou, X., Zhang, X., Lv, Y., & Xu, J. (2019). Impact of climate change on crops adaptation and strategies to tackle its outcome: A review. *Plants, 8(2),* 34.

Rehman, H., Majeed, B., Farooqi, M. A., Rasul, A. A., Sagheer, M., Ali, Q., & Akhtar, Z. R. (2021). Green synthesis of silver nitrate nanoparticles from *Camelina Sativa* (L.) and its effect to control insect pests of stored grains. *International Journal of Tropical Insect Science*, *41*(1), 3031–3039.

Rose, G. K., Soni, R., Rishi, P., & Soni, S. K. (2019). Optimization of the biological synthesis of silver nanoparticles using *Penicillium oxalicum* GRS-1 and their antimicrobial effects against common food-borne pathogens. *Green Processing Synthesis*, *8*, 144–156.

Salama, H. M. H. (2012). Effects of silver nanoparticles in some crop plants, common bean (*Phaseolus vulgaris* L.) and corn (*Zea mays* L.). *International Research Journal of Biotechnology, 3*(10), 190–197.

Salleh, A., Naomi, R., Utami, N. D., Mohammad, A. W., Mahmoudi, E., Mustafa, N., & Fauzi, M. B. (2020). The potential of silver nanoparticles for antiviral and antibacterial applications: A mechanism of action. *Nanomaterials*, *10*(8), 1566.

Salunkhe, R. B., Patil, S. V., Patil, C. D., & Salunkhe, B. K. (2011). Larvicidal potential of silver nanoparticles synthesized using fungus *Cochliobolus lunatus* against *Aedes aegypti* (Linnaeus, 1762) and *Anopheles stephensi* Liston (Diptera; Culicidae). *Parasitology Research, 109*(3), 823–831.

Sanghi, R., & Verma, P. (2009). Biomimetic synthesis and characterisation of protein capped silver nanoparticles. *Bioresource Technology, 100*(1), 501–504.

Sankar, M. V., & Abideen, S. (2015). Pesticidal Effect of green synthesized silver and lead nanoparticles using *Avicennia marina* against grain storage pest, *Sitophilus oryzae*. *Digest Journal of Nanomaterials and Biostructures, 5*(4), 32–39.

Saxena, J., Sharma, P. K., Sharma, M. M., & Singh, A. (2016). Process optimization for green synthesis of silver nanoparticles by *Sclerotinia sclerotiorum* MTCC 8785 and evaluation of its antibacterial properties. *Springerplus, 5*(1), 861.

Šebesta, M., Vojtková, H., Cyprichová, V., Ingle, A. P., Urík, M., & Kolenčík, M. (2022). Mycosynthesis of metal-containing nanoparticles-fungal metal resistance and mechanisms of synthesis. *International Journal of Molecular Sciences, 23*(22), 14084.

Shahzad, A., Saeed, H., Iqtedar, M., Hussain, S. Z., Kaleem, A., & Abdullah, R. (2019). Size-controlled production of silver nanoparticles by *Aspergillus fumigatus* BTCB10: likely antibacterial and cytotoxic effects. *Journal of Nanomaterials, 2019*(7), 5168698.

Shankar, S. S., Ahmad, A., Pasricha, R., & Sastry, M. (2003). Bioreduction of chloroaurate ions by geranium leaves and its endophytic fungus yields gold nanoparticles of different shapes. *Journal Materials of Chemistry, 13*(7), 1822–1826.

Sharma, B., Lakra, U., Sharma, R., & Sharma, S. R. (2022). A comprehensive review on nanopesticides and nanofertilizers-A boon for agriculture. In M. Ghorbanpour & M. A. Shahid (Eds.), *Nano-enabled agrochemicals in agriculture* (pp. 273–290). Academic Press.

Sharmin, S., Rahaman, M. M., Sarkar, C., Atolani, O., Islam, M. T., & Adeyemi, O. S. (2021). Nanoparticles as antimicrobial and antiviral agents: A literature-based perspective study. *Heliyon*, 7(3), e06456.

Sheikhloo, Z., & Salouti, M. (2011). Intracellular biosynthesis of gold nanoparticles by the fungus *Penicillium chrysogenum*. *International Journal Nanoscience and Nanotechnology, 7*(2), 102–105.

Shekhar, S., Sharma, S., Kumar, A., Taneja, A., & Sharma, B. (2021). The framework of nanopesticides: A paradigm in biodiversity. *Materials Advances, 2*, 6569–6588.

Shu, M., He, F., Li, Z., Zhu, X., Ma, Y., Zhou, Z., Yang, Z., Gao, F., & Zeng, M. (2020). Biosynthesis and antibacterial activity of silver nanoparticles using yeast extract as reducing and capping agents. *Nanoscale Research Letters, 15*(1), 14.

Shyam-Sundar, N., Karthi, S., Senthil-Nathan, S., Narayanan, K. R., Santoshkumar, B., Sivanesh, H., Chanthini, K. M. P., Stanley-Raja, V., Ramasubramanian, R., Abdel-Megeed, A., & Malafaia, G. (2023). Eco-friendly biosynthesis of TiO_2 nanoparticles using *Desmostachya bipinnata* extract: Larvicidal and pupicidal potential against *Aedes aegypti* and *Spodoptera litura* and acute toxicity in non-target organisms. *Science of The Total Environment, 858,* 159512.

Singh, P., Kim, Y. J., Zhang, D., & Yang, D. C. (2016). Biological synthesis of nanoparticles from plants and microorganisms. *Trends in Biotechnology*, *34*(7), 588–599.

Siva, C., & Kumar, M. S. (2015). Pesticidal activities of ecofriendly synthesized silver particles using *Aristolochia indica* extract against *Helicoverpa armigera* Hubner (Lepidoptera: Noctuidae). *International Journal of Advanced Scientific and Technical Research, 2*(5), 197–226.

Soltani Nejad, M., Samandari Najafabadi, N., Aghighi, S., Pakina, E., & Zargar, M. (2022). Evaluation of *Phoma* sp. biomass as an endophytic fungus for synthesis of extracellular gold nanoparticles with antibacterial and antifungal properties. *Molecules*, *27*, 1181.

Sonawane, H., Shelke, D., Chambhare, M., Dixit, N., Math, S., Sen, S., Borah, S. N., Islam, N. F., Joshi, S. J., Yousaf, B., Rinklebe, J., & Sarma, H. (2022). Fungi-derived agriculturally important nanoparticles and their application in crop stress management - Prospects and environmental risks. *Environmental Research, 212*(Pt D), 113543.

Soni, N., & Prakash, S. (2012a). Efficacy of fungus mediated silver and gold nanoparticles against *Aedes aegypti* larvae. *Parasitology Research, 110*(1), 175–184.

Soni, N., & Prakash, S. (2012b). Synthesis of gold nanoparticles by the fungus *Aspergillus niger* and its efficacy against mosquito larvae. *Reports of Parasitology, 2,* 1–7.

Soni, N., & Prakash, S. (2013). Possible mosquito control by silver nanoparticles synthesized by soil fungus (*Aspergillus niger* 2587). *Advances in Nanoparticles*, *2*(2), 125–132.

Sowndarya, P., Ramkumar, G., & Shivakumar, M. S. (2017). Green synthesis of selenium nanoparticles conjugated *Clausena dentata* plant leaf extract and their insecticidal potential against mosquito vectors. *Artificial Cells, Nanomedicine, and Biotechnology*, *45*(8), 1490–1495.

Subashini, G., & Bhuvaneswari, S. (2018). Nanoparticles from fungi (myconanoparticles). In P. Gehlot & J. Singh (Eds.), *Fungi and their Role in sustainable development: Current perspectives* (pp. 753–779). Springer, Singapore.

Sumanth. B., Lakshmeesha, T. R., Ansari, M. A., Alzohairy, M. A., Udayashankar, A. C., Shobha, B., Niranjana, S. R., Srinivas, C., & Almatroudi, A. (2020). Mycogenic synthesis of extracellular zinc oxide nanoparticles from *Xylaria acuta* and its nanoantibiotic potential. *International Journal of Nanomedicine, 15*, 8519–8536.

Sun, Y., Liang, J., Tang, L., Li, H., Zhu, Y., Jiang, D., Song, B., Chen, M., & Zeng, G. (2019). Nano-pesticides: A great challenge for biodiversity? *Nano Today*, *28*, 100757

Sundaravadivelan, C., & Padmanabhan, M. N. (2014). Effect of mycosynthesized silver nanoparticles from filtrate of *Trichoderma harzianum* against larvae and pupa of dengue vector *Aedes aegypti* L. *Environmental Science and Pollution Research*, *21*(6), 4624–4633.

Tarafdar, J. C., Raliya, R., Mahawar, H., & Rathore, I. (2014). Development of zinc nanofertilizer to enhance crop production in pearl millet (*Pennisetum americanum*). *Agricultural Research*, *3*, 257–262.

Tudi, M., Daniel, R. H., Wang, L., Lyu, J., Sadler, R., Connell, D., Chu, C., & Phung, D. T. (2021). Agriculture development, pesticide application and its impact on the environment. *International Journal of Environmental Research and Public Health, 18(3)*, 1112.

Umesha, S., Manukumar, H. M. G., & Chandrasekhar, B. (2018). Sustainable agriculture and food security. In R. L. Singh & S. Mondal (Eds.), *Biotechnology for sustainable agriculture: Emerging approaches and strategies* (pp. 67–92). Academic Press.

UN (2023). As the world's population surpasses 8 billion, what are the implications for planetary health and sustainability? https://www.un.org/en/un-chronicle/world-population-surpasses-8-billion-what-are-implications-planetary-health-and

UN (2017). World population prospects: the 2017 revision, key findings and advance tables. United Nations, Department of Economic and Social Affairs, Population Division, https://population.un.org/wpp/Publications/Files/WPP2017_KeyFindings.pdf

Urnukhsaikhan, E., Bold, B. E., Gunbileg, A., Sukhbaatar, N., & Mishig-Ochir, T. (2021). Antibacterial activity and characteristics of silver nanoparticles biosynthesized from *Carduus crispus*. *Scientific Reports, 11*(1), 21047.

Velhal, S. G., Kulkarni, S. D., & Latpate, R. V. (2016). Fungal mediated silver nanoparticle synthesis using robust experimental design and its application in cotton fabric. *International Nano Letters, 6*, 257–264.

Velusamy, P., Kumar, G. V., Jeyanthi, V., Das, J., & Pachaiappan, R. (2016). Bio-inspired green nanoparticles: Synthesis, mechanism, and antibacterial application. *Toxicological Research, 32*(2), 95–102.

Verma, K. K., Song, X. P., Joshi, A., Rajput, V. D., Singh, M., Sharma, A., Singh, R. K., Li, D. M., Arora, J., Minkina, T., & Li, Y. R. (2022a). Nanofertilizer possibilities for healthy soil, water, and food in future: An overview. *Frontiers in Plant Science, 13*, 865048.

Verma, K. K., Song, X. P., Joshi, A., Tian, D. D., Rajput, V. D., Singh, M., Arora, J., Minkina, T., & Li, Y. R. (2022b). Recent trends in nano-fertilizers for sustainable agriculture under climate change for global food security. *Nanomaterials (Basel), 12*(1), 173.

Verma, V. C., Kharwar, R. N., & Gange, A. C. (2010). Biosynthesis of antimicrobial silver nanoparticles by the endophytic fungus *Aspergillus clavatus*. *Nanomedicine (Lond), 5*(1), 33–40.

Vetchinkina, E. P., Loshchinina, E. A., Vodolazov, I. R., Kursky, V. F., Dykman, L. A., & Nikitina, V. E. (2017). Biosynthesis of nanoparticles of metals and metalloids by basidiomycetes. preparation of gold nanoparticles by using purified fungal phenol oxidases. *Applied Microbiology and Biotechnology,* 101(3), 1047–1062.

Vigneshwaran, N., Kathe, A. A., Varadarajan, P. V., Nachane, R. P., & Balasubramanya, R. H. (2006). Biomimetics of silver nanoparticles by white rot fungus *Phaenerochaete chrysosporium*. *Colloids and Surfaces B: Biointerfaces, 53*, 55–59.

Vivekanandhan, P., Swathy, K., Thomas, A., Kweka, E. J., Rahman, A., Pittarate, S., & Krutmuang, P. (2021). Insecticidal efficacy of microbial-mediated synthesized copper nano-pesticide against insect pests and non-target organisms. *International Journal of Environmental Research and Public Health, 18*(19), 10536.

Wang, D., Xue, B., Wang, L., Zhang, Y., Liu, L., & Zhou, Y. (2021). Fungus-mediated green synthesis of nano-silver using *Aspergillus sydowii* and its antifungal/antiproliferative activities. *Scientific Reports, 11*, 10356.

Win, T. T., Khan, S., & Fu, P. (2020). Fungus- (*Alternaria* Sp.) Mediated silver nanoparticles synthesis, characterization, and screening of antifungal activity against some phytopathogens. *Journal of Nanotechnology,* 2020, e8828878.

Xue, B., He, D., Gao, S., Wang, D., Yokoyama, K., & Wang, L. (2016). Biosynthesis of silver nanoparticles by the fungus *Arthroderma fulvum* and its antifungal activity against genera of *Candida*, *Aspergillus* and *Fusarium*. *International Journal of Nanomedicine*, *11*, 1899–1906.

Yadav, A., Kon, K., Kratosova, G., Duran, N., Ingle, A. P., & Rai, M. (2015). Fungi as an efficient mycosystem for the synthesis of metal nanoparticles: progress and key aspects of research. *Biotechnology Letters*, *37*(11), 2099–20120.

Yang, F. L., Li, X. G., Zhu, F., & Lei, C. L. (2009). Structural characterization of nanoparticles coded with garlic essential oil and their insecticidal activity against *Tribolium castaneum* (Herbst) (Coleoptera: Tenebrionidae). *Journal of Agricultural and Food Chemistry, 57*(21), 10156–10162.

Yassin, M. A., Elgorban, A. M., El-Samawaty, A. E. M. A., & Almunqedhi, B. M. A. (2021). Biosynthesis of silver nanoparticles using *Penicillium verrucosum* and analysis of their antifungal activity. *Saudi Journal of Biological Science*, *28*(4), 2123–2127.

Yousaf, M., Li, J., Lu, J., Ren, T., Cong, R., Fahad, S., & Li, X. (2017). Effects of fertilization on crop production and nutrient-supplying capacity under rice-oilseed rape rotation system. *Scientific Reports*, *7*, 1270.

Zahir, A. A., Bagavan, A., Kamaraj, C., Elango, G., & Rahuman, A. A. (2012). Efficacy of plant-mediated synthesized silver nanoparticles against *Sitophilus oryzae*. *Journal of Biopesticides*, *5,* 95–102.

Zaki, S. A., Ouf, S. A., Albarakaty, F. M., Habeb, M. M., Aly, A. A., & Abd-Elsalam, K. A. (2021). *Trichoderma harzianum*-mediated ZnO nanoparticles: A green tool for controlling soil-borne pathogens in cotton. *Journal of Fungi*, *7*, 952.

Zhang, X., He, X., Wang, K., & Yang, X. (2011). Different active biomolecules involved in biosynthesis of gold nanoparticles by three fungus species. *Journal of Biomedical* Nanotechnology, *7(2),* 245–254.

Zhao, X., Zhou, L., Riaz Rajoka, M. S., Yan, L., Jiang, C., Shao, D., Zhu, J., Shi, J., Huang, Q., Yang, H., & Jin, M. (2018). Fungal silver nanoparticles: synthesis, application and challenges. *Critical Reviews in Biotechnology*, *38*(6), 817–835.

CHAPTER 7

Role of Plants as Nanofactories for Nanomaterial Synthesis

AWANTIKA SINGH, AKANKSHA GAUTAM, ANU SINGH, and HEMANT RITTURAJ KUSHWAHA

School of Biotechnology, Jawaharlal Nehru University, New Delhi, India

ABSTRACT

Over the last 30 years, plants have gained attention as potential nanofactories for the synthesis of different metal nanoparticles (MNPs) such as gold, silver, zinc, copper, tin, and iron. The plant body consists of various phytochemicals that act both as reducing and stabilizing agents for the reduction of metal precursors and are nontoxic in nature. Plant-mediated synthesis of biogenic MNPs is a simple, convenient, economical, and environmentally friendly method that is suitable for large-scale production. There are three methods by which plants can be used for the synthesis of metal nanoparticles: intracellular synthesis, extracellular synthesis, and phytochemically mediated synthesis. The intracellular synthesis method involves whole plants or plant biomass (e.g., *M. sativa, B. juncea, Triticum aestivum,* etc.), where MNPs synthesize inside the plant and are recovered via cell rupturing. The extracellular synthesis method utilizes plant extract and is considered the most suitable approach due to its simple procedure of extract preparation, easy recovery of the end product, cost effectiveness, and high efficiency with rapid reaction rates. The phytochemically mediated synthesis method (e.g., synthesis via polyphenols, organic acids, and proteins) is more advanced, specific, and feasible. The limitations of the intracellular and extracellular synthesis methods are polydispersity and heterogeneity of the generated MNPs, while the phytochemical-mediated synthesis method is more complex due to its requirement for the isolation and purification of specific phytochemicals,

Nanobiotechnology for Agricultural Sciences: Nano-Agri-Input Products for Crop Production and Environmental Protection. J. C. Tarafdar, Deepak Kumar, Sanjay K. Singh, and Krishna Pal Singh (Eds.)

which is a time-consuming and costly affair. This chapter will focus on the principles of green chemistry behind plants as nanofactories for nanoparticle synthesis, their molecular and biochemical aspects of the mechanism of biogenic nanoparticle synthesis, and their recent applications and commercial usages in the biomedical, agricultural, and environmental sectors.

7.1 INTRODUCTION

Nanotechnology is a branch of advanced interdisciplinary sciences that is gaining attention throughout the globe and has the huge potential to begin a nanorevolution for humans. Nanoparticles are smaller particles at the scale of 1–100 nm in all three Cartesian dimensions (Jeevanandam et al., 2022), while nanomaterials are usually considered to be materials with at least one external dimension that measures 100 nm or less or with internal structures measuring 100 nm or less (Baig et al., 2021). It is stated that nanomaterials are advantageous over bulk materials due to their properties like surface plasmon light scattering (SP-LS), surface plasmon resonance (SPR), surface-enhanced Rayleigh scattering (SERS), and surface-enhanced Raman scattering (SERS) (Dauthal and Mukhopadhyay, 2016). These unique properties make the nanomaterials ideal building blocks for a wide range of applications such as agriculture equipment, industrial optoelectronics, optoelectronic and chemical and biological sensing, biomedical sectors, and so forth (Parmar and Sanyal, 2022; Kumar and Lal, 2014; Hu et al., 2020). The basic two approaches for the synthesis of desired nanomaterials or nanoparticles are the top-down and bottom-up approaches. The first approach involves the reduction of suitable bulk material into smaller particles at nanoscale size via adaptation of physical and/or chemical methods such as lithography, microwave irradiation, pyrolysis, thermolysis, radiation-induced methods, solvo-thermal methods, and precipitation methods (Hu et al., 2020; Jadoun et al., 2021). The second approach involves controlled accumulation or assembling of smaller atoms, ions, molecules, or clusters into a final nanosize material utilizing chemical and biological methods of synthesis (Kumar and Lal, 2014). Figure 7.1 shows different modes of synthesis of nanomaterials, that is physical, chemical, and biological. The physical approach is basically a dry synthesis process that is simple to perform, but it has limitations like the requirement of expensive, cumbersome instruments and high pressure and temperature, which in turn require and release lots of energy. On the other hand, the chemical method of synthesis is basically a wet synthesis procedure (e.g., chemical, electrochemical, sono-chemical, and polyol reduction)

(Dauthal and Mukhopadhyay, 2016), which involves the use of toxic chemicals, nonpolar organic solvents, and hazardous stabilizing and capping agents (Jadoun et al., 2021). Chemical approaches are actually not an ideal environmentally friendly method and suffer from the limitations of toxicity, as extensive exposure to these products and their byproducts could lead to hazardous environmental consequences and restrict their applications in the field of biomedical sciences. Replacement of these conventional methods by nontoxic, cost-effective, and cleaner approaches is an utter requirement for the synthesis of biocompatible nanomaterials (Hamdy et al., 2022; Dauthal and Mukhopadhyay, 2016). Since the last 30 years, nanotechnologists have been working to establish biological approaches to synthesis that are more environmentally friendly, simple, and cost-effective (Mustapha et al., 2022). The biological approach to synthesis utilizes living organisms as templates for the synthesis of nanomaterials/nanoparticles, for example, viruses, bacteria, lichen, fungi, algae, and plants (Mustapha et al., 2022). These bio-templates are rich in active biological entities (lipids, carbohydrates, proteins, nucleic acids, etc.), which play a role as reducing and stabilizing agents in the process of nanoparticle synthesis (Flieger et al., 2021; Begum et al., 2022). The biological method for nanoparticle synthesis is considered beneficial over other conventional synthesis processes due to its novel, nontoxic, and cost-effective approach, which involves sustainable organic capping and reducing agents and causes no harm to the environment or humans. In recent years, plant-mediated methods for nanoparticle synthesis have gained major interest among other biological approaches due to the bulk availability of plants rich in metabolites possessing reducing and capping abilities and a direct, safe, and economical approach (Nazneen et al., 2022).

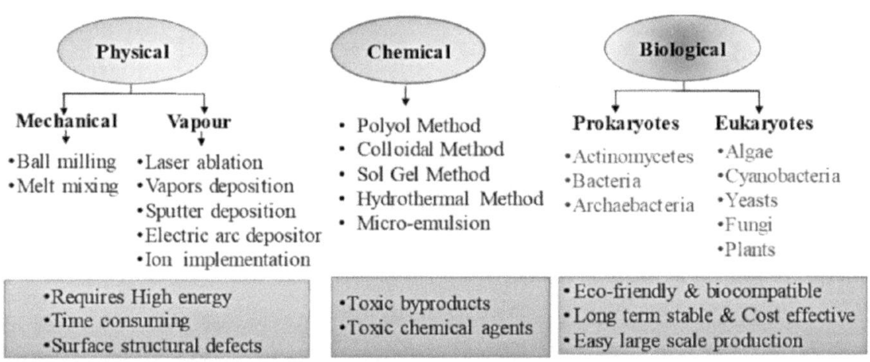

FIGURE 7.1 Diagrammatic representation of mode of synthesis of nanoparticles and their properties.

Due to the diverse physical and environmental factors across global geographical areas, the diversity in the plant kingdom is wide and divergent. Plants are considered ideal nanofactories because plant parts (fruit, flower, leaf, stem, and root) are highly rich in active phytochemicals (protein, carbohydrates, lipids, and metabolites), and therefore, all the plant parts can be used for nanoparticle synthesis (Raj et al., 2020). This mechanism depends on the action of phytochemicals as reducing agents and stabilizing or capping agents in a suitable solvent medium (Dauthal and Mukhopadhyay, 2016). The nanoparticle synthesis via plant nanofactories requires liquid medium (aqueous medium) and is considered a novel green method because it is nontoxic, simple, inexpensive, and environmentally beneficial, eliminating any use of intermediary base groups and the synthesis of toxic byproducts (Jadoun et al., 2021). The plant system is one of the best considered templates for nanoparticle synthesis as it helps in generating the most stable nanoparticles (Hamidi et al., 2019). Metal detoxification and accumulation via plant nanofactories are well studied, even at low concentrations (Hamidi et al., 2019). In comparison to other biological templates, plant nanofactories show better kinetics for nanoparticle synthesis procedures than chemical nanoparticle synthesis (Singh et al., 2021). Plant-mediated nanoparticles have been studied for their applications in various fields like agriculture, the environment, health care, and so forth, and some of the strategies have been successfully converted into commercial products. Figure 7.2 shows plant parts used in nanoparticle synthesis and possible fields of applications.

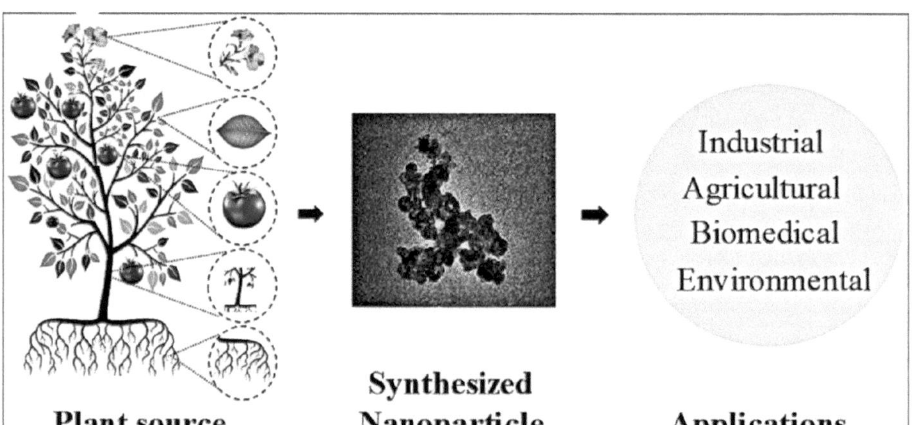

FIGURE 7.2 Diagrammatic representation of the mode of nanoparticle synthesis using plants as nanofactories and its generalized applications.

Therefore, considering the significance of plant-based nanomaterials, this chapter is focused on understanding the importance of plants as novel nanofactories for the fabrication of nanomaterials, methods of synthesis, their mechanistic pathways, large-scale production, and commercial applications in agriculture and environmental areas.

7.2 METHODS FOR SYNTHESIS OF PLANT-MEDIATED NANOPARTICLES

The methods of plant nanofactories-mediated nanomaterial synthesis are of three types: intracellular synthesis, extracellular synthesis, and phytochemical-mediated synthesis, as shown in Figure 7.3. Almost all the plant parts are used for nanomaterial synthesis, where active phytochemicals play a major role. The basic difference between these methods is the mode of usage of these phytochemicals: in the intracellular method, the whole plant is used, whereas in the extracellular method, plant extract is used, and in the phytochemically mediated method, these bioactive compounds are first extracted, purified, and then used.

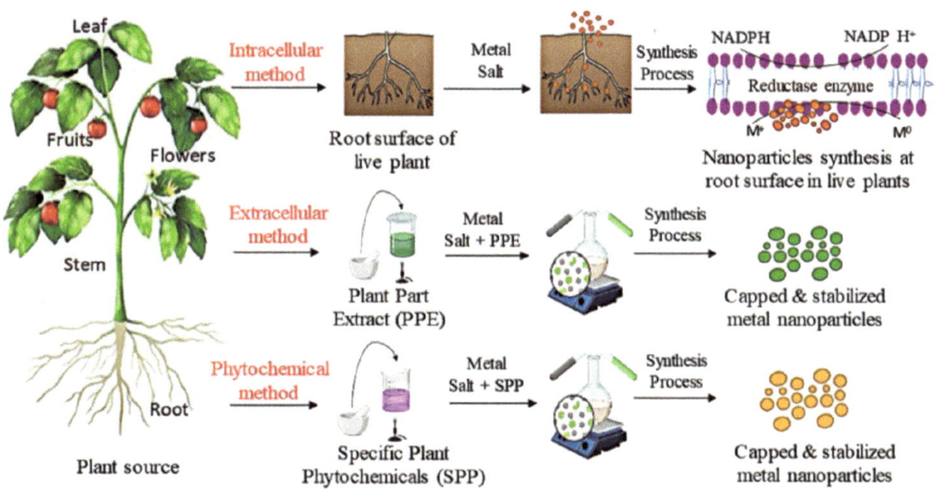

FIGURE 7.3 Plant nanofactories-mediated nanomaterial synthesis methods.

7.2.1 INTRACELLULAR MODE OF NANOPARTICLE SYNTHESIS

The intracellular approach to nanoparticle synthesis occurs *in vivo*, where a salt precursor is provided to the plant, which absorbs the metal precursor via different mechanisms like metal sequestration and biotransformation (Jadoun et al., 2021). Different metallophytes have shown the synthesis of a range of elemental metal nanoparticles and a rare combination of nanoalloys of Cu, Ag, and Au (Saim et al., 2021). Some examples of the intracellular method are as follows: The shoot parts of the plants *Brassica juncea and* Madia sativa have shown the metal accumulation and conversion into nanoparticles; beside these, the root exudates of *Arabidopsis thaliana* have shown the absorption and synthesis of gold nanoparticles (Saim et al., 2021). *Sesbania* seedlings have the potential to absorb gold salt precursors and produce stable AuNPs in the tissues of the plant. In *Vigna radiata,* the synthesized silver nanoparticles were noted to be found in cytoplasmic organelles like chloroplast cytoplasmic spaces, chloroplast vacuoles, and nucleolar plant regions. Other maize plants have also shown the production of silver nanoparticles along with nutrients at different plant parts (stems, roots, and leaves). The emerging sprouts of three plant beans, radish, and alfalfa showed the synthesis of silver nanoparticles intracellularly using different amino acids as reducing agents (Saim et al., 2021). Plant systems are rich in metal-interacting proteins and biomolecules that chelate or transform the metal salt precursor into elemental metal nanoparticles (Mustapha et al., 2022). Thus, various plant parts and biological entities have shown the capability of phytoextraction and detoxification of metal ions in the natural system. Scientists are focusing increasingly on investigating the potential for phytoextraction and phytoremediation in diverse plant species for metal nanoparticle synthesis (Dauthal and Mukhopadhyay, 2016).

7.2.2 EXTRACELLULAR MODE OF NANOPARTICLE SYNTHESIS

The extracellular approach to nanoparticle synthesis involves extracellular secretion of biochemical entities on the plant surface and the extraction of whole constituents present in plants, plant parts, or plant biomass, which help in the extracellular synthesis of nanoparticles (Jadoun et al., 2021). The extracted phytochemicals from plant sources act as stabilizing and reducing agents during the process of nanoparticle synthesis (Dauthal and Mukhopadhyay, 2016). Compared with the intracellular approach to synthesis, the extracellular approach is more common and easier and has been shown to have a wide range of applications (Saim et al., 2021). Different live plants have been used as nanofactories for extracellular nanoparticle synthesis, for example, *Tephrosia apollinea*, *Vigna unguiculata, Arachis hypogaea,* and *Vigna unguiculata.* The ease of the extracellular approach makes it more adapted by the nanotechnologist for scale-up and downstream processing (Jadoun et al., 2021). However, the extracellular mode of synthesis shows some major limitations, such as polydispersity, variation in shape and size, and the presence of high impurities in plant biomass or agricultural waste (Jadoun et al., 2021). The structural differences and complexities of biological systems are the major reasons for limitations. Plant extracts contain a complex soup of different biological entities, so identifying the precise bio-reducing and stabilizing agents in charge of synthesizing and stabilizing nanoparticles is a difficult task (Noah and Ndangili, 2022). To combat the limitations of plant extract-mediated nanoparticles and to gain ideal size and shape of biosynthesized nanoparticles, nano-biotechnologists are now focused on the pure mixture of phytochemical-based nanoparticle synthesis (Noah and Ndangili, 2022).

7.2.3 PHYTOCHEMICAL-MEDIATED NANOPARTICLE SYNTHESIS

The phytochemically mediated nanoparticle synthesis involves the *in vitro* isolation of a pure phytochemical soup of active biological entities from plant bodies (Abasi et al., 2022; Khan et al., 2021a; Naseer et al., 2020) and using these phytochemicals as reducing and/or capping agents to synthesize nanoparticles from metal salts. Researchers have isolated phytochemicals to produce nanoparticles with specific functionalities or for specific applications. For example, proteins act as reducing and/or capping agents, where the amide group and carboxyl group of enzymatic and nonenzymatic proteins act as binding templates or trapping centers for upcoming metal ions to convert them into elemental nanoforms (Kabir et al., 2022). The complexities found in the structure of proteins lead to variations in the size and shape of synthesized nanoparticles (Hamidi et al., 2019; Buntru et al., 2022).

Researchers have used various parts of plants to synthesize metal nanoparticles; however, all plant parts have not been explored yet for all three methods. Table 7.1 shows examples of plant-mediated synthesis of metal nanoparticles using the intracellular, extracellular, and phytochemical methods. In general, all the three methods are dependent on the mechanism of action of biological compounds present in different parts of the plants, but due to the differences in the methods and their feasibility, these methods offer some advantages and disadvantages, as shown in Table 7.2. This will guide researchers in planning their experiments for the generation of plant-mediated nanoparticles of their choice.

TABLE 7.1 Some of the Examples Show the Use of Different Parts of Plants to Synthesize Intracellular, Extracellular, and Phytochemical-Mediated Nanoparticles.

Plant Part	Plant Species	Type of Nanoparticles	Shape	Reference
		Intracellular Synthesis		
Root	Brassica juncea	Silver nanoparticles	Spherical	Marchiol et al., 2014
Leaf extract	Medicago sativa	Silver nanoparticles	Spherical	Marchiol et al., 2014
	Brassica juncea	Gold nanoparticles	-	Anderson et al., 2022
Stem/shoot	Brassica juncea	Gold nanoparticles	-	Anderson et al., 2022
		Extracellular Synthesis		
Root	Rumex hymenosepalus	Silver nanoparticles	Spherical	Rodríguez-León et al., 2017
	Linaria marocana	CdS quantum dots	Spherical	Boravaya et al., 2014
	Moringo oleifera	Nitrogen-doped Carbon quantum dots	Spherical	Wang et al., 2021
	Rhaphanus sativus	Cds QDs	Spherical	Gholami et al., 2020
Leaf	Ociman Sanctum (Tulsi)	Silver nanoparticles	Spherical	Jain and Mehata et al., 2017
	Pongamia pinnata (PLE)	Gold nanoparticles	Spherical	Khatua et al., 2020
	Stevia rebaudiana	Nickel nanoparticles	Spherical	Moghadam et al., 2020
	Gmelina arborea	Silver nanoparticles	Spherical	Chandrasekharan et al., 2022
Seeds	Cuminum cyminum	Iron nanoparticles	Spherical	Noor et al., 2022
Flowers	Cassia auriculata	Iron nanoparticles	Spherical	Sudhakar et al., 2022
Peel extract	Musa coocinea	Iron nanoparticles	Spherical	Chau et al., 2022
Fruit Extract	Ficus johannis	Cadmium telluride quantum dots	Spherical	Alvand et al., 2019
	Kei- apple	Silver nanoparticles	Spherical	Alrahji et al., 2022
	Yall pear (pynus Bretschelden)	Iron nanoparticles	Cubic	Rong K et al., 2022
		Phytochemical-Mediated Synthesis		
Leaf extract	Acalypha Indica	Silver nanoparticles		Krishnaraj et al., 2010
	Lantana camara	Silver nanoparticles	Spherical	Ajitha et al., 2015
	Ociman Sanctum (Tulsi)	Silver nanoparticles	Spherical	Jain and Mehata, 2017
	Mint	Copper sulfate	Spherical	Mahmoud et al., 2021
	Spinacia oleracea	Iron nanoparticles	Spherical	Fayyaz et al., 2022
	Abutilon indicum	NiO nanoparticles	spherical	Khan et al., 2021a
	Abutilon indicum	Chromium oxide nanoparticles	spherical	Khan et al., 2021b
Seed extract	Nigella arvensis	Silver nanoparticles	Spherical	Chahardoli et al., 2018

TABLE 7.1 *(Continued)*

Plant Part	Plant Species	Type of Nanoparticles	Shape	Reference
Root extract	Zingiber officinale	Silver/gold nanoparticles	-	Velmurugan et al., 2014
	Solanum indicum	Silver nanoparticles	-	Sengottaiyan et al., 2016
Fruit extract	Citrus X sinensis (orange peel extract)	Zinc oxide nanoparticles	Hexagonal	Chaudhuri and Malodia, 2017
	Orange	Copper sulfate	Spherical	Mahmoud et al., 2021
	Yali pear (*Pyrus bretschneideri*)	Iron nanoparticles	Spherical	Rong et al., 2022

TABLE 7.2 Advantages and Disadvantages of Plant-Mediated Nanoparticle Synthesis Methods.

S. No.	Methods	Advantages	Limitations
1.	Intracellular-mediated synthesis method	Simple but not specific, quick reduction of metal ions, to synthesize desired size and shape nanoparticle	Rare, expensive, time taken, low yield, and difficult scale-up and downstream processing
2.	Extracellular-mediated synthesis method	Simple but not specific, more common, easier cost effective, high yield, easy scale-up and downstream processing	Polydispersity, variation in shape and size, presence of high impurities of plant biomass or agricultural waste
3.	Phytochemical-mediated synthesis method	Specific, sustainability, renewability, biodegradable, and compatibility	Complex method, costly process, and additional purification procedures

7.3 MECHANISM OF ACTION FOR SYNTHESIS OF NANOPARTICLES AND THEIR CHARACTERISTICS

The three methods of plant-mediated nanoparticle synthesis are different on the basis of source material, but the mechanism of synthesis remains the same; that is the bioactive compounds or phytochemical (proteins, alkaloids, flavonoids, reducing sugars, polyphenols, etc.) present in the source material (plant part) act as reducing and capping agents. Reducing agents act to reduce metal ions to metal atoms, which, after nucleation, form nanoparticles (Alharbi et al., 2022a; Dikshit et al., 2021), stabilized by capping ligands to prevent further growth and aggregation (Singh et al., 2018). The functional groups (such as OH, -C-O-C-, -C-O-, -C=C-, and -C=O-) of phytochemicals help to generate metallic nanoparticles. Table 7.3 shows the different functional groups of phytochemicals leading to the synthesis of nanoparticles. Nanoparticle synthesis involved three stages where initially metal ions (M^+ or M^{2+}) were reduced to metal atoms (M^0), and then these atoms acted as a template (nucleation) for the growth of the nanoparticle via deposition of the reduced metal atoms on the nuclei with a simultaneous increase in thermodynamic stability, and finally the process got terminated, providing nanoparticles of final shape and size (Dikshit et al., 2021). It can be better understood by an example of silver nanoparticles (AgNPs) synthesized using Tulsi leaf extract, where the OH group of Quercetin (a phenolic compound) reacts with Ag^+ ions and reduces Ag^+ to Ag^0, which forms nanoparticles and stabilizes them against agglomeration (Alharbi et al., 2022b). Figure 7.4 shows the general mechanism of nanoparticle synthesis using phytochemicals.

Nanoparticles generated via plant-mediated methods possess different characteristics depending on the types of functional groups present in the phytochemicals, which play a major role in the reduction and capping of the metal salt precursor (Hamdy et al., 2022). These phytochemicals regulate the characteristics of the synthesized metal nanoparticles, such as stability, dispersion, structure, and functionalization, and guide their therapeutic applications. They can be explained as follows:

7.3.1 CARBOHYDRATES

Carbohydrates contain a wide range of side functional chains like hydroxyl, amino, sulfate, and carboxylic acid, which help in the formation of inter- and intramolecular hydrogen bonds (Shervani and Yamamoto, 2011). Carbohydrate-capped metal nanoparticles maintain the structural integrity and stability of nanoparticles, have drastically increased specificity and binding affinity toward different biological entities, trigger specific receptors for cellular uptake and responses, and regulate cellular processes. These also have ligand exchange (replace an already existing ligand with carbohydrates) abilities and mimic the cell surface by making multiple copies (Ndugire et al., 2021). A variety of applications have been achieved using these materials, including biosensing, bioimaging, targeted nanocarriers, and vaccines.

7.3.2 LIPIDS

Similarly, essential oils and fatty acids mostly help as capping and stabilizing agents for metallic nanoparticles due to the presence of diverse oxygenated functional groups (Dhifi et al., 2016; Omar et al., 2022; Pathania et al., 2022). Lipid capping has a major effect on the structure of nanoparticles and effectively controls the aggregation of lipid nanoparticles specifically prepared for nucleic acid delivery.

7.3.3 SECONDARY METABOLITES

Different bioactive compounds with various oxygenated functional groups help in the reduction and capping of metal salt precursors into nanoform (Pithawala and Jain, 2012; Uddin et al., 2022). These secondary metabolites are commonly classified into nitrogenous compounds, phenolic compounds, tannins, glycosides, and terpenes (Marslin et al., 2018) and offer the synthesis of nonagglomerated and controlled shape and size nanoparticles. Therefore, it is important to identify and understand the specific interaction between the phytochemicals present in the extracts and the metal salts in solution, which react to produce metal nanoparticles. Various studies have been conducted using plant-mediated synthesis of metal nanoparticles to increase their properties. For example, facile green synthesis of seed-extracted flavonoid-loaded bimetallic nanoparticles and statistically significant enhancement in wound-healing bioefficacy.

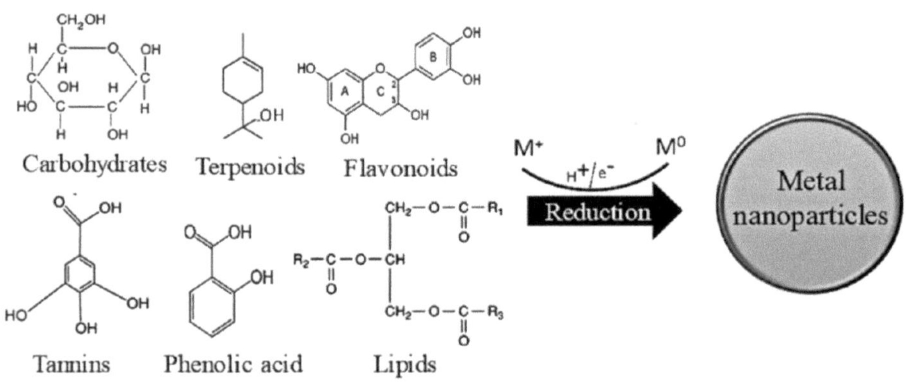

FIGURE 7.4 General reduction mechanism of phytochemicals.

7.4 PLANT-MEDIATED NANOPARTICLES AND THEIR APPLICATIONS

7.4.1 PHYTONUTRIENTS AND NANOFERTILIZERS

Phytogenic nanoparticles of gold, silver, zinc oxide, copper, and iron have been studied for their role as growth promoters for plants. Zinc oxide nanoparticles synthesized using *Calotropis* leaf extract were applied topically to seedlings (nursery stage) of neem *(Azadirachta indica)*, karanj *(Pongamia pinnata)*, and milkwood-*pine (Alstonia scholaris)* and showed noticeably accelerated growth (Chaudhuri and Malodia, 2017). Similarly, ZnO, MgO, and SiO_2 have

TABLE 7.3 Functional Groups and Structures of Various Phytochemicals Used for Plant-Mediated Nanoparticle Synthesis.

S. No	Phytochemical	Subclasses	Functional Group Involves in Nanoparticles Synthesis	Structure
1.	Carbohydrates	Sugar alcohols Monosaccharide	Hydroxyl, Carbonyl, (either aldehyde, ketone)	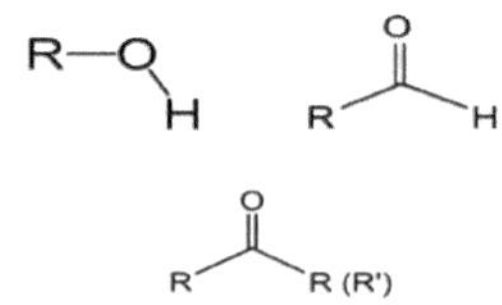
		Disaccharides Polysaccharides Oligosaccharides	Hydroxyl, carbonyl, (either aldehyde, ketone) and ether	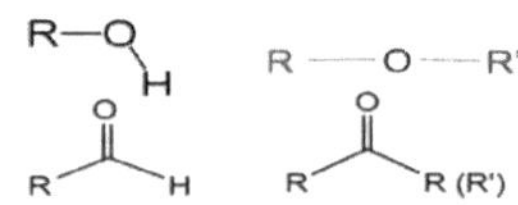
2.	Lipids	Unsaturated fat, saturated fat, fatty acids	Carbonyl, carboxylic, ester, and ether	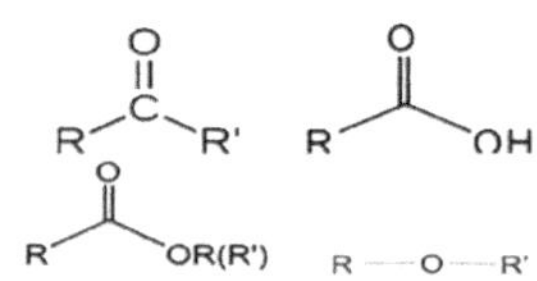
3.	Nitrogen-containing metabolites	Protein amino acids, purines, pyrimidine, peptides, amine	Carboxyl, amino, amide and sulfhydryl and hydroxyl	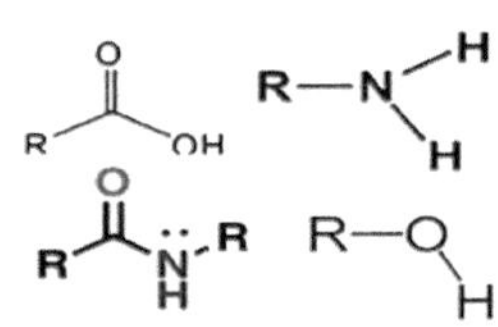
4.	Terpenoids	Carotenoids Hemiterpene Monoterpenoids Sesquiterpenooids Diterpenoids Sesterterpenoids Triterpenoids	Oxygen-containing functional group	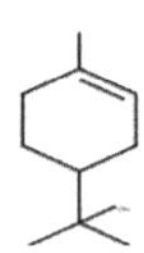
5.	Alkaloids	Indole Isoquinoline Pyrrolidone Piperidinequinoline Steroidal Cynogenic Glycosiode	Nitrogen-containing functional group	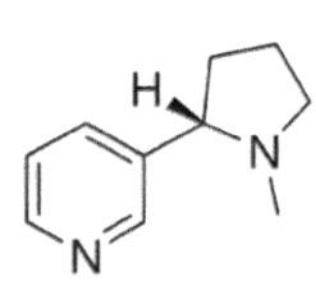

TABLE 7.3 *(Continued)*

S. No	Phytochemical	Subclasses	Functional Group Involves in Nanoparticles Synthesis	Structure
6.	Polyphenol compound	Cafeic acid	R_1=OH, R_2= H	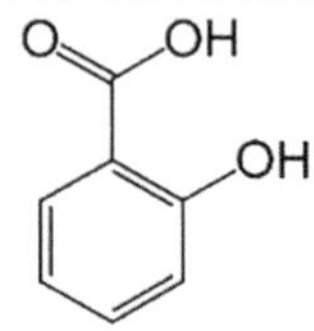
		P-Coumaric Asit	R_1=H, R_2= H	
	*(Phenolic acid)	Ferulik Asit	R_1=OCH_3, R_2= H	
		Gallic acid	R_1=OH, R_2=O H	
		Chlorogenic acid	ESTER	
	*(Flavonoids)	Kaempferol	R_1=R_2= H	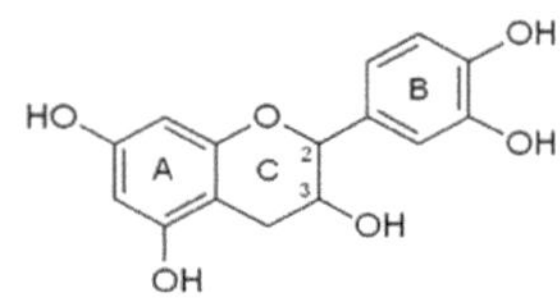
		Quercetin	R_1=OH, R_2= H	
		Quercitrin	R_1=OH, R_2= H, R_3=Rha	
		Rutin	R_1=OH, R_2= H, R_3=Rutinose	
		Hyperoside	R_1=OH, R_2= H, R_3=GAl	
		Cyanidin	R_1=OH, R_2= H	
		Delphinidin	R_1=R_2= H	
		Peonidin	R_1=OCH_3, R_2= H	
		Petunidin	R_1=OH, R_2= $0CH_3$	
	*(Tannin)		Hydroxyl	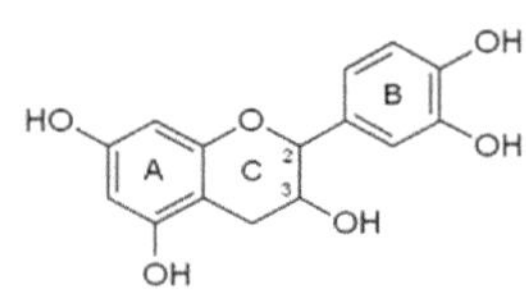
	(Lignin)		Oxygen-containing functional group	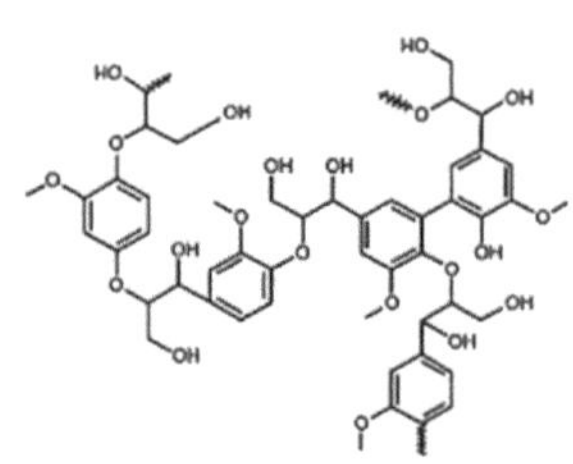
	*(Quinones)	Benzoquinone, Napthoquinone, and Arthraquinon	Oxygen-containing functional group	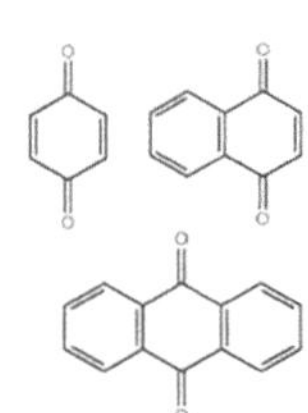

been synthesized using lemon peel and used as nanofertilizers to improve the absorption of Cu^{2+} and Fe^{3+} in the presence of NPK in the soil system and increase the productivity of *Origanum majorana* (marjoram).

7.4.2 ANTI-PHYTOPATHOGENIC

Fusarium species are associated with various plant diseases, such as cereal grain scab, head blight, and crown rot, as well as occasionally infecting animals. *Aspergillus flavus* infections are responsible for the ear rot in corn and yellow mold in peanuts pre- or postharvest. Similarly, *Didymella bryoniae* causes cucurbit-rot disease, gummy stem blight, and black rot. Plant-mediated nanoparticles have been studied to control these diseases and used as anti-phytopathogens. For example, the cubic-shaped silver nanoparticles synthesized using leaf extract of *Ipomoea obscura* have been used as anti-phytopathogenic agents against fungal species of *Didymella bryoniae, Fusarium oxysporum, Fusarium moniliforme,* and *Aspergillus flavus* (Pithawala and Jain, 2012). Zinc oxide nanoparticles synthesized using an aqueous extract of the seed coat of almond were studied on *Daucus carrota* (carrot plant) and found to have an inhibitory effect against the phytopathogen fungus "*Rhizoctonia solani*" (Khan et al., 2022). Similarly, silver nanoparticles (AgNPs) synthesized using an extract of strawberry waste showed significant control and growth inhibitory activity against the nematode *Meloidogyne incognita, the* plant pathogenic bacterium *Ralstonia solanacearum*, and the fungus *Fusarium oxysporum* (Khan et al., 2021b).

7.4.3 SOIL REMEDIATION

Agricultural fields are one of the major regions where heavy metals (cadmium and arsenic) are found in high concentrations. These metals get introduced into the biological system with daily diet intake from agriculture (Iravani and Varma, 2020). This problem can be resolved using plant-mediated nanoparticles; for example, phytogenic iron oxide nanoparticles (PION) decreased the concentration of Cadmium in flooded paddy soil via a redox reaction and helped in microbial enrichment (Lin et al., 2021).

7.4.4 WATER REMEDIATION

Plant-mediated nanoparticles have been studied as photocatalysts for the degradation of various organic pollutants present in water and the reduction or oxidation of inorganic pollutants to harmless substances. Phytogenic silver nanoparticles (PAgNPs) synthesized via *Parkia speciose* leaf aqueous extract have shown the properties of degradation of methylene blue (MB) at pH 11 under solar exposure (Ravichandran et al., 2019). Gold AuNPs synthesized using *Elaeis guineensis* leaves (EGL) extract showed methylene blue (MB) dye degradation under visible light irradiation at 70°C (Ahmad et al., 2020). Biosynthesized silver nanoparticles (AgNPs) using *caulifower (Brassica oleracea var. botrytis)* waste extract showed significant photocatalytic activity for methylene blue dye (Kadam et al., 2020). Silver nanoparticles synthesized using an aqueous extract of *Tribulus terrestris L.* shoots have also been used as photocatalysts for the degradation of methylene blue under UV light (Hamidi et al., 2019). Similarly, silver nanoparticles synthesized using *Kleinia grandiflora* leaf extract have shown a potential role in the photocatalytic degradation of norfloxacin under visible light (Kanagamani et al., 2019). Plant-mediated synthesis of CuO nanoparticles (NPs) by *Madhuca longifolia* plant extract has been studied for their photocatalytic properties and size scale, and it has been shown that phytogenic CuO is efficient in the degradation of methylene blue (MB) in the presence of visible light exposure (Das et al., 2018). Another example includes zinc oxide (ZnO-Np) and copper oxide (CuO-Np) nanoparticles prepared from *Ferulago angulata* extract that showed photocatalytic activity against the organic contaminant rhodamine B (Mehr et al., 2018).

7.4.5 BIOSENSING

Copper oxide nanoparticles (pm-CuO NPs) prepared from *Annona squamosa* aqueous seed extract are used as a platform for electro-catalytic determination of hydrogen peroxide (H_2O_2) produced during plant stress, scavenging elevated H_2O_2, and management of crops susceptibility toward oxidative damage (Singh et al., 2021). Silver nanoparticles (AgNPs) using cauliflower *(Brassica oleracea var. botrytis)* waste extract showed significant photocatalytic activity for methylene blue dye and detection of Hg^{2+} bio-sensing up to 0.1 mg/l Hg^{2+} ion (Kadam et al., 2020).

7.4.6 BIOMEDICAL APPLICATION

Biogenic-capped nanoparticles are used for different biomedical applications and the therapeutic improvement of nanoparticles. In antimicrobial studies, these types of metal nanoparticles elucidate their antimicrobial properties against pathogenic microorganisms (Naganthran et al., 2022; Simon et al., 2022). Similarly, biologically synthesized nanoparticles are also used for antitumor and cytotoxic activity (Naganthran et al., 2022; Simon et al., 2022). Anti-inflammatory drugs are more commonly used because chronic inflammatory diseases are becoming more common (Simon et al., 2022). It has been found that nanoparticles coated with biomolecules also help in the wound-healing mechanism (Naganthran et al., 2022). These types of nanoparticles are mainly used for drug delivery (Sidhu et al., 2022; Simon et al., 2022).

7.5 MARKET VALUE AND LARGE-SCALE PRODUCTION

Recent years have seen a significant rise in the importance of green nanoparticle synthesis because it is simple, economical, produces stable nanoparticles, does not produce toxic byproducts, is environmentally friendly, and can be scaled up for large-scale synthesis. Commercially, NPs are highly in demand because of their wide variety of applications, including industries, electronics, the environment, energy, and biomedicine, such as anticancer potential, anti-leishmanial potential, antimicrobial potential, agricultural applications (like nanofertilizers and nanopesticides), antioxidant action, and environmental remediation. Some of the commercially available products include gold nanoparticles or colloidal gold offered by TORSKAL, which are produced by crude and/or purified extracts from plants. The market value of these gold nanoparticles is in the range of €38–€1035 as per the different properties offered, e.g., self-resistance, biocompatibility, high purity, and stability. Next, Nano Max NPK Fertiliser, offered by JU Agri Sciences Pvt. Ltd., Janakpuri, New Delhi, is based on organic acid chelate with major nutrients, amino acids, organic carbon, and vitamins. Similarly, TAG NANO fertilizer offered by Tropical Agro Systems Pvt. Ltd., India, is synthesized from vitamins and seaweed extract.

The term "large-scale production" refers to the production of items on a large scale using highly specialized machines and processes. Commercialization is increasing through mainly two approaches: laboratory-scale and pilot-scale studies (Magalhães-Ghiotto et al., 2021). Nanoparticles are usually produced and evaluated in a laboratory or bench scale before they are applied in various scientific or industrial applications. In order to produce nanomaterials on a large scale, pilot studies would be the next step. Bench-scale systems are typically found in laboratories and can handle no more than 50 liters at a time (often much more per day). Using a pilot scale, we can demonstrate the concept by treating up to 50% of the total actual flow. In addition, a pilot plant is a valuable tool in estimating the economic viability, operation, and performance of this system's machinery. The majority of studies are still at bench-top level, so they should be continued for the pilot test. Scaling up the production of green nanoparticles will allow for greater accuracy, reproducibility, and efficiency, enabling cost-effectiveness and efficiency. Producing at a large scale involves a wide range of characteristics, including large quantities of raw materials, highly specialized machinery, a huge amount of capital, a large and highly skilled workforce, and efficient and effective planning (Magalhães-Ghiotto et al., 2021).

7.6 ACTS AND REGULATIONS

Every product on the market has to follow certain rules and regulations, and nanotechnology-based products are not unique. A public health agency uses scientific evidence to make regulations for the safe use of nanotechnology-based products. However, in terms of government regulation, nanotechnology and nanotechnology-based products are subject to considerable debate. Regulatory bodies such as the United States Environmental Protection Agency and the Food and Drug Administration in the U.S. or the Health and Consumer Protection Directorate of the European Commission have started dealing with the potential risks posed by nanoparticles. It is important to consider the institutional framework for nanotechnology research and development as well as product safety and regulation in terms of determining the regulatory space available to the industry. Applications of nanotechnology have been launched in several sectors (including electronics, textiles, health, and food packaging).

"The Food Safety and Standard Act 2006" was passed by the Indian parliament to consolidate the laws relating to food safety. The FSSA (Food Safety and Standards Authority) regulates the manufacture, storage, distribution, sale, and import of food. Environmental legislation such as the Environment Protection Act 1986 (for hazardous waste), the Water (Prevention and Control of Pollution) Act 1974, and the Air (Prevention and Control of Pollution) Act 1981, however, have a significant impact on nanomaterials' disposal in the environment.

The FDA has issued several guidelines on various aspects related to the use of nanotechnology in FDA-regulated products. The regulatory approach of the FDA is attributed to

1. The FDA continues to focus on product-based, science-based regulation.

2. Depending on the product class, the FDA respects varying legal standards.
3. In cases where statutory authority does not permit premarket review, consultation is encouraged to reduce unintended harm to human or animal health.
4. The FDA will continue to monitor the product after it is launched.
5. All legal requirements, including safety standards, must be met by industry.
6. On regulatory policy issues, the FDA will collaborate with domestic and international counterparts as appropriate.

The FDA will offer technical advice and guidance, as needed, to help the industry meet its regulatory and statutory obligations.

7.7 CONCLUSION AND FUTURE PROSPECTS

Nanomaterials and nanoparticles synthesized using chemical and physical approaches limit their current usage in wide applications, while biological approaches offer high biocompatibility to these materials and allow their usage in various fields like biomedical, agriculture, environmental, industrial applications, and so forth. Among biological approaches, that is microbial-mediated and plant-mediated, the latter offers a simple, easy, feasible, and cost-effective process compared to the first, where chances of contamination, the requirement of skilled manpower, and cumbersome infrastructure limit its usage. This chapter reveals exhaustive research and a detailed assessment of the mechanisms of biological entities present in plant nanofactories that act as reducing, capping, and stabilizing agents in nanoparticle synthesis. A detailed analysis of these biological entities and mechanisms involved in nanoparticle synthesis reveals a direct correlation between their roles in optimizing the quality and properties of synthesized nanoparticles. Plant-derived nanoparticles possess unique properties like stability, biocompatibility, and cost effectiveness. Therefore, it is concluded that plant nanofactories are a novel template for nanoparticle synthesis compared to other conventional approaches. However, despite the ease of synthesizing nanoparticles using plant nanofactories, a range of limitations are current challenges to overcome, such as specificity and cross-reactivity of synthesized nanoparticles, scaling up production, obtaining monodispersed nanoparticles, and post-use consequences. These limitations will need to be further analyzed in the future to achieve more success in the current field.

KEYWORDS

- **Biogenic synthesis**
- **metal nanoparticles (MNPs)**
- **nanofactories**
- **nanofertilizers**
- **nanoremediation**
- **phytochemicals**
- **phytonutrients**

REFERENCES

Abasi, F., Raja, N. I., Mashwani, Z. U. R., Amjad, M. S., Ehsan, M., Mustafa, N., Haroon, M., & Proćków, J. (2022). Biogenic silver nanoparticles as a stress alleviator in plants: A mechanistic overview. *Molecules*, *27*(11), 3378. https://doi.org/10.3390/molecules27113378

Ahmad, T., Bustam, M. A., Zulfiqar, M., Moniruzzaman, M., Idris, A., Iqbal, J., Asghar, H. M. A., & Ullah, S. (2020). Controllable phytosynthesis of gold nanoparticles and investigation of their size and morphology-dependent photocatalytic activity under visible light. *Journal of Photochemistry and Photobiology A: Chemistry*, *392*, 112429. https://doi.org/10.1016/j.jphotochem.2020.112429

Ajitha, B., Ashok Kumar Reddy, Y., & Sreedhara Reddy, P. (2015). Green synthesis and characterization of silver nanoparticles using Lantana camara leaf extract. *Materials Science and Engineering: C*, *49*, 373–381. https://doi.org/10.1016/J.MSEC.2015.01.035

Alharbi, N. S., Alsubhi, N. S., & Felimban, A. I. (2022a). Journal of radiation research and applied sciences green synthesis of silver nanoparticles using medicinal plants : characterization and application. *Journal of Radiation Research and Applied Sciences*, *15*(3), 109–124. https://doi.org/10.1016/j.jrras.2022.06.012

Alharbi, N. S., Alsubhi, N. S., & Felimban, A. I. (2022b). Green synthesis of silver nanoparticles using medicinal plants: characterization and application. *Journal of Radiation Research and Applied Sciences*, *15*(3), 109–124. https://doi.org/10.1016/J.JRRAS.2022.06.012

Baig, N., Kammakakam, I., & Falath, W. (2021). Nanomaterials: a review of synthesis methods, properties, recent progress, and challenges. *Materials Advances*, *2*(6), 1821–1871. https://doi.org/10.1039/D0MA00807A

Begum, S. J. P., Pratibha, S., Rawat, J. M., Venugopal, D., Sahu, P., Gowda, A., Qureshi, K. A., & Jaremko, M. (2022). Recent advances in green synthesis, characterization, and applications of bioactive metallic nanoparticles. *Pharmaceuticals (Basel)*, *15*(4), 455. https://doi.org/10.3390/ph15040455

Buntru, M., Hahnengress, N., Croon, A., & Schillberg, S. (2022). Plant-derived cell-free biofactories for the production of secondary metabolites. *Frontiers in Plant Science*, *12*, 1–10. https://doi.org/10.3389/fpls.2021.794999

Chahardoli, A., Karimi, N., & Fattahi, A. (2018). Nigella arvensis leaf extract mediated green synthesis of silver nanoparticles: their characteristic properties and biological efficacy. *Advanced Powder Technology*, *29*(1), 202–210. https://doi.org/10.1016/J.APT.2017.11.003

Chaudhuri, S. K., & Malodia, L. (2017). Biosynthesis of zinc oxide nanoparticles using leaf extract of calotropis gigantea: characterization and its evaluation on tree seedling growth in nursery stage. *Applied Nanoscience (Switzerland)*, *7*(8), 501–512. https://doi.org/10.1007/s13204-017-0586-7

Das, P., Ghosh, S., Ghosh, R., Dam, S., & Baskey, M. (2018). Madhuca longifolia plant mediated green synthesis of cupric oxide nanoparticles: A promising environmentally sustainable material for waste water treatment and efficient antibacterial agent. *Journal of Photochemistry and Photobiology B: Biology*, *189*, 66–73. https://doi.org/10.1016/j.jphotobiol.2018.09.023

Dauthal, P., & Mukhopadhyay, M. (2016). Noble metal nanoparticles: Plant-mediated synthesis, mechanistic aspects of synthesis, and applications. *Industrial and Engineering Chemistry Research*, *55*(36), 9557–9577. https://doi.org/10.1021/acs.iecr.6b00861

Dhifi, W., Bellili, S., Jazi, S., Bahloul, N., & Mnif, W. (2016). Essential oils' chemical characterization and investigation of some biological activities: A critical review. *Medicines, 3*(4), 25. https://doi.org/10.3390/medicines3040025

Dikshit, P. K., Kumar, J., Das, A. K., Sadhu, S., Sharma, S., Singh, S., Gupta, P. K., & Kim, B. S. (2021). Green synthesis of metallic nanoparticles : Applications and limitations. *Catalysts*, *11*(8), 902. https://doi.org/10.3390/catal11080902

Fayyaz, B., Zahra, M. B., Aftab, Z., & Haider, M. S. (2022). Screening of phenolic compounds from spinach *(Spinacia oleracea),* green synthesis of iron-nanoparticles and determination of its anti-microbial effect on *escherichia coli*, 1–20. https://doi.org/10.21203/rs.3.rs-1284313/v1

Flieger, J., Franus, W., Panek, R., Szymańska-Chargot, M., Flieger, W., Flieger, M., & Kołodziej, P. (2021). Green synthesis of silver nanoparticles using natural extracts with proven antioxidant activity. *Molecules*, *26*(16), 4986. https://doi.org/10.3390/molecules26164986

Hamdy, N. M., Boseila, A. A., Ramadan, A., & Basalious, E. B. (2022). Iron oxide nanoparticles-plant insignia synthesis with favorable biomedical activities and less toxicity, in the "Era of the-Green": A Systematic Review. *Pharmaceutics*, *14*(4). https://doi.org/10.3390/pharmaceutics14040844

Hamidi, A., Taghavizadeh Yazdi, M. E., Amiri, M. S., Hosseini, H. A., & Darroudi, M. (2019). Biological synthesis of silver nanoparticles in Tribulus terrestris L. extract and evaluation of their photocatalyst, antibacterial, and cytotoxicity effects. *Research on Chemical Intermediates*, *45*(5), 2915–2925. https://doi.org/10.1007/s11164-019-03770-y

Hu, X., Zhang, Y., Ding, T., Liu, J., & Zhao, H. (2020). Multifunctional gold nanoparticles: A novel nanomaterial for various medical applications and biological activities. *Frontiers in Bioengineering and Biotechnology*, *8*, 1–17. https://doi.org/10.3389/fbioe.2020.00990

Iravani, S., & Varma, R. S. (2020). Bacteria in heavy metal remediation and nanoparticle biosynthesis. *ACS Sustainable Chemistry and Engineering*, *8*(14), 5395–5409. https://doi.org/10.1021/acssuschemeng.0c00292

Jadoun, S., Arif, R., Jangid, N. K., & Meena, R. K. (2021). Green synthesis of nanoparticles using plant extracts: a review. *Environmental Chemistry Letters*, *19*(1), 355–374. https://doi.org/10.1007/s10311-020-01074-x

Jain, S., & Mehata, M. S. (2017). Medicinal plant leaf extract and pure flavonoid mediated green synthesis of silver nanoparticles and their enhanced antibacterial property. *Scientific Reports*, *7*(15867), 1–13. https://doi.org/10.1038/s41598-017-15724-8

Jeevanandam, J., Ling, J. K. U., Barhoum, A., Chan, Y. S., & Danquah, M. K. (2022). Bionanomaterials: definitions, sources, types, properties, toxicity, and regulations. *Fundamentals of Bionanomaterials*, 1–29. https://doi.org/10.1016/B978-0-12-824147-9.00001-7

Kabir, S. R., Islam, F., & Asaduzzaman, A. K. M. (2022). Biogenic silver/silver chloride nanoparticles inhibit human cancer cells proliferation *in vitro* and Ehrlich ascites carcinoma cells growth *in vivo*. *Scientific Reports*, *12*(8909), 1–14. https://doi.org/10.1038/s41598-022-12974-z

Kadam, J., Dhawal, P., Barve, S., & Kakodkar, S. (2020). Green synthesis of silver nanoparticles using cauliflower waste and their multifaceted applications in photocatalytic degradation of methylene blue dye and Hg^{2+} biosensing. *SN Applied Sciences*, *2*(738), 1–16. https://doi.org/10.1007/s42452-020-2543-4

Kanagamani, K., Muthukrishnan, P., Shankar, K., Kathiresan, A., Barabadi, H., & Saravanan, M. (2019). Antimicrobial, cytotoxicity and photocatalytic degradation of norfloxacin using kleinia grandiflora mediated silver nanoparticles. *Journal of Cluster Science*, *30*(6), 1415–1424. https://doi.org/10.1007/s10876-019-01583-y

Khan, M., Khan, A. U., Bogdanchikova, N., & Garibo, D. (2021a). Antibacterial and antifungal studies of biosynthesized silver nanoparticles against plant parasitic nematode meloidogyne incognita, plant pathogens ralstonia solanacearum and fusarium oxysporum. *Molecules*, *26*(9), 2462. https://doi.org/10.3390/molecules26092462

Khan, S. A., Shahid, S., Ayaz, A., Alkahtani, J., Elshikh, M. S., & Riaz, T. (2021b). Phytomolecules-coated NiO nanoparticles synthesis using abutilon indicum leaf extract: Antioxidant, antibacterial, and anticancer activities. *International Journal of Nanomedicine*, *16*, 1757–1773. https://doi.org/10.2147/IJN.S294012

Khan, A. U., Khan, M., Malik, N., Parveen, A., Sharma, P., Min, K., Gupta, M., & Alam, M. (2022). Screening of biosynthesized zinc oxide nanoparticles for their effect on Daucus carota pathogen and molecular docking. *Microscopy Research & Technique*, *85*(10), 3365–3373. https://doi.org/10.1002/jemt.24191

Kumar, R., & Lal, S. (2014). Synthesis of organic nanoparticles and their applications in drug delivery and food nanotechnology: A review. *Journal of Nanomaterials & Molecular Nanotechnology*, *3*(4). https://doi.org/10.4172/2324-8777.1000150

Krishnaraj, C., Jagan, E. G., Rajasekar, S., Selvakumar, P., Kalaichelvan, P. T., & Mohan, N. (2010). Synthesis of silver nanoparticles using Acalypha indica leaf extracts and its antibacterial activity against water borne pathogens. *Colloids and Surfaces B: Biointerfaces*, *76*(1), 50–56. https://doi.org/10.1016/j.colsurfb.2009.10.008

Lin, J., He, F., Owens, G., & Chen, Z. (2021). How do phytogenic iron oxide nanoparticles drive redox reactions to reduce cadmium availability in a flooded paddy soil? *Journal of Hazardous Materials*, *403*, 123736. https://doi.org/10.1016/j.jhazmat.2020.123736

Magalhães-Ghiotto, G. A. V., de Oliveira, A. M., Natal, J. P. S., Bergamasco, R., & Gomes, R. G. (2021). Green nanoparticles in water treatment: A review of research trends, applications, environmental aspects and large-scale production. *Environmental Nanotechnology, Monitoring & Management*, *16*, 100526. https://doi.org/10.1016/J.ENMM.2021.100526

Marslin, G., Siram, K., Maqbool, Q., Selvakesavan, R. K., Kruszka, D., Kachlicki, P., & Franklin, G. (2018). Secondary metabolites in the green synthesis of metallic nanoparticles. *Materials*, *11*(6), 1–25. https://doi.org/10.3390/ma11060940

Mustapha, T., Misni, N., Ithnin, N. R., Daskum, A. M., & Unyah, N. Z. (2022). A Review on Plants and Microorganisms Mediated Synthesis of Silver Nanoparticles, Role of Plants Metabolites and Applications. *International Journal of Environmental Research and Public Health*, *19*(2), 674. https://doi.org/10.3390/ijerph19020674

Naganthran, A., Verasoundarapandian, G., Khalid, F. E., Masarudin, M. J., Zulkharnain, A., Nawawi, N. M., Karim, M., Abdullah, C. A. C., & Ahmad, S. A. (2022). Synthesis, characterization and biomedical applications of silver nanoparticles. *Materials (Basel)*, *15*(2), 427. 10.3390/ma15020427

Naseer, M., Aslam, U., Khalid, B., & Chen, B. (2020). Green route to synthesize zinc oxide nanoparticles using leaf extracts of cassia fistula and melia azadarach and their antibacterial potential. *Scientific Reports*, *10*(9055), 1–10. https://doi.org/10.1038/s41598-020-65949-3

Nazneen, H., Rather, G. A., Ali, A., & Chakravorty, A. (2022). The role of plant-mediated biosynthesised nanoparticles in agriculture. In S. A. Bandh (Eds.), *Sustainable Agriculture* (pp. 97–117). Springer, Cham. https://doi.org/10.1007/978-3-030-83066-3_6

Ndugire, W., Liyanage, S. H., & Yan, M. (2021). Carbohydrate-presenting metal nanoparticles: synthesis, characterization and applications. *Comprehensive Glycoscience: Second Edition*, *4*, 380–405. https://doi.org/10.1016/B978-0-12-819475-1.00040-7

Noah, N. M., & Ndangili, P. M. (2022). Green synthesis of nanomaterials from sustainable materials for biosensors and drug delivery. *Sensors International*, *3*, 100166. https://doi.org/10.1016/J.SINTL.2022.100166

Omar, Z. A., Abduljabar, R. S., Sajadi, S. M., Mohmud, S. A., & Yahya, R. O. (2022). Recent progress in eco-synthesis of essential oil-based nanoparticles and their possible mechanisms. *Industrial Crops and Products*, *187*, 115322. https://doi.org/10.1016/J.INDCROP.2022.115322

Parmar, M., & Sanyal, M. (2022). Extensive study on plant mediated green synthesis of metal nanoparticles and their application for degradation of cationic and anionic dyes. *Environmental Nanotechnology, Monitoring & Management*, *17*, 100624. https://doi.org/10.1016/J.ENMM.2021.100624

Pathania, D., Kumar, S., Thakur, P., Chaudhary, V., Kaushik, A., Varma, R. S., Furukawa, H., Sharma, M., & Khosla, A. (2022). Essential oil-mediated biocompatible magnesium nanoparticles with enhanced antibacterial, antifungal, and photocatalytic efficacies. *Scientific Reports*, *12*(11431), 1–13. https://doi.org/10.1038/s41598-022-14984-3

Pithawala, N. A., & Jain, B. (2012). Journal of advanced scientific research. *Advancedment of Science, 1*(2), 19–23. http://www.sciensage.info/journal/1359303580JASR_3006121.pdf

Raj, S., Singh, H., Trivedi, R., & Soni, V. (2020). Biogenic synthesis of AgNPs employing Terminalia arjuna leaf extract and its efficacy towards catalytic degradation of organic dyes. *Scientific Reports*, *10*(9616). https://doi.org/10.1038/s41598-020-66851-8

Ravichandran, V., Vasanthi, S., Shalini, S., Shah, S. A. A., Tripathy, M., & Paliwal, N. (2019). Green synthesis, characterization, antibacterial, antioxidant and photocatalytic activity of Parkia speciosa leaves extract mediated silver nanoparticles. *Results in Physics*, *15*(102565). https://doi.org/10.1016/j.rinp.2019.102565

Rodríguez-León, E., Íñiguez-Palomares, R. A., Navarro, R. E., Rodríguez-Beas, C., Larios-Rodríguez, E., Alvarez-Cirerol, F. J., Íñiguez-Palomares, C., Ramírez-Saldaña, M., Hernández Martínez, J., Martínez-Higuera, A., Galván-Moroyoqui, J. M., & Martínez-Soto, J. M. (2017). Silver nanoparticles synthesized with Rumex hymenosepalus extracts: effective broad-spectrum microbicidal agents and cytotoxicity study. *Artificial Cells, Nanomedicine and Biotechnology*, *46*(6), 1194–1206. https://doi.org/10.1080/21691401.2017.1366332

Saim, A. K., Kumah, F. N., & Oppong, M. N. (2021). Extracellular and intracellular synthesis of gold and silver nanoparticles by living plants: a review. *Nanotechnology for Environmental Engineering*, *6*(1), 1–11. https://doi.org/10.1007/s41204-020-00095-9

Sengottaiyan, A., Mythili, R., Selvankumar, T., Aravinthan, A., Kamala-Kannan, S., Manoharan, K., Thiyagarajan, P., Govarthanan, M., & Kim, J. H. (2016). Green synthesis of silver nanoparticles using Solanum indicum L. and their antibacterial, splenocyte cytotoxic potentials. *Research on Chemical Intermediates*, *42*(4), 3095–3103. https://doi.org/10.1007/s11164-015-2199-7

Mehr, E. S., Sorbiun, M., Ramazani, A., & Fardood, S. T. (2018). Plant-mediated synthesis of zinc oxide and copper oxide nanoparticles by using ferulago angulata (schlecht) boiss extract and comparison of their photocatalytic degradation of Rhodamine B (RhB) under visible light irradiation. *Journal of Materials Science: Materials in Electronics*, *29*(2), 1333–1340. https://doi.org/10.1007/s10854-017-8039-3

Shervani, Z., & Yamamoto, Y. (2011). Carbohydrate-directed synthesis of silver and gold nanoparticles: effect of the structure of carbohydrates and reducing agents on the size and morphology of the composites. *Carbohydrate Research*, *346*(5), 651–658. https://doi.org/10.1016/J.CARRES.2011.01.020

Sidhu, A. K., Verma, N., & Kaushal, P. (2022). Role of biogenic capping agents in the synthesis of metallic nanoparticles and evaluation of their therapeutic potential. *Frontiers in Nanotechnology*, *3*, 1–17. https://doi.org/10.3389/fnano.2021.801620

Simon, S.; Sibuyi, N. R. S.; Fadaka, A. O.; Meyer, S.; Josephs, J.; Onani, M. O.; Meyer, M.; Madiehe, A. M. (2022). Biomedical applications of plant extract-synthesized silver nanoparticles. *Biomedicines, 10*(11), 2792. https://doi.org/10.3390/biomedicines10112792

Singh, J., Dutta, T., Kim, K. H., Rawat, M., Samddar, P., & Kumar, P. (2018). 'Green' synthesis of metals and their oxide nanoparticles : applications for environmental remediation. *Journal of Nanobiotechnology*, *16*(84), 1–24. https://doi.org/10.1186/s12951-018-0408-4

Singh, P., Singh, K. R. B., Singh, J., Das, S. N., & Singh, R. P. (2021). Tunable electrochemistry and efficient antibacterial activity of plant-mediated copper oxide nanoparticles synthesized by: *Annona squamosa* seed extract for agricultural utility. *RSC Advances*, *11*(29), 18050–18060. https://doi.org/10.1039/d1ra02382a

Uddin, S., Iqbal, J., Safdar, L. B., Ahmad, S., Abbasi, B. A., Capasso, R., Kazi, M., & Quraihi, U. M. (2022). Green synthesis of BPL-NiONPs using leaf extract of berberis pachyacantha: Characterization and multiple *in vitro* biological applications. *Molecules*, *27*(7), 2064. https://doi.org/10.3390/molecules27072064

Velmurugan, P., Anbalagan, K., Manosathyadevan, M., Lee, K. J., Cho, M., Lee, S. M., Park, J. H., Oh, S. G., Bang, K. S., & Oh, B. T. (2014). Green synthesis of silver and gold nanoparticles using Zingiber officinale root extract and antibacterial activity of silver nanoparticles against food pathogens. *Bioprocess and Biosystems Engineering*, *37*(10), 1935–1943. https://doi.org/10.1007/s00449-014-1169-6

CHAPTER 8

Physiochemical and Molecular Effects of Nanomaterials on Soil Fertility and Microbes

HIMANKI DABRAL[1], RAMESH KUMAR SINGH[2], AKANKSHA GAUTAM[3], AWANTIKA SINGH[3], SOURABH TYAGI[4], HEMANT RITTURAJ KUSHWAHA[3,4], and ANU SINGH[2,3]

[1]*Department of Genetics & Plant Breeding, School of Agricultural Sciences, Shri Guru Ram Rai University, Pathribagh, Dehradun, Uttarakhand, India*

[2]*Samvet Bharat Under the Aegis of Samvetan Society for Social and Scientific Research, Dehradun, Uttarakhand, India*

[3]*School of Biotechnology, Jawaharlal Nehru University, New Delhi, India*

[4]*Special Centre for Systems Medicine, Jawaharlal Nehru University, New Delhi, India*

ABSTRACT

During the past few decades, nanomaterial-based formulations have been extensively applied as nanocoated fertilizers, pesticides, herbicides, etc. for soil health improvement and growth of plant. Nanomaterials are defined as any functional device, structure, or system constructed on a nanometer scale. Due to their large surface area-to-volume ratio, higher reactivity, and diverse physical and chemical properties, they are more amenable to different applications in comparison to bulk materials. The excessive use of these nanomaterials in agriculture leads to disturbances in the physicochemical factors affecting soil fertility like soil texture, soil structure, water retention capacity,

Nanobiotechnology for Agricultural Sciences: Nano-Agri-Input Products for Crop Production and Environmental Protection. J. C. Tarafdar, Deepak Kumar, Sanjay K. Singh, and Krishna Pal Singh (Eds.)

electrical conductivity, bulk density, soil pH, cation exchange capacity, and plant nutrients (micronutrients and macronutrients). Similarly, excessive usage of nanomaterials can also affect biological factors, i.e., organic matter, microorganisms (bacteria and fungi), soil mineralogy, and biogeochemical cycles. Nanomaterials are capable of altering the microbiome of the soil by altering species richness and composition, which in turn affect plant communities. Reports on absorption of nanoparticles by the plant system show that (*Arabidopsis*) they change gene expression at the DNA level by upregulating the expression of harmful genes and downregulating the stress-responsive genes involved in metal detoxification. Nanomaterials also act as mutagens, inducing changes in chromosome structure and changes in cell cycle upon interaction with the plant cell. Due to the overproduction of reactive oxygen radicals by these nanomaterials, they are affecting all living organisms from microbes to human at molecular to whole organism level. This chapter focuses on the molecular behavior of the nanomaterials and their physiochemical effects on soil fertility and microbes.

8.1 INTRODUCTION

A rapidly increasing world population is generating a high demand for food supplies and, in turn, putting high pressure on the agriculture sector globally. This huge demand can be met by increasing the quality and production of food crops, for which various approaches have been employed in the agriculture sector, such as the adoption of hybrid seeds in traditional breeding methods, the use of chemicals, molecular breeding, and biotechnological approaches (Prasad et al., 2017). However, the desired result has not yet been obtained because the change in climate, rainfall pattern, temperature, and pest interaction has created huge challenges for the sustainable production of agricultural crops. For the past few decades, farmers have continuously used synthetic chemical fertilizers, weedicides, and pesticides to increase crop yield and crop production, but their excessive usage results in soil degradation, affecting both the soil microflora and fauna, which ultimately leads to environmental pollution (Nair, 2014). Therefore, it is important to develop and adopt new scientific advancements like nanoscale technologies as a sustainable approach for crop production. Nanotechnology makes use of materials synthesized in nanoscale dimensions to develop agricultural tools and smart delivery systems for the management of biotic and abiotic stresses and to overcome nutrient deficiency, uptake, and translocation into the plant system. Various types of materials used for nanomaterials and nanoparticles synthesis are metals, ceramics, biomolecules, semiconductors, and organic polymers. Some carbon-based nanoparticles involve single and multiwalled carbon nanotubes and fullerenes (Ghosh et al., 2015; Zaytseva and Neumann, 2016).

Nanomaterials possess varying physiochemical properties that make them amenable to various applications in the agriculture sector, such as the delivery of fertilizers, micronutrient supply, insect-pest management, nanoherbicides, nanofungicides, nanobiosensors, etc. Nanoparticles being used for crop protection on the other hand, nanocapsules containing active herbicide ingredients allow their effective penetration through the cuticle and their control release. Similarly, nanofertilizer has been reported to increase the bioavailability of nutrients to the plant system, especially yield-limiting micronutrients like iron, zinc, copper, and manganese, which are responsible for poor nutritional quality and are now available through nano-based fertilizer (Kumar et al., 2018). With their use, nutrients are efficiently targeted to the target crop plant, which ultimately contributes to sustainable crop production and food security. Nanomaterials are also helpful in generating disease- and stress-resistant crops. It is evident from recent research that nanotechnology is significantly contributing to the establishment of modern agriculture, and nanomaterials are going to be an important part of the sector. However, continuous application of nanomaterials in huge concentrations or their accidental release into the environment sometimes changes their fate, or they undergo various physical (aggregation, adsorption, deposition), chemical (redox reaction, photochemical reaction), and biological (biodegradation, biomodification) transformations that change their physiochemical properties and hence their efficiency and toxicity. Unfortunately, these transformed nanomaterials, when interacted with by the organisms in the soil microbiome, may have negative consequences for the respective ecosystem. The release of such transformed nanoparticles is a huge threat and concern to the soil microbiome by affecting the soil's physiochemical properties. Therefore, controlled application of nanomaterials or the synthesis of less toxic nanomaterials for agricultural purposes is required. This concern and awareness have guided researchers to opt for the green synthesis route for synthesizing biocompatible nanoparticles (Worrall et al., 2018). Bacteria-mediated and plant-mediated nanoparticles, where plants and their parts like leaves, stems, roots, flowers, seeds, bark, etc., as well as the phytochemicals and nutritional compounds extracted from plant parts, have been successfully used for the synthesis of metals and metal oxide nanoparticles, have been proven to be cost-effective, nontoxic, and eco-friendly approaches. This chapter provides detailed insights into the types of nanoparticles and their usage in sustainable agriculture and crop production, along with the effects of these nano-based agricultural inputs on soil physiochemical properties and the soil microbiome.

8.2 SOIL HEALTH AND SUSTAINABLE AGRICULTURE

Soil health is one of the important factors determining the sustainability of crop production in agriculture. Doran and Zeiss (2000) described the correlation between soil health and sustainability via improving soil quality through managing soil living component. The soil ecosystem, including both biotic and abiotic components, functions as a vital living system for the sustenance of plants and animals. Soil health can be assessed by its fitness to perform various functions as well as resist and recover from environmental degradation. A healthy soil should have rich organic matter, proper tillage and structure, high permeation and low compaction, a rich microbiome, resistance to erosion, the least trace of toxic chemicals, and low weed and disease pressure, as indicated in Table 8.1. Stable and dynamic soil properties are considered to be soil health indicators. Stable soil properties are affected by climate, land topography, organisms, and parent material. It includes properties like soil texture, soil type, and soil depth. While soil properties, which undergo change with land use and management practises, are dynamic in nature (e.g., pH, bulk density, etc.), it should be clearly distinguished that soil health affects crop productivity, while the gradual reduction and deterioration of agricultural land area due to natural and anthropogenic activities reduces land productivity. It is being expected that by the end of 2050, the world's population will rise to 8.9 billion, generating huge pressure on the agriculture sector (Lichtfouse et al., 2009). Under such circumstances, translation of scientific advancements into practical knowledge will be the only way to ensure the global sustainability of agricultural land. These sustainable approaches enhance the organic content of soil through plant diversity and reduce soil erosion through the adoption of conservation tillage practises that cause minimal disturbance to the soil. One of the important goals of sustainable agriculture is to increase crop production by overcoming the challenges of climate change and preserving the agroecosystem (Timsina, 2018). However, excessive use of synthetic fertilizer and pesticides in the past few decades in order to meet agricultural demand has increased degradation of land and environment causing harmful effect on the ecosystem. Similarly, some cultural practises, like monoculture in the case of wheat, subsequently reduced the soil quality and abundance of beneficial microorganisms with an increased incidence of pests and pathogens (Singh et al., 2011). To overcome these issues, sustainable agriculture is required, which is a kind of integrated approach that makes use of alternative resources to solve problems related to food production without harming our environment (Lal, 2008). The activity and diversity of soil microorganism forms the basics of sustainable agriculture, indicating the importance of soil health. Besides, several other physical and chemical properties of soil also act as key indicators for soil health assessment. Soil health indicators for sustainable agriculture are as follows:

TABLE 8.1 Soil Properties Indicators of Soil Health.

Physical	Chemical	Biological
Soil color, soil aggregate stability, water infiltration rate, water-holding capacity, bulk density, runoff and erosion, rooting depth	Organic carbon and nitrogen, active carbon, pH, cation exchange capacity, electrical conductivity, heavy metals	Soil respiration, microbial biomass, potential mineralizable nitrogen, soil enzymes, earthworms, crop condition, root growth, weeds, and disease pressure

Source: Adapted from Awale et al. (2017).

8.2.1 SOIL AGGREGATES

These are the primary soil particles that are held together by organic and inorganic materials. High soil aggregate stability is required for sustainable crop production because it protects the soil from disintegration through physical forces. The stability of soil clods can be measured by the proportion of different size aggregate classes, the percentage of stable aggregates upon wet or dry sieving, and the distribution of stable aggregates into different size ranges (Awale et al., 2017). Stable soil aggregates will have high infiltration rates, good aeration, resist erosion, enhance root growth, and protect soil organic matter from microbial decomposition.

8.2.2 SOIL BULK DENSITY AND COMPACTION

Bulk density is expressed as the ratio of the mass of oven-dry soil to its bulk volume which measures its compaction. Highly compact soil restricts plant root growth by hindering their penetration, due to which nutrient uptake and water permeability get reduced, resulting in nutrient deficiency, increased soil erosion, and reduced crop yield (Awale et al., 2017). Adoption of good crop management practises like the addition of organic matter through the application of crop residues, manures, and compost causes a decrease in bulk density and hence makes the soil less compact and suitable for crop production.

8.2.3 SOIL PH AND BUFFERING CAPACITY

It is a major determinant for soil condition and crop yield. Various biological, chemical, and physical characteristics of healthy soil depend on the pH of the soil, which regulates the availability of nutrients, cation exchange property, element toxicity, and finally crop yield. It has been reported that the availability of most macronutrients like calcium, potassium, nitrogen, magnesium, and sulfur is optimum at pH between 6 and 7, while it decreases beyond this range (Brown et al., 2008). Soil pH below 5.5 increases the solubility of aluminum and manganese, leading to increased toxicity to roots that affects plant development and crop yield. Microbial activity in soil is also reduced due to the low availability of organic substrate at low pH, causing low yield; however, this issue can be resolved by treating the acidic soil with lime, which results in a pH increase and facilitates the microbial activities of the soil, leading to increased yield (Awale et al., 2017). Therefore, it is utmost important to maintain a relatively stable pH, i.e., the buffering capacity of the soil, which generally affects factors like organic matter content, amount of clay, cation-exchange capacity, phosphate, carbonate–bicarbonate, organic acid, etc. (Schroder et al., 2011).

8.2.4 CATION EXCHANGE CAPACITY (CEC)

The ability of soil to hold cations by electrical attraction is known as its cation exchange capacity (CEC), which is an indicator of soil fertility and a measure of soil's ability to supply essential nutrients to plants. These nutrients in the soil solution include cations such as calcium (Ca^{2+}), magnesium (Mg^{2+}), potassium (K^{+}), aluminum (Al^{3+}), hydrogen (H+), and zinc (Zn^{2+}), which get exchanged with adsorbed cations on the clay and organic matter. Therefore, CEC is the sum of exchangeable cations that a soil can hold at a specific pH. The plants that grow in soil with low CEC generate deficiency symptoms as compared to those that grow in soil with high CEC (i.e., high nutrient retention capacity), thereby affecting soil buffering capacity.

8.2.5 SOIL ORGANIC MATTER

Soil organic matter is originally comprised of organic materials from plant and animal residues, excretion from plant roots, and soil microbes, which further undergo decomposition and return to the soil. It directly and indirectly affects soil physical, chemical, and biological activities and, hence, the agronomic yield of crops. Soil organic matter is one of the key indicators of soil health and affects soil structure, aggregate stability, water storage, water infiltration, nutrient storage and availability, soil pH, cation exchange capacity, electrical conductivity, etc. Crop sustainability depends upon the soil as well as the agronomic practises that increase the input and decay of organic matter in the soil. For example, conventional tillage practises incorporate crop residues at a higher rate with greater soil aeration, allowing more soil and residue contact and therefore rapid decay of organic matter, while no-till practises leave most of the crop residue on the soil surface, thereby favoring their accumulation. Soil organic matter consists of 58% carbon and 5% nitrogen; therefore, the soil organic matter is estimated by determining soil carbon and nitrogen content, which are assessed and strongly corelated with particulate organic matter, mineralizable carbon, microbial biomass carbon, dissolved organic carbon, permanganate oxidizable carbon, and potentially mineralizable nitrogen.

8.2.6 SOIL BIOTA

This soil health quality indicator comprises the microbial community, their diversity, their abundance, their activity, and their stability in the environment (Doran and Zeiss, 2000). Mineralization of plant residues and their bioavailability to plants for their growth and development are greatly enhanced by microbial community and structure (Meena et al., 2016). The nitrogen conversion between different forms by soil-borne bacteria and fungi has been reported to affect the composition, making, and absorption of nutrients by plant system (Van Der Heijden et al., 2008). These microbial communities contribute to the stability and productivity of any agroecosystem. For instance, the presence of arbuscular mycorrhizal fungi (AMF), bacteria, and useful nematodes plays a vital role in maintaining plant health and soil fertility by affecting the nutrient cycle, water retention, quality of fruit produce, and finally yield as shown in Figure 8.1 (Al-Karaki et al., 2007). Some of the agronomic practises, like organic farming and conservation tillage, significantly enhanced the soil biota in the case of watermelon grown in clay loam soils (Leskovar and Othman, 2018). As soil biota represent the largest reservoirs of biodiversity, their distribution and function are very important for advancing global agricultural sustainability.

Due to the global rise in population and rapidly changing lifestyles, agriculture has become exhaustive, and overutilization of natural resources has led to environmental degradation. In this aspect, sustainable agriculture is one of the solutions for agriculture and environmental sustainability. One of the sustainable approaches of the present time is the utilization of nanotechnology in agriculture. It has been effectively utilized for the targeted and efficient delivery of nutrients and pesticides to the crop plants, improving soil health by enhancing soil microfauna, gene editing for biotechnological applications, waste management, etc. (Tahat et al., 2020). Nanotechnology can be considered a holistic, sustainable approach toward precise and efficient agriculture to fulfil the requirement of present and future generations.

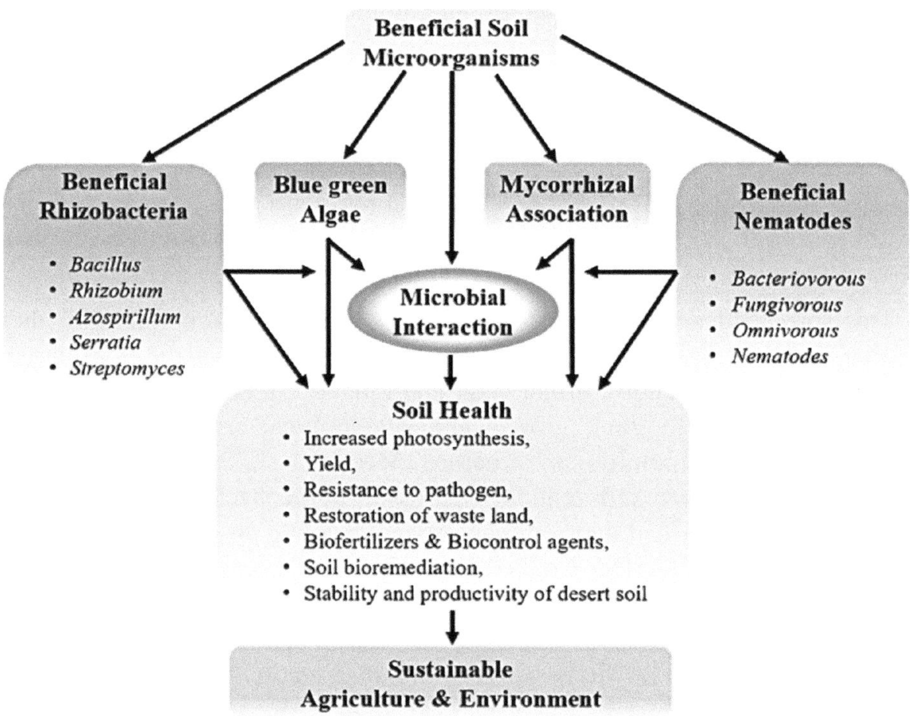

FIGURE 8.1 It indicates the role of soil microfauna and their interactions for the sustainability of agriculture and the environment.

8.3 NANOMATERIALS

The term "nano" suggests anything that is of the order 10^{-9} m in length. The word "nano" is derived from the Greek word "nano," which means extremely small or nanolevel. Nanomaterials can be defined as a set of substances that are built on the nanometer scale in the range of 1–100 nm (Baig et al., 2021). They have been broadly classified into three types: nanoparticles, nanoclays, and nanoemulsions, as shown in Figure 8.2.

8.3.1 TYPES OF NANOMATERIALS

8.3.1.1 NANOPARTICLES

Nanoparticles are constructed in the form of nanostructures or composites. These nanostructures are constructed in different dimensions; depending on that they can be classified as zero dimension as quantum dots in which movement of electrons is restricted in all three directions, one dimension in the form of quantum wires where free electrons can move in the X direction plane, two dimensions in the form of thin films where electrons can move freely in both the X and Y directions, and three dimensions in which free electrons can move in the X, Y, and Z directions (Jeevanandam et al., 2018; Khan et al., 2019; Siegel, 1993). Further, depending upon their physical and chemical properties, they are classified into carbon-based, metal and metal oxide-based, semiconductor, polymeric, and liquid-based nanoparticles (Jeevanandam et al., 2018; Khan et al., 2019).

i. **Carbon-based nanoparticles:** They are characterized into two types: carbon nanotubes and fullerene. Based on their wall number, carbon nanotubes are single-walled or multiwalled. On the other hand, fullerene carbon-based nanoparticles have a hollow cage-like structure made up in pentagonal and hexagonal patterns with 60 carbon atoms. Due to their high versatility and diverse chemical and physical properties like improved structure, high strength, good electrical conductivity, and affinity, they are more suited to diverse applications (Astefanei et al., 2015; Jeevanandam et al., 2018; Khan et al., 2019; Salem et al., 2023).

ii. **Metal and metal oxide-based nanoparticles:** Metal and metal oxide nanoparticles are synthesized from metal precursors obtained when metal salts are subjected to hydrolysis at room temperature or a temperature not more than 100 °C. Both metal and metal oxide nanoparticles have a surface plasmon resonance feature, due to which they show optoelectrical properties. This surface plasmon resonance is known as the collective oscillation of conduction band electrons, which get excited by the electromagnetic field of incident light. Metal NPs were first detected in solution by Faraday in 1857 (Daniel and Astruc, 2004), while the color-changing behavior of nanoparticles was first described quantitatively by Mie in 1908 (Link and El-Sayed, 2003). The most commonly used metal-based nanoparticles that are utilized in different areas are silver (Ag), gold (Au), titanium oxide (TiO_2), iron oxide (Fe_2O_3), zinc oxide (ZnO), or copper (Cu) (Khan et al., 2019; Salem et al., 2023).

iii. **Ceramic nanoparticles:** These are composed of metal oxides, carbides, phosphates, and carbonates of metalloids like calcium, silicon, titanium, etc. with wide applications due to their high heat resistance, and they are chemically inert. They have a porous, hollow, dense, amorphous structure (Sigmund et al., 2006), synthesized via uninterrupted heating and cooling. They are used in catalysis, photocatalysis, photodegradation of dyes, and imaging (Khan et al., 2019; Salem et al., 2023; Thomas et al., 2015).

iv. **Semiconductor nanoparticles**: Semiconductor nanoparticles are fluorescent materials with both metals and nonmetal characteristics. They show unique size-dependent material properties, which dramatically change their optical and electrical properties. For this reason they show wider applications in photocatalysis, photo-optics, and electronic devices and are utilized for analytes detection in biosensors (Khan et al., 2019; Salem et al., 2023).

v. **Polymeric nanoparticles:** Polymeric nanoparticles are organic-based materials such as polyvinyl alcohol, polyvinyl pyrrolidone, starch, and polylactic acid that have been utilized in delivering various agriculture inputs due to their improved characteristics like compatibility, nontoxic behavior, biodegradability, and cost effectiveness (Chen et al., 2018; Zhang et al., 2018). These polymeric nanoparticles can be prepared in the form of nanocapsules, where active compounds are confined and surrounded by a polymer shell, or nanospheres, where active compounds and polymers are uniformly scattered. These nanoparticles are synthesized by various processes like crosslinking and self-assembly that allow target and controlled release of the active compound into the plant system and decrease cellular and environmental toxicity (Khan et al., 2019; Oliveira et al., 2015; Salem et al., 2023).

vi. **Lipid-based nanoparticles:** Lipid-based nanoparticles are made up of a phospholipid bilayer with solid core fixed by surfactants and emulsifiers from outside with dimension between 50 and 100 nm. These nanoparticles are more stable and display site-specific delivery of the active compound. In agriculture, they have been utilized for delivering fungicide to target plant cells. Moreover, they have also been used in the field of nanomedicine as a carrier of drugs in RNA delivery for cancer therapy (Khan et al., 2019).

8.3.1.2 NANOCLAY

These are made up of phyllosilicates that comprise minerals like Mica, kaolin, montmorillonite, and serpentine ($Mg_3[Si_2O_5(OH)_4]$. They differ in the nature of ion present. Nanoclay is mainly found in the naturally occurring bentonite, which constitute 60 to 80% montmorillonite. These nanoclays improve soil fertility, especially by transforming desert land into fertile land and controlling soil erosion (Padidar et al., 2014).

8.3.1.3 NANOEMULSIONS

They are liquid-in-liquid dispersion with droplet dimension of 100 nm. Their small size contributes to their higher stability, high surface area per unit volume, and transparent appearance. These nanoemulsions have wider applicability in various areas of research and therefore are utilized in various areas of research like in food and cosmetic sectors, pharmaceutical, medical, and synthesis of material (Aswathanarayan and Vittal, 2019; Gupta et al., 2016; Wilson et al., 2022). They are prepared through high- and low-energy methods like phase inversion, emulsion inversion point, high-pressure homogenization, and ultrasonication.

8.4 NANOMATERIALS IN THE AGRICULTURE SYSTEM

Nanomaterials are built on a nanometer scale, with size in the range of 1 to 100 nm and this technology has touched different sector domains due to

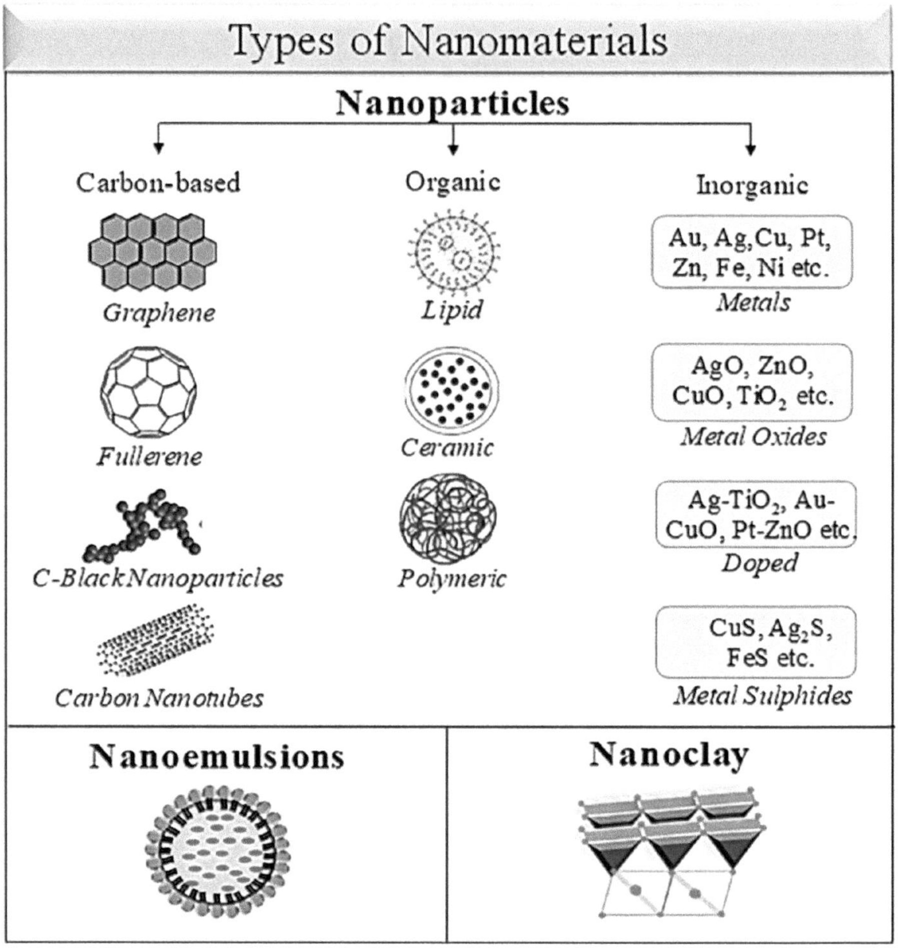

FIGURE 8.2 Types of nanomaterials.

its usefulness. Increasing demand for food with an increase in population is a global concern where adoption of modern technology such as nanotechnology is the only way to achieve food production and their security in a sustainable way. In recent times, nanotechnology has come out as an important tool in agriculture, especially its use as nanocarriers, nanobiosensors, nanopesticides, and nanofertilizers to improve plant growth and yield (Bratovcic et al., 2021; Hong et al., 2013; Singh et al., 2017). Nanomaterials are widely used in agriculture to detect and identify pathogens, help in controlling pest populations by controlling the release of pesticides and other agrochemicals, and improve the efficacy of nutrient distribution as nanofertilizers. A variety of materials, such as carbon-based, metal- and metal oxide-based, silica, clay, and biocomposites, in various forms like nanocapsules, nanoparticles, and nano-emulsions (Amenta et al., 2015) are being utilized for the synthesis of nanoagro inputs for crop production and protection, with agriculture expected to be the most benefited sector (Bouwmeester et al., 2009; Emamifar et al., 2010; Madhuri et al., 2010; Nair et al., 2010; Ghidan and Al Antary, 2019). It has been reported that the utilization of these nanomaterials as nanosensors to measure yield, soil condition, pest incidence, and weed population makes agriculture practises more efficient. Similarly, the use of nanocarriers as smart delivery systems will reduce the loss of active nutrients and optimize and sustain nutrient and water uptake by the plant system, which will make agriculture practises more efficient (Gogos et al., 2012; Sampathkumar et al., 2020). In the past few years, the utilization of nanomaterials in precision farming has revolutionized the agriculture sector. Due to their large surface area-to-volume ratio, nanomaterials offer improved reactivity and biochemical activity and are therefore amenable to their diverse utilization in crop production programmes. Every year, pest and disease outbreaks cause huge crop losses globally, around 20 to 40% (Daisley et al., 2022; Worrall et al., 2018), while the huge application of pesticides like herbicides, insecticides, and fungicides despite their fast action and reliability poses harmful effects on nontarget species as well as develops resistance in the pest population and environmental toxicity (Daisley et al., 2022; Stephenson, 2003). Therefore, it has created a need to develop cost-effective, efficient, fast-performing, and less harmful pesticides, which is now possible by employing nanotechnology. Major areas of using nanomaterials in agriculture are anti-phytomicrobial, nanobiosensors, nanopesticides, seed coating, genetic manipulation, nanofertilizers, etc., as shown in Figure 8.3.

8.4.1 NANOCARRIERS

Nanocarriers made up either of organic or inorganic nanoparticles constitute smart delivery systems. The efficacy of the treatment can be improved by the use of efficient nanocarriers such as polymeric nanocarriers, superparamagnetic nanoparticles, quantum dots, dendrimers, and lipid nanoparticles (Chamundeeswari et al., 2019). Nanocarriers provide the opportunity to increase crop yields in an environmentally friendly manner by delivering fertilizers and pesticides directly to plants while minimizing leaching. These have been utilized not only for delivering agro-inputs but also as carriers

FIGURE 8.3 Major areas of using nanomaterials in agriculture.

for targeting DNA into the plant genome, such as the use of gold and silica nanoparticles (Pereira et al., 2019; Torney et al., 2007). Pheromone traps also make use of nanogels to release chemicals to trap pest populations (Paula et al., 2011). Similarly, inorganic nanoparticles like clay, zeolites, ZnO, and SiO_2 have also been utilized in smart delivery systems (Mahmoodzadeh et al., 2013; Zhao et al., 2014).

8.4.2 NANOBIOSENSORS

These biosensors are at nanoscale size and combine a biological component with a physicochemical detector for the detection of analytes. Due to their high sensing ability and low limit of detection nanobiosensors hold wide scope in sustainable farming practices. Nano biosensors are being utilized in agriculture to detect variation in soil condition like pH, moisture content, mineral concentrations, and pest and pathogen detection (Mathivanan, 2018; Mondal et al., 2022). Appropriate and controlled use of nanobiosensors can support sustainable agriculture by improving crop productivity. Various reported applications of nanobiosensors in agriculture include crop stress management, smart farming, maintaining seed purity, and seed detection (Mathivanan, 2018). These biosensors are miniature in size and are used for precise data collection to detect variation in light, temperature, vibration, and chemicals through the identification of radiofrequency. Nanosensors based on TiO_2 are being used to detect variations in pH, gaseous oxygen, and metabolites due to microbial growth (Peters et al., 2016).

8.4.3 NANOFERTILIZERS

Nanofertilizers are made in nanoscale dimensions with nanoporous materials where nutrients are encapsulated. These nanofertilizers target the delivery of nutrients and agrochemicals in response to crop demand and environmental stimuli. Therefore, it helps in efficient nutrient distribution and utilization, reduces soil toxicity, and reduces the cost and need for frequent nutrient dose application (An et al., 2022; Kumar, 2021; Sampathkumar et al., 2020). Several organic nanoparticles like Chitosan, liposomes, and dendrimers are being used for nano-encapsulation of agrochemicals, nutrients, and vitamins for their efficient delivery and bioavailability to the plant system (Puri et al., 2009; Wani et al., 2019).

Deficiency of nutrients is the major problem affecting soil and plant health, thereby reducing the quality and quantity of farm produce and causing huge economic losses to farmers. Generally, chemical fertilizers are applied in large quantities, but only a small portion of them get taken up by the plants, or most macronutrients are insoluble and not available to plants (Connor et al., 2011; Sarbani and Yahaya, 2022), and the rest get leached or washed away by surface runoff, contributing to various types of pollution in the environment and also disturbing the soil microflora. It has been reported that due to volatile nature and other processes like leaching, 75% of the nitrogen was lost from the soil (Giordano and Petropoulos, 2021; Monreal et al., 1986). Therefore, nanofertilizers were found to be the sustainable alternative to chemical fertilizer and a need of an hour because nanoparticles are highly target-specific and make effective control release of the nutrient to address the nutrient deficiency in soils. Metal-based nanoparticles have been synthesized and used as commercial nanofertilizers. Table 8.2 shows some of the metal-based nanofertilizers and their roles in the plant system. Biological systems have also been utilized for the synthesis of these nanoparticle-based fertilizers (Huang et al., 2023). Nanoparticles in the form of nanofertilizers can be applied to the soil to improve and enhance the performance and optimum utilization of fertilizer by the plant (Kumari and Singh, 2019; Liu and Lal, 2015). Advance physiochemical properties of nanomaterials like small size, large surface area, high mobility, and high solubility help to maintain control release of encapsulated nanofertilizer (Kumari and Singh, 2019; Sasson et al., 2007), help to avoid several biological barriers (Kumari and Singh, 2019; Nair, 2014), facilitate easy translocation in plants (Reineke, 2012), and provide good health, resistance, and protection to plants (Kumari and Singh, 2019). Similarly, the slow discharge rate of nanomaterials maintains soil fertility and nutrient balance, thereby reducing surface runoff and environmental toxicity (Liu and Lal, 2015). For example, slow and controlled release of zeolites due to their high porosity selectivity and high surface area results in minimal nutrient loss and improved viability (Iavicolia et al., 2017). It has been reported that urea-coated hydroxyapatite NP designed as a nanofertilizer significantly enhances rice yield compared to conventional fertilizers (Kottegoda et al., 2017).

TABLE 8.2 The List of Metal-Based Nanofertilizers and Their Role in Plant Systems.

Nutrients	Nanoparticles	Mode of Application	Plant Species	Plant Response to Nanofertilizer
Titanium	TiO_2	Root	*Oryza sativa*	Increased phosphorous content in root, shoot, and grain with high amino acid and fatty acid content
		Root	*Hordeum vulgare*	Leads to increased content of P, Ca, Mg, Zn, Mn, amino acids
		Root, foliar	*Solanum lycopersicum*	Increased lycopene
Cerium	CeO_2	Root	*Oryza sativa*	Plant express increased amount of K, Ca, Na, protein albumin, total sugars
		Root	*Hordeum vulgare*	High fatty acid, amino acid content with increased P, K, Ca, Mg, S, Cu, Fe, Zn, Mn
		Root	Triticum aestivum	Expressed high content of P, K, Fe, amino acids, fatty acids, total sugars

TABLE 8.2 *(Continued)*

Nutrients	Nanoparticles	Mode of Application	Plant Species	Plant Response to Nanofertilizer
Zinc	ZnO	Root	*Pisum sativum*	Expressed high concentration of P, Fe, Zn, Mn, total sugars
		Root	*Cucumis sativus*	Showed high amount of I K, Mg, Fe, Mn, Zn, S, prolamin, globulin, glutelin
		Foliar	*Zea mays*	Increased Zn, germination, growth, yield
Copper	Cu	Root Foliar	*Solanum lycopersicum*	Increased K, total proteins, vitamin C, total phenols, flavonoids, lycopene, antioxidant
		Root	*Cucumis sativus*	Increased Cu, Fe, sugars, organic acids, amino acids, fatty acids
Iron	Fe_2O_3	Root	*Arachis hypogaea*	Showed increased growth, biomass, and zinc content
Calcium	CaO	Seed	*Vigna mungo*	Increased root and shoot growth, biomass
Magnesium	Mg	*Vigna unguiculata*	Foliar	Increased growth, yield
Silver	Ag	*Cucumis sativus*	Foliar	Increased growth, fruit
		Solanum lycopersicum	Root	Leads to higher yield, biomass and fruit total soluble solids
Gold	Au	*Arabidopsis thaliana*	Root	Increased seed germination, growth, yield
		Brassica juncea	Foliar	Increased germination, yield

Source: Adapted from Chugh et al. (2021).

Compared to other chemical fertilizers, the mode of application of nanoparticle-based nanofertilizers is generally divided into three methods: seed priming, directly applying to soil, and foliar application to the plants. Nanoparticles are passively taken up through the small openings of stomata, while they can be entered into plant cells through any injury or wound on the plant surface (Al-Salim et al., 2011). Lateral roots at the root tips in the rhizodermis have been reported to enhance the uptake of nanoparticles (Chichiriccò and Poma, 2015). Further, chitosan nanoparticles have been used for the delivery of nitrogen, phosphorous, and potassium and applied as foliar sprays in wheat (Abdel-Aziz et al., 2016). As compared to foliar spray, when applied directly to soil, nanoparticles affected the microbial community and also affected nanoparticle uptake by plant cells. It has also been reported that foliar sprays of nanonutrients are easily taken up by the plant system (Raliya et al., 2018). Moreover, it has also been reported that monodisperse particles produced through aerosol spray when applied through foliar application do not develop agglomerate. However, beyond the threshold level, the physiochemical properties of nanoparticles hinder their cellular uptake due to differences in contact sites, thus affecting the availability of free energy required for nanoparticle–cellular interaction.

8.4.4 NANOPESTICIDES

Nanopesticides contain pesticide active ingredients in a protective nanocarrier (Kookana et al., 2014; Xu et al., 2022). These active ingredients are highly stable inside these nanocarriers and allow their target host specificity without premature degradation or environmental toxicity (Choudhury et al., 2012; Wang et al., 2022). Nanopesticides are available in different formulations depending on their chemical components and pesticidal structure. Nanoparticles developed from clay, silica, chitosan, and polyesters in combination with biodegradable polymers are being used for the development of plant protection products (Dwivedi et al., 2016; Wani et al., 2019); even beeswax and corn oil are in demand for making eco-friendly matrices for nanopesticide development (Nguyen et al., 2012). Polymeric nanoparticles and microencapsulation are being used for the production of hydrophobic and other nanopesticides. Besides that, they have been reported to have toxic effects on fungi, bacteria, and insects (Feng et al., 2000; Wang et al., 2022). Several control-release formulations of nanopesticides, such as imidacloprid synthesized from polyethylene glycol produced through encapsulation techniques, proved beneficial against major pests in agriculture. Several metal-based nanopesticides, like silver nanopesticides, have been reported to control the mycelium growth of powdery mildew on cucurbits (Li et al., 2022). On the basis of their specific roles, these nanopesticides can be classified as nanoherbicides, nanofungicides, and nanoinsecticides applied in agriculture systems.

Nanopesticides have several advantages over conventional products; however, these nanopesticides also possess some limitations. Nanoparticles are more prone to degradation by exposure to sunlight due to their large surface area, which reduces the efficacy of the active ingredient in nanopesticides and causes early evaporation due to their small particle size. Similarly, it has also been reported that the delayed release of the active ingredient from the nanocarrier often affects nontarget organisms (Kah et al., 2013; Kumari and Singh, 2019). Several phytotoxic effects of these nanopesticides have also been reported in several crops, like decreased seed germination and seedling growth in rice and root elongation in tomatoes. TiO_2NP inhibits leaf area and respiration in maize seedlings (Asli and Neumann, 2009). Likewise, the application of nanoformulations has been reported to have an adverse effect on soil microbial community structure, which ultimately affects soil quality and health and hence the sustainability of agriculture. For example, application of silver nanoparticles to the soil affects the population of ectomycorrhizal fungi, which in turn reduces the root development and shoot biomass of pine trees grown in nanoparticle-treated soil (Sweet and Singleton, 2015). It has also been reported that the nanoparticle microbial interaction also varies with the soil type, like the application of silver nanoparticles to loamy soil, which does not have the same antibacterial effect on *Pseudomonas chlororaphis* as sandy soil (Calder et al., 2012; Kalwani et al., 2022). Therefore, these have serious effects on the soil microbial population and can also interfere with nitrogen fixation in leguminous crops. Likewise, they also damage plant genetic material if they enter the plant cell nucleus (Ali et al., 2021; Singh et al., 2009). Therefore, the development of more efficient nanopesticides is required to overcome these limitations and support sustainable agriculture.

8.4.5 NANOHERBICIDES

Nanoherbicides are formulated using nanoscale materials, allowing the effective delivery of chemical or biological pesticides, enhancing the solubility, maintaining the activity of their active ingredient, and reducing their environmental toxicity as compared to conventional herbicides (Manjunatha et al., 2016; Tang et al., 2022). A number of nanoherbicides have been developed using different types of nanoparticles, as given in Table 8.3. Nanocapsules made up of polymeric particles have been used to prepare nanoherbicides, e.g., polyepsilon caprolactone containing atrazine shows enhanced herbicidal activity and less mobility in the soil (Pereira et al., 2014). Alginate nanoparticles have also been used to encapsulate paraquat herbicide (Artusio et al., 2021; Silva et al., 2011). Similarly, inorganic silicon nanoparticles are also being used as an herbicide carrier, which allows easy deposition of herbicide on plant leaves and reduces their waste. Nanoparticle-encapsulated herbicides are applied to the plant roots, which then get translocated into the plant system and prevent glycolysis in the plant roots, leading to starvation and the death of weeds. These encapsulated nanoherbicides are of great use, especially in rained areas where chemical herbicides are lost through vaporization due to the low moisture content of the soil. Likewise, when the herbicides encapsulated in the hydrophobic nanocapsules are applied to the cuticles of leaves they were readily taken up by the target plant tissue (Yang et al., 2021). As compared to conventional chemical herbicides, nanoherbicides have a longer retention period on the weed foliage, which improves their adsorption, adhesion, and utilization efficiency as well as their performance. This causes upregulation of antioxidant-related genes and several other genes (Wang et al., 2022), changes chlorophyll content, changes lipid peroxidation, causes cell membrane damage, and induces leakage of the electrolyte into the extracellular environment (Jambunathan, 2010). Besides their advantages, these nanoherbicides also have an adverse effect on nontarget plant species and the bacterial rhizosphere communities over the long term of their exposure. Thus, plant–soil interaction and their relationship determine ways of safe use of nanoherbicides; therefore, their genotoxic studies need to be done (Yu et al., 2017) for effective and safe application of nanoherbicides.

TABLE 8.3 List of Nanoherbicides Effective Against Different Weed Species.

Nanoherbicides	Plant Species	Effect on Target Weed Species
DQ@MSN-SO_3	*Datura stramonium* L.	Stimulus-responsive delivery system. Good herbicidal activity
PCL_ATZ NPs	*Bidens pilosa* L.; *Amaranthus viridus* L.	Reduction in photosystem II activity and inhibition of plant growth
Nanoemulsion_garden sarvori (*Satureja ortensis*) essential oil	*Amaranthus retroflexus* L.; *Chenopodium album* L.	Leads to reduction in germination, growth of seedling, and their chlorophyll content
Nano-hydrogel glyphosate	*Oryza sativa*	Improved herbicidal activity
Herbicide@HTlcs	*Amaranthus retroflexus* L.	Potential herbicidal property

Source: Adapted from Mariana et al. (2022).

8.4.6 NANOFUNGICIDES

Nanofungicides are nanomaterials that are either effective against different fungal species themselves or used as a carrier molecule to generate nanocomposites for targeted delivery of fungicides. Several nanoparticle carriers like chitosan, silica, and polymer mixes have been utilized for the preparation of nanofungicides. Most of the common fungal species (*Fusarium spp.*, *Phtophthora spp.*, and *Aspergillus species*) causing maximum yield losses have been observed to be controlled by nanofungicides (Yadav et al., 2015). Several fungal diseases have been targeted using nanomaterial-based fungicides, as illustrated in Table 8.4. Nanomaterials like copper nanoparticles have been used for controlling diseases of fruit trees (Bhattacharyya et al., 2016); graphene oxide-silver nanoparticle composites have been used against *Fusarium graminearum* (Chen et al., 2016; Cobos et al., 2020); and chitosan-silver nanocomposites have been found to be effective against mango *Colletotrichum gloeosporioides* (Chowdappa et al., 2014). Zinc oxide nanoparticles were found to be effective against controlling postharvest diseases caused by the fungi *Botrytis cinerea and Penicillium expansum* (Sardella et al., 2017). Similarly, several essential oil-based nanoformulations have also been observed to be effective in controlling fungal diseases. Different types of nanomaterials utilized for the production of nanofungicides that are either mixed with seeds, applied to foliage, or applied to plant roots by drenching different crop species are shown in Table 8.5. However, applying high concentrations of these nanofungicides may show phytotoxic effects as well as contaminate the soil microbial community. As the long-term effect of nanofungicides on health and the environment is a major concern (Mustafa et al., 2014; Rajput et al., 2021), it is therefore important to assess the impact of inhaling nanofungicides during foliar spray (Maluin et al., 2020). The study conducted on mice using nanopesticide chlorfenapyr shows that it is less toxic than conventional formulation; therefore, it is stated that nanoformulation could lessen the adverse effect on humans and the environment as compared to conventional fungicides (Maluin et al., 2020).

8.4.7 NANOINSECTICIDES

Insecticides are a group of chemicals and biological compounds that are used to control destructive insect populations. Insects reach pest status when their population reaches beyond the threshold level and becomes damaging and destructive to crops. Chemical-based conventional insecticide formulations have a nonspecific delivery system, and their repeated application sometimes makes insects more resistant to the concerned insecticides and also increases the chemical toxicity in the environment. On the other hand, nanoinsecticides

TABLE 8.4 Nanofungicides Effective Against Different Fungal Species with their Advantages and Disadvantages for the Plant and Microbial Community.

Fungicides	**Nanoparticles**	**Plant Species**	**Target Fungi**	**Advantages**	**Disadvantages**
Chitosan–Dazomet-hexaconozole	Hexaconozole, Dazomet and chitosan	Oil palm	*Ganoderma boninense*	Control release of the actives	High concentration may cause phytotoxicity
Chitosan–hexaconozole	Chitosan and hexaconozole	Oil palm	*G. boninense*	Effective nano delivery system	Phytotoxicity at higher concentration
Carbendazim	Chitosan/Pectin	Cucumber Maize Tomato	*A. parasiticus* and *F. oxysporum*	Control release	Phytotoxicity
Prochloraz	PHSN	Cucumber	*B. cinerea*	Control release of nutrients with increased crop protection	Phytotoxicity at higher concentration
Tebuconazole	Bacterial ghosts	Wheat, cucumber, and barley	*L. nodorum*, *P. teres*, *S. fuliginea*, and *E. graminis*	It possesses a multifunctional delivery platform	No phytotoxicity
Tebuconazole Chlorothalonil	PVP and PVP copolymer	Southern yellow pine	*G. trabeum*	Enhanced solubility in solvents of different polarity, improved binding ability, and stabilizing effect	Cause microbial contamination due to absorption of moisture with high hygroscopicity and hydrophilicity of PVP.

Source: Adapted from Kutawa et al. (2021).

TABLE 8.5 Nanomaterials Used as Nanofungicides for Plant Protection, their Mode of Application, and Effect.

Nanomaterials Type	Crop Species	Application Method	Effect
Silver nanoparticles	*Vigna unguiculata*	Foliar application	Inhibit *Xanthomonas axonopodis pv. malvacearum* and *Xanthomonas campestris pv. campestris* under vitro condition with no phtotoxicity
CuO	*Solanum lycopersicum*	Foliar application	Controlled late blight caused by *Phytophthora infestans*
MgO	*Solanum lycopersicum*	Drenching	Suppression of *Ralstonia solanacearum* leading to bacterial wilt disease control
A12O3 NPs	*Solanum lycopersicum*	Foliar application	Provided protection against *Fusarium* root rot in tomato
Copper oxide and silver nanoparticles	*Prunus domestica* fruits	Fruit spray	Put down the symptom caused by *B. cinerea* and soil-borne diseases

Source: Adapted from Shang et al. (2019).

developed by encapsulation of insecticides in different forms like nanocapsules, nanogels, nanospheres, micelles, and liposomes using polymers like chitosan, polyethylene glycol, lignin, starch-based polymers, and clay-based nanomaterials offer better efficiency and less toxicity. These nanoformulations of encapsulated nanoinsecticides act as a protective shield and protect the active component of the insecticides from degradation processes such as degradation by light, temperature, and hydrolysis, as well as some amount lost through leaching in the environment which helps them provide control and target delivery of the insecticides to the host insect pest without harming the natural enemies of the insect pest present in the environment. Further, the usage of nanoinsecticides reduces the repeated application of insecticides as compared to conventional insecticides, thereby reducing the cost of crop production and environmental pollution through rapid decomposition in soil and plant systems without having a harmful effect on human health (Gavrilescu, 2005). The effectiveness of nanoinsecticides depends upon the size and shape of the nanoparticles as well as the type of carrier used for their synthesis. As compared to commercial insecticides, these nanoformulations have been reported to have a higher lethal effect and residual activity. Metal nanoparticles like silver and aluminum have been used for preparing nanoformulations of insecticides. The insecticidal properties of nanostructured alumina have been reported against the stored grain pests *Rhyzopertha dominica* and

Sitophilus oryzae (Stadler et al., 2018). Chlorantraniliprole, thiocyclam, and nanoinsecticides were found to be effective against the black cutworm, Agrotisipsilon, affecting its development and reproduction ability (Awad et al., 2022). These nanoformulations can provide future effective methods for insect pest management as well as for environmental sustainability.

8.5 TRANSFORMATION OF NANOMATERIALS IN THE ENVIRONMENT

Engineered nanoparticles are released in the environment from different sources, either intentionally or unintentionally, from point and nonpoint sources from the nanoenabled products (Abbas et al., 2020) (Figure 8.4). The effectiveness of nanoparticles is determined by their physiochemical properties, which is further influenced by environmental factors after their discharge into the environment. Relative to soil and water compartments, the atmosphere also receives an annual 8100 metric tonnes of engineered nanoparticles globally (Abbas et al., 2020; Keller and Lazareva, 2013). After release, these nanoparticles undergo several changes or transformation processes, which include sedimentation, homo and hetero agglomeration processes, oxidation, reduction, sulfidation, photochemical and biologically mediated processes, and finally altering their mobility and bioavailability to different organisms in the ecosystem (Bundschuh et al., 2018). Agglomeration processes affect the mobility of the nanoparticles in the environment; for instance, with more agglomeration, particle flock size increases, which reduces the concentration of engineered nanoparticles in soil and water and favors the deposition of large agglomerated nanoparticles, thereby reducing their mobility and bioavailability in the environment compartments (Mitrano and Nowack, 2017). The agglomeration process of metal nanoparticles is greatly influenced by pH and ionic strength, which enhanced the agglomeration kinetics of zinc oxide and copper oxide nanoparticles in soil (Abbas et al., 2020; Bundschuh et al., 2018; Conway et al., 2015). On the other hand, natural organic matter adsorbs on the surface of nanoparticles and modifies their surface chemistry, structure, and potential. Likewise, redox reactions decreased the stability of engineered nanoparticles through the formation of an active surface layer, which facilitates their interaction with other components in the environment (Amde et al., 2017). Nanoparticles also act as carriers of contaminants and target them in different organisms, like the bioaccumulation of atrazine and hexachlorobenzene in fungi by the adsorption of $TiO2$ nanoparticles (Zhang et al., 2017). Interaction of carbon, silver, and aluminum nanomaterials with organochlorine pesticides in soil promoted their uptake and bioaccumulation in *Zea mays*, as reported (Wu et al., 2017).

Light-induced chemical reactions like photooxidation reduction, photolysis, and photocatalysis affect the rate at which reactive oxygen species are generated by nanoparticles and their persistence in the environment (Mitrano and Nowack, 2017). Toxicity of graphene TiO2 has been reported in *Daphnia magna* and *Oryzias latipes* (Li et al., 2014). Under UV light, zinc oxide nanoparticles undergo chemical transformation in the environment and become toxic to human keratocytes (Gao et al., 2016; Tu et al., 2013). Biological transformation of nanoparticles also changes their fate; white-rot basidiomycete fungi have been reported to enzymatically degrade carbon nanoparticles through oxidation (Abbas et al., 2020). Horseradish-peroxidase (HRP) phytoenzyme present in the rhizospheric zone of horseradish causes biological degradation of carboxylate and nitrogen-doped MWCNTs (Raghu et al., 2014; Zhao et al., 2011). The organic coating present on the surface of nanomaterials provides them with stability for various applications or in order to achieve their desired properties (Khan et al., 2019). In the environment, these coatings undergo microbial degradation, which induces agglomeration and deposition processes. Since the natural macromolecules are abundant in the environment, their utilization as a coating on the nanoparticles further hinders their microbial degradation by altering their bioavailability (Batley et al., 2012; Lead et al., 2018).

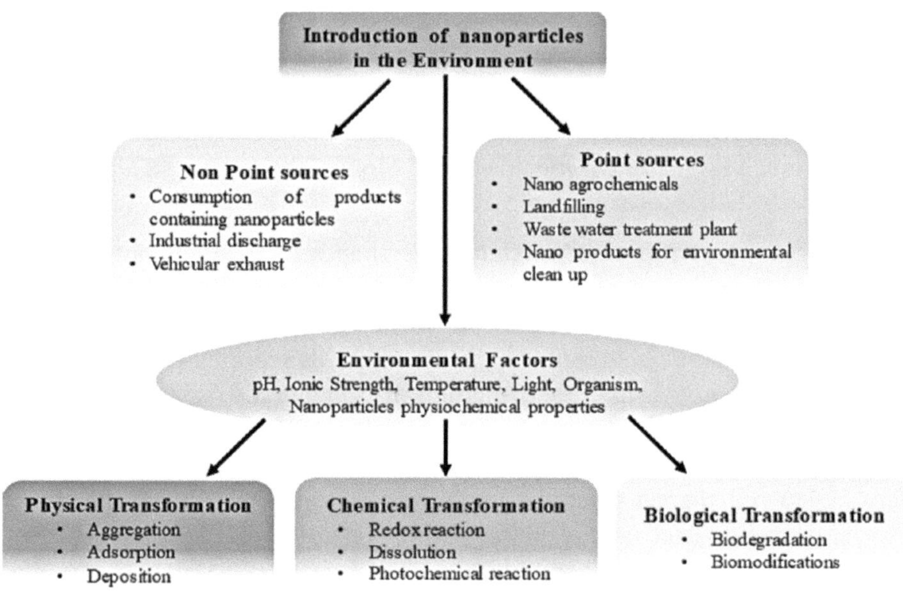

FIGURE 8.4 The schematic pathway determining the environmental fate of nanoparticles.

8.6 PHYSIOCHEMICAL AND MOLECULAR EFFECTS OF NANOMATERIALS

8.6.1 PHYSIOCHEMICAL PROPERTIES OF NANOMATERIALS

In addition to electronics, cosmetics, household appliances, energy storage devices, semiconductor devices, and biological sensors, nanomaterials are effectively utilized in biomedical sciences as a carrier for drugs, as bioprobes and implants due to their unique size and properties. Due to their diverse physicochemical properties such as surface properties, surface charge, size, shape, structure, composition, crystallinity, and other parameters have a major impact on wide applications of nanomaterials.

i. **Surface area and size of the nanoparticle:** The size and surface area of the nanoparticle affect their interaction with the biological system in which they are applied. With a decrease in size, the nanomaterial's surface area increases exponentially relative to its volume, thereby making its surface more reactive. It was reported that mechanisms like cellular uptake and endocytosis of nanoparticles are affected by the particle size of the nanomaterial (Aillon et al., 2009; Gatoo et al., 2014; Nel et al., 2006). On the other hand, the amount of reactive oxygen radical generated by nanoparticles are directly proportional to radical's size; the smaller the nanoparticle, the greater the production of ROS, which leads to DNA damage. It has also been reported that a decrease in size as compared to large size particles having same mass dose will have increased surface area, which increases oxidation and DNA damaging ability at a much higher rate (Donaldson and Stone, 2003; Gatoo et al., 2014; Li et al., 2015; Risom et al., 2005).

ii. **Shape of the nanoparticle:** The shapes of the nanoparticles vary from spherical, cubical, triangular, oval, hexagonal, rod-shaped, and helical-shaped. It has been reported that the shape of the nanoparticles is associated with their toxic effect, as reported in some carbon-based, gold, silver, and titanium nanomaterials (Ispas et al., 2009; Petersen and Nelson, 2010), and also affects the process of their delivery to the target biological system. For example, as compared to rod-shaped spherical nanoparticles, they have been reported to have a faster rate of endocytosis (Champion and Mitragotri, 2006), as well as being relatively less toxic in nature as compared to other shapes. Blockage of K^+ ion channels by rod-shaped nanoparticles has been reported to be more efficient than spherical carbon fullerenes (Gatoo et al., 2014; Li et al., 2015; Park et al., 2003).

iii. **Surface charge:** It has been reported that surface charge on nanomaterials affects their various behaviors, like colloidal behavior, selective adsorption, and transmembrane permeability (Hoshino et al., 2004). Surface charge determines the colloidal property of the nanomaterials, affecting organism's response to nanoparticles in response to their change in shape and size which in turn affects the interaction between nanoparticles and their surrounding environment. However, positively charged nanoparticles show a higher cellular uptake by plasma protein due to enhanced opsonization than negatively and neutrally charged nanoparticles (Gatoo et al., 2014).

iv. **Composition and crystalline structure:** According to studies, nanoparticle toxicity also depends on their composition and crystalline structure (Gatoo et al., 2014).

v. **Aggregation and concentration:** Nanoparticles' aggregation states also influence their toxicities. However, size, surface charge, composition, and other factors affect the aggregation states of NPs. Nanoparticles' toxicity decreases at higher concentrations as their concentration increases (Gatoo et al., 2014; Li et al., 2015).

vi. **Surface coating and roughness**: Nanoparticles' surface properties play an important role in determining the toxicity of nanoparticles because they determine the outcome of their interactions with cells and other biological components. Cytotoxic property of nanoparticles is affected by their surface coating which in turn alters their magnetic, electric, and optical properties and chemical reactivity which also affects their distribution, accumulation, and toxicity. In order to stabilize nanoparticulate systems in biological system several surfactant copolymers like polyethylene and hydrophilic and flexible polyethylene glycol have been used in the area of nanotechnology. Further, the uptake of nanoparticles by the cell is highly influenced by their properties like surface roughness, charge, hydrophobicity which in turn affect nanoparticles' toxicity (Gatoo et al., 2014).

8.6.2 EFFECTS ON SOIL HEALTH AND FERTILITY

Soil acts as a major sink or receptor for the nanoparticles, as they are constantly being introduced from various sources in the agriculture field. As the soil profile or matrix already has natural nanoparticles, the artificial entry of nanoparticles may have a significant effect on soil fertility, especially their microscopic properties (Ben-Moshe et al., 2013). On the other hand, the degree of effect of nanoparticles on soil varies with the soil type and enzymatic activity of the soil. The negative effect on the dehydrogenase activity of the enzyme is observed at high nanoparticle concentrations (Jośko et al., 2014). Nanoparticles influence the rate of soil self-cleaning, i.e., they affect the balance of nutrients in the soil, thereby affecting soil fertility. Soil properties affect plant–soil interactions and the sustainability of crop growth and production; also, the introduction of engineered nanomaterials can act as soil conditioners by improving their physiochemical properties (Rafiq et al., 2017). Application of oxidized orthorhombic iron oxalate-capped Fe_3O_4 (OCIO) nanomaterials in alluvial soil results in reduced soil bulk density and increased soil organic carbon (Das et al., 2016). Similarly, application of nano-submicron mineral-based soil conditioners results in improved pH, bulk density, and soil porosity (Liu et al., 2017). Another universal approach for improving most of the soil is the application of nanoparticle bio-chars because these have the ability to increase soil porosity, carbon sequestration, absorptive and catalytic properties, increase soil water and nutrient retention capacity, reduce nutrient loss, and prevent soil erosion (Hagemann et al., 2017). Ca-alginate/ball-milled biochar composite is utilized as a soil amendment that increases soil water holding capacity and nutrient retention capacity (Wang et al., 2018). Similarly, application of green silver nanoparticles with a shift in soil pH from acidic to neutral shows significant improvement in water-holding capacity, nitrogen, phosphorous, and organic carbon content (Das et al., 2018).

8.6.3 EFFECTS ON MICROBIAL DIVERSITY

Soil is a natural habitat where microbial species diversity exists and forms a favorable ecosystem for the interaction of soil and microbes with plants. These microorganisms play a vital role in influencing crop growth and yield. Sometimes, these microorganisms act as an indicator of the soil condition if any alteration is made to the soil, like introducing nanoparticles in the form of nanofertilizers or nanochemicals (Rajput et al., 2018). Nanoparticles in the form of nanofertilizers affect soil microorganisms in various ways: first, they induce a direct toxicity effect, alter the accessibility of essential elements, or have an antagonistic effect (Haris and Ahmad, 2017). In some cases, the symbiotic relationship is affected by the concentration of nanoparticles as well as other species present and other soil properties. For example, application of silver nanoparticles coated with polyvinyl pyrrolidone in tomatoes reduced colonization by arbuscular mycorrhizal fungi, while noncoated Ag_2SNPs showed no impact. Similarly, application of Ag nanoparticles against *Bacillus cereus* and *Pseudomonas stutzeri* shows antibacterial effects and morphological alterations (Fajardo et al., 2014). The effect of silver nanoparticles on microbial richness is also affected by their exposure time, concentration, and soil texture. Silver nanoparticles are known to reduce microbial enzymatic activity, while exposure to copper- and zinc-based nanoparticles reduces growth and biomass (Frenk et al., 2013). Further, soil organic matter content also influences the release of toxic ions from nanomaterials and alters the level of toxicity for microorganisms (Klaine et al., 2008). For instance, the toxic effect of ZnO nanoparticles on *Folsomia candida* is increased in acidic soil (Waalewijn-Kool et al., 2014). In the case of soil with low pH and low organic content, it hinders Ag absorption by soil colloids as compared to soil with higher pH and high organic content, thereby increasing the toxicity to microorganisms (Benoit et al., 2013). These nanoparticles also have toxic effects on soil annelids and arthropods (Rajput et al., 2018), which together affect community structure, composition, and function.

Nanoparticles are also reported to regulate the expression of genes involved in several metabolic pathways in microorganisms. Like silver nanoparticles, exposure to *Articulospora tetracladia* shows significant changes in hierarchical protein clusters (Barros et al., 2019). The toxic effect of silver nanoparticles against *Fusarium solani* species has been elucidated at the RNA level using transcriptome sequencing (Shen et al., 2020), which regulates the differential expression of 2503 genes involved in several metabolic pathways. These nanoparticles inhibited the growth of fungal mycelium by regulating genes involved in signal transduction and substance metabolism. In the same way, these nanoparticles also regulate m-RNA expression in several bacterial species, such as *Bacillus cereus*, where the expression of genes related to the biosynthesis of the catalase enzyme increases on exposure to aluminum oxide nanoparticles. Similarly, the exposure of copper oxide nanoparticles to *Desulfovibrio vulgaris* upregulates the expression of genes involved in ATP synthesis, ferric uptake regulator, and protein synthesis (Chen et al., 2019) and downregulates the expression of genes involved in electron transfer, cellular respiration, and flagellar movement. The molecular changes in bacterial and plant cells that induce their cytotoxicity are represented in Figure 8.5. Various nanoparticles used in agriculture systems to treat different microorganisms and their effects on these organisms are shown in Table 8.6.

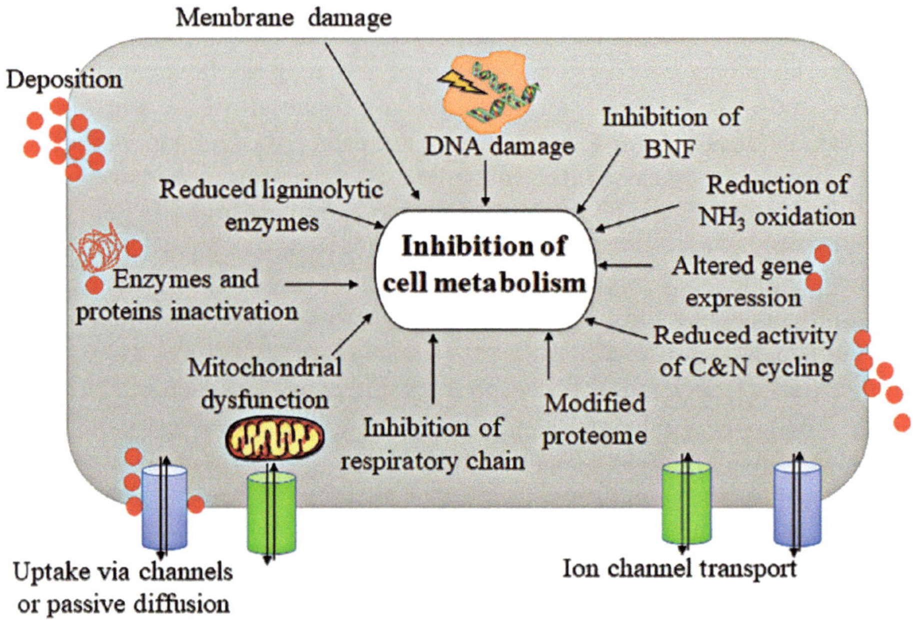

FIGURE 8.5 Nanoparticles' toxicity affects gene expression in microorganisms.

TABLE 8.6 Effects of Various Nanoparticles on Microbial Diversity.

Nanoparticles	Microorganism Affected	Effects
CuO, Cu	Fungi	Decreased fungi population with nano-Cu, while increased with nano-CuO
Ag, ZnO	*Arbuscularmycorrhizal* fungi	Negative effects on colonization with tomato plants
ZnO	*Rhizobium leguminosarum*	Rhizobial symbiosis negatively affected with pea
ZnO	*Pseudomonas chlororaphis* O6	Caused disruption in cell-signaling Reduced phenazine production in dose-dependent manner which results in reduced quorum sensing molecules, acyl-homoserine lactones
TiO_2, ZrO_2	*Bacillus megaterium*, *Azotobactervinelandii*, *Pseudomonas fluorescens*, *Bacillus brevis*	TiO_2 showed toxicity toward PGPRs and *B. brevis* showed highest susceptibility followed by *B. megaterium, P. fluorescens*, and *A. vinelandii*, respectively
Ag	*Bacillus*	Bacterial transformation in chloroplasts due to cell-damage, thereby causing leakage of proteins and sugars

8.6.4 ETHICS OF NANOMATERIALS' APPLICABILITY

Before developing any nanofertilizer, its environmental impact should be assessed. Appropriate formulations and dosages with the desired concentration should be tested and applied (Hasler et al., 2015). The toxicity level of nanomaterials and the risk associated with their release and bioaccumulation in the environment or in the food chain should be studied first to help in framing policies for their commercialization (Tilman et al., 2002). Biocompatibility studies of the nanoparticles, especially the metal core or protein coating, with the surrounding environment are required to explain the cellular response generated upon invasion by nanoparticles (Mubarak Ali et al., 2015). More advanced tools of genomics can be used to conduct nanotoxicological studies, like c-DNA microarrays, a powerful tool for analyzing cell responses to nanoparticles as well as toxicological aspects of nanoparticles (Xu et al., 2011). Such types of nanotoxicological and risk assessment studies will form the basis for the development of safe nanoproducts for agriculture and enhance their commercialization. The risk associated with nanoparticles can also be determined by studying their environmental fate, which leads to their transformation, as reported in the environmental assessment of agrochemicals (Kah et al., 2013). To increase the commercial acceptance of nanoparticles in agriculture systems (soil and plant systems), risk assessments and their cytotoxicity and phytotoxicity studies need to be performed on a priority basis. Further, the integration of various scientific branches with strict laws and social development will speed up the nanoproducts' utilization and acceptance by consumers (Kah, 2015). Therefore, several barriers present in the polices and regulations which have not been yet enforced need to overcome for the successful utilization of nano-based agriculture inputs in agriculture sector.

8.7 CONCLUSION AND FUTURE PROSPECTIVE

In order to feed and sustain the increasing global population, the integration of engineered nanomaterials (ENM) with conventional farming practises has the huge potential to overcome several global challenges, including climate change, soil health security, the sustainability of crop production, and food safety. The use of nanomaterials for the management of soil health and fertility served as a driving force toward sustainable green farming and a soil restoration approach. On the other hand, due to the risk associated with the use of nano-agri-inputs in the soil ecosystem, these are still under study and need to be further explored. The physiochemical properties of nanomaterials such as their surface charge, size, and reactivity affect their fate in the environment which influence the beneficial interaction between soil and plant microbes. Therefore, considering the toxicity of nanoparticles and their environmental transformation, it is important to develop approaches for their target delivery and safe disposal in the environment without affecting the beneficial soil microbiome. Nanoformulations affect DNA at the molecular level, which has both positive effects (i.e., providing resistance to plants against biotic and abiotic stresses) and a negative effect on the expression of certain protein-coding genes. Similarly, their intake through food and different environmental sources has also adversely affected human health. The evaluation of engineered nanomaterials is a complex process, and it is hard to generalize their impact on soil biota. This highlights the need for more research on the interaction of ENMs with soil components and quantitative assessment of the aggregation, dispersion, adsorption, desorption, precipitation, dissolution, decomposition, and mobility of ENMs in the soil environment. Researchers should also initiate top-down or field research in addition to lab-based experiments for a complete understanding of the interaction of ENMs and to observe the unanticipated ecological risks and benefits before delivering the technology to the farm gate for the sustainable development of the agriculture sector.

KEYWORDS

- **Nanomaterials**
- **reactive oxygen radicals**
- **soil mineralogy**
- **soil health**
- **microbes**
- **metal detoxification**
- **gene expression**

REFERENCES

Abbas, Q., Yousaf, B., Amina, Ali, M. U., Munir, M. A. M., El-Naggar, A., Rinklebe, J., & Naushad, M. (2020). Transformation pathways and fate of engineered nanoparticles (ENPs) in distinct interactive environmental compartments: A review. *Environment International, 138*, 105646. https://doi.org/10.1016/J.ENVINT.2020.105646

Abdel-Aziz, H. M. M., Hasaneen, M. N. A., & Omer, A. M. (2016). Nano chitosan-NPK fertilizer enhances the growth and productivity of wheat plants grown in sandy soil. *Spanish Journal of Agricultural Research, 14*(3), 1–9.

Aillon, K. L., Xie, Y., El-Gendy, N., Berkland, C. J., & Forrest, M. L. (2009). Effects of nanomaterial physicochemical properties on in vivo toxicity. *Advanced Drug Delivery Reviews, 61*(6), 457–466. https://doi.org/10.1016/J.ADDR.2009.03.010

Ali, S. S., Al-Tohamy, R., Koutra, E., Moawad, M. S., Kornaros, M., Mustafa, A. M., Mahmoud, Y. A. G., Badr, A., Osman, M. E. H., Elsamahy, T., Jiao, H., & Sun, J. (2021). Nanobiotechnological advancements in agriculture and food industry: Applications, nanotoxicity, and future perspectives. *Science of the Total Environment, 792*, 148359. https://doi.org/10.1016/j.scitotenv.2021.148359

Al-Karaki, G. N. O., Othman, Y., & Al-Ajmi, A. (2007). Effects of mycorrhizal fungi inoculation on landscape turf establishment under Arabian Gulf region conditions. *Arab Gulf Journal of Scientific Research, 25*(3), 147–152.

Al-Salim, N., Barraclough, E., Burgess, E., Clothier, B., Deurer, M., Green, S., Malone, L., & Weir, G. (2011). Quantum dot transport in soil, plants, and insects. *Science of the Total Environment, 409*(17), 3237–3248. https://doi.org/10.1016/j.scitotenv.2011.05.017

Amde, M., Liu, J. fu, Tan, Z. Q., & Bekana, D. (2017). Transformation and bioavailability of metal oxide nanoparticles in aquatic and terrestrial environments. A review. *Environmental Pollution, 230*, 250–267. https://doi.org/10.1016/J.ENVPOL.2017.06.064

Amenta, V., Aschberger, K., Arena, M., Bouwmeester, H., Botelho Moniz, F., Brandhoff, P., Gottardo, S., Marvin, H. J. P., Mech, A., Quiros Pesudo, L., Rauscher, H., Schoonjans, R., Vettori, M. V., Weigel, S., & Peters, R. J. (2015). Regulatory aspects of nanotechnology in the agri/feed/food sector in EU and non-EU countries. *Regulatory Toxicology and Pharmacology, 73*(1), 463–476. https://doi.org/10.1016/J.YRTPH.2015.06.016

An, C., Sun, C., Li, N., Huang, B., Jiang, J., Shen, Y., Wang, C., Zhao, X., Cui, B., Wang, C., Li, X., Zhan, S., Gao, F., Zeng, Z., Cui, H., & Wang, Y. (2022). Nanomaterials and nanotechnology for the delivery of agrochemicals: Strategies towards sustainable agriculture. *Journal of Nanobiotechnology, 20*(1), 11. https://doi.org/10.1186/s12951-021-01214-7

Artusio, F., Casà, D., Granetto, M., Tosco, T., & Pisano, R. (2021). Alginate nanohydrogels as a biocompatible platform for the controlled release of a hydrophilic herbicide. *Processes, 9*(9), 1–14. https://doi.org/10.3390/pr9091641

Asli, S., & Neumann, P. M. (2009). Colloidal suspensions of clay or titanium dioxide nanoparticles can inhibit leaf growth and transpiration via physical effects on root water transport. *Plant, Cell and Environment, 32*(5), 577–584. https://doi.org/10.1111/j.1365-3040.2009.01952.x

Astefanei, A., Núñez, O., & Galceran, M. T. (2015). Characterisation and determination of fullerenes: A critical review. *Analytica Chimica Acta, 882*, 1–21. https://doi.org/10.1016/J.ACA.2015.03.025

Aswathanarayan, J. B., & Vittal, R. R. (2019). Nanoemulsions and their potential applications in food industry. *Frontiers in Sustainable Food Systems, 3*, 1–21. https://doi.org/10.3389/fsufs.2019.00095

Awad, M., Ibrahim, E. D. S., Osman, E. I., Elmenofy, W. H., Mahmoud, A. W. M., Atia, M. A. M., & Moustafa, M. A. M. (2022). Nano-insecticides against the black cutworm Agrotis ipsilon (Lepidoptera: Noctuidae): Toxicity, development, enzyme activity, and DNA mutagenicity. *PLoS ONE, 17*(2). https://doi.org/10.1371/journal.pone.0254285

Awale, R., Machado, S., Ghimire, R., & Bista, P. (2017). Soil health. *Sustainable Landscaping* (pp. 169–188). https://doi.org/10.1201/b15717-11

Baig, N., Kammakakam, I., Falath, W., & Kammakakam, I. (2021). Nanomaterials: A review of synthesis methods, properties, recent progress, and challenges. *Materials Advances, 2*(6), 1821–1871. https://doi.org/10.1039/d0ma00807a

Barros, D., Pradhan, A., Mendes, V. M., Manadas, B., Santos, P. M., Pascoal, C., & Cássio, F. (2019). Proteomics and antioxidant enzymes reveal different mechanisms of toxicity induced by ionic and nanoparticulate silver in bacteria. *Environmental Science: Nano, 6*(4), 1207–1218. https://doi.org/10.1039/c8en01067f

Ben-Moshe, T., Frenk, S., Dror, I., Minz, D., & Berkowitz, B. (2013). Effects of metal oxide nanoparticles on soil properties. *Chemosphere, 90*(2), 640–646. https://doi.org/10.1016/j.chemosphere.2012.09.018

Benoit, R., Wilkinson, K. J., & Sauvé, S. (2013). Partitioning of silver and chemical speciation of free Ag in soils amended with nanoparticles. *Chemistry Central Journal, 7*(1), 1–7. https://doi.org/10.1186/1752-153X-7-75

Bhattacharyya, A., Duraisamy, P., Govindarajan, M., Buhroo, A. A., & Prasad, R. (2016). *Nano-Biofungicides: Emerging Trend in Insect Pest Control* (pp. 307–319). https://doi.org/10.1007/978-3-319-42990-8_15

Bouwmeester, H., Dekkers, S., Noordam, M. Y., Hagens, W. I., Bulder, A. S., de Heer, C., ten Voorde, S. E. C. G., Wijnhoven, S. W. P., Marvin, H. J. P., & Sips, A. J. A. M. (2009). Review of health safety aspects of nanotechnologies in food production. *Regulatory Toxicology and Pharmacology, 53*(1), 52–62. https://doi.org/10.1016/J.YRTPH.2008.10.008

Bratovcic, A., Hikal, W. M., Said-Al Ahl, H. A. H., Tkachenko, K. G., Baeshen, R. S., Sabra, A. S., & Sany, H. (2021). Nanopesticides and nanofertilizers and agricultural development: Scopes, advances and applications. *Open Journal of Ecology, 11*(04), 301–316. https://doi.org/10.4236/oje.2021.114022

Brown, T. T., Koenig, R. T., Huggins, D. R., Harsh, J. B., & Rossi, R. E. (2008). Lime effects on soil acidity, crop yield, and aluminum chemistry in direct-seeded cropping systems. *Soil Science Society of America Journal, 72*(3), 634–640. https://doi.org/10.2136/sssaj2007.0061

Bundschuh, M., Filser, J., Lüderwald, S., McKee, M. S., Metreveli, G., Schaumann, G. E., Schulz, R., & Wagner, S. (2018). Nanoparticles in the environment: Where do we come from, where do we go to? *Environmental Sciences Europe, 30*(1). https://doi.org/10.1186/s12302-018-0132-6

Calder, A. J., Dimkpa, C. O., McLean, J. E., Britt, D. W., Johnson, W., & Anderson, A. J. (2012). Soil components mitigate the antimicrobial effects of silver nanoparticles towards a beneficial soil bacterium, Pseudomonas chlororaphis O6. *Science of the Total Environment, 429*, 215–222. https://doi.org/10.1016/j.scitotenv.2012.04.049

Champion, J. A., & Mitragotri, S. (2006). Role of target geometry in phagocytosis. *Proceedings of the National Academy of Sciences of the United States of America, 103*(13), 4930–4934. https://doi.org/10.1073/pnas.0600997103

Chamundeeswari, M., Jeslin, J., & Verma, M. L. (2019). Nanocarriers for drug delivery applications. *Environmental Chemistry Letters, 17*(2), 849–865. Springer. https://doi.org/10.1007/s10311-018-00841-1

Chen, B. Z., Ashfaq, M., Zhang, X. P., Zhang, J. N., & Guo, X. D. (2018). In vitro and in vivo assessment of polymer microneedles for controlled transdermal drug delivery. *Journal of Drug Targeting, 26*(8), 720–729. https://doi.org/10.1080/1061186X.2018.1424859

Chen, J., Sun, L., Cheng, Y., Lu, Z., Shao, K., Li, T., Hu, C., & Han, H. (2016). Graphene oxide-silver nanocomposite: Novel agricultural antifungal agent against fusarium graminearum for crop disease prevention. *ACS Applied Materials and Interfaces, 8*(36), 24057–24070. https://doi.org/10.1021/acsami.6b05730

Chen, Z., Gao, S. hong, Jin, M., Sun, S., Lu, J., Yang, P., Bond, P. L., Yuan, Z., & Guo, J. (2019). Physiological and transcriptomic analyses reveal CuO nanoparticle inhibition of anabolic and catabolic activities of sulfate-reducing bacterium. *Environment International, 125*, 65–74. https://doi.org/10.1016/j.envint.2019.01.058

Chichiriccò, G., & Poma, A. (2015). Penetration and toxicity of nanomaterials in higher plants. *Nanomaterials, 5*(2), 851–873. https://doi.org/10.3390/nano5020851

Choudhury, S. R., Pradhan, S., & Goswami, A. (2012). Preparation and characterisation of acephate nano-encapsulated complex. *Nanoscience Methods, 1*(1), 9–15. https://doi.org/10.1080/17458080.2010.533443

Chowdappa, P., Gowda, S., Chethana, C. S., & Madhura, S. (2014). Antifungal activity of chitosan-silver nanoparticle composite against Colletotrichum gloeosporioides associated with mango anthracnose. *African Journal of Microbiology Research, 8*(17), 1803–1812. https://doi.org/10.5897/ajmr2013.6584

Chugh, G., Siddique, K. H. M., & Solaiman, Z. M. (2021). Nanobiotechnology for agriculture: Smart technology for combating nutrient deficiencies with nanotoxicity challenges. In *Sustainability (Switzerland)* (Vol. 13, Issue 4, pp. 1–20). MDPI AG. https://doi.org/10.3390/su13041781

Cobos, M., De-La-pinta, I., Quindós, G., Fernández, M. D., & Fernández, M. J. (2020). Graphene oxide–silver nanoparticle nanohybrids: Synthesis, characterization, and antimicrobial properties. *Nanomaterials, 10*(2). https://doi.org/10.3390/nano10020376

Connor, D. J., Loomis, R. S., & Cassman, K. G. (2011). Productivity and management in agricultural systems. *Nature, 166*(4235), 1097–1098. https://doi.org/10.1038/1661097d0

Conway, J. R., Adeleye, A. S., Gardea-Torresdey, J., & Keller, A. A. (2015). Aggregation, dissolution, and transformation of copper nanoparticles in natural waters. *Environmental Science and Technology, 49*(5), 2749–2756. https://doi.org/10.1021/es504918q

Daisley, B. A., Chernyshova, A. M., Thompson, G. J., & Allen-Vercoe, E. (2022). Deteriorating microbiomes in agriculture - the unintended effects of pesticides on microbial life. *Microbiome Research Reports, 1*(1), 6. https://doi.org/10.20517/MRR.2021.08

Daniel, M. C., & Astruc, D. (2004). Gold nanoparticles: Assembly, supramolecular chemistry, quantum-size-related properties, and applications toward biology, catalysis, and nanotechnology. *Chemical Reviews, 104*(1), 293–346. https://doi.org/10.1021/cr030698

Das, P., Barua, S., Sarkar, S., Karak, N., Bhattacharyya, P., Raza, N., Kim, K. H., & Bhattacharya, S. S. (2018). Plant extract–mediated green silver nanoparticles: Efficacy as soil conditioner and plant growth promoter. *Journal of Hazardous Materials, 346*, 62–72. https://doi.org/10.1016/j.jhazmat.2017.12.020

Das, P., Sarmah, K., Hussain, N., Pratihar, S., Das, S., Bhattacharyya, P., Patil, S. A., Kim, H. S., Khazi, M. I. A., & Bhattacharya, S. S. (2016). Novel synthesis of an iron oxalate capped iron oxide nanomaterial: A unique soil conditioner and slow-release eco-friendly source of iron sustenance in plants. *RSC Advances, 6*(105), 103012–103025. https://doi.org/10.1039/c6ra18840k

Donaldson, K., & Stone, V. (2003). Current hypotheses on the mechanisms of alcoholism. *Annali dell'Istituto Superiore di Sanità, 3*, 405–410. https://doi.org/10.1016/B978-0-444-62619-6.00027-6

Doran, J. W., & Zeiss, M. R. (2000). Soil health and sustainability: Managing the biotic component of soil quality. *Applied Soil Ecology, 15*(1), 3–11. https://doi.org/10.1016/S0929-1393(00)00067-6

Dwivedi, S., Saquib, Q., A. Al-Khedhairy, A., & Musarrat, J. (2016). Understanding the role of nanomaterials in agriculture. *Microbial Inoculants in Sustainable Agricultural Productivity: Vol. 2: Functional Applications* (pp. 1–308). https://doi.org/10.1007/978-81-322-2644-4

Batley, G., K. Kirby, J., & J. McLaughlin, M. (2012). Fate and risks of nanomaterials in aquatic and terrestrial environments. *Accounts of Chemical Research, 46*(3), 854–862. https://doi.org/10.1021/ar2003368

Emamifar, A., Kadivar, M., Shahedi, M., & Soleimanian-Zad, S. (2010). Evaluation of nanocomposite packaging containing Ag and ZnO on shelf life of fresh orange juice. *Innovative Food Science and Emerging Technologies, 11*(4), 742–748. https://doi.org/10.1016/j.ifset.2010.06.003

Fajardo, C., Saccà, M. L., Costa, G., Nande, M., & Martin, M. (2014). Impact of Ag and Al2O3 nanoparticles on soil organisms: In vitro and soil experiments. *Science of the Total Environment, 473–474*, 254–261. https://doi.org/10.1016/j.scitotenv.2013.12.043

Feng, Q. L., Wu, J., Chen, G. Q., Cui, F. Z., Kim, T. N., & Kim, J. O. (2000). A mechanistic study of the antibacterial effect of silver ions on Escherichia coli and Staphylococcus aureus. *Journal of Biomedical Materials Research, 52*(4), 662–668. https://doi.org/10.1002/1097-4636(20001215)52:4<662::AID-JBM10>3.0.CO;2-3

Forini, M. M. L., Pontes, M. S., Antunes, D. R., Lima, P. H. C. de, Santos, J. S., Santiago, E. F., & Grillo, R. (2022). Nano-enabled weed management in agriculture: From strategic design to enhanced herbicidal activity. *Plant Nano Biology, 1*, 100008. https://doi.org/10.1016/j.plana.2022.100008

Frenk, S., Ben-Moshe, T., Dror, I., Berkowitz, B., & Minz, D. (2013). Effect of metal oxide nanoparticles on microbial community structure and function in two different soil types. *PLoS ONE, 8*(12), 1–12. https://doi.org/10.1371/journal.pone.0084441

Gao, F., Ma, N. J., Zhou, H., Wang, Q., Zhang, H., Wang, P., Hou, H. L., Wen, H., & Li, L. J. (2016). Zinc oxide nanoparticles-induced epigenetic change and G2/M arrest are associated with apoptosis in human epidermal keratinocytes. *International Journal of Nanomedicine, 11*, 3859. https://doi.org/10.2147/IJN.S107021

Gatoo, M. A., Naseem, S., Arfat, M. Y., Mahmood Dar, A., Qasim, K., & Zubair, S. (2014). Physicochemical properties of nanomaterials: Implication in associated toxic manifestations. *BioMed Research International, 2014*, 498420. https://doi.org/10.1155/2014/498420

Gavrilescu, M. (2005). Fate of pesticides in the environment and its bioremediation. In *Engineering in Life Sciences* (Vol. 5, Issue 6, pp. 497–526). Wiley-VCH Verlag. https://doi.org/10.1002/elsc.200520098

Ghidan, A., & M. Al Antary, T. (2019). *Applications of Nanotechnology in Agriculture*. September, 2016–2017.

Ghosh, M., Bhadra, S., Adegoke, A., Bandyopadhyay, M., & Mukherjee, A. (2015). MWCNT uptake in Allium cepa root cells induces cytotoxic and genotoxic responses and results in DNA hyper-methylation. *Mutation Research/Fundamental and Molecular Mechanisms of Mutagenesis, 774*, 49–58. https://doi.org/10.1016/J.MRFMMM.2015.03.004

Giordano, M., & Petropoulos, S. A. (2021). The fate of nitrogen from soil to plants: influence of agricultural practices in modern agriculture. *Agriculture, 11*, 944.

Gogos, A., Knauer, K., & Bucheli, T. D. (2012). Nanomaterials in plant protection and fertilization: Current state, foreseen applications, and research priorities. *Journal of Agricultural and Food Chemistry, 60*(39), 9781–9792. https://doi.org/10.1021/jf302154y

Gupta, A., Eral, H. B., Hatton, T. A., & Doyle, P. S. (2016). Nanoemulsions: Formation, properties and applications. *Soft Matter, 12*(11), 2826–2841. https://doi.org/10.1039/c5sm02958a

Hagemann, N., Joseph, S., Schmidt, H. P., Kammann, C. I., Harter, J., Borch, T., Young, R. B., Varga, K., Taherymoosavi, S., Elliott, K. W., McKenna, A., Albu, M., Mayrhofer, C., Obst, M., Conte, P., Dieguez-Alonso, A., Orsetti, S., Subdiaga, E., Behrens, S., & Kappler, A. (2017). Organic coating on biochar explains its nutrient retention and stimulation of soil fertility. *Nature Communications, 8*(1), 1–11. https://doi.org/10.1038/s41467-017-01123-0

Haris, Z., & Ahmad, I. (2017). Impact of metal oxide nanoparticles on beneficial soil microorganisms and their secondary metabolites. *International Journal of Life-Sciences Scientific Research, 3*(3), 1020–1030. https://doi.org/10.21276/ijlssr.2017.3.3.10

Hasler, K., Bröring, S., Omta, S. W. F., & Olfs, H. W. (2015). Life cycle assessment (LCA) of different fertilizer product types. *European Journal of Agronomy, 69*, 41–51. https://doi.org/10.1016/j.eja.2015.06.001

Hong, J., R. Peralta-Videa, J., & L. Gardea-Torresdey, J. (2013). Nanomaterials in agricultural production: Benefits and possible threats? In *ACS Symposium Series*. https://doi.org/10.1021/bk-2013-1124.ch005

Hoshino, A., Fujioka, K., Oku, T., Suga, M., Sasaki, Y. F., Ohta, T., Yasuhara, M., Suzuki, K., & Yamamoto, K. (2004). Physicochemical properties and cellular toxicity of nanocrystal quantum dots depend on their surface modification. *Nano Letters, 4*(11), 2163–2169. https://doi.org/10.1021/nl048715d

Huang, T., Li, X., Maier, M., O'Brien-Simpson, N. M., Heath, D. E., & O'Connor, A. J. (2023). Using inorganic nanoparticles to fight fungal infections in the antimicrobial resistant era. *Acta Biomaterialia, 158*, 56–79. https://doi.org/10.1016/J.ACTBIO.2023.01.019

Iavicolia, I., Veruscka, L., Beezholdb, D. H., & Shvedova, A. A. (2017). Nanotechnology in agriculture: Opportunities, toxicological implications, and occupational risks. *Toxicology and Applied Pharmacology, 176*(3), 139–148. https://doi.org/10.1016/j.taap.2017.05.025. Nanotechnology

Ispas, C., Andreescu, D., Patel, A., V. Goia, D., Andreescu, S., & N. Wallace, K. (2009). Toxicity and developmental defects of different sizes and shape nickel nanoparticles in Zebrafish. *Environmental Science & Technology, 43*(16), 6349–6356. https://doi.org/10.1021/es9010543

Jambunathan, N. (2010). Determination and detection of reactive oxygen species (ROS), lipid peroxidation, and electrolyte leakage in plants. *NIH Public Access, 639*, 1–14. https://doi.org/10.1007/978-1-60761-702-0

Jeevanandam, J., Barhoum, A., Chan, Y. S., Dufresne, A., & Danquah, M. K. (2018). Review on nanoparticles and nanostructured materials: History, sources, toxicity and regulations. *Beilstein Journal of Nanotechnology, 9*(1), 1050–1074. https://doi.org/10.3762/bjnano.9.98

Jośko, I., Oleszczuk, P., & Futa, B. (2014). The effect of inorganic nanoparticles (ZnO, Cr2O3, CuO and Ni) and their bulk counterparts on enzyme activities in different soils. *Geoderma, 232–234*, 528–537. https://doi.org/10.1016/j.geoderma.2014.06.012

Kah, M. (2015). Nanopesticides and nanofertilizers: Emerging contaminants or opportunities for risk mitigation? *Frontiers in Chemistry, 3*(NOV), 1–6. https://doi.org/10.3389/fchem.2015.00064

Kah, M., Beulke, S., Tiede, K., & Hofmann, T. (2013). Nanopesticides: State of knowledge, environmental fate, and exposure modeling. *Critical Reviews in Environmental Science and Technology, 43*(16), 1823–1867. https://doi.org/10.1080/10643389.2012.671750

Kalwani, M., Chakdar, H., Srivastava, A., Pabbi, S., & Shukla, P. (2022). Effects of nanofertilizers on soil and plant-associated microbial communities: Emerging trends and perspectives. *Chemosphere, 287*(P2), 132107. https://doi.org/10.1016/j.chemosphere.2021.132107

Keller, A., & Lazareva, A. (2013). Predicted releases of engineered nanomaterials: From global to regional to local. *Environmental Science & Technology Letters, 1*(1), 65–70. https://doi.org/10.1021/ez400106t

Khan, I., Saeed, K., & Khan, I. (2019). Nanoparticles: Properties, applications and toxicities. *Arabian Journal of Chemistry, 12*(7), 908–931. https://doi.org/10.1016/J.ARABJC.2017.05.011

Klaine, S. J., Alvarez, P. J. J., Batley, G. E., Fernandes, T. F., Handy, R. D., Lyon, D. Y., Mahendra, S., McLaughlin, M. J., & Lead, J. R. (2008). Nanomaterials in the environment: Behavior, fate, bioavailability, and effects. *Environmental Toxicology and Chemistry, 27*(9), 1825–1851. https://doi.org/10.1897/08-090.1

Kookana, R. S., Boxall, A. B. A., Reeves, P. T., Ashauer, R., Beulke, S., Chaudhry, Q., Cornelis, G., Fernandes, T. F., Gan, J., Kah, M., Lynch, I., Ranville, J., Sinclair, C., Spurgeon, D., Tiede, K., & Van Den Brink, P. J. (2014). Nanopesticides: Guiding principles for regulatory evaluation of environmental risks. *Journal of Agricultural and Food Chemistry, 62*(19), 4227–4240. https://doi.org/10.1021/jf500232f

Kottegoda, N., Sandaruwan, C., Priyadarshana, G., Siriwardhana, A., Rathnayake, U. A., Berugoda Arachchige, D. M., Kumarasinghe, A. R., Dahanayake, D., Karunaratne, V., & Amaratunga, G. A. J. (2017). Urea-hydroxyapatite nanohybrids for slow release of nitrogen. *ACS Nano, 11*(2), 1214–1221. https://doi.org/10.1021/acsnano.6b07781

Kumar, J., Das, S., & Mawlong, L. G. (2018). Nanotechnology and its importance in micronutrient fertilization. *International Journal of Current Microbiology and Applied Sciences, 7*(5), 2306–2325. https://doi.org/10.20546/ijcmas.2018.705.267

Kumari, R., & Singh, D. P. (2019). Nano-biofertilizer : An emerging eco-friendly approach for sustainable agriculture. *Proceedings of the National Academy of Sciences, India Section B: Biological Sciences, 90*, 733–741. https://doi.org/10.1007/s40011-019-01133-6

Kumar, Y. (2021). Nanofertilizers and their role in sustainable agriculture. *Annals of Plant and Soil Research, 23*(3), 238–255. https://doi.org/10.47815/apsr.2021.10067

Kutawa, A. B., Ahmad, K., Ali, A., Hussein, M. Z., Wahab, M. A. A., Adamu, A., Ismaila, A. A., Gunasena, M. T., Rahman, M. Z., & Hossain, M. I. (2021). Trends in nanotechnology and its potentialities to control plant pathogenic fungi: A review. *Biology,* 10(9), 881. https://doi.org/10.3390/biology10090881

Lal, R. (2008). Soils and sustainable agriculture. A review. *Agronomy for Sustainable Development, 28*(1), 57–64. https://doi.org/10.1051/agro:2007025

Lead, J. R., Batley, G. E., Alvarez, P. J. J., Croteau, M. N., Handy, R. D., McLaughlin, M. J., Judy, J. D., & Schirmer, K. (2018). Nanomaterials in the environment: Behavior, fate, bioavailability, and effects—An updated review. *Environmental Toxicology and Chemistry, 37*(8), 2029–2063. https://doi.org/10.1002/ETC.4147

Leskovar, D., & Othman, Y. A. (2018). Organic and conventional farming differentially influenced soil respiration, physiology, growth and head quality of artichoke cultivars.

Journal of Soil Science and Plant Nutrition, 18(3), 865–880. https://doi.org/10.4067/S0718-95162018005002502

Lichtfouse, E., Navarrete, M., Debaeke, P., Souchère, V., Alberola, C., & Ménassieu, J. (2009). Agronomy for sustainable agriculture. A review. *Agronomy for Sustainable Development, 29*(1), 1–6. https://doi.org/10.1051/agro:2008054

Link, S., & El-Sayed, M. A. (2003). Optical properties and ultrafast dynamics of metallic nanocrystals. *Annual Review of Physical Chemistry, 54*, 331–366. https://doi.org/10.1146/annurev.physchem.54.011002.103759

Li, S., Pan, X., Wallis, L. K., Fan, Z., Chen, Z. L., & Diamond, S. A. (2014). Comparison of TiO2 nanoparticle and graphene-TiO2 nanoparticle composite phototoxicity to Daphnia magna and Oryzias latipes. *Chemosphere, 112*, 62–69. https://doi.org/10.1016/j.chemosphere.2014.03.058

Liu, R., & Lal, R. (2015). Potentials of engineered nanoparticles as fertilizers for increasing agronomic productions. *Science of the Total Environment, 514*(2015), 131–139. https://doi.org/10.1016/j.scitotenv.2015.01.104

Liu, S., Qi, X., Han, C., Liu, J., Sheng, X., Li, H., Luo, A., & Li, J. (2017). Novel nano-submicron mineral-based soil conditioner for sustainable agricultural development. *Journal of Cleaner Production, 149*, 896–903. https://doi.org/10.1016/j.jclepro.2017.02.155

Li, X., Liu, W., Sun, L., Aifantis, K. E., Yu, B., Fan, Y., Feng, Q., Cui, F., & Watari, F. (2015). Effects of physicochemical properties of nanomaterials on their toxicity. *Journal of Biomedical Materials Research - Part A, 103*(7), 2499–2507. https://doi.org/10.1002/jbm.a.35384

Li, Y., Zhang, P., Li, M., Shakoor, N., Adeel, M., Zhou, P., Guo, M., Jiang, Y., Zhao, W., Lou, B., & Rui, Y. (2022). Application and mechanisms of metal-based nanoparticles in the control of bacterial and fungal crop diseases. *Pest Management Science, 79*, 21–36. https://doi.org/10.1002/PS.7218

Madhuri, S., Kr. Choudhary, A., & Kumar, R. (2010). Nanotechnology in agricultural diseases and food safety. *Journal of Phytology, 2*(4), 83–92.

Mahmoodzadeh, H., Nabavi, M., & Kashefi, H. (2013). Effect of nanoscale titanium dioxide particles on the germination and growth of canola (Brassica napus). *Journal of Ornamental and Horticultural Plants, 3*(1), 25–32. www.SID.ir

Maluin, F. N., Hussein, M. Z., & Idris, A. S. (2020). An overview of the oil palm industry: Challenges and some emerging opportunities for nanotechnology development. *Agronomy, 10*(3). https://doi.org/10.3390/agronomy10030356

Manjunatha, S. B., Biradar, D. P., & Aladakatti, Y. R. (2016). Nanotechnology and its applications in agriculture: A review. *Journal Farm Science, 29*(1), 1–13. https://www.researchgate.net/publication/303665019

Mathivanan, S. (2018). *Perspectives of Nano-Materials and Nanobiosensors in Food Safety and Agriculture*. www.intechopen.com

Meena, V. S., Maurya, B. R., Verma, J. P., & Meena, R. S. (2016). Potassium solubilizing microorganisms for sustainable agriculture. In *Potassium Solubilizing Microorganisms for Sustainable Agriculture* (pp. 1–331). https://doi.org/10.1007/978-81-322-2776-2

Mitrano, D. M., & Nowack, B. (2017). The need for a life-cycle based aging paradigm for nanomaterials: Importance of real-world test systems to identify realistic particle transformations. *Nanotechnology, 28*(7). https://doi.org/10.1088/1361-6528/28/7/072001

Mondal, R., Dam, P., Chakraborty, J., Paret, M. L., Katı, A., Altuntas, S., Sarkar, R., Ghorai, S., Gangopadhyay, D., Mandal, A. K., & Husen, A. (2022). Potential of nanobiosensor in

sustainable agriculture: the state-of-art. *Heliyon, 8*(12), e12207. https://doi.org/10.1016/J.HELIYON.2022.E12207
Monreal, C., McGill, W. B., & Nyborg, M. (1986). Spatial heterogeneity of substrates: Effects on hydrolysis' immobilization and nitrification of urea-N. *Canadian Journal of Soil Science, 66*, 499–511.
Mubarak Ali, D., Arunkumar, J., Pooja, P., Subramanian, G., Thajuddin, N., & Alharbi, N. S. (2015). Synthesis and characterization of biocompatibility of tenorite nanoparticles and potential property against biofilm formation. *Saudi Pharmaceutical Journal, 23*(4), 421–428. https://doi.org/10.1016/j.jsps.2014.11.007
Mustafa, M. A., Ali, A., Manickam, S., & Siddiqui, Y. (2014). Ultrasound-assisted chitosan-surfactant nanostructure assemblies: Towards maintaining postharvest quality of tomatoes. *Food and Bioprocess Technology, 7*(7), 2102–2111. https://doi.org/10.1007/s11947-013-1173-x
Nair, P. K. R. (2014). Grand challenges in agroecology and land use systems. *Frontiers in Environmental Science, 2*, 1–4. https://doi.org/10.3389/fenvs.2014.00001
Nair, R., Varghese, S. H., Nair, B. G., Maekawa, T., Yoshida, Y., & Kumar, D. S. (2010). Nanoparticulate material delivery to plants. *Plant Science, 179*(3), 154–163. https://doi.org/10.1016/j.plantsci.2010.04.012
Nel, A., Xia, T., Mädler, L., & Li, N. (2006). Toxic potential of materials at the nanolevel. *Science, 311*(5761), 622–627. https://doi.org/10.1126/science.1114397
Nguyen, H. M., Hwang, I. C., Park, J. W., & Park, H. J. (2012). Photoprotection for deltamethrin using chitosan-coated beeswax solid lipid nanoparticles. *Pest Management Science, 68*(7), 1062–1068. https://doi.org/10.1002/ps.3268
Sarbani, N. M. M., & Yahaya, N. (2022). Advanced development of bio-fertilizer formulations using microorganisms as inoculant for sustainable agriculture and environment: A review. *Malaysian Journal of Science Health & Technology, 8*(1), 92–101. https://doi.org/10.33102/mjosht.v8i1.228
Oliveira, H. C., Stolf-Moreira, R., Martinez, C. B. R., Grillo, R., De Jesus, M. B., & Fraceto, L. F. (2015). Nanoencapsulation enhances the post-emergence herbicidal activity of atrazine against mustard plants. *PLoS ONE, 10*(7), 1–12. https://doi.org/10.1371/journal.pone.0132971
Padidar, M., Jalalian, A., Abdouss, M., Najafi, P., & Honarjoo, N. (2014). *Effect of Nanoclay on Soil Erosion Control* (pp. 5–8).
Park, K. H., Chhowalla, M., Iqbal, Z., & Sesti, F. (2003). Single-walled carbon nanotubes are a new class of ion channel blockers. *Journal of Biological Chemistry, 278*(50), 50212–50216. https://doi.org/10.1074/jbc.M310216200
Paula, H. C. B., Sombra, F. M., Cavalcante, R. D. F., Abreu, F. O. M. S., & De Paula, R. C. M. (2011). Preparation and characterization of chitosan/cashew gum beads loaded with Lippia sidoides essential oil. *Materials Science and Engineering C, 31*(2), 173–178. https://doi.org/10.1016/j.msec.2010.08.013
Pereira, A. do E. S., Oliveira, H. C., & Fraceto, L. F. (2019). Polymeric nanoparticles as an alternative for application of gibberellic acid in sustainable agriculture: A field study. *Scientific Reports, 9*(1), 1–10. https://doi.org/10.1038/s41598-019-43494-y
Pereira, A. E. S., Grillo, R., Mello, N. F. S., Rosa, A. H., & Fraceto, L. F. (2014). Application of poly(epsilon-caprolactone) nanoparticles containing atrazine herbicide as an alternative technique to control weeds and reduce damage to the environment. *Journal of Hazardous Materials, 268*, 207–215. https://doi.org/10.1016/j.jhazmat.2014.01.025

Petersen, E. J., & Nelson, B. C. (2010). Mechanisms and measurements of nanomaterial-induced oxidative damage to DNA. *Analytical and Bioanalytical Chemistry, 398*(2). https://doi.org/10.1007/s00216-010-3881-7

Peters, R. J. B., Bouwmeester, H., Gottardo, S., Amenta, V., Arena, M., Brandhoff, P., Marvin, H. J. P., Mech, A., Moniz, F. B., Pesudo, L. Q., Rauscher, H., Schoonjans, R., Undas, A. K., Vettori, M. V., Weigel, S., & Aschberger, K. (2016). Nanomaterials for products and application in agriculture, feed and food. *Trends in Food Science and Technology, 54*, 155–164. https://doi.org/10.1016/j.tifs.2016.06.008

Prasad, R., Bhattacharyya, A., & Nguyen, Q. D. (2017). Nanotechnology in sustainable agriculture: Recent developments, challenges, and perspectives. *Frontiers in Microbiology, 8*, 1–13. https://doi.org/10.3389/fmicb.2017.01014

Puri, A., Loomis, K., Smith, B., Lee, J. H., Yavlovich, A., Heldman, E., & Blumenthal, R. (2009). Lipid-based nanoparticles as pharmaceutical drug carriers: From concepts to clinic. *Critical Reviews in Therapeutic Drug Carrier Systems, 26*(6), 523–580. https://doi.org/10.1615/CritRevTherDrugCarrierSyst.v26.i6.10

Rafiq, M. K., Joseph, S. D., Li, F., Bai, Y., Shang, Z., Rawal, A., Hook, J. M., Munroe, P. R., Donne, S., Taherymoosavi, S., Mitchell, D. R. G., Pace, B., Mohammed, M., Horvat, J., Marjo, C. E., Wagner, A., Wang, Y., Ye, J., & Long, R. J. (2017). Pyrolysis of attapulgite clay blended with yak dung enhances pasture growth and soil health: Characterization and initial field trials. *Science of the Total Environment, 607–608*, 184–194. https://doi.org/10.1016/j.scitotenv.2017.06.186

Raghu, P., Reddy, T. M., Gopal, P., Reddaiah, K., & Sreedhar, N. Y. (2014). A novel horseradish peroxidase biosensor towards the detection of dopamine: A voltammetric study. *Enzyme and Microbial Technology, 57*, 8–15. https://doi.org/10.1016/j.enzmictec.2014.01.002

Rajput, V. D., Minkina, T. M., Behal, A., Sushkova, S. N., Mandzhieva, S., Singh, R., Gorovtsov, A., Tsitsuashvili, V. S., Purvis, W. O., Ghazaryan, K. A., & Movsesyan, H. S. (2018). Effects of zinc-oxide nanoparticles on soil, plants, animals and soil organisms: A review. *Environmental Nanotechnology, Monitoring and Management, 9*, 76–84. https://doi.org/10.1016/j.enmm.2017.12.006

Rajput, V. D., Singh, A., Minkina, T., Rawat, S., Mandzhieva, S., Sushkova, S., Shuvaeva, V., Nazarenko, O., Rajput, P., Komariah, Verma, K. K., Singh, A. K., Rao, M., & Upadhyay, S. K. (2021). Nano-enabled products: Challenges and opportunities for sustainable agriculture. *Plants, 10*(12), 1–12. https://doi.org/10.3390/plants10122727

Raliya, R., Saharan, V., Dimkpa, C., & Biswas, P. (2018). Nanofertilizer for precision and sustainable agriculture: Current state and future perspectives [Review-article]. *Journal of Agricultural and Food Chemistry, 66*(26), 6487–6503. https://doi.org/10.1021/acs.jafc.7b02178

Reineke, J. (2012). Nanotoxicity methods and protocols. In *Methods in Molecular Biology*, (pp. 1064–3745). https://doi.org/https://doi.org/10.1007/978-1-62703-002-1

Risom, L., Møller, P., & Loft, S. (2005). Oxidative stress-induced DNA damage by particulate air pollution. *Mutation Research - Fundamental and Molecular Mechanisms of Mutagenesis, 592*(1–2), 119–137. https://doi.org/10.1016/j.mrfmmm.2005.06.012

Salem, S. S., Hammad, E. N., Mohamed, A. A., & El-Dougdoug, W. (2023). A comprehensive review of nanomaterials: Types, synthesis, characterization, and applications. *Biointerface Research in Applied Chemistry, 13*(1). https://doi.org/10.33263/BRIAC131.041

Sampathkumar, K., Tan, K. X., & Loo, S. C. J. (2020). Developing nano-delivery systems for agriculture and food applications with nature-derived polymers. *IScience, 23*(5), 101055. https://doi.org/10.1016/j.isci.2020.101055

Sardella, D., Gatt, R., & Valdramidis, V. P. (2017). Physiological effects and mode of action of ZnO nanoparticles against postharvest fungal contaminants. *Food Research International, 101*, 274–279. https://doi.org/10.1016/j.foodres.2017.08.019

Sasson, Y., Levy-Ruso, G., Toledano, O., and Ishaaya, I. (2007). Nanosuspensions: emerging novel agrochemical formulations. In I. Ishaaya, R. Nauen, R. Horowitz (Eds.), *Insecticides Design using Advances Technologies* (pp. 1–32). Springer-Verlag, Berlin Heidelberg.

Schroder, J. L., Zhang, H., Girma, K., Raun, W. R., Penn, C. J., & Payton, M. E. (2011). Soil acidification from long-term use of nitrogen fertilizers on winter wheat. *Soil Science Society of America Journal, 75*(3), 957–964. https://doi.org/10.2136/sssaj2010.0187

Shang, Y., Kamrul Hasan, M., Ahammed, G. J., Li, M., Yin, H., & Zhou, J. (2019). Applications of nanotechnology in plant growth and crop protection: A review. *Molecules, 24*(14), 2558. https://doi.org/10.3390/molecules24142558

Shen, T., Wang, Q., Li, C., Zhou, B., Li, Y., & Liu, Y. (2020). Transcriptome sequencing analysis reveals silver nanoparticles antifungal molecular mechanism of the soil fungi Fusarium solani species complex. *Journal of Hazardous Materials, 388*, 122063. https://doi.org/10.1016/J.JHAZMAT.2020.122063

Siegel, R. W. (1993). Synthesis and properties of nanophase materials. *Materials Science and Engineering: A, 168*(2), 189–197. https://doi.org/10.1016/0921-5093(93)90726-U

Sigmund, W., Yuh, J., Park, H., Maneeratana, V., Pyrgiotakis, G., Daga, A., Taylor, J., & Nino, J. C. (2006). Processing and structure relationships in electrospinning of ceramic fiber systems. *Journal of the American Ceramic Society, 89*(2), 395–407. https://doi.org/10.1111/j.1551-2916.2005.00807.x

Silva, M. dos S., Cocenza, D. S., Grillo, R., Melo, N. F. S. de, Tonello, P. S., Oliveira, L. C. de, Cassimiro, D. L., Rosa, A. H., & Fraceto, L. F. (2011). Paraquat-loaded alginate/chitosan nanoparticles: Preparation, characterization and soil sorption studies. *Journal of Hazardous Materials, 190*(1–3), 366–374. https://doi.org/10.1016/j.jhazmat.2011.03.057

Singh, D. M., Gautam, C., Prakash, P. O., Mohan, M. H., Prakasha, G., & Vishwajith, V. (2017). Nano-fertilizers is a new way to increase nutrients use efficiency in crop production. *International Journal of Agriculture Sciences Citation, 9*(7), 3831–3833. http://www.bioinfopublication.org/jouarchive.php?opt=&jouid=BPJ0000217

Singh, J. S., Pandey, V. C., & Singh, D. P. (2011). Efficient soil microorganisms: A new dimension for sustainable agriculture and environmental development. *Agriculture, Ecosystems and Environment, 140*(3–4), 339–353. https://doi.org/10.1016/j.agee.2011.01.017

Singh, N., Manshian, B., Jenkins, G. J. S., Griffiths, S. M., Williams, P. M., Maffeis, T. G. G., Wright, C. J., & Doak, S. H. (2009). NanoGenotoxicology: The DNA damaging potential of engineered nanomaterials. *Biomaterials, 30*(23–24), 3891–3914. https://doi.org/10.1016/j.biomaterials.2009.04.009

Stadler, T., Buteler, M., Valdez, S. R., & Gitto, J. G. (2018). Particulate nanoinsecticides: A new concept in insect pest management. In *Insecticides - Agriculture and Toxicology*. InTech. https://doi.org/10.5772/intechopen.72448

Stephenson, G. R. (2003). *Pesticide Use and World Food Production: Risks and Benefits*. 261–270. https://doi.org/10.1021/bk-2003-0853.ch015

Sweet, M. J., & Singleton, I. (2015). Soil contamination with silver nanoparticles reduces Bishop pine growth and ectomycorrhizal diversity on pine roots. *Journal of Nanoparticle Research, 17*(11), 1–8. https://doi.org/10.1007/s11051-015-3246-4

Tahat, M. M., Alananbeh, K., Othman, Y., & Leskovar, D. I. (2020). *Sustainability*. November. https://doi.org/10.3390/su121248590

Tang, G., Tian, Y., Gao, Y., Zhou, Z., Chen, X., Li, Y., Yu, X., Wang, H., Li, X., & Cao, Y. (2022). Supramolecular self-assembly of herbicides with reduced risks to the environment. *ACS Nano, 16*(3), 4892–4904. https://doi.org/10.1021/acsnano.2c00539

Thomas, S., Harshita, B. S. P., Mishra, P., & Talegaonkar, S. (2015). Ceramic nanoparticles: fabrication methods and applications in drug delivery. *Current Pharmaceutical Design, 21*(42), 6165–6188. https://doi.org/10.2174/1381612821666151027153246

Tilman, D., Cassman, K. G., Matson|, P. A., Naylor, R., & Polasky, S. (2002). Agricultural sustainability and intensive production practices. *Nature, 418*, 671–677. https://doi.org/10.1038/nature01014.

Timsina, J. (2018). Can organic sources of nutrients increase crop yields to meet global food demand? *Agronomy, 8*(10), 1–20. https://doi.org/10.3390/agronomy8100214

Torney, F., Trewyn, B. G., Lin, V. S. Y., & Wang, K. (2007). Mesoporous silica nanoparticles deliver DNA and chemicals into plants. *Nature Nanotechnology, 2*(5), 295–300. https://doi.org/10.1038/nnano.2007.108

Tu, M., Abbood, H. A., Zhu, Z., Li, H., & Gao, Z. (2013). Investigation of the photocatalytic effect of zinc oxide nanoparticles in the presence of nitrite. *Journal of Hazardous Materials, 244–245*(2), 311–321. https://doi.org/10.1016/j.jhazmat.2012.11.048

Van Der Heijden, M. G. A., Bardgett, R. D., & Van Straalen, N. M. (2008). The unseen majority: Soil microbes as drivers of plant diversity and productivity in terrestrial ecosystems. *Ecology Letters, 11*(3), 296–310. https://doi.org/10.1111/j.1461-0248.2007.01139.x

Waalewijn-Kool, P. L., Rupp, S., Lofts, S., Svendsen, C., & van Gestel, C. A. M. (2014). Effect of soil organic matter content and pH on the toxicity of ZnO nanoparticles to Folsomia candida. *Ecotoxicology and Environmental Safety, 108*, 9–15. https://doi.org/10.1016/j.ecoenv.2014.06.031

Wang, B., Gao, B., Zimmerman, A. R., Zheng, Y., & Lyu, H. (2018). Novel biochar-impregnated calcium alginate beads with improved water holding and nutrient retention properties. *Journal of Environmental Management, 209*, 105–111. https://doi.org/10.1016/j.jenvman.2017.12.041

Wang, D., Saleh, N. B., Byro, A., Zepp, R., Sahle- Demessie, E., P. Luxton, T., T. Ho, K., M. Burgess, R., Flury, M., C. White, J., & Su, C. (2022). Nano-enabled pesticides for sustainable agriculture and global food security. *Nature, 544*(2). https://doi.org/10.1038/s41565-022-01082-8.Nano-enabled

Wani, T. A., Masoodi, F. A., Baba, W. N., Ahmad, M., Rahmanian, N., & Jafari, S. M. (2019). Nanoencapsulation of agrochemicals, fertilizers, and pesticides for improved plant production. In *Advances in Phytonanotechnology: From Synthesis to Application*. Elsevier Inc. https://doi.org/10.1016/B978-0-12-815322-2.00012-2

Wilson, R. J., Li, Y., Yang, G., & Zhao, C. X. (2022). Nanoemulsions for drug delivery. *Particuology, 64*, 85–97. https://doi.org/10.1016/J.PARTIC.2021.05.009

Worrall, E. A., Hamid, A., Mody, K. T., Mitter, N., & Pappu, H. R. (2018). Nanotechnology for plant disease management. *Agronomy, 8*(12), 1–24. https://doi.org/10.3390/agronomy8120285

Wu, J., Zhao, Y., Qi, H., Zhao, X., Yang, T., Du, Y., Zhang, H., & Wei, Z. (2017). Identifying the key factors that affect the formation of humic substance during different materials composting. *Bioresource Technology, 244*, 1193–1196. https://doi.org/10.1016/j.biortech.2017.08.100

Xu, L., Takemura, T., Xu, M., & Hanagata, N. (2011). Toxicity of silver nanoparticles as assessed by global gene expression analysis. *Materials Express, 1*(1), 74–79. https://doi.org/10.1166/mex.2011.1010

Xu, Z., Tang, T., Lin, Q., Yu, J., Zhang, C., Zhao, X., Kah, M., & Li, L. (2022). Environmental risks and the potential benefits of nanopesticides: A review. *Environmental Chemistry Letters, 20*(3), 2097–2108. https://doi.org/10.1007/s10311-021-01338-0

Yadav, A., Kon, K., Kratosova, G., Duran, N., Ingle, A. P., & Rai, M. (2015). Fungi as an efficient mycosystem for the synthesis of metal nanoparticles: Progress and key aspects of research. *Biotechnology Letters, 37*(11), 2099–2120. https://doi.org/10.1007/s10529-015-1901-6

Yang, C.-W., Hu, Y., Yuan, L., Zhou, H.-Z., & Sheng, G.-P. (2021). Selectively tracking nanoparticles in aquatic plant using core–shell nanoparticle-enhanced Raman spectroscopy imaging. *ACS Nano, 15*(12), 19828–19837. https://doi.org/10.1021/acsnano.1c07306

Yu, M., Yao, J., Liang, J., Zeng, Z., Cui, B., Zhao, X., Sun, C., Wang, Y., Liu, G., & Cui, H. (2017). Development of functionalized abamectin poly(lactic acid) nanoparticles with regulatable adhesion to enhance foliar retention. *RSC Advances, 7*(19), 11271–11280. https://doi.org/10.1039/c6ra27345a

Zaytseva, O., & Neumann, G. (2016). Differential impact of multiwalled carbon nanotubes on germination and seedling development of *Glycine max, Phaseolus vulgaris* and *Zea mays*. *European Chemical News, 5*(5), 202–210.

Zhang, J. N., Chen, B. Z., Ashfaq, M., Zhang, X. P., & Guo, X. D. (2018). Development of a BDDE-crosslinked hyaluronic acid based microneedles patch as a dermal filler for anti-ageing treatment. *Journal of Industrial and Engineering Chemistry, 65*, 363–369. https://doi.org/10.1016/J.JIEC.2018.05.007

Zhang, S., Deng, R., Lin, D., & Wu, F. (2017). Distinct toxic interactions of TiO2 nanoparticles with four coexisting organochlorine contaminants on algae. *Nanotoxicology, 11*(9–10), 1115–1126. https://doi.org/10.1080/17435390.2017.1398358

Zhao, L., Peralta-Videa, J. R., Rico, C. M., Hernandez-Viezcas, J. A., Sun, Y., Niu, G., Servin, A., Nunez, J. E., Duarte-Gardea, M., & Gardea-Torresdey, J. L. (2014). CeO2 and ZnO nanoparticles change the nutritional qualities of cucumber (Cucumis sativus). *Journal of Agricultural and Food Chemistry, 62*(13), 2752–2759. https://doi.org/10.1021/jf405476u

Zhao, Y., Allen, B. L., & Star, A. (2011). Enzymatic degradation of multi-walled carbon nanotubes. *The Journal of Physical Chemistry A, 23*(1), 9536–9544. https://doi.org/10.1021/jp112324d.Enzymatic

CHAPTER 9

The Role of Nanomaterials for Stress Tolerance and Gum Production

J. C. TARAFDAR[1], AMITA GHARU DHARIWAL[2], and DEEPAK KUMAR[3,4]

[1]*ICAR – Central Arid Zone Research Institute, Jodhpur, India*

[2]*Department of Chemistry, Jai Narain Vyas University, Jodhpur, India*

[3]*Nextnode Bioscience Pvt Ltd., Opposite GEB Office, Kadi-Kalol Road, Kadi, India*

[4]*Shri Ram Solvent Extraction Pvt Ltd., Kashipur Road, Jaspur, India*

ABSTRACT

Nanoparticles like Zn, Si, Ag, CeO_2, TiO_2, Se, and chitosan may efficiently be used to overcome plant abiotic stresses such as heat, cold, drought, salinity, heavy metals, flooding, and ultraviolet rays, as well as plant biotic stresses due to pathogenic attack, insect attack, and herbivores. They may be used either as foliage, soil, and seed soaks or through irrigation water. The main impacts of nanomaterials are the elevation of antioxidant enzyme activities, photosynthetic pigments, and reduced H_2O_2 and MDA levels. They help improve plant tolerance and defense mechanisms after retaining important physiological and biochemical functions. The net results are increased seed germination, vigor index, plant biomass, plant tolerance, chlorophyll level, hormonal control, photosynthetic rate, electrolytic level, as well as overall plant development and yield attributes. They may help to alter plant metabolic pathways, regulate energy dissipation, improve membrane permeability and stability, improve overall plant health and growth kinetic traits, and ultimately decrease the negative impact of different stresses. Nanoparticles like Zn, Fe, and Mg can help release exopolysaccharides of many folds from

Nanobiotechnology for Agricultural Sciences: Nano-Agri-Input Products for Crop Production and Environmental Protection. J. C. Tarafdar, Deepak Kumar, Sanjay K. Singh, and Krishna Pal Singh (Eds.)

the polysaccharide-releasing fungi, bacteria, algae, and so forth. The important microbial polysaccharides are xanthan gum, gellan, pollulan, dextran, curdlan, and alginates. This microbial polysaccharide can be used for soil aggregation, soil moisture retention, soil carbon buildup, texture modifier, thickening agent, dispersant, emulsion stabilizer, heat retention, stabilizing agent, controlling viscosity, and gelling agent.

9.1 INTRODUCTION

Nanomaterials are rapidly gaining interest for their versatile uses, including agriculture, because of their large surface area-to-volume ratio and greater triggering capacity. Due to their unique properties, they can be used as plant stress tolerance materials and trigger microbial gum production. There are two main categories of plant stresses. They are abiotic stresses that are mainly caused by heat (temperature), cold, drought, salt, metals, rain (flooding), and ultraviolet rays that primarily affect plant metabolism and ultimately plant development. Similarly, the biotic stresses are mainly caused by pathogenic attack, insect attack, and herbivore attack, which ultimately affect phytohormone production and break the plant's defense. It has been observed that plants acquire stress very easily and cannot break out of it. In fact, at every stage of the plant life cycle (from seed germination to fruit ripening), they suffer stress. Nanomaterials and various nanotechnology approaches have been found to be effective in mitigating different plant stresses (Tarafdar, 2021). The different abiotic and biotic stresses the plants are suffering from are presented in Figure 9.1.

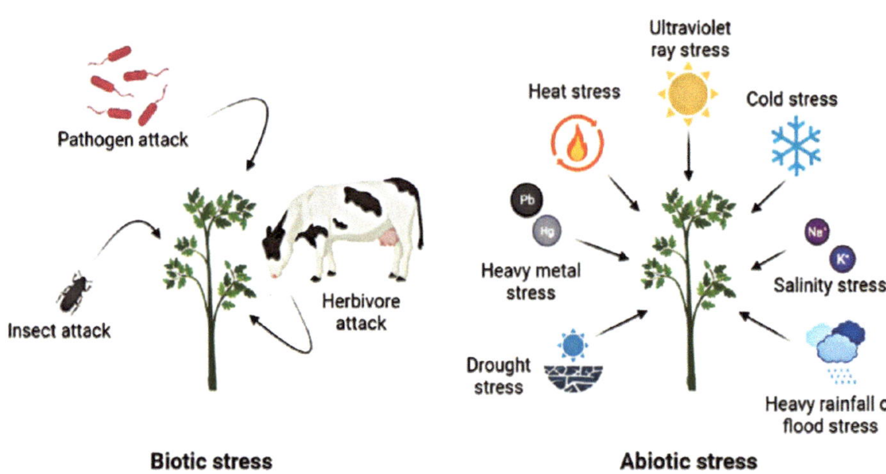

FIGURE 9.1 Different plants are suffering different stresses during their growth period.

Microorganisms produce gum as a metabolite, which is mainly polysaccharides. It can act as secondary carbon sources and protect the soil and environment after soil aggregation, moisture retention, and carbon buildup. It can also take part in cell aggregation, cell adhesion, and biofilm formation, which can help protect the cell from hostile environments. It has been noted that nanomaterials have the capability to enhance microbial polysaccharide production (Raliya et al., 2014), which can be well utilized under drought conditions. The important microbial gums are xanthan, curdlan, pollulan, and gellan. Their production can be enhanced by 12–15 times through nanomaterial intervention. The unique property of microbial gums is their ability to form gel after coming into contact with water. Recently, nanotechnology has gained tremendous focus for plant and soil stress management (Tarafdar et al., 2018; Rady et al., 2021; Abd El-Maged et al., 2022). The present chapter demonstrated the use of nanomaterials for plant stress management and microbial gum production for sustainable agriculture and global farming.

9.2 ABIOTIC STRESSES

Plants suffer different environmental stresses such as heat due to sunlight, cold weather, drought, soil salinity, heavy metals, heavy rainfall (flooding), and ultraviolet rays.

9.2.1 SUNLIGHT OR HEAT STRESS

Sunlight is essential for plant growth and development as it supplies the energy for the plants to translate carbon dioxide and water into carbohydrate and oxygen through the process of photosynthesis. This process of carbohydrate production is utilized for the vegetative and reproductive growth of plants. When the temperature goes up, plants suffer heat stress, which has a negative effect by interrupting the stability of various cytoskeleton structures, membranes, and proteins as well as affecting pollen grain formation. It may also cause physiological damage to plants like leaf scorching, leaf abscission, senescence, and a reduction in root and shoot growth that results in a reduction in yield.

9.2.2 COLD STRESS

Due to the very low temperature, the plant cannot protect its body at a normal temperature, which may cause ice formation in the plant tissues, resulting in cellular dehydration and permanent tissue damage or death. Plants may show symptoms like abnormal curling as well as crinkling of leaves. There is every possibility to change membrane structure and composition, decrease protoplasmic streaming and electrolyte leakage, and increase plasmolysis. It can also cause disorder in plant metabolism and enhance the production of ROS (El-Saadory et al., 2022). It may change physiological and biochemical reactions in the plant cells, resulting in wilting, leaf chlorosis, and sometimes necrosis of the plant cells (Ruelland and Zachowski, 2010).

9.2.3 DROUGHT STRESS

It is a major factor that limits crop production. It enhances the leaves' ageing and drooping, including rolling and brittleness, closed flowering, and premature fall. It affects water relations with plants and drastically reduces water use efficiency (Farooq et al., 2010). Drought influences membrane damage and the activity of ATP synthesis enzymes. The photosynthesis rate is reduced due to stomatal closure. Drought conditions in plants can modify carbon and nitrogen cycles, affecting water and mineral uptake by roots. The litter decomposition is reduced under drought conditions, and a biogenic release of CO_2 was observed from the soil.

9.2.4 SALINITY

The salt stress on the plant may be due to more salt accumulation in the soil, which affects plant growth and development. It normally enhances intracellular osmotic pressure as well as the accumulation of sodium to toxic levels. It affects plant's ability to take the water they need and ultimately influences plant growth and reproduction. Legume may suffer oxidative stresses due to sodium chloride, resulting in a reduction in growth and seed nutrients as well as lower nodulation (Hernandez et al., 2000; Ahmad et al., 2008). Soil salinity is reported to inhibit seed germination due to the low osmotic pressure created around the seeds (Taha et al., 2021), which also prevents water uptake. Generally, plants are applying both enzymatic and nonenzymatic antioxidant defence mechanisms to overcome the salinity stress. The important damages due to salinity stresses are membrane damage, nutrient imbalance, a decrease in photosynthetic activity, and impairment in detoxifying ROS.

9.2.5 HEAVY METALS

The toxic levels of heavy metals such as Al, Pb, Cd, As, Cu, and so forth may affect seed germination, root growth, and development, including dry biomass.

A higher level of metal concentration may cause chlorosis in young leaves as well as decoupling of the photosynthetic enzymes. It may inhibit ATP production and damage DNA. Heavy metals also help to overproduce ROS and damage proteins and carbohydrates, as well as chlorophyll production. The toxic levels of different heavy metals unfavorably or negatively influence various metabolic processes like degradation or displacement of protein structures (Hall, 2002), disruption of cytoplasmic membrane stability (Farid et al., 2013), as well as repression of photosynthesis, respiration, and enzymatic actions (Hossain et al., 2012). Soil contamination is the main cause of heavy metal accumulation.

9.2.6 HEAVY RAINFALL OR FLOODING

More rainfall causes flooding and stresses plants in many ways, such as submerging and water logging, depending on the soil depth and water table. Under submerged conditions, photosynthesis is badly affected, along with flowering and pollination. It may cause hypoxia, which results in plants going from aerobic respiration to anaerobic fermentation with the accumulation of ethanol. It also inhibits ATP synthesis. The condition also dictates the species composition of the ecosystem. It has been noticed that the concentration of oxygen, carbon dioxide, ROS, and ethylene production changes upon flooding and has a disastrous impact on crop growth and survival (Normila, 2008).

9.2.7 ULTRAVIOLET RAYS

It is a nonionizing radiation released by the sun or artificial sources like tanning beds. It may damage DNA, protein polymerization, inactivate enzymes, and enhance the permeability of the cell membrane. It affects photosynthesis in many species and may change plant color. It may damage cell membranes as well as almost all organelles within the cell, like mitochondria and chloroplasts.

9.3 ROLE OF NANOPARTICLES TO CONTROL DIFFERENT ABIOTIC STRESSES

Nanosized particles (1–100 nm) may be used effectively to control different abiotic plant stresses. Normally, foliage application is very effective. But they can also be applied directly to the soil or through drip and sprinkler systems. People are also applying in hydroponics, aquaponics, or aeroponics. During foliage application, they may enter through the cuticle, stomata, epidermis, hydathodes, root tips, cortex, lateral roots, or any wounds. The effect of nanoparticles depends on the plant species, application methods, exposure time, and doses applied (Rizwan et al., 2017; Tarafdar, 2022). After entering the nanoparticles through the stomata or cuticle, they may move within the plant by the phloem and then be carried by pressure differences between the leaves and roots (Turgeon, 2010). During transport, they may trigger different plant coenzyme systems and make the plants more active to release the different beneficial enzymes (Tarafdar and Rathore, 2018). Nanomaterials have the capability to alter numerous plant processes such as seed germination, antioxidant activity, photosynthesis, chlorophyll content, nutrient mobilisation in plants, and chloroplast number (Tarafdar et al., 2015; Cinisli et al.,2019). Nanoparticles can also affect ethylene production (Syu et al., 2014). Mainly, nanoparticles are transported by osmotic pressure, capillary forces, or the symplastic route (Usman et al., 2020). It has been observed that nanomaterials may tie up to carrier proteins through ion channels, aquaporin, and endocytosis and derange the plasma membrane to inspire the formation of pores for crossing into the cells. The effective action of nanoparticles and nanomaterials depends on the size, shape, and structure of the nanoparticles (Tarafdar et al., 2012a; Tunca, 2015).

Silver nanoparticles applied through irrigation water improved heat resistance in wheat crops (Iqbal et al., 2019) by increasing leaf number and plant development. Spraying tomato leaves with TiO_2 nanoparticles can overcome heat stress (Qi et al., 2013) and increase photosynthesis after regulating energy dissipation and leaf cooling by increasing stomatal opening. Seed dormancy can be averted in wheatgrass by applying SiO_2 nanoparticles (Azimi et al., 2014), which resulted in improved seed germination and seedling weight. Silver nanoparticles can overcome cold stress in *Arabidopsis thaliana* after activating and increasing the expression of antioxidant genes (Kohan-Baghkheirati and Geisler-Lee, 2015). Chickpea crops can avoid cold stress with the application of TiO_2 nanoparticles (Mohammadi et al., 2013; 2014), which help improve antioxidative enzyme activities and reduce H_2O_2 levels and electrolyte leakage. A similar observation with TiO_2 nanoparticles on chickpea was noticed by Hasanpour et al. (2015), who found increased expression of genes encoding Rubisco- and chlorophyll-binding proteins, enhanced activity of phosphoenolpyruvate carboxylase, and decreased H_2O_2 levels. Amini et al. (2017) concluded that TiO_2 nanoparticles effectively controlled the cold stress of chickpea by altering metabolic pathways, which were noticed through transcription profiling. TiO_2 nanoparticles were noticed to overcome the cold stress of Licorice plants after decreasing lipid peroxidation and H_2O_2 levels and simultaneously increasing phenolics, total protein, and osmolyte contents.

Sedghi et al. (2013) observed that soybean crops can overcome drought stress with the application of nanozinc, which helps increase seed germination and reduce seed residual weight. A similar observation with Zn nanoparticles was reported by Dimkpa et al. (2019) with a barley crop and reported that Zn nanoparticles assist in stimulating growth, improving yield, and N acquisition under drought conditions. By enhancing melatonin synthesis and improving the functioning of the antioxidant system in maize, Zn nanoparticles can overcome drought conditions (Sun et al., 2020). Application of SiO_2 nanoparticles can overcome drought stress in hawthorn plants and improve plant tolerance by retaining critical physiological and biochemical functions as well as exhibiting nonsignificant effects on chlorophyll and carotenoid contents (Ashkavand et al., 2015). Silica nanoparticles can help overcome drought stress in higher plants by enhancing antioxidant enzyme activities that promote plant stress resistance (Liang et al., 2007). Application of Ag nanoparticles to lentils helps overcome drought stress by decreasing the germination ratio and increasing the root length and dry weight (Hojjat and Ganjali, 2016). Nanosize Ag particles applied to flax can defeat drought stress (Aghdam et al., 2016) by improving chlorophyll and carotenoid content and reducing H_2O_2 and MDA levels, resulting in improved flax growth and yield attributes. Foliar application of Ag nanoparticles to cotton plants under drought conditions can increase total phenolics, total soluble proteins, total free amino acids, total antioxidant capacity, and CAT, POD, and SOD activities (Shallan et al., 2016). The wheat crop can overcome drought stress with the foliar application of TiO2 nanoparticles (Jaberzadeh et al., 2013; Faraji and Sepehri, 2020) that help to improve development, productivity, seed gluten, and starch content. It also helps dry weight, chlorophyll and carotenoid content, transpiration rate, RWC, and stomatal conductance. A similar observation with TiO_2 nanoparticles was reported for linseed (Aghdam et al., 2016), basil (Kiapour et al., 2015), and moldavian dragonhead (Mohammadi et al., 2016) to overcome drought stress. They found that TiO_2 nanoparticles were very effective in increasing photosynthetic pigments, reducing oxidative stress, and reducing H_2O_2 and MDA levels in drought situations. Foliar and soil application of chitosan nanoparticles was able to overcome drought stress in wheat and barley crops after increasing leaf area, RWC, chlorophyll content, photosynthetic rate, CAT, and SOD activities, resulting in more biomass and grain yield. Fe nanoparticles were found to be very effective in nullifying drought stress in wheat after enhancing Fe uptake and improving growth parameters and photosynthetic activities (Adrees et al., 2020). Rice crops were able to overcome drought stress with the foliar and soil application of Fe nanoparticles that enhanced nutrient uptake and photosynthetic parameters as well as growth and plant biomass (Ahmed et al., 2021). Kim et al. (2015) observed that Fe nanoparticles are very effective in defeating drought stress in *Arabidopsis thaliana*, which promoted H^+-ATPase activity, maintained stomatal opening and closure, increased biomass and photosynthetic pigments, as well as internal CO_2. Application of Se nanoparticles to wheat was able to conquer drought stress (El-Saadony et al., 2021) after maintaining leaf water status as well as chlorophyll and carotenoid contents, which ultimately enhanced plant growth and biomass content.

Zn nanoparticles were reported to be very effective in controlling salinity stress in *Lupinus termis* (Abdel Latef et al., 2017), which can influence plant growth and restore the levels of photosynthetic pigments, organic solutes, total phenols, ascorbic acid, and antioxidant enzymes. Foliar application of nano-Zn particles has been reported to overcome the salinity stress experienced by wheat (Fathi et al., 2017) and tomato (Faizan et al., 2021) after enhancing antioxidant enzyme levels and promoting root and shoot growth as well as photosynthetic pigments. Enhanced protein and antioxidant activity, as well as reduced H_2O_2 and MDA levels, were also noticed in *Trigonella foenum-graecum* after foliar application of Zn nanoparticles to overcome salinity stress (Noohpisheh et al., 2021). A similar observation was also made by Eisheery et al. (2020) in *Mangifera indica,* where Zn nanoparticle application increased the amount of antioxidant enzymes as well as total sugar to overcome the salinity effect. Foliar application of Si nanoparticles can overcome salinity stress in common beans (*Phaseolus vulgaris*) after mitigating salinity-induced adverse effects and improving antioxidant enzyme activities. Heghighi et al. (2012) reported that the application of Si nanoparticles to tomato plants can improve germination, root length, dry matter, chlorophyll content, leaf water content, and photosynthetic rate under saline soil conditions. A similar observation was also noticed with lentil (Sabaghnia and Janmohammadi, 2014), basil (Kalteh et al., 2014), squash (Siddiqui et al., 2014), broad bean (Qados, 2015), maize (Geo et al., 2006), cucumber (Alsaeedi et al., 2018), soybean (Farhangi-Abriz and Torabian, 2018), and many other crops. In general, it has been observed that Si nanoparticles can be used to overcome salinity stress in higher plants, which promotes plant stress resistance, including a tremendous increase in antioxidant enzyme activities (Liang et al., 2007).

Seed soaking with Ag nanoparticles has been reported to overcome salinity stress in tomato (Almutairi, 2016) and wheat (Mohamed et al., 2017) after mitigating the damaging impact due to salinity by increasing germination ratio, root length, dry matter production, and gene regulation. After spraying with CeO_2 nanoparticles, mouse-ear cress plants were reported to overcome salinity stress (Wu et al., 2018) due to improved leaf mesophyll K^+ retention, chlorophyll content, biomass production, and photosynthetic activity. Similarly, with the application of CeO_2 nanoparticles in soil, canola plants can overcome salinity (Rossi et al., 2016) after shortening root apoplastic barriers, which allows more Na^+ transport to shoots, thereby reducing Na^+ accumulation and photosynthetic apparatus efficiency. Foliage application of TiO_2 nanoparticles to broad beans may overcome salinity stress after increasing the levels of soluble sugars, amino acids, metabolites, and antioxidant enzyme activities, resulting in osmoprotection for crops (Abdel Latief et al., 2018). Under saline conditions, bell peppers improved seed germination and root growth as well as altered gene expression with the foliar application of Mn nanoparticles (Ye et al., 2020). The cucumber crop was found to be very responsive under salinity conditions after foliage application of Mn_3O_4 nanoparticles (Lu et al., 2020), which helped increase photosynthetic pigment content and net photosynthesis. Application of Fe nanoparticles to grapes under saline conditions helps increase total protein content and K^+ content as well as decrease Na^+ levels (Mozafari et al., 2018). Moldavian balm plants overcome salinity stress with the foliar application of Fe nanoparticles (Moradbeygi et al. 2020) after an increase in the gene expression of TAT, RAS, and RA, as well as affecting the amino acid concentration, polyphenyl oxidase, phenylalanine ammonia-lipase, and superoxiDe dismutase activities. Sorghum crops can overcome salinity after seed soaking with Fe nanoparticles (Maswada et al., 2018), which help increase photosynthetic rate, chlorophyll index, photosystem II efficiency, relative water content, and lipid peroxidation activity. The positive effect of Fe nanoparticles to overcome salinity was also noticed in wheat crops after soil application (Manzoor et al., 2021), which helped increase photosynthetic pigments, NPK uptake, and antioxidant enzyme activities that assisted increased growth and plant biomass, including yield.

Supplication of Zn nanoparticles in nutrient solutions can get rid of heavy metal toxicity in rice crops (Yan et al., 2021) after increasing phytochelatin content, which ultimately promotes the growth of the plant. Foliar application of Si nanoparticles to rice was also found to reduce heavy metal toxicity and encourage development by reducing the bioaccumulation and translocation of heavy metals in plants. After applying Si nanoparticles through irrigation water in soybean (Li et al., 2020), the Hg toxicity of the crop can be immobilized and inactivated. Nano-Si application in irrigation water or through nutrient solution can decrease Cr(VI) absorption by pea crops, promote antioxidant defence mechanisms, and increase overall growth potential (Delfani et al., 2014). Seed soaking by Si nanoparticles may overcome arsenic (As) toxicity to maize crops and develop Increased arsenic (As) resistance (Tripathi et al., 2017). Application of Si nanoparticles as foliage also overcomes Cd toxicity to wheat after increasing superoxide dismutase and peroxidase activity (Ali et al., 2019) and mitigates oxidative stress. Silver (Ag) nanoparticles applied with irrigation water to Bok choy plants reduce Cd toxicity after improving biomass, chlorophyll content, and vitamin C levels and reducing malondialdehyde (MDA) levels (Li and Huang, 2014). TiO_2 nanoparticles also help to get rid of heavy metal toxicity in mung bean plants (Katiyar et al., 2020), which induced antioxidant activities and reduced MDA levels that resulted in more growth and improved biomass. Foliar as well as soil application of Fe nanoparticles to wheat crops can conquer heavy metal toxicity after enhancing SOD and guaiacol peroxidase (POX) activity (Konate et al., 2017).

Cerium dioxide (CeO_2) nanoparticle application in the nutrient solution of soybean can overcome flooding stress (Mustafa et al., 2015) after reducing the formation of cytotoxic byproducts of glycolysis and increasing the production of stress-related proteins, resulting in better growth of the seedling. Seed soaking with CeO_2 on saffron plants has been reported to overcome the adverse effects of flooding by blocking ethylene signaling and promoting root growth as well as root length (Rezvani et al., 2012). Al_2O_3 soaking of soybean seeds helps to overcome flooding stress after regulating the ascorbate–glutathione pathway and increasing membrane permeability as well as the activity of the tricarboxylic acid cycle (Mustafa and Komatsu, 2016).

Foliar application of Si nanoparticles to wheat crops may help to overcome the stress due to ultraviolet radiation (Tripathi et al., 2017), mainly due to nitric oxide-mediated antioxidant defence as well as counterbalancing toxic ROS generation, resulting in improved protection of the plants. Application of silver nanoparticles as foliage to spinach plants may get rid of the adverse effects caused by UV radiation on the crops after reducing reactive oxygen species (ROS) and MDA levels and inducing oxygen evolution rate as well as antioxidant enzyme activities (Lei et al., 2007). The summarized effect of different nanoparticles on various abiotic stresses is presented in Table 9.1. In a nutshell, different nanoparticles so far have been found to be effective against plant abiotic stresses after improving growth and crop metabolism and increasing the production of different hormones as well as antioxidants.

TABLE 9.1 Impacts of Various Nanoparticles on Plant Abiotic Stresses.

S. No.	Abiotic Stresses	Effective Nanoparticles	Most Effective Plants
1.	Heat stress	Ag	Wheat
		TiO_2	Tomato
2.	Cold stress	SiO_2	Wheatgrass
		Ag	*Arabidopsis thaliana*
		TiO_2	Chickpea, Licorice
3.	Drought stress	Zn	Soybean, Barley, Maize
		Ag	Cotton, Lentil, Flax
		SiO_2	Howthorn
		Si	Higher plants
		TiO_2	Basil, Wheat, Linseed, Moldavian dragonhead
		Chitosan	Wheat, Barley, Rice
		Fe	Wheat, Rice, *Arabidopsis thaliana*
		Se	Wheat
4.	Salinity stress	Zn	Sunflower, Soybean, *Lupinus termis*, Wheat, Tomato, *Trigonella foemum-graecum, Mangifera indica* L.
		Si	Maize, Broad bean, Common bean, Squash, Tomato, Lentil, Basil, Soybean, Cucumber, Sweet pepper, Peregrina, *Prunus mahaleb*
		Ag	Tomato, Wheat
		CeO_2	Canola, Mouse-ear cress
		TiO_2	Broad bean
		Mn	Bell peppers
		Mn_3O_4	Cucumber
		Fe	Grape, Moldavian balm, Sorghum, Wheat
5.	Heavy metal stress	Zn	Rice
		Si	Rice, Pea, Maize, Wheat
		Si	Bok choy
		TiO_2	Mung bean
		Fe	Wheat

6.	Flooding stress	CeO_2	Soybean, Saffron
		Al_2O_3	Soybean
7.	UV rays stress	Si	Wheat
		Ag	Spinach

9.4 BIOTIC STRESSES

It is the plant stress that is induced by living organisms, notably bacteria, fungus, insects, viruses, nematodes, birds, and weeds. It may deplete the host's nutrients and lead to pre- and postharvest losses. It may cause huge damage to agricultural production. Biotic stress can affect plant phytohormones and ultimately plant defence. The best way to overcome biotic stresses is to develop resistant varieties. Nanoparticles have excellent antimicrobial properties to control plant biotic stresses or manage stress. It can be used as a potential material to alleviate the damage caused by biotic stress to plants.

9.4.1 PATHOGENIC STRESS

Pathogenic stress in plants is mainly due to bacteria, fungi, antinobacteria, viruses, and viroids, which cause plant diseases and affect plants' normal growth and development. Sometimes, the losses are very severe, destroying the entire plant. The pathogens are invading the plants and affecting the host immune system. They may affect the cells and produce toxins.

9.4.2 INSECT STRESS

Under stressful conditions, plants are more vulnerable to attack by different insects or pests. This may be due to the fact that plants may be more suitable as food or have less ability to defend themselves. Normally, they attack the plants after cutting the root, stem, and leaf or sucking the cell sap from different parts of the plant. They may also bore the stems of the plants or fruits, affecting the plant health and yield of the crops. There are many chewing insects that do pit feeding on leaves called leaf beetles, flea beetles, and young and larger caterpillars, weevils, and grasshoppers that slit along the edges of leaves and cause damage to plants. They may also alter the plant's volatile emission rate. To overcome insect stress, plants can go for cell wall modifications, through which they may receive signals and activate their immune system.

9.4.3 HERBIVORE STRESS

It is a major plant stress, and sometimes that affects the plant very severely, decreasing the plant's fitness and affecting the development of leaves. Herbivores can put a stop to shoot growth and may start off more branches, which affects shoot-to-root ratios. Plants also become more susceptible to infection following an herbivore attack. Some plants try to overcome herbivore stress by producing allele-chemicals as secondary metabolites as a defence mechanism that can act as repellents or toxins to herbivores to minimize their plant digestibility. Normally, herbivores initiate enzymes to overcome and minimize the adverse effects of toxic secondary metabolites produced by plants. The major enzyme groups are called mixed-function oxidases, which detoxify dangerous plant compounds after catalyzing oxidative reactions.

9.5 USE OF NANOMATERIAL TO CONTROL BIOTIC STRESSES

To control wilt disease caused by *Fusarium species,* chitosan nanoparticles were found to be very effective. They can inspire upregulation of pathogenesis-related proteins and antioxidant genes that help improve plant defence against pathogenic attack (Chun and Chandrasekaran, 2019). Chitosan nanoparticles may also help to overpower rust diseases caused by *Puccinia striiformis* after increasing the antioxidant enzyme activities, reducing reactive oxygen species formation, and activating the transcription levels of PR1-PR5 and PR10 genes (Elsharkawy et al., 2022). Silver nanoparticles can be used to overcome bacterial leaf blight caused by *Xanthomonas oryzae* and help increase antioxidant enzyme levels to regulate the adverse effects of reactive oxygen species and promote nutrient uptake as well as the cellular antioxidative system (Ahmed et al., 2020). MgO nanoparticles may control black shank disease caused by *Phytophthora nicotianae* and black root rot disease due to *Thielaviopsis basicola.* The application of nanomagnesium oxide inspires reactive oxygen species production and destroys membrane integrity after altering morphological characteristics through pathogen cell uptake. It also participates in many physiological and biological processes to boost plant defence (Chen et al., 2020). The yellow stripe rust disease caused by *Puccinia striiformis* on wheat crops can very well be controlled with the application of TiO_2 nanoparticles, which can regulate the proteins provoking defence-related responses, which in turn are implicated in facilitating the activity of NADPH-oxidase, which is the important ROS-producing enzyme during infections by pathogens (Satti et al., 2022). Seed priming with ZnO nanoparticles can get rid of fusarium wilt caused by *Fusarium oxysporum* on chickpea crops (Farhana et al., 2022). It has been found that nano ZnO particles may assist plants in accumulating more quantities of sugars, phenol, and total proteins, as well as help activate plant defence enzymes like SOD, PO, and CAT, ultimately generating resistance against the pathogen. Si nanoparticles are found to be very effective against root-knot nematodes, which inhibit egg hatching as well as cause nematode mortality.

Insect mortality can be increased with the application of CuO nanoparticles, which help plants after an increase in the activity of enzymes like SOD, POD, and APX as well as an increase in the concentration of leaf pigments that play a major role in scavenging reactive oxygen species that protect the plants from stress (Badawy et al., 2021). The chitosan nanoparticles loaded with the essential oil peppermint may be able to control insects like *Sitophilus oryzae* and *Tribolium castaneum* after inhibiting essential detoxification enzymes (AChE) of insect organization (Rajkumar et al., 2020). Moreover, ZnO, Cu, Ag, and SiO_2 nanoparticles are reported to be effective in controlling many insects singly or in combination with other insecticides (Chhipa, 2017). Generally, metal nanoparticles bind to S and P in proteins and nucleic acids, resulting in decreases in membrane permeability that help in organelle and enzyme denaturation and ultimately cell death. Besides this, nanoparticles of Al_2O_3 and TiO_2 have proven to have insecticidal properties and can be used to control various insects.

Herbivore attacks on plants may be possible to control with the application of TiO_2, Ag, and multiwall carbon nanotubes after increasing plant resistance. Silicon nanodots may encourage plant resistance against herbivores after concurrently activating both plant physical and chemical defences. Normally, nanoparticles help to enhance herbivore-induced plant volatile (HIPV) production, which activates the plant's defences against herbivores. The nanoparticles' application decreases herbivore attack and consumption rates either directly or indirectly. It may generally act as a repellent for herbivores. The herbivore-induced plant volatiles may be classified into three groups: terpenoids (isoprene, monoterpenes, homoterpenes, and sesquiterpenes), lipoxygenases such as (Z)-3-hexenyl acetate and (Z)-3-hexenal, and

TABLE 9.2 Nanomaterials Responsible for Controlling Plant Biotic Stresses.

S. No.	Type of Biotic Stress	Effective Nanoparticles	Control Agents
1.	Pathogenic stress	Chitosan	Wilt disease caused by *Fusarium* sp.
			Rust disease caused by *Puccinia* sp.
		Ag	Bacterial leaf blight caused by *Xanthomonas* sp.
		MgO	Black shank disease caused by *Phytopthora* sp.
			Black root rot disease caused by *Thielaviopsis* sp.
		TiO_2	Yellow stripe rust disease caused by *Puccinia* sp.
		Si	Nematodes
2.	Insect stress	CuO	Stored grain insects
		Chitosan with peppermint oil	Weevils and Beetles
		ZnO, Ag, SiO_2	Common insects
		Al_2O_3 and TiO_2	Insecticidal properties
3.	Herbivore stress	TiO_2, Ag	Herbivores
		Multiwalled carbon nanotubes	Herbivores
		Silicon nanorods	Herbivores

volatile aromatic compounds (methyl salicylate, indole). Table 9.2 represents the different nanoparticles effective at controlling various biotic stresses.

9.6 MICROBIAL GUM PRODUCTION

Microorganisms can release polysaccharides, which are high-molecular-weight, environmentally friendly natural polymers. They are mostly surface active and can be used as metal chelators, emulsifiers, and flocculants in industrial and environmental fields, as well as retaining soil moisture after aggregating soil particles. Polysaccharides, in their true sense, are rather complex carbohydrates composed of multiple monosaccharides attached together with glycosidic linkages. Generally, bacteria, fungi, algae, and yeast may produce polysaccharides that have various applications in agriculture, food industries, cosmetics, oil industries, and pharmaceuticals. The cofactors in polysaccharide structure are normally Zn, Fe, and Mg. Therefore, it has been noticed that the polysaccharide production by the polysaccharide-releasing organisms can be manipulated to a large extent by the application of Zn and Fe nanoparticles. The important microbial polysaccharides are dextran, curdlan, pollulan, chitosan, and arabic gum. They have immense commercial

TABLE 9.3 Important Polysaccharide-Releasing Microorganisms' Positive Response to Nanoparticles.

Microorganisms Name	Major Polysaccharide Produced	Type of Nanoparticle Responded	Quantum of Increase in Production
Bacillus subtilis	Dextron	Zn	6–9 folds
Agrobacterium tumefaciens	Xanthan and Curdlan	Zn and Fe	8–11 folds
Caulobacter vibrioides	Curdlan	Zn and Mg	5–8 folds
Paenibacillus illinoisensis	Curdlan and Dextran	Zn and Fe	7–9 folds
Stenotrophomonas maltophilia	Curdlan and Chitosan	Zn and Fe	8–12 folds
Sporosarcina ginsengisoli	Xanthan and Dextran	Zn	6–10 folds
Aspergillus terreus	Pollulan	Zn and Mg	8–11 folds
Aspergillus flavus	Pollulan	Zn	7–12 folds
Aspergillus fumigates	Pollulan and Dextran	Zn and Fe	8–14 folds
Chaetomium globossum	Pollulan and Xanthan	Zn	9–16 folds
Curvularia lunata	Pollulan and Dextran	Zn and Fe	5–12 folds
Sinarhizobium sp.	Xanthan and Curdlan	Zn	4–8 folds

importance. A list of selected important polysaccharide-producing organisms that responded to nanoparticles and the types of polysaccharides produced are listed below (Table 9.3).

9.7 FACTORS AFFECTING THE PRODUCTION OF MICROBIAL GUMS

Microbial polysaccharide release is influenced by an excess supply of carbon, but the release is affected by a higher nitrogen supply. The best release was noticed when the C:N ratio was maintained at 10:1. Temperature, moisture level, pH, nutrient content, and oxygen level are the other factors affecting microbial polysaccharide production. Temperature has a significant effect on polysaccharide release by microorganisms. More microbial release of polysaccharides was observed between 30 and 35°C, but their secretion was reduced considerably below 25°C and above 50°C. For the release of polysaccharides, pH plays an important role, as bacteria were found to be more comfortable releasing polysaccharides between pH 6 and 7.5, while most polysaccharides produced by fungi were between pH 5 and 6. In general, marginal polysaccharide release by the organisms was noted below pH 4 and above pH 10. Microbes are unable to release polysaccharides under moisture-stress conditions. The higher release was observed between 50 and 75% moisture levels. Nutrients like N, Mg, Zn, and Fe can manipulate the polysaccharide secretion. Less than 20% oxygen level affected the production of microbial polysaccharides.

9.8 ROLE OF NANOPARTICLES FOR MICROBIAL GUM PRODUCTION

Microbial gum production can be triggered by nanoparticles like Zn, Fe, and Mg. Nanoparticles can enhance extracellular polysaccharide production, which is mainly associated with the cell surfaces, between 4 and 16 folds. Tarafdar et al. (2012b) found that the application of ZnO nanoparticles helped increase polysaccharide release by 8–9 folds when applied to the fungal cultures of *Aspergillus terreus* CZR1 and *Aspergillus flavus* CZR 2. A simultaneous release of internal polysaccharides between 5-8 folds was also recorded. The increased production of exo-polysaccharides may help in minimizing soil erosion, increasing moisture retention, and encouraging carbon buildup in the system. Microbial polysaccharides, focused as essential biopolymers for soil application, have been found to be an emerging economic competitor with synthetic polymers. The important microbial polysaccharides induced by nanoparticles are alginate, xanthan, pollulan, curdlan, dextran, levan, and gellan. Alginate is mainly found in the cell walls of brown algae, whose production can be triggered by Zn nanoparticles. Different nanoformulations can be made with alginate for nutraceutical delivery. Many fungi and bacteria can produce curdlan; the noted genera are *Alcaligenus, Rhizobium, Bacillus, Agrobacterium,* and *Cellulomonas*. It is actually a homopolysaccharide of glucose connected by β-1,3 glycosidic linkages. Nanoparticles like Fe and Zn may trigger curdlan production (Tarafdar, 2021). Dextran is formed from D-glucose with α-1,6 glycosidic linkages between the monomers and α-1,2, α1,3, and α-1,4 linkages in the branching. It is a neutral polysaccharide and can trigger its production through the introduction of Zn and Fe nanoparticles. Gellan is a natural anionic polysaccharide mainly produced by *Pseudomonas* species, basically a tetrasaccharide formulated of repeating units of α-L-rhamnose, β-D-glucose, and β-D-glucuronate in a ratio of 1:2:1. Zn nanoparticles are found to be effective in triggering their production. It can be used as a very good coating material due to its bioavailability and targeting of the receptor. Levan is mainly produced by the bacteria *Bacillus, Acinetobacter, Acetobacter, Pseudomonas, and Zymomonas*. It is mainly composed of β-(2→6) linked β-D-fructofuranose residues that can be linear or branched with β-(2→1) linkages. The production of levan can be induced with the application of both Zn and Fe nanoparticles. The most exciting properties of levan are to self-assemble into nanoparticles in water. Nano-induced levan-polysaccharide can be used as a stable biopolymer. Pollulan is a neutral polysaccharide that contains repeating maltotriose units attached by α-1,4 glycosidic bonds, and trimers are linked together via α-1,6 glycosidic bonds. The chain of pollulan contains a high percentage of hydroxyl groups. Many strains of *Aureobasidium* produce pollulan along with some other fungi. Zn, Fe, and Mg nanoparticles may trigger pollulan production. Besides its use in soil, this polysaccharide can be used in the food, biomedical, and pharmacological industries. During application, pollulan can be transformed into multiple forms like nanospheres, nanogels, and nanoparticles. The main source of xanthan is the Gram-negative bacteria *Xanthomonas campestris*. It is generally made of a linear backbone of glucose units connected by β-1,4 glycosidic bonds. Some fungi can also release xanthan. It has a high molecular weight and hydrogen bonding. Zn nanoparticles are found to be very effective in enhancing xanthan production. It is a helpful material to use as a polysaccharide-based delivery system.

9.9 USES OF NANOINDUCED MICROBIAL GUM

The microbial gum produced by the organisms and induced through the action of nanoparticles may be purified in sucrose solution followed by drying at 100°C (Tarafdar, 2021). The gum pellet can then be powdered for agricultural applications. It has been observed that they are very effective in arid and semiarid areas for soil aggregation, moisture retention, and carbon buildup. The major components of microbial gum are xanthan, gellan, dextran, pollulan, alginate, curdlan, and glycan. The polysaccharide powder is capable of building the moisture retention capacity of arid soil after incorporation into it (Table 9.4).

TABLE 9.4 Nanoparticle-Induced Polysaccharide Powder on Improvement in Soil Moisture Retention in Arid Soils (Modified After Tarafdar, 2021).

Nano-Induced Polysaccharide Powder (%)	Improvement in Moisture Retention (%)
1.0	10–12
2.0	12–13
3.0	13–14
4.0	14–15
6.0	15–16
8.0	15–17
10.0	16–17

Four important nanoparticle-induced exo-polysaccharides such as xanthan, dextran, curdlan, and pollulan have been found to be very effective in playing an important role in soil aggregation. Polysaccharide concentration was found to be an important factor for aggregation in soil. Normally, 1–2% concentration was found to be efficient enough, although the soil aggregation percent varies with the type of polysaccharide-producing organisms as well as the nanoparticle used for triggering the polysaccharide production. Table 9.5 displays the efficiency of aggregate formation after 30 days of application by different nano-induced microbial polysaccharides.

TABLE 9.5 Improvement of Soil Aggregation Percent of 0.5 Mm and 1.0 Mm Soil Aggregates After 1.0% (W/V) Application of Nanoparticle-Induced Polysaccharides.

Nano-Induced Polysaccharide from Microorganisms	Nanoparticle Used for Triggering	Percent Improvement of 0.5 Mm Soil Aggregate	Percent Improvement of 1.0 Mm Soil Aggregate
Aspergillus faecalis	Zn	35–42	77–79
Aspergillus terreus	Zn	45–50	80–84
Bacillus coagulans	Fe	52–57	67–73
	Zn	30–33	80–86
Alcaligens faecalis	Fe	23–25	69–71
	Zn	63–65	83–85
	Mg	23–24	47–49

The percent improvement in soil aggregation may also depend on the size of the nanoparticles used for triggering, soil condition, and polysaccharide origin. But the results indicated the definite role of nano-induced microbial polysaccharides in the process of soil aggregation. It was observed that the entire process of nanoparticle-induced soil aggregation is mainly guided by the density of the particle, the frequency of collision, and the binding capacity of nanoparticle-induced polysaccharides. The buildup of soil carbon with the application of nano-induced microbial polysaccharides was reported (Tarafdar, 2021). It was found that 3–5% carbon builds up in arid soil with the application of 1–5% nanoparticle-induced microbial polysaccharides. The four most important microbial polysaccharides that have a role in soil aggregation are xanthan, dextran, curdlan, and pollulan. For good soil aggregate formation due to the nanoparticle-induced polysaccharide, it may take one month's time. Nano-induced soil aggregate may prevent wind erosion and encourage moisture retention. The best size of nanoparticles to trigger microbial polysaccharide production was noted as 40 nm, and in general, 40 ppm concentration is the safer limit for application.

The nanoparticle-induced microbial polysaccharides may also be used as emulsifiers, binders, stabilizers, and coagulants in the food industry, where their diversity of structure and physicochemical characteristics may help. There is also a possibility to use them as food packaging material, wastewater treatment material, and wound dressing due to their nontoxicity, low cost, and both biocompatibility and biodegradability. For example, nano-induced xanthan polysaccharides can be used in salad dressings as a dispersant and emulsion stabilizer. It may also be used in beverages, milk products, baked goods, and frozen foods as a stabilizing agent. Similarly, nano-induced gellan polysaccharides may be used in jellies and jams as a gelling material and for texture modification. Moreover, pollulan and dextran may be used in bakery products as well as frozen and dried foods. Curdlan may be used in processed meat or freeze-dried foods for modification of texture and improving the mixture's rehydration.

9.10 FUTURE PROSPECTS

The use of nanomaterials in agriculture may provide a chance for sustainability in global food production. Important benefits of nanoparticles are their penetration power, triggering the co-enzyme systems, enhancing the nutrient use efficiency, regulating the availability of nutrients in crops, helping to overcome abiotic and biotic plant stresses, and enhancing the microbial polysaccharide production, which results in more soil aggregate formation, moisture retention, and soil carbon buildup. Moreover, nanomaterials are needed in small quantities; therefore, soil is not loaded with salts. It may also be used as a bionanoformulation to overcome the problem of shelf life and delivery of biofertilizers. Nanomaterials may be used in all areas of crop production, from seed treatment to grain storage. It may help to minimize the important inputs for crop production like fertilizers, pesticides, and herbicides. It can also be used to precisely locate the problems crops are facing and minimize environmental pollution. Nanomaterials may also be used as nanosensors to monitor grain storage, crop health, and the removal of contaminants. There is also enormous scope to work on biosensors with metal nanoparticles based on surface plasmon resonance (SPR). Nanoparticles can be used in an eco-friendly way to remove weeds, plant pathogens, and insects. But there is a need to work on safe doses for different crops. Due to the high surface area and reactivity of nanomaterials, there is scope to use them for water management and reducing run-off in agriculture.

Incorporation of nanoscale particles into membranes or filters may help in water purification with low operating energy. To maximize clean energy production while reducing emissions, nanoparticles are a good candidate. Nano-induced microbial polysaccharides have massive commercial importance. It has great importance in the oil, food, and pharmaceutical industries. But we must be careful with the exact doses of nanoparticles to trigger the most polysaccharide secretion from the microbes. There is scope to measure the level of exopolysaccharide that can help minimize the erodibility of soil. More work is also needed on the carbon buildup under different soil types with the application of nano-induced exopolysaccharides that are produced by various organisms. It is also needed to assess the quantity of drought stress that may be overcome through nano-induced microbial polysaccharides.

9.11 CONCLUSION

Nanoparticles or nanomaterials can be very well prepared through different physical, chemical, biological, and aerosol techniques that may be used to overcome plant abiotic (heat, cold, drought, salt, metals, flooding, ultraviolet rays) and biotic (pathogen, insect, herbivore) stresses. Nanoparticles like Zn, Fe, and Mg can effectively be used to trigger microbial polysaccharide production to improve soil conditions (aggregate formation, moisture retention, carbon buildup), food stabilization, and other industrial and pharmaceutical production. Nanomaterials have the potential for more stress-tolerant bio-nanoformulations, plant growth regulation, and soil health maintenance at a minimum cost. It has the potential for precise nutrient management in agricultural crops as well as enormous commercial importance. With the help of nanomaterials, precise application, and safe doses, it is possible to revolutionize global agricultural farming. Nanoscale or nanostructured materials can be used as carriers of fertilizers and pesticides for control release to reduce cost and increase efficiency. Nanoparticles also influence crop physiological parameters after modifying the reactive oxygen species, catalase, peroxidase, and superoxide dismutase activities, as well as chlorophyll, leaf protein, and phenol contents. Nano-induced polysaccharide application may improve root hydraulic conductivity as well as water stress tolerance. Further, the information gathered so far must be disseminated and demonstrated so that farmers can be convinced to use it.

KEYWORDS

- **Agricultural application**
- **microbial polysaccharides**
- **nano induced**
- **nanoparticles**
- **plant abiotic and biotic stress**
- **soil and plant health**

REFERENCES

Abd El-Mageed, T. A., Rady, M. O. A., Abd El-Wahed, M. H., Abd El-Mageed, S. A., Omran, W. M., Aljuaid, B. S., El-Shehawi, A. M., El-Tahan, A. M., El-Saadony, M. T., & Abdou, N. M. (2022). Consecutive seasonal effect on yield and water productivity of drip deficit irrigated sorghumin saline soils. *Saudi Journal of Biological Sciences, 29*(4), 2683–2690.

Abdel Latef, A. A. H., Alhmad, M. F., & Abdelfattah, K. E. (2017). The possible roles of priming with ZnO nanoparticles in mitigation of salinity stress in lupine (Lupinus termis) plants. *Journal of Plant Growth Regulation, 36*(1), 60–70.

Abdel Latef, A. A. H., Srivastava, A. K., El Sadek, M. S. A., Kordrostami, M., & Tran, L. S. P. (2018). Titanium dioxide nanoparticles improve growth and enhance tolerance of broad bean plants under saline soil conditions. *Land Degradation and Development, 29*(4), 1065–1073.

Adrees, M., Khan, Z. S., Ali, S., Hafeez, M., Khalid, S., Rehman, M. Z. U., Hussain, A., Hussain, K., Chatha, S. A. S., & Rizwan, M. (2020). Simultaneous mitigation of cadmium and drought stress in wheat by soil application of iron nanoparticles. *Chemosphere, 238*, 124681.

Aghdam, M. T. B., Mohammadi, H., & Ghorbanpour, M. (2016). Effects of nanoparticulate anatase titanium dioxide on physiological and biochemical performance of *Linum usitatissimum* (Linaceae) under well-watered and drought stress conditions. *Brazilian Journal of Botany, 39*(1), 139–146.

Ahmad, P., John, R., Sarwat, M., & Umar, S. (2008). Responses of proline, lipid peroxidation and antioxidative enzymes in two varieties of *Pisum sativum* L. under salt stress. *International Journal of Plant Production, 2*(4), 353–366.

Ahmed, T., Shahid, M., Noman, M., Niazi, M. B. K., Mahmood, F., Manzoor, I., Zhang, Y., Li, B., Yang, Y., Yan, C., & Chen, J. (2020). Silver nanoparticles synthesized by using bacillus cereus sztl ameliorated the damage of bacterial leaf blight pathogen in rice. *Pathogens, 9*(3), 160.

Ahmed, T., Noman, M., Manzoor, N., Shahid, M., Abdullah, M., Ali, L., Wang, G., Hashem, A., Al-Arjani, A. B. F., Alqarawi, A. A., Allah, E. F. A., & Li, B. (2021). Nanoparticle-based amelioration of drought stress and cadmium toxicity in rice via triggering the stress responsive genetic mechanisms and nutrient acquisition. *Ecotoxicology and Environmental Safety, 209*, 111829.

Ali, S., Rizwan, M., Hussain, A., Ur Rehman, M. Z., Ali, B., Yousaf, B., Wijaya, L., Alyemeni, M. N., & Ahmad, P. (2019). Silicon nanoparticles enhanced the growth and reduced the cadmium accumulation in grains of wheat (*Triticum aestivum* L.). *Plant Physiology and Biochemistry*, *140*, 1–8.

Almutairi, Z. M. (2016). Influence of silver nano-particles on the salt resistance of tomato (*Solanum lycopersicum*) during germination. *International Journal of Agriculture and Biology*, *18*(2), 449–457.

Alsaeedi, A., El-Ramady, H., Alshaal, T., El-Garawani, M., Elhawat, N., & Al-Otaibi, A. (2018). Exogenous nanosilica improves germination and growth of cucumber by maintaining K^+/Na^+ ratio under elevated Na^+ stress. *Plant Physiology and Biochemistry, 125*, 164–171.

Amini, S., Maali-Amiri, R., Mohammadi, R., & Kazemi-Shahandashti, S. S. (2017). cDNA AFLP analysis of transcripts induced in chickpea plants by TiO_2 nanoparticles during cold stress. *Plant Physiology and Biochemistry*, *111*, 39–49.

Ashkavand, P., Tabari, M., Zarafshar, M., Tomášková, I., & Struve, D. (2015). Effect of SiO_2 nanoparticles on drought resistance in hawthorn seedlings. *Forest Research Papers, 76*(4), 350–359.

Azimi, R., Borzelabad, M. J., Feizi, H., & Azimi, A. (2014). Interaction of SiO_2 nanoparticles with seed prechilling on germination and early seedling growth of tall wheatgrass (*Agropyron elongatum* L.). *Polish Journal of Chemical Technology, 16*(3), 25–29.

Badawy, A. A., Abdelfattah, N. A. H., Salem, S. S., Awad, M. F., & Fouda, A. (2021). Efficacy assessment of biosynthesized copper oxide nanoparticles (CuO-NPs) on stored grain insects and their impacts on morphological and physiological traits of wheat (*Triticum aestivum* L.) plant. *Biology*, *10*(3), 233.

Chen, J., Wu, L., Lu, M., Lu, S., Li, Z., & Ding, W. (2020). Comparative study on the fungicidal activity of metallic MgO nanoparticles and macroscale MgO against soilborne fungal phytopathogens. *Frontiers in Microbiology*, *11*, 365.

Chhipa, H. (2017). Nanofertilizers and nanopesticides for agriculture. *Environmental Chemistry Letters*, *15*, 15–22.

Chun, S. C., & Chandrasekaran, M. (2019). Chitosan and chitosan nanoparticles induced expression of pathogenesis-related proteins genes enhances biotic stress tolerance in tomato. *International Journal of Biological Macromolecules*, *125*, 948–954.

Cinisli, K. T., Uçar, S. M., & Dikba, S. N. (2019). Use of nanomaterials in agriculture. *Yuzuncu Yil University Journal of Agriculture and Science*, *29*, 817–831.

Delfani, M., Firouzabadi, M. B., Farrokhi, N., & Makarian, H. (2014). Some physiological responses of black-eyed pea to iron and magnesium nanofertilizers. *Communications in Soil Science Plant Analysis*, *45*, 530–540.

Dimkpa, C. O., Singh, U., Bindraban, P. S., Elmer, W. H., Gardea-Torresdey, J. L., & White, J. C. (2019). Zinc oxide nanoparticles alleviate drought-induced alterations in sorghum performance, nutrient acquisition, and grain fortification. *Science Total Environment*, *688*, 926–934.

El-Saadony, M. T., Saad, A. M., Najjar, A. A., Alzahrani, S. O., Alkhatib, F. M., Shafi, M. E., Selem, E., Desoky, E. S. M., Fouda, S. E. E., El-Tahan, A. M., & Hassan, M. A. A. (2021). The use of biological selenium nanoparticles to suppress *Triticum aestivum* L. crown and root rot diseases induced by *Fusarium* species and improve yield under drought and heat stress. *Saudi Journal of Biological Sciences, 28*(8), 4461–4471.

El-Saadony, M. T., Saad, A. M., Soliman, S. M., Salem, H. M., Desoky, E. S. M., Babalghith, A. O., El-Tahan, A. M., Ibrahim, O. M., Ebrahim, A. M. M., El-Mageed, T. A. A., Elrys, A. S., Elbadawi, A. A., El-Tarabily, K. A., & AbuQamar, S. F. (2022). Role of nanoparticles

in enhancing crop tolerance to abiotic stress: A comprehensive review. *Frontiers in Plant Science, 13*, 946717.
Elsharkawy, M. M., Omara, R. I., Mostafa, Y. S., Alamri, S. A., Hashem, M., Alrumman, S. A., & Ahmad, A. A. (2022). Mechanism of wheat leaf rust control using chitosan nanoparticles and salicylic acid. *Journal of Fungi, 8*(3), 304.
Faizan, M., Bhat, J. A., Chen, C., Alyemeni, M. N., Wijaya, L., Ahmad, P., & Yu, F. (2021). Zinc oxide nanoparticles (ZnO-NPs) induce salt tolerance by improving the antioxidant system and photosynthetic machinery in tomato. *Plant Physiology and Biochemistry, 161*, 122–130.
Faraji, J., & Sepehri, A. (2020). Exogenous nitric oxide improves the protective effects of TiO_2 nanoparticles on growth, antioxidant system, and photosynthetic performance of wheat seedlings under drought stress. *Journal of Soil Science Plant Nutrition, 20*(2), 703–714.
Farhana, Munis, M. F. H., Alamer, K. H., Althobaiti, A. T., Kamal, A., Liaquat, F., Haroon, U., Ahmed, J., Chaudhary, H. J., & Attia, H. (2022). ZnO Nanoparticle-mediated seed priming induces biochemical and antioxidant changes in chickpea to alleviate fusarium wilt. *Journal of Fungi, 8*(7), 753.
Farhangi-Abriz, S., & Torabian, S. (2018). Nano-silicon alters antioxidant activities of soybean seedlings under salt toxicity. *Protoplasma, 255*(3), 953–962.
Farid, M., Shakoor, M. B., Ehsan, S., Ali, S., Zubair, M., & Hanif, M. A. (2013). Morphological, physiological and biochemical responses of different plant species to Cd stress. *International Journal of Chemical and Biochemical Science, 3*(1), 53–60.
Farooq, M., Hussain, M., Wahid, A., & Siddique, K. H. (2010). Drought stress in plants: an overview. In R. Aroca (Ed.), *Plant responses to drought stress* (pp. 1–33). Springer Publication.
Fathi, A., Zahedi, M., Torabian, S., & Khoshgoftar, A. (2017). Response of wheat genotypes to foliar spray of ZnO and Fe_2O_3 nanoparticles under salt stress. *Journal of Plant Nutrition, 40*(10), 1376–1385.
Gao, X., Zou, C., Wang, L., & Zhang, F. (2006). Silicon decreases transpiration rate and conductance from stomata of maize plants. *Journal of Plant Nutrition, 29*(9), 1637–1647.
Hall, J. L. (2002). Cellular mechanisms for heavy metal detoxification and tolerance. *Journal of Experimental Botany, 53*(366), 1–11.
Hasanpour, H., Maali-Amir, R., & Zeinali, H. (2015). Effect of TiO_2 nanoparticleson metabolic limitations to photosynthesis under cold in chickpea. *Russian Journal of Plant Physiology, 62*, 779–787.
Hernandez, J. A., Jimenez, A., Mullineaux, P., & Seviela, F. (2000). Tolerance of pea (*Pisum sativum* L.) to long-term salt stress is associated to induction of antioxidant defences. *Plant, Cell and Environment, 23*(8), 853–862.
Hojjat, S. S., & Ganjali, A. (2016). The effect of silver nanoparticle on lentil seed germination under drought stress. *International Journal of Farming and Allied Sciences, 5*, 208–212.
Hossain, M. A., Piyatida, P., da Silva, J. A. T., & Fujita, M. (2012). Molecular mechanism of heavy metal toxicity and tolerance in plants: central role of glutathione in detoxification of reactive oxygen species and methylglyoxal and in heavy metal chelation. *Journal of Botany*, 872875.
Iqbal, M., Raja, N. I., Mashwani, Z. U. R., Hussain, M., Muhammad, E., & Yasmeen, F. (2019). Effect of silver nanoparticles on growth of wheat under heat stress. *Iranian Journal of Science and Technology. Transaction A, Science, 43*(2), 387–395.

Jaberzadeh, A., Moaveni, P., Tohidi Moghadam, H. R., & Zahedi, H. (2013). Influence of bulk and nanoparticles titanium foliar application on some agronomic traits, seed gluten and starch contents of wheat subjected to water deficit stress. *Notulae Botanicae Horti Agrobotanici Cluj-Napoca, 41*(1), 201–207.

Kalteh, M., Alipour, Z. T., Shahram, A., Aliabadi, M. M., & Nosratabadi, A. F. (2014). Effect of silica nanoparticles on basil (*Ocimum basilicum*) under salinity stress. *Journal of Chemical Health Risks*, *4*(3), 49–55.

Katiyar, P., Yadu, B., Korram, J., Satnami, M. L., Kumar, M., & Keshavkant, S. (2020). Titanium nanoparticles attenuates arsenic toxicity by up-regulating expressions of defensive genes in *Vigna radiata* L. *Journal of Environmental Sciences*, *92*, 18–27.

Kiapour, H., Moaveni, P., Habibi, D., & Sani, B. (2015). Evaluation of the application of gibbrellic acid and titanium dioxide nanoparticles under drought stress on some traits of basil (*Ocimum basilicum* L.). *International Journal of Agronomy and Agricultural Research*, *6*(4), 138–150.

Kim, J. H., Oh, Y., Yoon, H., Hwang, I., & Chang, Y.-S. (2015). Iron nanoparticle-induced activation of plasma membrane H^+-ATPase promotes stomatal opening in *Arabidopsis thaliana*. *Environmental Science and Technology*, *49*(2), 1113–1119.

Kohan-Baghkheirati, E., & Geisler-Lee, J. (2015). Gene expression, protein function and pathways of Arabidopsis thaliana responding to silver nanoparticles in comparison to silver ions, cold, salt, drought, and heat. *Nanomaterials (Basel)*, *5*(2), 436–467.

Konate, A., He, X., Zhang, Z., Ma, Y., Zhang, P., Alugongo, G. M., & Rui, Y. (2017). Magnetic (Fe_3O_4) nanoparticles reduce heavy metals uptake and mitigate their toxicity in wheat seedling. *Sustainability*, *9*(5), 790.

Lei, Z., Mingyu, S., Chao, L., Liang, C., Hao, H., Xiao, W., Xiaoqing, L., Fan, Y., Fengqing, G., & Fashui, H. (2007). Effects of nanoanatase TiO_2 on photosynthesis of spinach chloroplasts under different light illumination. *Biological Trace Element Research*, *119*(1), 68–76.

Li, Z., & Huang, J. (2014). Effects of nanoparticle hydroxyapatite on growth and antioxidant system in pakchoi (*Brassica chinensis* L.) from cadmium-contaminated soil. *Journal of Nanomaterials*, *2014*(470962), 7.

Li, Y., Zhu, N., Liang, X., Bai, X., Zheng, L., Zhao, J., Li, Y.-F., Zhang, Z., & Gao, Y. (2020). Silica nanoparticles alleviate mercury toxicity via immobilization and inactivation of Hg(II) in soybean (*Glycine max*). *Environmental Science: Nano*, *7*(6), 1807–1817.

Liang, Y., Sun, W., Zhu, Y.-G., & Christie, P. (2007). Mechanisms of silicon mediated alleviation of abiotic stresses in higher plants: a review. *Environmental Pollution*, *147*(2), 422–428.

Lu, L., Huang, M., Huang, Y., Corvini, P. F.-X., Ji, R., & Zhao, L. (2020). Mn_3O_4 nanozymes boost endogenous antioxidant metabolites in cucumber (*Cucumis sativus*) plant and enhance resistance to salinity stress. *Environmental Science: Nano*, *7*(6), 1692–1703.

Manzoor, N., Ahmed, T., Noman, M., Shahid, M., Nazir, M. M., Ali, L., Alnusaire, T. S., Li, B., Schulin, R., & Wang, G. (2021). Iron oxide nanoparticles ameliorated the cadmium and salinity stresses in wheat plants, facilitating photosynthetic pigments and restricting cadmium uptake. *Science of the Total Environment*, *769*, 145221.

Maswada, H. F., Djanaguiraman, M., & Prasad, P. V. V. (2018). Seed treatment with nano-iron (III) oxide enhances germination, seeding growth and salinity tolerance of sorghum. *Journal of Agronomy and Crop Science*, *204*(6), 577–587.

Mohamed, A. K. S. H., Qayyum, M. F., Abdel-Hadi, A. M, Rehman, R. A., Ali, S., & Rizwan, M. (2017). Interactive effect of salinity and silver nanoparticles on photosynthetic and biochemical parameters of wheat. *Archives of Agronomy and Soil Science*, *63*, 1736–1747.

Mohammadi, R., Maali-Amiri, R., & Abbasi, A. (2013). Effect of TiO_2 nanoparticles on chickpea response to cold stress. *Biological Trace Element Research*, *152*, 403–410.

Mohammadi, R., Maali-Amiri, R., & Mantri, N. L. (2014). Effect of TiO_2 nanoparticles on oxidative damage and antioxidant defense systems in chickpea seedlings during cold stress. *Russian Journal of Plant Physiol*ogy, *61*, 768–775.

Mohammadi, H., Esmailpour, M., & Gheranpaye, A. (2016). Effects of TiO_2 nanoparticles and water-deficit stress on morpho-physiological characteristics of dragonhead (*Dracocephalum moldavica* L.) plants. *Acta Agriculturae Slovenica*, *107*(2), 385–396.

Moradbeygi, H., Jamei, R., Heidari, R., & Darvishzadeh, R. (2020). Investigating the enzymatic and non-enzymatic antioxidant defense by applying iron oxide nanoparticles in *Dracocephalum moldavica* L. plant under salinity stress. *Scientia Horticulturae*, *272*(1), 109537.

Mozafari, A.-A., Asl, A. G., & Ghaderi, N. (2018). Grape response to salinity stress and role of iron nanoparticle and potassium silicate to mitigate salt induced damage under *in vitro* conditions. *Physiology and Molecular Biology of Plants*, *24*(1), 25–35.

Mustafa, G., & Komatsu, S. (2016). Insights into the response of soybean mitochondrial proteins to various sizes of aluminum oxide nanoparticles under flooding stress. *Journal of Proteome Research*, *15*(12), 4464–4475.

Mustafa, G., Sakata, K., Hossain, Z., & Komatsu, S. (2015). Proteomic study on the effects of silver nanoparticles on soybean under flooding stress. *Journal of Proteomics*, 122, 100–118.

Noohpisheh, Z., Amiri, H., Mohammadi, A., & Farhadi, S. (2021). Effect of the foliar application of zinc oxide nanoparticles on some biochemical and physiological parameters of Trigonella foenum-graecum under salinity stress. *Plant Biosystems - An International Journal Dealing with all Aspects of Plant Biology*, *155*(2), 267–280.

Normila, D. (2008). Reinventing rice to feed the world. *Science*, 321, 330–333.

Qados, A. M. A. (2015). Mechanism of nanosilicon-mediated alleviation of salinity stress in faba bean (*Vicia faba* L.) plants. *American Journal of Experimental Agriculture*, *7*(2), 78–95.

Qi, M., Liu, Y., & Li, T. (2013). Nano-TiO_2 improve the photosynthesis of tomato leaves under mild heat stress. *Biological Trace Element Research*, *156*(1–3), 323–328.

Rady, M. M., Boriek, S. H. K., Abd El-Mageed, T. A., Seif El-Yazal, M. A., Ali, E. F., Hassan, F. A. S., & Abdelkhalik, A. (2021). Exogenous gibberellic acid ordilute bee honey boosts drought stress tolerance in *Vicia faba* by rebalancing osmoprotectants, antioxidants, nutrients, and phytohormones. *Plants*, *10*(4), 748.

Rajkumar, V., Gunasekaran, C., Paul, C. A., & Dharmaraj, J. (2020). Development of encapsulated peppermint essential oil in chitosan nanoparticles: Characterization and biological efficacy against stored-grain pest control. *Pesticide Biochemistry and Physiology*, *170*, 104679.

Raliya, R., Tarafdar, J. C., Chaudhary, K., Mal, P., Raturi, A., & Singh, S. K. (2014). Synthesis of MgO nanoparticles using *Aspergillus tubingensis* TFR-3. *Journal of Bionanoscience*, *8*(1–5), 34–38.

Rezvani, N., Sorooshzadeh, A., & Farhadi, N. (2012). Effect of nano-silver on growth of saffron in flooding stress. *World Academy of Science, Engineering and Technology*, *6*(1), 517–522.

Rizwan, M., Ali, S., Qayyum, M. F., Ok, Y. S., Adrees, M., Ibrahim, M., Zia-ur-Rehman, M., Farid, M., & Abbas, F. (2017). Effect of metal and metal oxide nanoparticles on growth and physiology of globally important food crops: A critical review. *Journal of Hazardous Materials*, *322*, 2–16.

Rossi, L., Zhang, W., Lombardini, L., & Ma, X. (2016). The impact of cerium oxide nanoparticles on the salt stress responses of *Brassica napus* L. *Environmental Pollution*, *219*, 28–36.

Ruelland, E., & Zachowski, A. (2010). How plants sense temperature. *Environmental and Experimental Botany*, *69*(3), 225–232.

Sabaghnia, N., & Janmohammadi, M. (2014). Effect of nano-silicon particles application on salinity tolerance in early growth of some lentil genotypes. *Annales UMCS Biologia*, *69*(2), 39–55.

Satti, S. H., Raja, N. I., Ikram, M., Oraby, H. F., Mashwani, Z.-U., Mohamed, A. H., Singh, A., & Omar, A. A. (2022). Plant-Based Titanium Dioxide Nanoparticles Trigger Biochemical and Proteome Modifications in *Triticum aestivum* L. under Biotic Stress of *Puccinia striiformis*. *Molecules*, *27*(13), 4274.

Sedghi, M., Mitra, H., & Sahar, T. (2013). Effect of nano zinc oxide on the germination of soybean seeds under drought stress. *Annales of West University of Timisoara. Series of Biology*, *16*(2), 73–78.

Shallan, M. A., Hassan, H. M. M., Namich, A. A. M., & Ibrahim, A. A. (2016). Biochemical and physiological effects of TiO_2 and SiO_2 nanoparticles on cotton plant under drought stress. *Research Journal of Pharmaceutical, Biological and Chemical Sciences*, *7*(4), 1540–1551.

Siddiqui, M. H., Al-Whaibi, M. H., Faisal, M., & Al-Sahli, A. A. (2014). Nano silicon dioxide mitigates the adverse effects of salt stress on *Cucurbita pepo* L. *Environmental Toxicology and Chemistry*, *33*(11), 2429–2437.

Sun, L., Song, F., Guo, J., Zhu, X., Liu, S., Liu, F., & Li, X. (2020). Nano-ZnO-induced drought tolerance is associated with melatonin synthesis and metabolism in maize. *International Journal of Molecular Science*, *21*(3), 782.

Syu, Y. Y., Hung, J. H., Chen, J. C., & Chuang, H. W. (2014). Impacts of size and shape of silver nanoparticles on Arabidopsis plant growth and gene expression. *Plant Physiology and Biochemistry*, *83*, 57–64.

Taha, R. S., Seleiman, M. F., Shami, A., Alhammad, B. A., & Mahdi, A. H. A. (2021). Integrated application of selenium and silicon enhances growth and anatomical structure, antioxidant defense system and yield of wheat grown in salt-stressed soil. *Plants*, *10*(6), 1040.

Tarafdar, J. C. (2021). *Nanofertilizers: Challenges and prospects*. Scientific publisher, India.

Tarafdar, J. C. (2022). Nanotechnology application in agriculture: current developments of nanofertilizers. In P. Singh, Anam, T. K. Srivastava, & R. R. Verma (Eds.), *Nanoparticles application in agriculture, chapter 2* (pp. 19–29). Scientific Publishers.

Tarafdar, J. C., & Rathore, I. (2018). Nano-nutrient from fungal protein: future prospects on crop production. In H. B. Singh, S. Mishra, L. F. Fraceto, & R. D. Lima, (Eds.), *Emerging trends in agri-nanotechnology, fundamental and applied aspects* (pp. 37–55). CAB International, U. K.

Tarafdar, J. C., Raliya, R., & Praveen-Kumar. (2018). Nanoinduced bacterial polysaccharide production. Indian Patent No. 304253.

Tarafdar, J. C., Rathore, I., & Thomas, E. (2015). Enhancing nutrient use efficiency through nanotechnological interventions. *Indian Journal of Fertilizers*, *11*(12), 46–51.
Tarafdar, J. C., Xiang, Y., Wang, W. N., Dong, Q., & Biswas, P. (2012a). Standardization of size, shape and concentration of nanoparticle for plant application. *Applied Biological Research*, *14*(2), 138–144.
Tarafdar, J. C., Agrawal, A., Raliya, R., Kumar, P., Burman, U., & Kaul, R. K. (2012b). ZnO nanoparticles induced synthesis of polysaccharides and phosphatases by *Aspergillus* fungi. *Advanced Science Engineering and Medicine, 4*(4), 1–5.
Tripathi, D. K., Singh, S., Singh, V. P., Prasad, S. M., Dubey, N. K., & Chauhan, D. K. (2017). Silicon nanoparticles more effectively alleviated UV-B stress than silicon in wheat (*Triticum aestivum*) seedlings. *Plant Physiology and Biochemistry*, *110*, 70–81.
Tunca, E. (2015). Nanoparticles as the base of nanotechnology and phytoremediation of nanoparticles. *Ordu University Journal of Science Technology*, *5*, 23–34.
Turgeon, R. (2010). The puzzle of phloem pressure. *Plant Physiology*, *154*(2), 578–581.
Usman, M., Farooq, M., Wakeel, A., Nawaz, A., Cheema, S. A., ur Rehman, H., Ashraf, I., & Sanaullah, M. (2020). Nanotechnology in agriculture: current status, challenges and future opportunities. *The Science of The Total Environment*, *721*, 137778.
Wu, H., Shabala, L., Shabala, S., Shabala, S., & Giraldo, J. P. (2018). Hydroxyl radical scavenging by cerium oxide nanoparticles improves *Arabidopsis* salinity tolerance by enhancing leaf mesophyll potassium retention. *Environmental Science: Nano*, *5*(7), 1567–1583.
Yan, S., Wu, F., Zhou, S., Yang, J., Tang, X., & Ye, W. (2021). Zinc oxide nanoparticles alleviate the arsenic toxicity and decrease them accumulation of arsenic in rice (*Oryza sativa* L.). *BMC Plant Biology*, *21*(150).
Ye, Y., Cota-Ruiz, K., Hernandez-Viezcas, J. A., Valdes, C., Medina-Velo, I. A., Turley, R. S., Peralta-Videa, J. R., & Gardea-Torresdey, J. L. (2020). Manganese nanoparticles control salinity-modulated molecular responses in Capsicum annuum L. through priming: A sustainable approach for agriculture. ACS Sustainable Chemistry & Engineering, *8*(3), 1427–1436.

CHAPTER 10

Biogenic Synthesis of Liquid Carrier-Based Nanonutrient Formulations

AYON TARAFDAR

Livestock Production and Management Section, ICAR-Indian Veterinary Research Institute, Bareilly, Uttar Pradesh, India

ABSTRACT

This chapter explores the emerging field of biogenic synthesis for the preparation of liquid carrier-based nanonutrient formulations. Biogenic synthesis, utilizing natural sources like plants, microbes, and biopolymers, offers a sustainable and eco-friendly approach for producing nanoparticles. This chapter highlights the significance of nanotechnology in agriculture and nutrition and provides an in-depth analysis of the synthesis techniques, characterization methods, and potential applications of liquid carrier-based nanonutrient formulations. Furthermore, it discusses the challenges and future prospects of this innovative technology in enhancing crop productivity and human health.

10.1 INTRODUCTION

Nanotechnology has made considerable strides in recent years, revolutionizing a number of industries, including agriculture. Nano-agri-input products, which use nanomaterials for agricultural applications, have a great deal of promise to boost crop output, lessen environmental effects, and encourage sustainable farming methods. Among such inputs, the role of liquid-carrier-based nanonutrients is immense. Adequate and balanced nutrition is essential for the optimal growth and development of plants and humans alike. In the agricultural sector, nutrient deficiencies in crops can lead to reduced yields,

Nanobiotechnology for Agricultural Sciences: Nano-Agri-Input Products for Crop Production and Environmental Protection. J. C. Tarafdar, Deepak Kumar, Sanjay K. Singh, and Krishna Pal Singh (Eds.)

decreased quality, and increased susceptibility to pests and diseases. Likewise, human health is heavily dependent on the intake of essential nutrients, and deficiencies can lead to various health disorders. To address these challenges, novel and efficient delivery systems for nutrients are required, and nanotechnology has emerged as a promising approach to meeting these demands. Liquid carrier-based nanonutrient formulations have gained significant attention due to their potential to enhance nutrient delivery and bioavailability. This scientific exploration delves into the reasons why such formulations are needed and how they can revolutionize both agriculture and human nutrition.

In traditional agricultural practises, nutrients are typically supplied to crops through soil amendments or foliar applications. However, conventional methods have several limitations. Nutrient losses due to leaching and volatilization are common, leading to inefficiencies in nutrient utilization. Additionally, poor nutrient uptake by plants limits the full potential of nutrient supplementation. Similarly, in the context of human nutrition, conventional supplementation methods such as oral pills may suffer from low bioavailability and poor absorption, hindering their effectiveness. Nanotechnology offers unique advantages in addressing the challenges of nutrient delivery. By manipulating matter at the nanoscale, nanoparticles can be engineered to have distinct properties, including enhanced stability, controlled release, and improved bioavailability. Liquid carrier-based nanonutrient formulations utilize biocompatible carriers, such as liposomes, emulsions, micelles, and nanogels, to encapsulate nutrients. These carriers protect the nutrients from degradation, facilitate controlled release, and enable targeted delivery to specific plant tissues or human cells.

Enhancing nutrient absorption and utilization is one of the main advantages of liquid carrier-based nanonutrient formulations. In agriculture, nanoparticles can get through plant barriers such as cell walls and membranes to help plants absorb nutrients more efficiently. This results in increased nutrient absorption by plant tissues, which enhances crop output and nutritional quality. Similar to this, nanonutrient formulations improve the absorption of crucial vitamins, minerals, and phytonutrients in human nutrition, resulting in more successful supplementing methods. The adoption of liquid carrier-based nanonutrient formulations in agriculture can significantly reduce the environmental impact of nutrient application. By increasing nutrient uptake efficiency, less fertilizer is required to achieve the same results, leading to decreased nutrient runoff and minimized pollution of water bodies. This can contribute to sustainable agricultural practises and reduce the adverse effects of nutrient leaching on ecosystems. Nanonutrient formulations offer the potential for precision agriculture, where nutrients can be delivered with spatial and temporal precision based on specific crop requirements. This approach optimizes resource utilization and reduces waste. Similarly, in personalized nutrition, nanonutrient formulations can be tailored to individual nutritional needs, considering factors such as age, gender, health conditions, and genetic predispositions, leading to more effective and personalized supplementation.

This chapter discusses the different methods of nanoparticle synthesis for application in nanoformulations, along with the factors affecting the development of such formulations and their environmental and regulatory aspects. This chapter is intended to provide the readers with an overview of the biogenic synthesis strategies of nanoparticles, which can later be implemented in research on nanoformulations.

10.2 MICROBIAL SYNTHESIS OF METALLIC NANONUTRIENTS

The creation of metallic nanoparticles (NPs) in an aqueous metal salt (precursor) solution is made possible by reducing agents that come from biological sources (Salem and Fouda, 2020). Microorganisms, plants, and algae are examples of biological sources that are efficient, accessible, and eco-friendly (Ahmad et al., 2019). Because it is straightforward, affordable, and uses less energy-intensive machinery to physically convert bulk materials to nanomaterials, the biological technique is preferred for the synthesis of NPs (Ahmad et al., 2019). Figure 10.1 shows the mechanism of synthesis of metal nanoparticles using microorganisms, while the advantages and disadvantages of different biological synthesis methods have been described in Table 10.1. The composition, shape, and size distribution of NPs determine their usefulness, stability, and physicochemical characteristics. From a thermodynamic perspective, the Van der Waals interaction potential rises with a drop in NP size. As a measure of the stability of the surface chemistry of NPs, changes in the surface potential of NPs in solution may be calculated using the zeta potential. However, the kind of solvent employed and the solution's pH can significantly influence these findings (Phan and Haes, 2019). Nanoparticles are thought to be stable at a zeta potential higher than +30 mV or lower than –30 mV. According to Singh et al. (2016), high yields of homogeneous, stable NPs may be made by biological synthesis by modifying factors including pH, temperature, mixing rate, incubation time, concentration of the metal precursor, and biological source.

Due to their metabolic capacity to endure high metal concentrations through the reduction of metal ions, metal-resistant microorganisms such as bacteria, fungi, and algae are suited for metallic NP production (Salunke et al., 2016). Microorganisms generate a variety of reductase enzymes, including nitrate reductase and NADH-dependent reductases, which function as reducing agents for the production of metallic NP (Singh et al., 2016). Nevertheless, research has demonstrated that other biomolecules, including sugars, cofactors, and secondary metabolites, may also convert metal ions from dead microbial cells to NPs (Binupriya et al., 2010). According to Salem and Fouda (2020), microorganisms are thought of as nanofactories that produce metallic NPs either intracellularly or extracellularly.

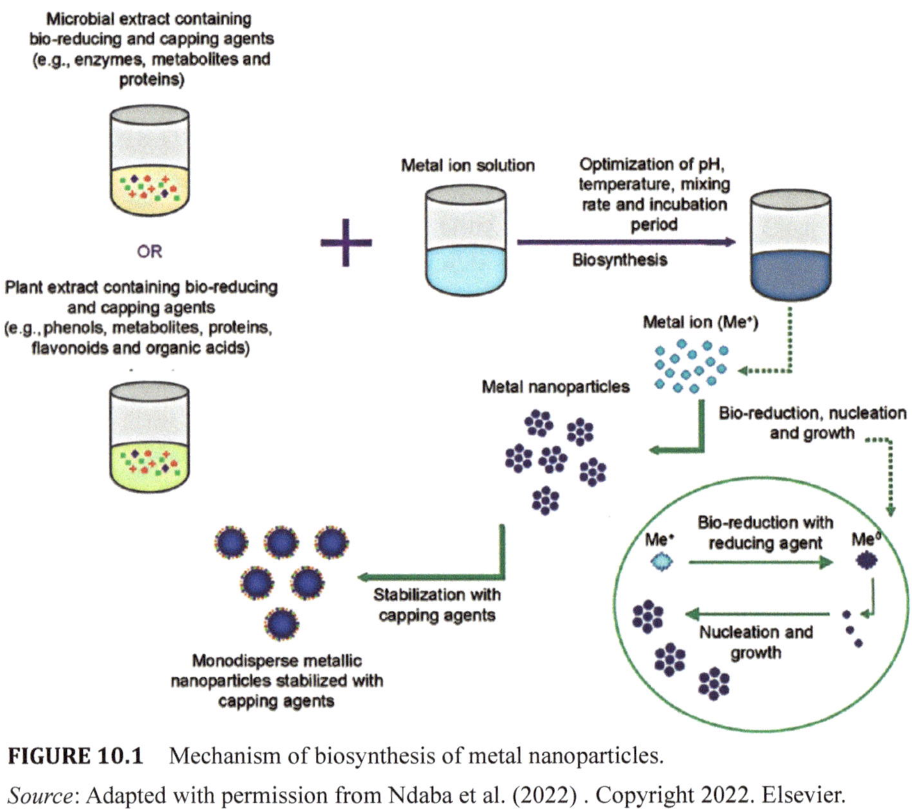

FIGURE 10.1 Mechanism of biosynthesis of metal nanoparticles.

Source: Adapted with permission from Ndaba et al. (2022) . Copyright 2022. Elsevier.

In order to produce extracellular synthesis from a cell-free extract, microorganisms must first be cultured before being extracted by centrifugation. The obtained supernatant is then added to a metal salt solution. For the purpose of producing metallic NPs, the reaction mixture is further incubated (Singh et al., 2016). In contrast, the pelleted biomass is cleaned and suspended in a metal salt solution during intracellular synthesis. Metal ions are carried into the microbial cells during incubation, when they are converted to NPs (Salunke et al., 2016). To dismantle microbial cell walls and remove accumulated NPs, the incubated mixture is repeatedly exposed to ultrasonication. Because extracellular synthesis requires fewer downstream ultrasonication stages than intracellular synthesis does, it is more cost-effective (Singh et al., 2016). This section will cover the different biogenic synthesis methods of NPs and derived nanoformulations.

TABLE 10.1 Pros and Cons of Biosynthesis Methods.

Method	Pros	Cons	Reference
Bacterial-mediated synthesis	Occur in high abundance, grow quickly, and are easily accessible. Have unique metal-binding proteins	Lower metabolic capability and bio-reducing agents produced as unicellular microorganisms	Ovais et al. (2018); Mukherjee and Nethi (2019)
Fungal-mediated synthesis	More effective for large-scale production of NPs due to higher metabolic capability and amounts of bio-reducing agents produced as multicellular microorganisms. High-yield synthesis of monodisperse NPs	Require more time to grow	Jeevanandam et al. (2016); Ovais et al. (2018)
Algal-mediated synthesis	Occur in abundance and are easily culturable	Limited studies on algal-mediated synthesis	Ali et al. (2020)
Yeast-mediated synthesis	Low-cost and easily cultivable with lower environmental impact	Limited scalability and complex process for purification	Koul et al. (2021)
Virus-mediated synthesis	High degree of specificity and self-assembly capabilities, lower agglomeration of resulting particles	Problems with large-scale application, safety issues, requirement of a host for protein expression and under-defined methods for synthesis	Koul et al. (2021)
Plant-mediated synthesis	Easily available, scalable, produce high amounts of bio-reducing agents compared to microorganisms	Produce polydisperse NPs	Rajeshkumar and Bharath (2017); Nasrollahzadeh et al. (2019); Solgi and Taghizadeh (2020)

Source: Adapted from Ndaba et al. (2022) with permission.

10.2.1 BACTERIA-MEDIATED SYNTHESIS

Numerous culturable bacteria exist in great abundance and are readily available (Mukherjee and Nethi, 2019). These bacteria are able to grow quickly and adapt to severe environments. Several studies have demonstrated the bacterial-mediated synthesis of diverse metallic NPs that varied in form, size, and yield by different bacterial groups (Pathak et al., 2019; Ameen et al., 2020). Current suggested processes for bacterial NP synthesis are covered here since the mechanism is still not completely understood. Metal-resistant bacteria possess a variety of distinctive biomolecules that allow metal ions to be transported from the external environment to the cytoplasm of bacterial cells (internally) via metal-binding proteins (Marooufpour et al., 2019; Nasrollahzadeh et al., 2019). Externally, negatively charged polyanionic functional groups that make up the structural elements of bacterial cell walls bind to positively charged metal ions (Ali et al., 2020). Metal binding happens by electrostatic interactions, ion exchange, physical adsorption, or complexation during this process, which is referred to as biosorption (Mukherjee and Nethi, 2019). On cell surfaces or in the cytoplasm, biomolecular reducing or stabilizing agents such as enzymes (reductases) convert metal ions to stable NPs (Ali et al., 2020).

10.2.2 FUNGAL-MEDIATED SYNTHESIS

Fungi are a varied category of multicellular microorganisms that are widely distributed and have better metabolic capacities than bacteria. As a result, large-scale fungal production of metallic NPs is preferred over bacterial synthesis of NPs. Due to the greater quantity of biomolecular reducing agents that this multicellular group of microorganisms produces, fungi may manufacture monodisperse metallic NPs far more effectively and at higher yields than bacteria (Ovais et al., 2018). Metabolites that help in electron transport and are involved in the reduction of metal clusters to NPs by extracellular enzymes include hydroxyquinoline, napthoquinones, anthraquinones, and other quinones (Nasrollahzadeh et al., 2019). Due to their stiff cell wall structure, which is made up of glycoproteins and polysaccharides that permit high metal-binding capacity and intracellular absorption of metal ions, many fungal species are metal-tolerant and have significant bioaccumulation capacities (Nasrollahzadeh et al., 2019). Similar to bacterial production, metallic NP is synthesized through fungal-mediated processes.

According to several studies (Jeevanandam et al., 2016), various species of fungus produce metallic NPs that differ in size and form either extracellularly by biosorption in the mycelia or intracellularly by a variety of intracellular enzymes and complex proteins. Since extracellular synthesis does not need the extraction and purification of NPs from fungal biomass, it is a more favorable method. As contrasted to bacterial synthesis, fungal production of metallic NPs requires a longer time (between 24 and 120 h) for culture reasons (Jeevanandam et al., 2016).

10.2.3 ALGAE-MEDIATED SYNTHESIS

Algae are a kind of multicellular or unicellular eukaryotic photosynthetic organism. Microalgae are smaller groupings of algae, whereas macroalgae are bigger clumps of algae. A new word, phyconanotechnology, has appeared to describe the production of NPs by algae. According to Ali et al. (2020), algae are plentiful, simple to cultivate, and show promise as a source of biological, nontoxic reducing agents. Both live and dead microalgal biomass can be used to make metallic NPs. Alternately, boiling or heating cultures can be used to extract biomolecules from microalgal cells. The extract is then combined with precursor metal salt solutions and incubated to create NPs. By using the bio-reduction pathway, which involves algal biomolecules including pigments, proteins, and secondary metabolites, microalgae may synthesize metallic NPs from metal ions (Ali et al., 2020). Additionally, capping agents such as the amino groups/cysteine groups of proteins and/or the carboxyl, hydroxyl, or sulfate groups of polysaccharides that are naturally formed by algae are used to decrease and stabilize the metallic NPs that are created by algae. According to LewisOscar et al. in 2016, the kind and quantity of algal biomass employed determine the size, shape, and yield of NPs generated. Many different algal groups have recently been demonstrated to synthesize metallic NPs, but studies on the synthesis of NPs by algae are relatively scarce in comparison to other biological methods of NP synthesis. Further research should examine these tiny organisms' capacity to synthesize metallic NPs since they are eco-friendly sources of reducing agents.

10.2.4 YEAST-MEDIATED SYNTHESIS

The yeast strains have greater advantages than bacteria due to the bulk production of NPs, ease of regulating yeasts in lab settings, creation of various enzymes, and quick growth using basic nutrients. Some research has been done to look at the use of yeast in the manufacture of metallic nanoparticles. But one of the main strategies for using biological material was achieved for this goal by using the eukaryotic systems *Candida glabrata* and *S. pombe* (Moghaddam et al., 2015).

Several studies have demonstrated the potential use of NPs produced by yeast. S. pombe's intracellularly synthesized sulfur NPs, which operated at low voltage and had a high forward current value, were used to create a diode made of cadmium. It is presumable that these characteristics can create a flawless artificial diode (Kowshik et al., 2003).

Additionally, yeast strains have been used to create silver and gold nanoparticles. Additionally, it was noted that extracellular production of silver nanoparticles using the silver-tolerant yeast strain MKY3 was successful, with hexagonal silver nanoparticles (2–5 nm) being synthesized at the log stage of development. The standardization and recording of the proper conditions for the synthesis of a large number of silver nanoparticles were also carried out in accordance with the various warming situations. Cell-related gold NPs and nanoplates were produced during the incubation of Yarrowia lipolytica cells with various doses of chloroauric acid. Additionally, it has been discovered that the number of cells and the salt concentrations used can influence NP size (Pimprikar et al., 2009). Similar to this, zirconium phosphate was synthesized using yeast as a biotemplate to create an air electrode with exceptional electrocatalytic activity for oxygen decrease (ORR) (Tian et al., 2010). Therefore, yeasts were used as bio-based templates in the manufacture of zinc phosphate nanopowders. Yan et al. demonstrated the synthesis of $Zn_{3(}PO_4)_2$ powders with butterfly-like microstructures that ranged in size from 10 to 80 nm in width and 80 to 200 nm in length.

10.2.5 VIRUS-MEDIATED SYNTHESIS

It is a novel strategy to generate synthetic nanocrystals like cadmium sulfide, silicon dioxide, ferrous oxide, and zinc sulfide using viruses. Green chemistry is interested in semiconductor nanoparticles like zinc sulfide and cadmium sulfide, and the electronics sector is actively looking for ways to produce them. The creation of nanomaterials using whole viruses has been studied for a while now (Zeng et al., 2013). By creating a highly reactive metallic ion-binding surface, the virus' outer capsid protein aids in the creation of nanoparticles (Pokorski and Steinmetz, 2011). Approximately, 2130 capsid proteins may be found on the tobacco mosaic virus (TMV) surface. For a number of medicinal applications, all peptides may be employed as notched connectors to deposit or create three-dimensional vessels (Royston et al., 2008). The size of the synthesized nanoparticles decreased when Au and Ag salts were added to TMV in small amounts prior to adding *Nicotiana benthamiana* or Hordeum vulgare plant extracts. It increased their amounts relative to people who did not get the viral supplement because, at higher TMV concentrations, they produced proportionally fewer free nanoparticles. For the metallization of nanowires, TMV was also used as a bio-template (Love et al., 2014). In contrast to the absence of a virus, the presence of a pathogen not only shortened the length of biosynthesized NPs but also markedly enhanced their production.

10.3 PLANT-BASED SYNTHESIS OF METALLIC NANONUTRIENT

Plants may swiftly and easily produce stable, biocompatible, and affordable metallic NPs via a method known as phytosynthesis (Solgi and Taghizadeh, 2020). In comparison to microbes, plants are easily accessible biological sources that generate more biomolecular-reducing agents. For the synthesis of metallic NPs, extracts from a variety of plant components, including stems, leaves, roots, flowers, and fruits, have been used (Rajeshkumar and Bharath, 2017; Solgi and Taghizadeh, 2020). Defence mechanisms against hazardous metal concentrations and reduction by secondary metabolites and other phytochemicals are two key bio-reduction processes that have been found in recent years (Solgi and Taghizadeh, 2020). Under circumstances of metal stress, plants bioaccumulate metals by producing chelating substances such as phytochelatins, metallothioneins, and peptides that form complexes with metal ions. Phenolic substances, such as flavonoids, are secondary metabolites that are also created when a metal stressor is present in order to chelate metal ions by creating less toxic, stable complexes with hydroxyl or carbonyl groups for detoxification. Metallic NPs can arise both within and outside of plant cells as a result of the interaction between accumulated metal ions and phytochemicals. Plant material is cleaned, dried, boiled, and then filtered to make plant extract. The plant filtrate is then added to a solution of metal salts, which is then further incubated to produce metallic nanoparticles (Rajeshkumar and Bharath, 2017). According to Marslin et al. in 2018, plant extracts contain a variety of secondary metabolites, proteins, polysaccharides, and other phytochemicals such as ketones, aldehydes, terpenoids, flavones, quinones, and organic acids that function as electron donors to convert metal ions into NPs in aqueous solutions. Due to their high surface energy state, metallic NPs are often extremely unstable and need to aggregate to stabilize. According to Rajeshkumar and Bharath (2017), several of the reducing agents found in plant extracts also function as capping agents, stabilizing metallic NPs as they develop. The shape and size of NPs are influenced by a number of variables, including pH, temperature, incubation time, agitation, metal ion concentration, extract content, and composition. The formation of polydisperse NPs as a result of extract composition change is a significant impediment to phytosynthesis by plant extract (Nasrollahzadeh et al., 2019). Recent investigations have employed a variety of plants to synthesize metallic NPs. The antimicrobial qualities of the biosynthesized NPs have, nevertheless, been the focus of the majority of these investigations. Future research should look at the use of the biosynthesized NPs as nanofertilizers in agriculture.

10.4 FACTORS AFFECTING THE SYNTHESIS OF NANOFORMULATIONS

When employing nanoparticles for biomedical purposes, toxicity is a crucial consideration to make sure they are both safe and effective (Fadeel and Garcia-Bennett, 2010). Metal nanoparticles have been demonstrated in several studies to be harmful to humans. Size and surface load both influence how hazardous metal nanoparticles are. Recognizing the possible health risks associated with nanoparticles is essential. In order to rule out potential genotoxicity, it is crucial to look at minute cellular changes in DNA damage and oxidative stress in human tissues. When using nanoparticles for implants, chemical inertness is a crucial factor as well. Iron, zinc, and titanium oxides are frequently used, and they may all be created and modified using carbon derivatives to enable them to attach to antibodies, medications, and ligands for attention (Amiri et al., 2019). Noncovalent interactions between hydroxyl groups and surface metal ions or ligands may help nanoparticles operate (Arranz-Mascarós et al., 2020). The stability of nanoparticles is closely related to the heat needed for their creation (Yu and Xie, 2012). The morphological characteristics of nanoparticles may be controlled by a variety of variables, including reaction duration, reactant concentrations, pH, and temperature. These traits are essential for comprehending how environmental factors affect NP synthesis since they may be used to optimize the production of metallic NP using biological techniques (Roy and Bharadvaja, 2019). But according to earlier studies, the more extract volume utilized, the quicker the rate of synthesis since more chemical components are available in the solution to interact with the precursor and hasten bio-reduction and nanoparticle stabilization. To achieve the optimal conditions for the green manufacturing of nanoparticles, the volume of green extract and concentration of nanoparticle precursor employed must correlate. The amount of extract utilized during the manufacturing process substantially influences the yield of nanoparticles. The results demonstrate that the amount and kind of extract used during the production of nanoparticles significantly affect both their morphological characteristics and biological activity.

10.5 ENVIRONMENTAL CONCERNS AND REGULATION

Nano-agricultural inputs have a number of advantages, but there are also some possible dangers to the environment and human health that need to be considered. Concerns include the dispersion of nanoparticles in the environment, their buildup in soils and water bodies, and their possible effects on creatures other than their intended targets. It is crucial to create suitable regulatory frameworks in order to guarantee the safe and responsible use of nano-agri-input products. To analyze the possible risks related to nanomaterials, robust risk assessment processes must be devised. To lessen any possible negative consequences, efforts should also be made to improve openness, educate stakeholders, and encourage the adoption of best practises. Through the use of these nanoscale materials, the use of nanotechnological developments in the purification of water and wastewater has been established, and it is related to various hazardous consequences for human health and environmental deterioration. Reactive oxygen species (ROS) and free radicals are produced as a result of the high biological and chemical reactivity of nanoparticles, which leads to damage to biological membranes and DNA. NPs are so tiny that they may enter the human body by inhalation or ingestion and then transfer to other organelles.

10.6 CONCLUSIONS AND PERSPECTIVES

Biogenic synthesis of liquid carrier-based nanonutrient formulations represents a promising avenue in agriculture and nutrition. This chapter has emphasized the significance of utilizing natural sources for nanoparticle synthesis and discussed various approaches, such as plant-mediated, microbial, and biopolymer-assisted methods. Furthermore, the characterization techniques, potential applications, and safety considerations of these formulations have been elucidated. As research in this field continues to progress, biogenic synthesis is expected to play a crucial role in addressing global challenges related to food security and human health while ensuring environmental sustainability. As a substantial improvement in agricultural practises, nano-agri-input products have the potential to increase crop output, lessen environmental effects, and support sustainable farming systems. Technologies such as nanofertilizers, nanopesticides, and nano-cleanup show promise for enhancing nutrient management, pest management, and environmental remediation.

KEYWORDS

- **Biogenic synthesis**
- **environmental and regulatory concerns**
- **liquid carrier**
- **nanomaterials**
- **nanonutrient formulations**

REFERENCES

Ahmad, F., Ashraf, N., Ashraf, T., Zhou, R. B., & Yin, D. C. (2019). Biological synthesis of metallic nanoparticles (MNPs) by plants and microbes: Their cellular uptake, biocompatibility, and biomedical applications. *Applied Microbiology and Biotechnology*, *103*, 2913–2935.

Ali, M. A., Ahmed, T., Wu, W., Hossain, A., Hafeez, R., Islam Masum, M. M., Wang, Y., An, Q., Sun, G., & Li, B. (2020). Advancements in plant and microbe-based synthesis of metallic nanoparticles and their antimicrobial activity against plant pathogens. *Nanomaterials*, *10*(6), 1146.

Ameen, F., AlYahya, S., Govarthanan, M., ALjahdali, N., Al-Enazi, N., Alsamhary, K., ... & Alharbi, S. A. (2020). Soil bacteria Cupriavidus sp. mediates the extracellular synthesis of antibacterial silver nanoparticles. *Journal of Molecular Structure*, *1202*, 127233.

Amiri, M., Salavati-Niasari, M., & Akbari, A. (2019). Magnetic nanocarriers: Evolution of spinel ferrites for medical applications. *Advances in Colloid and Interface Science*, *265*, 29–44.

Arranz-Mascarós, P., Godino-Salido, M. L., López-Garzón, R., García-Gallarín, C., Chamorro-Mena, I., López-Garzón, F. J., Fernández-García, E., & Gutiérrez-Valero, M. D. (2020). Non-covalent functionalization of graphene to tune its band gap and stabilize metal nanoparticles on its surface. *ACS Omega*, *5*(30), 18849–18861.

Binupriya, A. R., Sathishkumar, M., & Yun, S. I. (2010). Myco-crystallization of silver ions to nanosized particles by live and dead cell filtrates of Aspergillus oryzae var. viridis and its bactericidal activity toward Staphylococcus aureus KCCM 12256. *Industrial & Engineering Chemistry Research*, *49*(2), 852–858.
Fadeel, B., & Garcia-Bennett, A. E. (2010). Better safe than sorry: Understanding the toxicological properties of inorganic nanoparticles manufactured for biomedical applications. *Advanced Drug Delivery Reviews*, *62*(3), 362–374.
Jeevanandam, J., Chan, Y. S., & Danquah, M. K. (2016). Biosynthesis of metal and metal oxide nanoparticles. *ChemBioEng Reviews*, *3*(2), 55–67.
Kowshik, M., Ashtaputre, S., Kharrazi, S., Vogel, W., Urban, J., Kulkarni, S. K., & Paknikar, K. M. (2002). Extracellular synthesis of silver nanoparticles by a silver-tolerant yeast strain MKY3. *Nanotechnology*, *14*(1), 95.
Koul, B., Poonia, A. K., Yadav, D., & Jin, J. O. (2021). Microbe-mediated biosynthesis of nanoparticles: Applications and future prospects. *Biomolecules*, *11*(6), 886.
Love, A. J., Makarov, V., Yaminsky, I., Kalinina, N. O., & Taliansky, M. E. (2014). The use of tobacco mosaic virus and cowpea mosaic virus for the production of novel metal nanomaterials. *Virology*, *449*, 133–139.
Marooufpour, N., Alizadeh, M., Hatami, M., & Asgari Lajayer, B. (2019). Biological synthesis of nanoparticles by different groups of bacteria. In *Microbial Nanobionics: Vol 1, State-of-the-Art* (pp 63–85).
Moghaddam, A. B., Namvar, F., Moniri, M., Md. Tahir, P., Azizi, S., & Mohamad, R. (2015). Nanoparticles biosynthesized by fungi and yeast: A review of their preparation, properties, and medical applications. *Molecules*, *20*(9), 16540–16565.
Mukherjee, S., & Nethi, S. K. (2019). Biological synthesis of nanoparticles using bacteria. In *Nanotechnology for Agriculture: Advances for Sustainable Agriculture* (pp 37–51).
Nasrollahzadeh, M., Sajadi, S. M., Issaabadi, Z., & Sajjadi, M. (2019). Biological sources used in green nanotechnology. In *Interface Science and Technology* (Vol. 28, pp. 81–111), Elsevier.
Ndaba, B., Roopnarain, A., Haripriya, R. A. M. A., & Maaza, M. (2022). Biosynthesized metallic nanoparticles as fertilizers: An emerging precision agriculture strategy. *Journal of Integrative Agriculture*, *21*(5), 1225–1242.
Ovais, M., Khalil, A. T., Ayaz, M., Ahmad, I., Nethi, S. K., & Mukherjee, S. (2018). Biosynthesis of metal nanoparticles via microbial enzymes: A mechanistic approach. *International Journal of Molecular Sciences*, *19*(12), 4100.
Pathak, J., Ahmed, H., Singh, D. K., Pandey, A., Singh, S. P., & Sinha, R. P. (2019). Recent developments in green synthesis of metal nanoparticles utilizing cyanobacterial cell factories. *Nanomaterials in Plants, Algae and Microorganisms*, *2*, 237–265.
Pimprikar, P. S., Joshi, S. S., Kumar, A. R., Zinjarde, S. S., & Kulkarni, S. K. (2009). Influence of biomass and gold salt concentration on nanoparticle synthesis by the tropical marine yeast Yarrowia lipolytica NCIM 3589. *Colloids and Surfaces B: Biointerfaces*, *74*(1), 309–316.
Phan, H. T., & Haes, A. J. (2019). What does nanoparticle stability mean? *The Journal of Physical Chemistry C, 123*(27), 16495–16507.
Pokorski, J. K., & Steinmetz, N. F. (2011). The art of engineering viral nanoparticles. *Molecular Pharmaceutics*, *8*(1), 29–43.
Rajeshkumar, S., & Bharath, L. V. (2017). Mechanism of plant-mediated synthesis of silver nanoparticles–A review on biomolecules involved, characterisation and antibacterial activity. *Chemico-Biological Interactions*, *273*, 219–227.

Roy, A., & Bharadvaja, N. (2019). Silver nanoparticle synthesis from Plumbago zeylanica and its dye degradation activity. *Bioinspired, Biomimetic and Nanobiomaterials*, *8*(2), 130–140.
Royston, E., Ghosh, A., Kofinas, P., Harris, M. T., & Culver, J. N. (2008). Self-assembly of virus-structured high surface area nanomaterials and their application as battery electrodes. *Langmuir, 24*(3), 906–912.
Salem, S. S., & Fouda, A. (2021). Green synthesis of metallic nanoparticles and their prospective biotechnological applications: an overview. *Biological Trace Element Research*, *199*, 344–370.
Salunke, B. K., Sawant, S. S., Lee, S. I., & Kim, B. S. (2016). Microorganisms as efficient biosystem for the synthesis of metal nanoparticles: Current scenario and future possibilities. *World Journal of Microbiology and Biotechnology, 32*, 1–16.
Singh, P., Kim, Y. J., Zhang, D., & Yang, D. C. (2016). Biological synthesis of nanoparticles from plants and microorganisms. *Trends in Biotechnology*, *34*(7), 588–599.
Solgi, M., & Taghizadeh, M. (2020). Biogenic synthesis of metal nanoparticles by plants. In *Biogenic Nano-Particles and their Use in Agro-ecosystems* (pp. 593–606). Springer Singapore. http://dx.doi.org/10.1007/978-981-15-2985-6_27.
Tian, X., He, W., Cui, J., Zhang, X., Zhou, W., Yan, S., Sun, X., Han, X., Han, S., & Yue, Y. (2010). Mesoporous zirconium phosphate from yeast biotemplate. *Journal of Colloid Interface Science, 343*(1), 344–249. DOI: 10.1016/j.jcis.2009.11.037. Epub 2009 Nov 23. PMID: 20031146.
Yu, W., & Xie, H. (2012). A review on nanofluids: Preparation, stability mechanisms, and applications. *Journal of Nanomaterials*, *2012*, 1–17.
Zeng, Q., Wen, H., Wen, Q., Chen, X., Wang, Y., Xuan, W., Liang, J., & Wan, S. (2013). Cucumber mosaic virus as drug delivery vehicle for doxorubicin. *Biomaterials*, *34*(19), 4632–4642.

CHAPTER 11

Nanomaterials in Controlling Plant Pests, Pathogens, and Associated Risks

DEEPTI SRIVASTAVA[1], MD. SHAMIM[2], DEEPAK KUMAR[3,4], PRAMILA PANDEY[5], MD ABU NAYYER[1], SABA SIDDIQUI[1], and MOHAMMED HARIS SIDDIQUI[1]

[1]*Department of Agriculture, Integral Institute of Agricultural Science and Technology (IIAST), Integral University, Dasauli, Uttar Pradesh, India*

[2]*Department of Molecular Biology and Genetic Engineering, Dr. Kalam Agricultural College, Kishanganj, Bihar Agricultural University, Sabour, Bhagalpur, Bihar, India*

[3]*Nextnode Bioscience Pvt Ltd., Kadi, Gujarat, India*

[4]*Shri Ram Solvent Extraction Pvt Ltd., Jaspur, Uttarakhand, India*

[5]*Department of Botany, BSNVPG College, Charbagh, Lucknow, India*

ABSTRACT

Worldwide crop yield loss is significantly influenced by insect pests and plant diseases. Their management is tedious and challenging because of the limitations in early diagnosis, broad host range, and soil stability. Therefore, it is necessary to establish an innovative and efficient method of control to prevent the losses caused by insect pests and diseases that affect plants. The basis of present plant disease management strategies, which have the potential to upset the balance of nature, is the use of chemicals. An effective solution to the difficulties in recognizing and managing soil-borne plant disease has been offered by nanotechnology. Presently, use of nanotechnology for the control of diseases and insect pests involves utilizing a range of methods,

Nanobiotechnology for Agricultural Sciences: Nano-Agri-Input Products for Crop Production and Environmental Protection. J. C. Tarafdar, Deepak Kumar, Sanjay K. Singh, and Krishna Pal Singh (Eds.)

including protective nanoparticles, transporters of actives such as pesticides, fertilizers, antimicrobials, and microorganisms, or tactics that enhance plant growth and development. To develop an effective management strategy, plant diseases and insect pests can also be precisely and accurately detected using nanotechnology. Nanoparticles' distinctive physicochemical characteristics enable higher penetration and interaction with biological membranes, improving their effectiveness and release ability. However, agricultural nanotechnology, in particular, is immobile in its formative years, and to reach its occupied apparent, general field experiments, the use of plant-host interaction, and toxicological studies are necessary to address the fundamental questions connected to the creation of commercial nanoformulations.

11.1 INTRODUTION

In recent years, nanotechnology has emerged as a promising field with vast potential for various applications. One such area is the use of nanomaterials in agriculture, particularly in controlling plant pests and pathogens. Nanotechnology offers innovative solutions that can enhance crop protection while minimizing environmental risks and reducing the need for traditional chemical pesticides. This chapter explores the applications of nanomaterials in managing plant pests and pathogens and addresses the associated risks and challenges. Traditional agricultural practices have been altered by the use of nanoscale agrochemicals in agriculture, including nanopesticides, nanofertilizers, nanosensors, and nanoformulations (Aouada and De Moura, 2015; Panpatte et al., 2016; Scott et al., 2018; Pirzadah et al., 2019; Singh et al., 2021a, 2021b). Plant pathogens are a wide set of microorganisms that live in the soil and comprise fungi, bacteria, viruses, and nematodes that cause significant economic damage by damaging the root and collar regions of plants. Since chemical pesticides hardly ever penetrate the soil system and their excessive use disturbs the health of the soil and ecosystem, these infections are challenging to control using traditional management techniques (Ramezani et al., 2019). According to Ivo et al. (2017), using pesticides has some disadvantages, including an upsurge in pesticide resistance, pesticide residues in food, unexpected side effects on beneficial species, disruption of animal feed, and disruption of the ecosystem. A major challenge to establishing food safety, sustainable agriculture, and long-term human health has been the high cost of employing pesticides. Therefore, biological control agents that can eliminate diseases without harming the environment are urgently needed (Adetunji et al., 2017). Recent advances in nanoscale engineering have introduced a new type of particulate bionanotechnology that makes use of biological imitation to better blend adjuvant and antigen. These particles may come from totally synthetic platforms or biologically derived and self-assembling systems, among other sources. Recent studies that used pathogen-like particles as platforms of vaccine delivery have demonstrated great potential against unsolved, high-impact vaccine targets ranging from viral and bacterial pathogens to addiction and cancer by using molecularly engineered targeting and stimulating key immune cells (Rosenthal et al. 2014). Nanoparticles (1–100 nm in diameter) can be used as effective insecticides against harmful weeds, plant diseases, and insect pests, reported by Barik et al. (2008), Gajbhiye et al. (2009), Goswami et al. (2010), and Thabet et al. (2021). These nanoparticles can be used for the developments of new insecticide and insect repellent formulations just as effectively as conventional synthetic pesticides, but without the same risks.

Nanocides, in comparison to traditional hydrophobic pesticides, may be water-soluble (hydrophilic), increasing bioactivity and uniformity of application (Zhang et al., 2008). Nanocides can help target pests acquire resistance more slowly since they can be sprayed in small amounts and are swiftly absorbed by cells (Ayoub et al., 2017). One potential management tactic to reduce the threat of biotic stressors is nanotechnology. Due to their diverse antibacterial capabilities, which include the production of ROS, cell membrane, organelle, and other macromolecule instability and toxicity due to nanoparticles, nanomaterials can be efficiently used to manage soil-borne diseases and insect pests (Dutta et al., 2023). Due to their smaller dimensions and greater surface area-to-volume ratio, nanoparticles are more effectively suited to permeate soil and interact with bacteria that reside there. Nanomaterials can be utilized to create intelligent delivery systems for chemical fertilizers, insecticides, and fungicides for reducing their adverse impacts on the environment. A newly emerging scientific area that can assist in resolving the issues with soil-borne pathogen detection is nanodiagnostics. On the basis of the qualitative and quantitative detection of particular metabolites released by soil-borne diseases, nanotechnology has been used to develop cheap biosensor devices. It has been difficult to manage pests and study plant pathology using nanobiotechnology (Ghormade et al., 2011).

Plant defence mechanisms are activated, and plant development is encouraged, as part of the indirect management strategy for soil-borne diseases. For the successful application of nanomaterials in the control of soil-borne diseases, one must be aware of their ecotoxicity, phytotoxicity, cytotoxicity, genotoxicity, and interactions with the soil system and its inhabitants. The nanomaterials interact with soil particles as soon as they penetrate the soil, altering their physicochemical properties, fertility, and beneficial organisms. Analyzing the effects of nanoparticles on soil structure, soil function, organic matter, siderophore production, nitrogen fixation, phosphate and potassium solubilization, and related processes is important to gain a thorough understanding of the behavior and fate of nanoparticles in soil. Additionally, the effects of nanoparticle use on beneficial bacteria must be investigated. It is necessary to further clarify the process by which nanoparticles interact with the rhizosphericmicrobiota. The use of biological substances, organic fertilizers, and nanomaterials can all be used to help minimize soil-borne illnesses and boost crop output. Considering that the technology is still in its infancy, there must be a great deal more research to be done in this field. Nanocomposites can also be created utilizing fertilizers and active chemicals for the complete development of agricultural crops. It would take more than 10 years for nanopesticides to reach the market and their final consumers. Fewer items have so far made it to the market, and the majority of materials developed with the help of nanotechnology are presently in the testing or start-up stages.

The global market for nano-based materials was valued at 8.5 billion US dollars in 2019 and was projected to continue to expand at a rate of 13.1% each year from 2020 to 2027. The rise in social acceptance, demand, and adoption of nanomaterials in a variety of fields, including the medical, food, sports, aerospace, and energy industries, among others, is credited with the increase in market share. However, because they are so directly linked to food and human health, nanomaterials in agriculture are not widely accepted. The performance of nanotechnology for agricultural related point is disputed due to the high initial production cost and complicated production method. It is common knowledge that nanoparticles created by physical and chemical processes are extremely expensive and dangerous for the environment. The long-term profitability and economic effectiveness of green synthesis techniques are comparable. They also pose less of a threat to the environment and react more quickly. Global efforts are being undertaken to control the safe production and application of nanomaterials and nanodevices through oversight, recommendations, and legislation (Choi et al., 2009).

There is currently no single law that is entirely dedicated to offering guidelines regarding the usage of nanotechnology in any nation on the planet. There is a need for rules and instructions to analyze looming threats and for proposals to assure safe use of nanotechnology, whether they come from the US Food and Drug Administration (FDA), the Organisation for Economic Cooperation and Development (OCED), or both (Coles and Frewer, 2013). It is urgently necessary to carefully evaluate the advantages of using nanoparticles throughout their lifecycle, taking into account potential risks to occupational safety and health as well as environmental, social, and financial consequences.

The impact of nanoparticles on workers in the agricultural, industrial, and nonindustrial sectors, as well as the assessment of worker exposure, are crucial. Additionally, it is important to identify the toxicological characteristics of nanoparticles and collect data that can be easily accessed by researchers. Safety precautions should be given top priority in nanoparticle manufacturing facilities to eliminate workplace dangers. Additionally, safety rules for the handling, application, and disposal of nanoparticles and associated waste products should be created in laboratories. To ensure that there are no negative effects from the use of nanoparticles in agricultural fields, proper precautions should be taken while promoting their advantages. In order to validate these expertise in compared to the existing technologies, more study is required to ascertain the possibility, sustainability, capability, applicability, and release ability of nanotechnology-based goods in field conditions (Dutta et al., 2023). Additionally, more in vivo and field tests of pesticide-loaded nanomaterials or nanopesticides will be carried out. Long-term trial data is necessary, although it is currently limited, in order to acquire a thorough understanding of the efficacy of these pesticides. Contrary to chemical pesticides, regulatory agencies have not provided a comprehensive definition of nanomaterials and nanopesticides.

In contrast to conventional pesticides, the effectiveness of nanopesticides may be influenced by the absorption, bioavailability, quantity, and lethality of the nanoparticles as well as the quantity of the active ingredient linked to them (Kookana et al., 2014). Furthermore, little is known about the issue of pesticide resistance and possible strategies for using nanoparticles to reduce its incidence. Without the application of sophisticated analytical tools, it is impossible to develop regulatory criteria for risk assessment. There need to be a wide variety of nanoformulations accessible for future uses because the use of chemical pesticides in the field is regularly restricted and banned by regulatory organizations. It could be necessary to employ cutting-edge instruments and techniques to generate reliable data for exploration, categorization, and menace quantification. Therefore, a full understanding of the structural properties of the nanoparticles, jointly with their outline, amount, useful faction, and dynamic adsorption/loading ability, is required for the logical selection of relevant nanomaterials and nanopesticides. It is also essential to pick a reliable and reproducible system in order to conduct biocompatibility and efficacy studies at the cell, individual organisms, and pest-host ecosystem concentration under as close-to-field conditions as is practical.

Additionally, for the full development of nanoscience, it is necessary to incorporate experts from a variety of fields, including biologists, agricultural engineers, plant pathologists, biotechnologists, and soil microbiologists. As a result, importance needs to be given for developing a long-lasting, secure, efficient, and ecologically friendly soil disease control strategy. To reduce the spread of soil-borne diseases, farmers apply chemical fungicides and synthetic fumigants at regular intervals during the cropping season. However, frequent use of fungicides and fumigants like methyl bromide disturbs the ecological balance, poses a risk to human and animal health, degrades the aquatic ecosystem, and harms beneficial soil organisms (Panth et al., 2020). Farmers mostly use cultural techniques such as crop rotation, biofumigation, anaerobic soil disinfestation, soil solarization, and soil steam sterilization to reduce losses, although these techniques are less reliable and successful than chemical control options. The necessity for an alternative, sustainable, and successful management plan is highlighted by the growing environmental restrictions and the inefficient management solutions available. One of the sciences of the twenty-first century with the greatest rate of advancement is nanotechnology. There are numerous disciplines, such as environmental science, health care, information technology, food security and safety, and pest and disease management, and each of these have been found to benefit from the diverse application of nanotechnology in future.

According to the perspective of agriculture, the main issues with soil and environmental health include more pesticide contamination of soil and water sources, a decline in beneficial soil organisms, changes to the physical and chemical properties of soil, pesticide resistance in pathogens, several other reports have been studied (Sharma et al., 2019). Most of these issues can be resolved by nanotechnology, which can also bring about revolutionary breakthroughs. It has wonderful applications in targeted drug delivery, high-sensitivity disease detection, and antimicrobial and therapeutic compounds. As a result, it is likely to increase agricultural productivity due to a decrease in the cost of agricultural production practises (Dutta et al., 2021; Dutta et al., 2022).

11.2 NANOPARTICLES AS PROTECTANTS

Plant pathogens, such as bacteria, fungi, and viruses, pose significant threats to crop yields and food security. Nanomaterials have shown promising results in managing plant diseases (Figure 11.1). For example, nanoformulations of antimicrobial agents can penetrate plant tissues more effectively, combating pathogens at the cellular level. Additionally, nanoscale carriers for gene-editing tools have enabled precise and efficient modification of plant genes to enhance disease resistance. Nanotechnology has shown great potential in plant disease management, offering innovative and sustainable approaches to combat plant pathogens and improve agricultural practises (Gogos et al., 2012; Kah and Hofmann, 2014). The direct use of nanoparticles as antimicrobial agents has been proven successful against a variety of soil-borne diseases. It can be used to protect soil, seed, roots, and leaves from pests and pathogens such as fungi, bacteria, and viruses. When nanoparticles enter a plant's system, they either directly combat the pathogen or act as elicitor molecules, triggering both local and systemic defence responses in the plant. The most thoroughly investigated nanoparticles are metallic ones, which include gold, silver, titanium oxide, zinc oxide, and copper oxide (Kim et al., 2018). These particles are also known to have antifungal, antibacterial, and antiviral activities. According to several studies, nanoparticles are a powerful antibacterial agent that can reduce the threat posed by soil-borne phytopathogens. The nanoparticles have been tested at a concentration of 100 ppm on the soil-borne phytopathogens, namely, *Sclerotium rolfsii*, *Rhizoctonia solani*, *Fusarium oxysporum*, and *Sclerotinia sclerotiorum* for

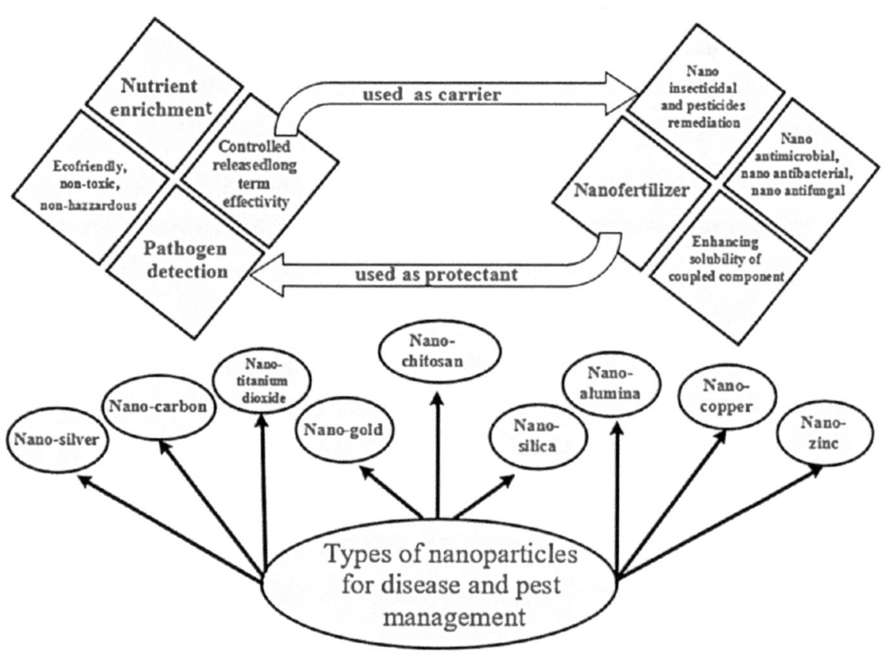

FIGURE 11.1 Types of different nanoparticles and their use as plant protectants.

their antifungal efficacy (Kaman and Dutta, 2019). AgNPs' antifungal effectiveness against the disease *Sclerotium rolfsii* in wheat plants was reported by Desai et al. (2021). At a dosage of 100 ppm AgNP, they noticed a 100% suppression of mycelial growth and a 100% inhibition of sclerotial germination. However, at this concentration, wheat plant roots displayed phytotoxic effects. As a result, it was concluded that plant growth and disease management were optimal at a 50 ppm AgNP concentration.

Another investigation by Zaki et al. (2022) found that soil-borne pathogens *R. solani*, *M. phseolina*, and *F. fujikuroi* were resistant to the antifungal activity of mycogenically produced Zinc oxide nanoparticles (ZnONPs) through *Trichoderma* spp. Both on cotton seedlings and *in vitro*, a significant antifungal impact was noted. Additionally, *R. solani*, *S. sclerotiorum*, and *S. minor*mycelial growth were shown to be inhibited by AgNPs in a dose-dependent manner (Min et al., 2009). Microscopic findings indicate that the route of action of AgNPs is fungal cell disintegration, layer separation of the hyphal wall, and finally hyphal collapse and death.

It is clear that a great deal of investigation has been done on nanoparticles' potential antimicrobial effects against soil-borne diseases, particularly fungi. There is little research on how soil-borne bacterial and viral diseases are affected by nanoparticles. The control of the majority of soil-borne diseases has been found to benefit from the use of metallic nanoparticles like Ag, ZnO, and CuO NPs. In contrast to their bulk counterparts, positively charged metallic nanoparticles are more readily adsorbed on surfaces and can enter cells (Gupta and Rai, 2017). Additionally, prior studies have proven that nanoparticles function through contact, accumulation, cell wall disintegration, membrane leakage, suppression of RNA and protein synthesis, reticence of ATPase activity, intracellular ROS production, and a variety of other mechanisms (Dakal et al., 2016). According to Alkubaisi and Aref (2016), the antiviral activity includes the direct interaction of nanoparticles with the capsid protein and the destruction of virus particles. To ascertain its precise mode of action, additional research is required, particularly with regard to soil-borne bacteria and viruses. Additionally, *in vivo* tests are required to verify the outcomes of *in vitro* testing. However, using heavy metal nanoparticles raises questions about their toxicity, persistence, and buildup in the food chain and ecology. To arrive at reliable results in this regard, much research is required. Natural polymeric nanoparticles with biodegradable and environmentally beneficial properties, like chitosan, lignin, and cellulose, may prove to be a good replacement.

11.3 NANOMATERIALS AS ANTIMICROBIAL AGENTS' TRANSPORTERS

By lowering the amount of chemical fungicides and pesticides in the environment, nanotechnology can help with the requirement for sustainable agriculture. The effectiveness of the product is increased by the use of nanoparticles as a carrier molecule that enables the delayed, suitable, and embattled discharge of dynamic substances in the environment (Table 11.1). To protect the crop throughout the emergent period, a consistent and delayed release of active ingredients is necessary because soil-borne diseases or their propagules do not simultaneously infect the roots of the plant (Khan et al., 2011). Because the chemical truly resourceful aligned with the pathogen is 10–15% less than the chemical required with standard formulations, nanocarriers minimize the rate of application. Smaller nanoparticles with higher kinetic stability, lower viscosity, and optical transparency can be more effective and intelligent (Xu et al., 2010).

TABLE 11.1 Nanomaterials and Carriers for Plant Disease Management.

S. No.	Type of Nanoparticle	Most Common Biological and Chemically Active Ingredients	Action and Benefits in Plant Systems
1	Nanoemulsions	Neem oil	Improved absorption and dispersal of nanoparticles by organisms
2	Nanocapsule	Lansiumamide B	Enhanced absorption and, hence, improved pest-control efficiency
3	Nanogel	Copper (antifungal activity) Essential oils (various pests)	Elevated absorption and consequential boost in pest-control effectiveness
4	Nanofibers	Thiametoxam (insecticide)	Enhanced dispersion and gradual active ingredient release
5	Liposomes	Etofenprox (insecticide) Pyrifluquinazon (insecticide)	Slow release of active ingredient
6	Lipid nanoparticles	Deltamethrin (insecticide)	Slow release of the active ingredient, protection from degradation
7	Inorganic nanoparticles	Silicon dioxide Titanium dioxide SilverCopper Nanoclays	Improved distribution, high effectiveness of active ingredient against pests
8	Inorganic nanoparticles linked to an organic dynamic part	Silicon dioxide Titanium dioxide	Allocate the active part to be released gradually

The waterproofing, bioavailability, solubility, and dispersion of chemical fungicides and pesticides can all be improved by using nanoformulations as carriers of active antimicrobial agents (Bergeson, 2010a). Nanomaterials have special qualities that are necessary for the creation of nanopesticides, things like biodegradability, permeability, thermal stability, solubility, and crystallinity (Bordes et al., 2009; Bouwmeester et al., 2009). Nanoencapsulates, nanotubes, nanowires, nanoemulsions, and nanocages are typical nano-based delivery systems that may be useful in plant protection techniques (Bouwmeester et al., 2009; Lyons and Scrinis, 2009; Bergeson, 2010b). The studies on the application of nano-based elegant deliverance arrangements for agrochemicals, such as insecticides, fertilizers, and plant growth boosters, were compiled by Ghormade et al. (2011). Agrochemicals can be enclosed in nanoencapsulates, notably nanoclay, which offers interaction surfaces with an aspect ratio. The primary concern with regard to soil health is the leaching of pollutants through the soil; however, there has only been a little research done on this topic. Metalaxyl was encapsulated by Wanyika (2013) using MSNs, and over the course of 30 days, she compared the amounts of free and encapsulated metalaxyl that leached into the soil (11.5%) and 76%, respectively. Encapsulated metalaxyl discharged 47% faster into water than into soil, illustrated the importance of conducting experiments in a farming context. In their 2015 study, Campos et al. looked into the cytotoxicity of carbendazim and/or tebuconazole attached to two different kinds of solid lipid or polymeric nanoparticles. It was discovered that the toxicity of the insecticides with nanoparticles was decreased in preosteoblast and fibroblast mouse cell lines. The majority of research has been done to create fungicides that are nano-encapsulated; however, there is little literature on encapsulating antibiotics and antiviral medicines to fight phytopathogens. Therefore, it suggests that more study is required in this area.

11.4 NANOTECHNOLOGY IN INSECT PEST MANAGEMENT

Additionally, novel formulations can be made using nanoparticles that can be alternative to pesticides, insecticides, and insect repellents that are effective against insects and other pests (Barik et al., 2008; Ramezani et al., 2019; Jasrotia et al., 2022). Insecticides, pesticides, and insect repellents have all been developed using nanoparticles by scientists (Barik et al., 2008; Gajbhiye et al., 2009; Goswami et al., 2010). According to Bhattacharyya et al. (2010), nanotechnology will soon revolutionize agriculture, especially pest control. Additionally, they forecast that during the next 20 years, nanotechnology will hasten the "green revolution." One use of this technology is nanoencapsulation. It offers the best chance of protecting host plants from insect pests.

Conversely, herbicides can be properly absorbed into plants through nanoencapsulation with nanoparticles (Scrinis and Lyons, 2007). Oil in water (nanoemulsions) is excellent for creating pesticides and extremely beneficial in a number of crucial pests in agriculture, according to Wang et al. (2007). Similarly, Liu et al. (2006) found that solid lipid nanoparticles loaded with essential oils may be used to create nanopesticides. According to Yang et al. (2009), mature *Tribolium castaneum* insects detected in stored goods can be killed by polyethylene glycol-coated nanoparticles loaded with garlic essential oil. According to reports, control efficacy against adult *T. castaneum* is about 80%, presumably as a result of the constant and progressive release of active ingredients from nanoparticles.

An experiment was carried out by Goswami et al. (2010), by using the applications of different types of nanoparticles, including silver nanoparticles (SNP), zinc oxide, aluminum oxide (ANP), and titanium dioxide, which were tested for controlling grasses and rice weevil disease in silkworms (*Bombyx mori*) caused by the baculovirus BmNPV (*B. mori* nuclear polyhedrosis virus) and *Sitophilus oryzae*, respectively. They created the aforementioned nanoparticles in liquid and solid formulations for their study's bioassay, put them on rice, and then stored the rice in a plastic box with 20 *S. oryzae* adults while observing the results for seven days. There have been several formulations created with various modes of action and applications, including nanoemulsions, nanosuspensions, controlled-release formulations, and solid-based nanopesticides. There are many ways to control insects, including physical, mechanical, chemical, and biological methods. However, fumigation is the most widely used approach because it can be used in a wide range of storage environments, including bags, silos, warehouses, and mills (Nguyen et al., 2015; Nayak et al., 2020). However, there are commercially available synthetic pesticides for the management of storage pests; however, these pesticides are expensive, ineffective, and hazardous to both human and environmental features (Jallow et al., 2017; Poudel et al., 2020).

By increasing pesticide effectiveness, improving active ingredient stability, lowering the required pesticide dose, and conserving agricultural inputs, the application of nanotechnology will be able to solve the drawbacks of conventional pesticides (Jasrotia et al., 2018; Rikta and Rajiv, 2021). Though, fresh improved nano-based formulations are anticipated to have a novel mechanism of action, be target-specific, cost-effective, stable, and have the capacity to remain active in a variety of application conditions. Different shapes of nanoparticles can be designed, such as a capsule that serves as a physical shell and is better able to survive environmental deterioration. Therefore, compared to traditional pesticides, nanoparticles provide greater long-lasting protection (Sushil et al., 2021). Nanoformulations, in contrast to conventional pesticide formulations, are specifically designed to release the biocide in a regulated and targeted manner while improving the solubility of insoluble or poorly soluble active components. As a result, a lesser amount of the active ingredient per area is needed for application, and this method can supply the active ingredient over a longer period of time while still ensuring that it is effective (Singh et al., 2020). Therefore, the cost of application is also lowered as a result of the lower dose requirement. Additionally until the substance's active component is released, the controlled-release formulations must remain inactive.

11.5 NANOPARTICLES FOR INDUCTION OF PLANT DEFENCE MECHANISM

11.5.1 MODE OF ACTION

A lot of scientific interest has recently been shown in the development of novel insecticides using nanoparticles created by diverse synthesis processes. Although numerous studies have been carried out to assess their poisonous potential against a variety of insect pests, it is still unclear exactly how they work against insects (Rai et al., 2014; Athanassiou et al., 2018). Since nanopesticides are relatively new substances that have not been thoroughly explored, only a small number of studies have been done to examine the toxicokinetics or toxicodynamics of nanopesticides against storage grain insect pests. The concept of "toxicodynamics" applies to the mechanisms by which chemicals act as well as their physiological, biochemical, and molecular impacts, whereas toxicokinetics primarily refers to the movement and changes an insecticide undergoes inside an organism: absorption, distribution, metabolism, and excretion (Alzogaray & Zerba, 2017). Several studies have examined the mechanism of action of silica, alumina, silver, and graphene oxide nanoparticle-based nanopesticides against insects (Benelli, 2018). By influencing antioxidants and detoxifying enzymes, silver nanopesticides reduce acetylcholinesterase function resulting in cell death and oxidative stress.

By up- or downregulating important insect genes, likewise it has been shown that they decrease protein synthesis and gonadotropin release, resulting in developmental harm and reproductive failure (Nair and Choi, 2011). By attaching to S and P in proteins and nucleic acids, respectively, metal nanoparticles decrease membrane permeability, causing organelle and enzyme denaturation, which is followed by cell death. Furthermore, there is an impact on growth and reproduction, gold nanoparticles can also operate as trypsin inhibitors (Patil et al., 2016; Small et al., 2016). According to Debnath et al. (2011) and Arumugam et al. (2016), ANP and silicon dioxide nanopesticides work by adhering to the insect cuticle and then physically absorbing waxes and lipids, which causes the insect to become dehydrated. According to Ibrahim and Salem (2019), after adhering to the body of *T. confusum*, the nanozeolite formulation produced scratching and cuticle splitting, which ultimately caused the insect's dehydration and death. The dangers posed by nanostructured alumina and how it works to protect against *S. oryzae* were described by Stadler et al. (2017). According to the experiments, triboelectric forces caused charged nanostructured alumina to adhere to the beetle cuticle, which caused the insect to become dehydrated by sorbing away its wax covering due to surface area phenomena.

11.5.2 ANTIOXIDANT SYSTEM

Numerous soil-borne diseases, such as fungus, bacteria, and viruses, seriously affect plant health. Plants have evolved complex defence mechanisms, including both inducive and constitutive responses, to respond to a wide range of biotic and abiotic stressors. The pathogen is first protected by constitutive mechanisms. In order to evade constitutive defence mechanisms, the pathogen must enter the roots of plants, which triggers the activation of induced defensive mechanisms. The induction of plant defences is a key component of any management strategy. One aspect of the plant defence response to various stresses is the creation of ROS, which prevents the spread of pathogens and sets off local and systemic defence reactions, such as the release of pathogenesis-related (PR) proteins. When ROS levels rise over the cutoff, oxidative products are generated, and the equilibrium between antioxidants and ROS is disturbed. Plants' antioxidant systems function to mitigate the impacts of oxidants. Superoxide dismutase (SOD), ascorbate peroxidase (APX), catalase (CAT), and guaiacol peroxidase (GPX) are a few of the enzymes that comprise up the mechanism of antioxidants (Tan et al., 2018). By raising or decreasing oxidative stress, nanoparticles change the cellular redox balance. According to earlier research, nanoparticles can either promote the creation of ROS or decrease the oxidative burst by producing antioxidant enzymes and secondary metabolites, depending on the needs of the host plant. Silicon dioxide nanoparticles (SiO_2 NPs) were found by Abdelrhim et al. (2021) to activate wheat seedlings' innate defense mechanisms and antioxidant systems in response to the pathogen *R. solani.*

SiO_2 NP application induced the accumulation of defense-related stimulants (in particular salicylic acid), increased the amount of photosynthetic pigments (chlorophylls and carotenoids), and reduced oxidative stress by creating each enzymatic (POD, SOD, APX, CAT, and PPO) and nonenzymatic (phenolics and flavonoids) antioxidant defence processes. SOD, a catalyzing enzyme for the detoxification of superoxide, is oxygen into either regular molecular oxygen (O2) or H_2O_2, and APX, which detoxifies peroxides like H_2O_2 using ascorbic acid (Asc) as a substrate, were both expressed at higher levels in plants exposed to nanoparticles (Fu et al., 2014). Dehydroascorbate reductase (DHAR) and monodehydroascorbate reductase (MDAR), the enzymes that regulate the cellular Asc redox state, were decreased in their expression levels (Fu et al., 2014). Using proteome analysis, SOD, APX, and glutathione-S-transferase (GST) were discovered at higher concentrations in AgNP-treated *O. sativa* roots (Mirzajani et al., 2014).

11.5.3 PHYTOHORMONES AND SIGNALING MOLECULES

The correct defensive processes are initiated by the signaling chemicals ethylene (ET), jasmonic acid (JA), and salicylic acid (SA). Other plant hormones that influence plant growth include gibberellin (GA), cytokinin (CK), auxin [indole-3-acetic acid (IAA)], abscisic acid (ABA), brassinosteroids (BRs), and strigolactone (SL). The balance between a plant's defences and growth is controlled by crosstalk between different plant hormones. The balance of plant hormones is known to be impacted by nanoparticles (Rastogi et al., 2017). Strawberry plants treated with selenium nanoparticles (SeNPs) show a rise of the stress signaling molecules IAA and ABA. After treatments, increased levels of organic acids (such as malic, citric, and succinic acids) and sugars (such as glucose, fructose, and sucrose) in the fruits of strawberry were reported. Further treatment with Se-NPs under saline conditions also demonstrated the benefits of Se-NPs on the improvement of fruit quality and nutritional values (Zahedi et al., 2019). Azhar et al. (2021) investigated phytohormone signaling after exposing Arabidopsis to several metallic nanoparticles (ZnO, SiO_2, and ZnO/SiO_2 composite NPs). They found that nanoparticle accumulation in plant tissue changed the level of expression of genes connected to the CK signalling system (ARR7 and ARR15), suggesting the importance of CK in the plant's response to nanoparticles. They discovered that exposure to nHA markedly increased the total phenolic content (40–68%) and phenylalanine ammonialyase activity (30–80%) in infected plants. A connection among nHA antioxidant systems and phytohormones boosted defence against the fungal disease in the host plant, as evidenced by the 10–45% increase in SA levels in shoots. PR proteins, which are used as diagnostic biomarkers of plant defence signaling pathways, are another vital part of a plant's immune system.

The increase in the SA signaling pathway is shown by the activation of the PR1, PR2, and PR5 genes (Ali et al., 2018). SiO_2 NPs, ZnONPs, and treatment with magnetite nanoparticles (Fe_3O_4NPs) all increased the expression of the SA-inducible PR1 and PR2 genes in healthy tobacco plants (*N. benthamiana*) (Cai et al., 2020). Therefore, the altered concentrations of phytohormones and PR proteins entail activation of the plant's defence mechanism when plants are exposed to nanoparticles. In Grodetskaya et al.'s (2022) study, soil-borne infections *Fusarium oxysporum* and *Fusarium avenaceum* were used to infect downy birch micro-clones. They also measured the expression of genes linked to plant defence responses brought on by microbes. The issues with disease management require a clear, unambiguous remedy offered by nanotechnology. The scientific and industrial spheres are being revolutionized by these breakthroughs. The development of potent formulations that ensure optimal distribution of agrochemicals, nutrients, pesticides, insecticides, and even growth regulators for enhanced usage efficiency is particularly promising for nano-based instruments in agriculture.

The scientific and industrial fields are being revolutionized by these innovations. The development of potent formulations that ensure optimal distribution of agrochemicals, nutrients, pesticides, insecticides, and even growth regulators for enhanced usage efficiency is particularly promising for nano-based instruments in agriculture. Nanotechnology offers a clear, unambiguous answer to the problems with illness management. These breakthroughs are revolutionizing the scientific and industrial fields. For nano-based agricultural instruments, the creations of potent formulations that ensure appropriate distribution of agrochemicals, nutrients, pesticides, insecticides, and even growth regulators for better usage efficiency are particularly promising. The most often utilized nanoparticles for the prevention and treatment of plant diseases are alumino-silicates, carbon, silver, silica, and nonmetal oxides. Consequently, we demonstrate collectively the function of nanocomposites in the treatment of fungus infections in this review.

11.5.4 PLANT RESPONSES TO NANOPARTICLES

The majority of the ways that NPs reach a plant's system are through its leaves and roots. After entering plants, NPs engage in cellular and subcellular interactions that influence morphological and physiological changes (Khan et al., 2019). Depending on the NPs and the plant species, these interactions could be advantageous or harmful. Effects of NPs on plant systems may depend on the chemical make-up, reactivity, size, and concentration of NPs in or on the plant (Ma et al., 2010). In order to investigate the effects of nanoparticles on seed germination or plant growth, the researchers used a variety of NPs application methods, such as soil application, foliar spraying, or seed treatment. Different NPs can stimulate seed germination (Hatami, 2017) and plant growth and development (Aslani et al., 2014; Pallavi et al., 2016; Nair, 2016) at concentrations below specific limitations, according to the available research. Most of these examinations were conducted in pots or plate growth media, which are artificial treatment settings. We explore the beneficial impacts of nanoparticles on plant growth and seed germination as well as the advantages of NPs in enhancing plant stress tolerance in order to comprehend how nanoparticles affect plant growth.

11.5.5 REGULATIONS OF NANOPESTICIDES

In today's high-tech environment, the use of nanomaterials in a variety of sectors is on the rise. However, it is also true that before a product can be sold all over the world, it must first undergo assessments of the environmental risks posed by nanomaterials like nanopesticides. Prior to their release, nanomaterials should undergo a risk assessment to ascertain their persistence, behavior, and fate in the environment. Global regulatory and legislative agencies are necessary to guarantee that they take into account safe use and effective risk assessment. The size, surface area, and catalytic characteristics of nanoparticles necessitate additional testing beyond that required to determine their toxicity limits (Singh et al., 2020). To assure the safety of nanomaterials, numerous national and international regulatory authorities are working. Heterogeneous approaches are sometimes employed to regulate nanoproducts in specific countries, such as Asia, Africa, and

Oceania (Kookana et al., 2014). As an illustration, criteria for the assessment of nanomaterials as either nanopesticides or nanofertilizers have just been developed in India (Naqvi and Flora, 2020). Prior to their release on the market, potential health and environmental risks associated with such nanomaterials should be taken into account, according to the Scientific Advisory Panel of the US Federal Insecticide, Fungicide, and Rodenticide Act (FIFRA) (Grillo et al., 2021).

Additionally, Regulation (EC) No. 1107/2009, which mandates previous government approval, governs the majority of plant protection products (PPPs) in Europe (Villaverde et al., 2014; Côa et al., 2020). The REACH Regulation (EC) No. 1907/2006, which informs manufacturers and importers about the risks associated with their use, regulates the official control of nanomaterial residues in food and protects workers from exposure to these chemicals (European Food Safety Authority et al., 2019). According to Bowman and Hodge (2009) and Walker et al. (2017), the Australian Pesticides and Veterinary Medicines Authority (APVMA) also created rules for the discharge of nanomaterials. Priorities in the R&D sectors of nations all over the world must center on reducing the ecotoxicity of nanopesticides through standardization of testing protocols for nanopesticides, improved knowledge of the dangers posed by nanopesticides and their degradation products, lengthening the time of exposure of nanopesticide tests against specific organisms, and identification of nanopesticides that fall within boundaries for regulatory coverage.

11.6 FUTURE PERSPECTIVES

The field of nanomaterials in agricultural pest and disease management is still evolving. Continued research and development are necessary to explore new nanotechnologies, optimize the existing applications, and address the identified risks. Collaborations between scientists, policymakers, and industry stakeholders are vital to ensuring the responsible and sustainable integration of nanomaterials in agriculture. Environmentally friendly management and sustainable strategies become dominant as a result of nanopesticides' dissemination through plants, microorganisms, and their derivatives. Future nanotechnology research should focus on the development of smart nanopesticide formulations to address problems with conventional formulations, environmentally friendly green nanopesticide chemistry, technologies for the cost-effective and commercial production of nanopesticides, and nanoformulations with conventional analogues to determine their practical utility and ecotoxicological evaluation. With very few exceptions, nanomaterials have a significant potential to transform crop protection practises from ones that harm the environment to ones that benefit it. Additionally, it lowers the cost of agriculture.

11.7 CONCLUSION

Nanotechnology holds great promise for revolutionizing plant pest and pathogen management in agriculture. The use of nanomaterials can lead to safer and more sustainable agricultural practises while minimizing environmental impacts. Nevertheless, the potential risks associated with nanomaterials must not be overlooked, and it is crucial to implement a precautionary approach that prioritizes human health, environmental protection, and ethical considerations. Finding the ideal balance will allow us to fully utilize the nanomaterials to safeguard global food security and pave the way for a greener and more sustainable future.

KEYWORDS

- **Nanofertilizers**
- **nanoformulations**
- **nanopesticides**
- **nanoscale carriers**
- **plant protection**

REFERENCES

Abdelrhim, A. S., Mazrou, Y. S. A., Nehela, Y., Atallah, O. O., El-Ashmony, R. M., & Dawood, M. F. A. (2021). Silicon dioxide nanoparticles induce innate immune responses and activate antioxidant machinery in wheat against Rhizoctonia solani. *Plants*, *10*(12), 2758. DOI: 10.3390/plants10122758

Adetunji, C., Oloke, J., Kumar, A., Swaranjit, S., & Akpor, B. (2017). Synergetic effect of rhamnolipid from Pseudomonas aeruginosa C1501 and phytotoxic metabolite from *Lasiodiplodiapseudotheobromae* C1136 on *Amaranthushybridus* L. and *Echinochloacrus-galli* weeds. *Environment Science and Pollution Research*, *24*(15), 13700–13709.

Ali, S., Ganai, B. A., Kamili, A. N., Bhat, A. A., Mir, Z. A., Bhat, J. A., et al. (2018). Pathogenesis-related proteins and peptides as promising tools for engineering plants with multiple stress tolerance. *Microbiological Research*, *212*, 29–37. DOI: 10.1016/j.micres.2018.04.008

Alkubaisi, N. A., & Aref, N. M. A. (2016). Dispersed gold nanoparticles potentially ruin gold barley yellow dwarf virus and eliminate virus infectivity hazards. *Applied Nanoscience*, *7*(1), 31–40. DOI: 10.1007/s13204-016-0540-0

Alzogaray, R. A., & Zerba, E. N. (2017). *Rhodniusprolixus* intoxicated. *Journal of Insect Physiology*, *97*, 93–113. DOI: 10.1016/j.jinsphys.2016.04.004

Aouada, F. A., & DeMoura, M. R. (2015). Nanotechnology applied in agriculture: Controlled release of agrochemicals. In *Nanotechnologies in food and agriculture* (pp. 103–118). Springer. DOI: 10.1007/978-3-319-14024-7_5

Arumugam, G., Velayutham, V., Shanmugavel, S., & Sundaram, J. (2016). Efficacy of nanostructured silica as a stored pulse protector against the infestation of Bruchid Beetle, *Callosobruchus maculatus* (Coleoptera: Bruchidae). *Applied Nanoscience*, *6*(3), 445–450. DOI: 10.1007/s13204-015-0446-2

Aslani, F., Bagheri, S., Muhd Julkapli, N., Juraimi, A. S., Hashemi, F. S. G., & Baghdadi, A. (2014). Effects of engineered nanomaterials on plants growth: An overview. *The Scientific World Journal*, *2014*, 641759.

Athanassiou, C. G., Kavallieratos, N. G., Benelli, G., Losic, D., Usha Rani, P., & Desneux, N. (2018). Nanoparticles for pest control: Current status and future perspectives. *Journal of Pest Science*, *91*, 1–15. DOI: 10.1007/s10340-017-0898-0

Ayoub, H. A., Khairy, M., Rashwan, F. A., & Abdel-Hafez, H. F. (2017). Synthesis and characterization of silica nanostructures for cotton leaf worm control. *Journal of Nanostructure in Chemistry*, 7, 91–100.

Azhar, B. J., Noor, A., Zulfiqar, A., Zeenat, A., Ahmad, S., Chishti, I., et al. (2021). Effect of ZnO, SiO2 and composite nanoparticles on Arabidopsis thaliana and involvement of ethylene and cytokinin signaling pathways. *Pakistan Journal of Botany*, *53*(2), 437–446. DOI: 10.30848/ PJB2021-2(40)

Barik, T. K., Sahu, B., & Swain, V. (2008). Nanosilica- from medicine to pest control. *Parasitology Research*, *103*, 253–258.

Benelli, G. (2018). Mode of action of nanoparticles against Insects. *Environmental Science and Pollution Reseach*, *25*(13), 12329–12341. DOI: 10.1007/s11356-018-1850-4

Bergeson, L. L. (2010a). Nanosilver pesticide products: What does the future hold? *Environmental Quality Management*, *19*(4), 73–82. DOI: 10.1002/tqem.20263

Bergeson, L. L. (2010b). Nanosilver: US EPA's pesticide office considers how best to proceed. *Environmental Qualality Management*, *19*, 79–85. DOI: 10.1002/tqem.20255

Bhattacharyya, A., Bhaumik, A., Pathipati, U. R., Mandal, S., & Epidi, T. T. (2010) Nanoparticles – a recent approach to insect pest control. *African Journal of Biotechnology*, *9*(24), 3489–3493.

Bordes, P., Pollet, E., & Averous, L. (2009). Nano-biocomposites: Biodegradable polyester/ nanoclay systems. *Progress in Polymer Science*, *34*(2), 125–155. DOI: 10.1016/j.progpolymsci.2008.10.002

Bouwmeester, H., Dekkers, S., Noordam, M. Y., Hagens, W. I., Bulder, A. S., de Heer, C., et al. (2009). Review of health safety aspects of nanotechnologies in food production. *Regulatory Toxicology and Pharmacology*, *53*(1), 52–62. DOI: 10.1016/j.yrtph.2008.10.008

Bowman, D., & Hodge, G. (2009). Nanotechnology products in Australia: Chemicals, cosmetics and regulatory character. In *New global frontiers in regulation: The age of nanotechnology* (pp. 239–264). Edward Elgar Publishing Ltd.

Cai, L., Jia, H., Liu, C., Wang, D., & Sun, X. (2020). Foliar exposure of Fe_3O_4 nanoparticles on *Nicotiana benthamiana*: Evidence for nanoparticles uptake, plant growth promoter and defense response elicitor against plant virus. *Journal of Hazardous Materials*, *393*, 1224. DOI: 10.1016/j.jhazmat.2020.122415

Campos, E. V. R., De Oliveira, J. L., Da Silva, C. M. G., Pascoli, M., Pasquoto, T., Lima, R., et al. (2015). Polymeric and solid lipid nanoparticles for sustained release of carbendazim and tebuconazole in agricultural applications. *Scientific Reports*, *5*, 13809. DOI: 10.1038/srep13809

Choi, J. Y., Ramachandran, G., & Kandlikar, M. (2009). The impact of toxicity testing costs on nanomaterial regulation. *Environmental Science & Technology*, *43*, 3030–3034. DOI: 10.1021/ es802388s

Côa, F., Bortolozzo, L. S., Petry, R., Da Silva, G. H., Martins, C. H., de Medeiros, A. M., et al. (2020). Environmental toxicity of nanopesticides against non-target organisms: The state of the art. In *Nanopesticides* (pp. 227–279).

Coles, D., & Frewer, L. J. (2013). Nanotechnology applied to European food production – A review of ethical and regulatory issues. *Trends in Food Science and Technology*, *34*, 32–43. DOI: 10.1016/j.tifs.2013.08.006

Dakal, T. C., Kumar, A., Majumdar, R. S., & Yadav, V. (2016). Mechanistic basis of antimicrobial actions of silver nanoparticles. *Frontiers in Microbiology*, *7*. DOI: 10.3389/fmicb.2016.01831

Debnath, N., Das, S., Seth, D., Chandra, R., Bhattacharya, S. C., & Goswami, A. (2011). Entomotoxic effect of silica nanoparticles against *Sitophilus oryzae* (L.). *Journal of Pest Science*, *84*(1), 99–105. DOI: 10.1007/s10340-010-0332-3

Desai, P., Jha, A., Markande, A., & Patel, J. (2021). Silver nanoparticles as a fungicide against soil-borne *Sclerotium rolfsii*: A case study for wheat plants. In *Biobased nanotechnology for green applications, nanotechnology in the life sciences* (pp. 1–29). Springer Nature. DOI: 10.1007/978-3-030-61985-5_18

Dutta, P., Kumari, A., Mahanta, M., Upamanya, G. K., Heisnam, P., Borua, S., Kaman, P. K., Mishra, A. K., Mallik, M., Muthukrishnan, G., Sabarinathan, K. G., Puzari, K. R., & Vijayreddy, D. (2023). Nanotechnological approaches for management of soil-borne plant pathogens. *Frontiers in Plant Science*, *14*, 1136233. DOI: 10.3389/fpls.2023.1136233

Dutta, P., Das, G., Boruah, S., Kumari, A., Mahanta, M., Yasin, A., Sharma, A., Deb, L. (2021). Nanoparticles as nano-priming agent for antifungal and antibacterial activity against plant pathogens. *Biol. Forum Int*. J. 13, 476–482.

Dutta, P., Kumari, A., Mahanta, M., Biswas, K. K., Dudkiewicz, A., Thakuria, D., et al. (2022). Advances in nanotechnology as a potential alternative for plant viral disease management. *Frontiers in Microbiology*, *13*, 935193. DOI: 10.3389/fmicb.2022.935193

European Food Safety Authority (Efsa), Anastassiadou, M., Brancato, A., Brocca, D., Carrasco Cabrera, L., Ferreira, L., et al. (2019). Reporting data on pesticide residues in food and feed according to regulation (EC) No 396/2005 (2018 data Collection). *EFSA Journal*, *17*(4), e05655. DOI: 10.2903/j.efsa.2019.5655

Fu, P. P., Xia, Q., Hwang, H. M., Ray, P. C., & Yu, H. (2014). Mechanisms of nanotoxicity: Generation of reactive oxygen species. *Journal of Food and Drug Analysis*, *22*, 64–75. DOI: 10.1016/j.jfda.2014.01.005

Gajbhiye, M., Kesharwani, J., Ingle, A., Gade, A., & Rai, M. (2009) Fungus-mediated synthesis of silver nanoparticles and their activity against pathogenic fungi in combination with fluconazole. *Nanomedicine, 5*(4), 382–386.

Ghormade, V., Deshpande, M. V., & Paknikar, K. M. (2011). Perspectives for nano biotechnology enabled protection and nutrition of plants. *Biotechnology Advances*, *29*(6), 792–803. DOI: 10.1016/j.biotechadv.2011.06.007

Gogos, A., Knauer, K., & Bucheli, T. D. (2012). Nanomaterials in plant protection and fertilization: Current state, foreseen applications, and research priorities. *Journal of Agricultural and Food Chemistry*, *60*, 9781–9792. DOI: 10.1021/jf302154y

Goswami, A., Roy, I., Sengupta, S., & Debnath, N. (2010). Novel applications of solid and liquid formulations of nanoparticles against insect pests and pathogens. *Thin Solid Films*, *519*(3), 1252–1257.

Grillo, R., Fraceto, L. F., Amorim, M. J., Scott-Fordsmand, J. J., Schoonjans, R., & Chaudhry, Q. (2021). Ecotoxicological and regulatory aspects of environmental sustainability of nanopesticides. *Journal of Hazardous Materials*, *404*, 124148. DOI: 10.1016/j.jhazmat.2020.124148

Grodetskaya, T. A., Evlakov, P. M., Fedorova, O. A., Mikhin, V. I., Zakharova, O. V., Kolesnikov, E. A., et al. (2022). Influence of copper oxide nanoparticles on gene expression of birch clones In vitro under stress caused by phytopathogens. *Nanomaterial (Basel), 12*, 864. DOI: 10.3390/nano12050864

Gupta, R., & Rai, B. (2017). Effect of size and surface charge of gold nanoparticles on their skin permeability: A molecular dynamics study. *Scientific Report*, *7*, 45292. DOI: 10.1038/srep45292

Hatami, M. (2017). Stimulatory and inhibitory effects of nanoparticulates on seed germination and seedling vigor indices. In *Nanoscience and plant–soil systems* (pp. 357–385). Springer, Cham.

Ibrahim, S. S., & Salem, N. Y. (2019). Insecticidal efficacy of nano zeolite against *Triboliumconfusum* (Col., Tenebrionidae) and *Callosobruchus maculatus* (Col., Bruchidae). *Bulletin of the National Research Centre*, *43*(1), 1–8. DOI: 10.1186/s42269-019-0128-4

Ivo, I., Veruscka, L., Donald, H. B., & Anna, A. S. (2017) Nanotechnology in agriculture: Opportunities, toxicological implications, and occupational risks. *Toxicology and Applied Pharmacology*, *329*, 96–111.

Jallow, M. F., Awadh, D. G., Albaho, M. S., Devi, V. Y., & Thomas, B. M. (2017). Pesticide risk behaviors and factors influencing pesticide use among farmers in Kuwait. *Science of the Total Environment*, *574*, 490–498. DOI: 10.1016/j.scitotenv.2016.09.085

Jasrotia, P., Nagpal, M., Mishra, C. N., Sharma, A. K., Kumar, S., Kamble, U., Bhardwaj, A. K., Kashyap, P. L., Kumar, S., & Singh, G. P. (2022). Nanomaterials for postharvest management of insect pests: Current state and future perspectives. *Frontiers in Nanotechnology*, *3*, 811056. DOI: 10.3389/fnano.2021.811056

Jasrotia, P., Kashyap, P. L., Bhardwaj, A. K., Kumar, S., & Singh, G. P. (2018). Scope and applications of nanotechnology for wheat production: A review of recent advances. *Wheat Barley Research*, *10*(1), 1–14. DOI: 10.25174/2249-4065/ 2018/76672

Kah, M., & Hofmann, T. (2014). Nanopesticide research: Current trends and future priorities. *Environment International, 63*, 224–235. DOI: 10.1016/j.envint.2013.11.015

Kaman, P. K., & Dutta, P. (2019). Synthesis, characterization and antifungal activity of biosynthesized silver nanoparticle. *Indian Phytopathology*, *72*, 79–88. DOI: 10.1007/s42360- 018-0081-4

Khan M. R., Adam V., Rizvi T. F., Zhang, B., Ahamad, F., Josko, I., Zhu, Y., Yang, M., & Mao, C. (2019). Nanoparticle–plant interactions: Two-way traffic. *Small*, *15*(37), 1901794.

Khan, M. R., Majid, S., Mohidin, F. A., & Khan, N. A. (2011). A new bioprocess to produce low-cost powder formulations of biocontrol bacteria and fungi to control fusarial wilt and root-knot nematode of pulses. *Biological Control*, *59*(2), 130–140. DOI: 10.1016/j.biocontrol.2011.04.007

Kim, D. Y., Kadam, A., Shinde, S., Saratale, R. G., Patra, J., & Ghodake, G. (2018). Recent developments in nanotechnology transforming the agricultural sector: A transition replete with opportunities. *Journal of the Science of Food and Agriculture*, *98*, 849–864. DOI: 10.1002/ jsfa.8749.

Kookana, R. S., Boxall, A. B. A., Reeves, P. T., Ashauer, R., Beulke, S., Chaudhry, Q., et al. (2014). Nanopesticides: Guiding principles for regulatory evaluation of environmental risks. *Journal of Agriculture and Food Chemistry*, *62*, 4227–4240. DOI: 10.1021/jf500232f

Liu, F., Wen, L.-X., Li, Z.-Z., Yu, W., Sun, H.-Y., Chen, J.-F. (2006). Porous hollow silica nanoparticles as controlled delivery system for water-soluble pesticide. *Materials Research Bulletin*, *41*(12), 2268–2275.

Lyons, K., & Scrinis, G. (2009). Under the regulatory radar nanotechnologies and their impacts for rural Australia. In F. Merlan (Ed.), *Tracking rural change: Community, policy and technology in Australia, New Zealand and Europe* (pp. 151–171). Australian National University E Press.

Ma, Y., Kuang, L., He, X., Bai, W., Ding, Y., Zhang, Z., Zhao, Y., & Chai, Z. (2010). Effects of rare earth oxide nanoparticles on root elongation of plants. *Chemosphere*, *78*(3), 273–279.

Min, J. S., Kim, K., Kim, S. W., Jung, J., Lamsal, K., Kim, S., Jung, M., Lee, Y. S. (2009). Effects of colloidal silver nanoparticles on sclerotium-forming phytopathogenic fungi. *The Plant Pathology Journal*, *25*, 376–380. DOI: 10.5423/PPJ.2009.25.4.376

Mirzajani, F., Askari, H., Hamzelou, S., Schober, Y., Rompp, A., Ghassempour, A., Spengler, B. (2014). Proteomics study of silver nanoparticles toxicity on *Oryza sativa* l. *Ecotoxicology and Environmental Safety*, *108*, 335–339. DOI: 10.1016/j.ecoenv.2014.07.013

Nair, R. (2016). Effects of nanoparticles on plant growth and development. In C. Kole, D. Kumar & M. Khodakovskaya (Eds.), *Plant nanotechnology* (pp. 95–118). Springer, Cham.

Nair, P. M. G., & Choi, J. (2011). Identification, characterization and expression profiles of *Chironomus Riparius* glutathione S-transferase (GST) genes in response to cadmium and silver nanoparticles exposure. *Aquatic Toxicology*, *101*(3), 550–560. DOI: 10.1016/j.aquatox.2010.12.006

Naqvi, S., & Flora, S. J. S. (2020). Nanomaterial's toxicity and its regulation strategies. *Journal of Environmental Biology*, *41*(4), 659–671. DOI: 10.22438/jeb/41/4/mrn-1394

Nayak, M. K., Daglish, G. J., Phillips, T. W., & Ebert, P. R. (2020). Resistance to the fumigant phosphine and its management in insect pests of stored products: A global perspective. *Annual Review of Entomology*, *7*(65), 333–350. DOI: 10.1146/annurev-ento-011019-025047. Epub 2019 Oct 14. PMID: 31610132.

Nguyen, T. T., Collins, P. J., & Ebert, P. R. (2015). Inheritance and characterization of strong resistance to phosphine in *Sitophilus oryzae* (L.). *PLoS One*, *10*(4), e0124335. DOI: 10.1371/journal.pone.0124335

Pallavi, M. C., Srivastava, R., Arora, S., & Sharma, A. K. (2016). Impact assessment of silver nanoparticles on plant growth and soil bacterial diversity. *3 Biotech*, *6*(2), 254.

Panpatte, D. G., Jhala, Y. K., Shelat, H. N., & Vyas, R. V. (2016). Nanoparticles: The next generation technology for sustainable agriculture. In *Microbial inoculants in sustainable agricultural productivity* (pp. 289–300). Springer, New Delhi. DOI: 10.1007/978-81-322-2644-4_18

Panth, M., Hassler, S. C., & Baysal-Gurel, F. (2020). Methods for management of soilborne diseases in crop production. *Agriculture*, *10*(1), 16. DOI: 10.3390/agriculture10010016

Patil, C. D., Borase, H. P., Suryawanshi, R. K., & Patil, S. V. (2016). Trypsin inactivation by latex fabricated gold nanoparticles: A new strategy towards insect control. *Enzyme and Microbial Technology*, *92*, 18–25. DOI: 10.1016/ j. enzmictec.2016.06.005

Pirzadah, T. B., Malik, B., Maqbool, T., & Rehman, R. U. (2019). Development of nanobioformulations of nutrients for sustainable agriculture. In *Nanobiotechnology in bioformulations* (pp. 381–394). Springer, Cham. DOI: 10.1007/978-3-030-17061-5_16

Poudel, S., Poudel, B., Acharya, B., & Poudel, P. (2020). Pesticide use and its impacts on human health and environment. *Environmental & Ecosystem Science*, *4*, 47–51. DOI: 10.26480/ees.01.2020.47.51

Rai, M., Kon, K., Ingle, A., Duran, N., Galdiero, S., & Galdiero, M. (2014). Broad-spectrum bioactivities of silver nanoparticles: The emerging trends and future prospects. *Applied Microbiology and Biotechnology*, *98*(5), 1951–1961. DOI: 10.1007/s00253-013-5473-x

Ramezani, M., Ramezani, F., & Gerami, M. (2019). Nanoparticles in pest incidences and plant disease control. In D. Panpatte & Y. Jhala (Eds.), *Nanotechnology for agriculture: Crop production and protection* (pp. 233–272). Springer, Singapore.

Rastogi, A., Zivcak, M., Sytar, O., Kalaji, H., Xiaolan, H., Mbarki, S., et al. (2017). Impact of metal and metal oxide nanoparticles on plant: A critical review. *Frontiers in Chemistry*, *5*, 78. DOI: 10.3389/fchem.2017.00078

Rikta, S. Y., & Rajiv, P. (2021). Applications of silver nanomaterial in agricultural pest control. In *Silver Nanomaterials for Agri-Food Applications* (pp. 453–470). Elsevier. DOI: 10.1016/b978-0-12-823528-7.00002-0

Rosenthal, J. A., Chen, L., Baker, J. L., Putnam, D., & DeLisa, M. P. (2014). Pathogen-like particles: Biomimetic vaccine carriers engineered at the nanoscale. *Current Opinion in Biotechnology*, *28*, 51–58.

Scott, N. R., Chen, H., & Cui, H. (2018). Nanotechnology applications and implications of agrochemicals toward sustainable agriculture and food systems. *Journal of Agricultural and Food Chemistry*, *66*, 6451–6456. DOI: 10.1021/acs.jafc.8b00964

Scrinis, G., & Lyons, K. (2007). The emerging nano-corporate paradigm: Nanotechnology and the transformation of nature, food and agri-food systems. *IJAFR*, *15*(2), 22–44.

Sharma, A., Kumar, V., Shahzad, B., Tanveer, M., Sidhu, G. P. S., Handa, N., Kohli, S. K., Yadav, P., Bali, A. S., Parihar, R. D., Dar, O. I., Singh, K., Jasrotia, S., Bakshi, P., Ramakrishnan, M., Kumar, S., Bhardwaj, R., & Thukral, A. K. (2019). Worldwide pesticide usage and its impacts on ecosystem. *SN Applied Sciences*, *1*, 1446. DOI: 10.1007/s42452-019-1485-1

Singh, A., Dhiman, N., Kar, A. K., Singh, D., Purohit, M. P., Ghosh, D., & Patnaik, S. (2020). Advances in controlled release pesticide formulations: Prospects to safer integrated pest management and sustainable agriculture. *Journal of Hazardous Materials*, *385*, 121525. DOI: 10.1016/j.jhazmat.2019.121525

Singh, R. P., Handa, R., & Manchanda, G. (2021a). Nanoparticles in sustainable agriculture: An emerging opportunity. *Journal of Controlled Release*, *329*, 1234–1248. DOI: 10.1016/j.jconrel.2020.10.051

Singh, H., Sharma, A., Bhardwaj, S. K., Arya, S. K., Bhardwaj, N., & Khatri, M. (2021b). Recent advances in the applications of nano-agrochemicals for sustainable agricultural development. *Environmental Science: Processes & Impacts*, *23*(2), 213–239. DOI: 10.1039/D0EM00404

Small, T., Ochoa-Zapater, M. A., Gallello, G., Ribera, A., Romero, F. M., Torreblanca, A., & Garcera, M. D. (2016). Gold-nanoparticles ingestion disrupts reproduction and development in the German cockroach. *Science of the Total Environment, 565*, 882–888. DOI: 10.1016/j.scitotenv.2016.02.032

Stadler, T., Lopez Garcia, G. P., Gitto, J. G., & Buteler, M. (2017). Nanostructured alumina: Biocidal properties and mechanism of action of a novel insecticide powder. *Bulletin of Insectology, 70*(1), 17–25.

Sushil, A., Kamla, M., Nisha, K., Karmal, M., & Sandeep, A. (2021). Nano-enabled pesticides in agriculture: Budding opportunities and challenges. *Journal of Nanoscience Nanotechnology, 21*(6), 3337–3350. DOI: 10.1166/jnn.2021.19018

Tan, B. L., Norhaizan, M. E., Liew, W. P. P., & Sulaiman, R. H. (2018). Antioxidant and oxidative stress: A mutual interplay in age-related diseases. *Frontiers in Pharmacology, 9*, 1162. DOI: 10.3389/fphar.2018.01162

Thabet, A. F., Boraei, H. A., Galal, O. A. El-Samahy, M. F. M., Mousa, K. M., Zhang, Y. Z., Tuda, M., Helmy, E. A., Wen, J., & Nozaki, T. (2021). Silica nanoparticles as pesticide against insects of different feeding types and their non-target attraction of predators. *Scientific Report, 11*, 14484. https://doi.org/10.1038/s41598-021-93518-9

Villaverde, J. J., Sevilla-Morán, B., Sandín-España, P., López-Goti, C., & Alonso-Prados, J. L. (2014). Biopesticides in the framework of the European Pesticide Regulation (EC) No. 1107/2009. *Pest Management Science, 70*(1), 2–5. DOI: 10.1002/ps.3663

Walker, G. W., Kookana, R. S., Smith, N. E., Kah, M., Doolette, C. L., Reeves, P. T., & Navarro, D. A. (2017). Ecological risk assessment of nano-enabled pesticides: A perspective on problem formulation. *Journal of Agricultural and Food Chemistry, 66*(26), 6480–6486. DOI: 10.1021/acs.jafc.7b02373

Wang, L., Li, X., Zhang, G., Dong, J., & Eastoe, J. (2007). Oil-in-water nanoemulsions for pesticide formulations. *Journal of Colloid and Interface Science, 314*(1), 230–235.

Wanyika, H. (2013). Sustained release of fungicide metalaxyl by mesoporous silica nanospheres. *Journal ofNanoparticle Research, 15*, 1831. DOI: 10.1007/s11051-013-1831-y

Xu, L., Liu, Y., Bai, R., & Chen, C. (2010). Applications and toxicological issues surrounding nanotechnology in the food industry. *Pure and Applied Chemistry, 82*(2), 349–372. DOI: 10.1351/PAC-CON-09-05-09

Yang, F.-L., Li, X.-G., Zhu, F., & Lei, C.-L. (2009) Structural characterization of nanoparticles loaded with garlic essential oil and their insecticidal activity against *Triboliumcastaneum* (Herbst) (Coleoptera: Tenebrionidae). *Journal of Agricultural and Food Chemistry, 57*(21), 10156–10162.

Zahedi, S. M., Abdelrahman,M., Hosseini, M. S., Hoveizeh, N. F., & Tran, L. S. P. (2019). Alleviation of the effect of salinity on growth and yield of strawberry by foliar spray of selenium nanoparticles. *Environmental Pollution, 253*, 246–258. DOI: 10.1016/j.envpol.2019.04.078

Zaki, S. A., Ouf, S. A., Abd-Elsalam, K. A., Asran, A. A., Hassan, M. M., Kalia, A., & Albarakaty, F. M. (2022). Trichogenic silver-based nanoparticles for suppression of fungi involved in damping-off of cotton seedlings. *Microorganisms, 10*, 344. DOI: 10.3390/microorganisms10020344

Zhang, H., Wang, D., Butler, R., Campbell, N. L., Long, J., Tan, B., Duncalf, D. J., Foster, A. J., Hopkinson, A., Taylor, D., Angus, D., Cooper, A. I., & Rannard, S. P. (2008). Formation and enhanced biocidal activity of water-dispersible organic nanoparticles. *Nature Nanotechnology, 3*, 506–511.

CHAPTER 12

Nanotechnology Interventions for Improved Pesticide and Fertilizer Delivery

VANDANA GHORMADE, VAIDEHI BHAGWAT, DEEPALI CHOUDHARY, MAHESHWARI G., and SHIVANGI SINGH

Nanobioscience Group, Agharkar Research Institute, Pune, Maharashtra, India

ABSTRACT

Although agriculture relies on pesticides and fertilizer applications, their excessive usage often leads to environmental pollution, biodiversity loss, and the upsurge of insect pests and fungal pathogens. Nanotechnology, by virtue of its nanomaterial-related properties, can contribute to the controlled delivery and release of agrochemicals for the mitigation of these problems. In addition, nano-scale pesticides and fertilizers can promote improved efficacy and lower the toxicity of agrochemicals. Since the last decade, nanotechnology applications for plant and soil improvement have demonstrated that nanomaterials have potential for agri-biotechnological applications. The translation of these benefits can provide the required boost to agriculture. An understanding of the regulatory requirements will further assist the development of nano-scale pesticides and fertilizers for sustainable agriculture.

12.1 INTRODUCTION

Intensive pesticide and fertilizer applications have contributed to increased food production to feed the spiraling global population. However, agricultural intensification has led to the indiscriminate application of pesticides

Nanobiotechnology for Agricultural Sciences: Nano-Agri-Input Products for Crop Production and Environmental Protection. J. C. Tarafdar, Deepak Kumar, Sanjay K. Singh, and Krishna Pal Singh (Eds.)

and fertilizers, with environmental fallout (Pretty et al., 2018). Around 4 million metric tons of pesticides are used annually, with a small proportion reaching the target organism (25%), releasing a large proportion into the environment as a potential hazard (Zhang et al., 2018). Increasing resistance is reported to the existing groups of fungicides and pesticides. Plant improvement faces challenges with genetic modifications and plant nutrition. Plant improvement by conventional breeding is another approach to increasing yield and nutrient density. However, plant breeding is labor-intensive, time-consuming, genetically non-targeted, and could potentially introduce undesirable traits. Plant nutrition by fertilizer application also faces the problem of poor efficiency. A large proportion of the fertilizers are lost as run-off, with severe environmental consequences like the eutrophication of water bodies. Sustainable agricultural practices consider the long-term effects of contamination by agrochemicals.

Recently, nanotechnology applications have gained importance for efficient agrochemical applications through the development of delivery systems for pesticides, genetic material, and fertilizers. Such approaches are intended to improve application efficiency, reduce toxicity, and enhance the agrochemical performance of the active ingredients (AI). The negative impacts of excessive agrochemicals on ecosystems and human health are curtailed. Nanomaterials have the potential for pesticide and fertilizer deployment attributed to their nano-scale properties and large surface-to-volume ratio. A variety of inorganic and organic nanomaterials, such as metal oxides, ceramics, silicates, magnetic materials, lipids, polymers, dendrimers, and emulsions are reported as agrochemical carriers (Pouci et al., 2008; Ghormade et al., 2011) (Figure 12.1). Nanotechnology has improved the toxicity profile of chemical insecticides and fungicides and reduced their applicable field concentrations for crop protection against various pests and pathogens. The incursion of nanomaterials has improved the possibility of the application of natural products like hydrolytic enzymes, essential oils, and other metabolites for the control of pests and pathogens. The nanoformulation complex is targeted to pollen grains, protoplasts, cell lines, pathogens, and leaves by biolistic gun-aided gene integration, injection, or spraying on plants.

Conventionally, plant transformation is facilitated by *Agrobacterium*-mediated and biolistic gene-gun transformation. However, the host specificity of *Agrobacterium* and tissue damage by the gene-gun method cause limitations. Applications of nanomaterials have expanded to include genetic material delivery into cells for plant transformation and crop improvement. Nanomaterials enable the targeted delivery and release of cargo to

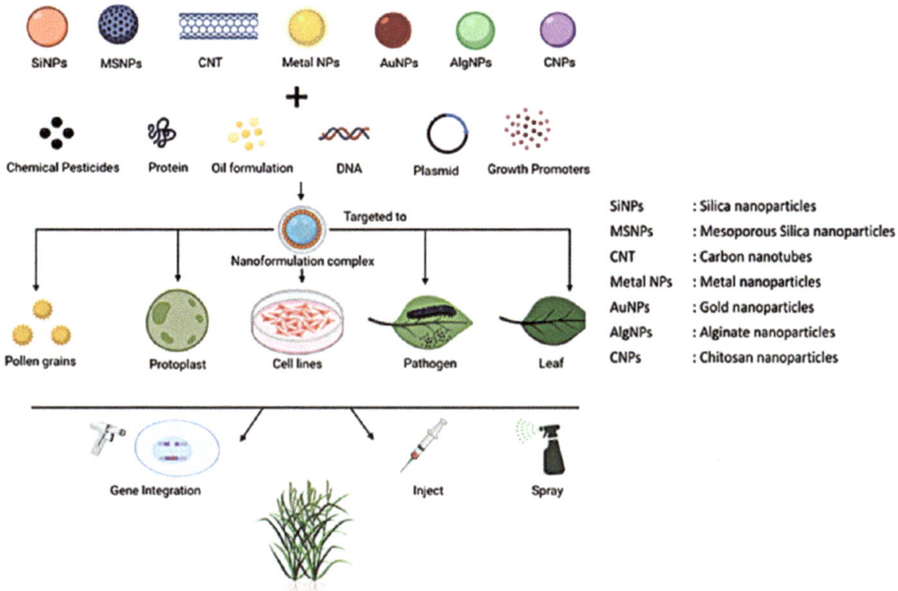

FIGURE 12.1 Nanoparticles in pesticide and fertilizer delivery: Schematic representation of different nanoparticles as delivery agents or carriers for various chemical pesticides, proteins, oil formulations, genetic material, plasmids, fertilizers, and growth promoters.

plant cells in a species-independent manner and protect the genetic material from degradation. Recent approaches for plant protection involve the non-transformative delivery of RNA interference (RNAi). The siRNAs, which are protected and delivered with the aid of nanovehicles, give plants protection against viruses and pests.

Pesticide and fertilizer agrochemical delivery is attempted with nano-composites, nanocapsules, and nanoemulsions that improve their delivery. The concerns for improved delivery can also affect the ecosystem, and the AI in the nanoparticles need to be evaluated systematically for their possible effects. The regulatory approaches are outlined to help the development of future nano-scale pesticides and fertilizers for sustainable agriculture.

12.2 NANO-MEDIATED CHEMICAL PESTICIDES/FUNGICIDES DELIVERY

Pesticides are broadly defined as compounds used for the protection of plants from pests and pathogens. The term "pesticide" includes different chemicals used as insecticides, herbicides, fungicides, antimicrobials, nematicides, rodenticides, and molluscicides in agriculture (Mali et al., 2020). Pesticides are classified into different groups depending on their chemical structure, such as organochlorine pesticides, organophosphorus pesticides, carbamate pesticides, triazines, synthetic pyrethroids, and neonicotinoids (Nicolopoulou-Stamati et al., 2016).

Crop protection chemicals are required to remain active in various environmental conditions (sun, rain, heat, or cold) and reach the targeted organism effectively in an eco-friendly manner. However, several pesticides face the requirement of organic solvents or co-formulants due to their hydrophobic nature (diesel, acetonitrile, N, N-dimethylformamide, etc.) for their field application. Additionally, a considerable fraction of the pesticide (> 90%) is lost before it reaches its target area due to factors such as evaporation, leaching, photolysis, and washing away. Due to these limitations, large amounts of pesticides are applied, leading to hazardous effects on human and plant health and the environment as well (Nuruzzaman et al., 2016). Excessive usage of chemical pesticides has numerous ill effects on health, which include dermatological, carcinogenic, respiratory, endocrine, and gastrointestinal effects (Nicolopoulou-Stamati et al., 2016). Indiscriminate use of chemical pesticides also leads to the development of pesticide-resistant strains of insects and pathogens, biomagnification of chemicals in the food chain, soil, and groundwater pollution, reduced soil diversity, etc. (Nuruzzaman et al., 2016).

Currently, nanotechnology is being explored for its application to pesticide delivery in agriculture. Using nanomaterials as a carrier or as a coating for the pesticide can protect the pesticide from heat and other abiotic factors (Nuruzzaman et al., 2016). Liu et al. (2022) reported the pesticidal activity of aqueous abamectin/graphene oxide (Abm/GO) nanoformulation for *Plutella xylostella* (diamondback moth) control. Abm is a broad-spectrum pesticide but has poor water solubility and gets rapidly degraded under UV light. A nanoformulation was synthesized by simple adsorption of Abm with GO. The Abm/GO suspension showed enhanced water solubility as compared with only Abm. The foliar spray of Abm/GO nanoformulation achieved improved pesticidal activity (LC_{50} = 0.996) as compared with only Abm (LC_{50} = 1.503). Moreover, nanoformulation displayed higher stability under UV exposure (Liu et al., 2022).

Shoaib et al. (2018) employed three different nanoformulations of emamectin benzoate for the control of diamondback moths (*Plutella xylostella*). The emamectin benzoate polymeric nanocapsules (EB+PNC), emamectin benzoate silicon dioxide nanoparticles (EB+SNP), and emamectin benzoate mesoporous nanosilica (EB+MCM-48) nanoformulations improved the UV stability of the agrochemical. First, the nanoformulations were prepared, and their encapsulation/entrapment efficiency and UV stability were determined. The entrapment efficiency of the nanoformulations was EB+PNC at 92.84%, EB+SNPs – 87.45%, EB+MCM-48 – 71.19%, EB+PNC and EB+MCM-48 were more stable under UV light than EB+SNP. However, EB+SNP showed higher pesticidal activity than EB+PNC and EB+MCM-48. The LC_{50} for EB, EB+PNC, EB+SNP, and EB+MCM-48 was 4.22 mgL^{-1}, 4.52 mgL^{-1}, 0.18 mgL^{-1}, and 1.8 mgL^{-1}, respectively (Shoaib et al., 2018).

Avermectin (Av) is a natural macrolide compound secreted by *Streptomyces avermitilis* and is widely used as an insecticide and nematicide in agriculture. It displays efficient, nontoxic, and selective insecticidal activity. Sodium lignosulfonate was cross-linked by p-phenylenediamine diazonium salt and embedded with a pesticide molecule for the development of an Av nano-delivery system (Av-NDs) of 80–150 nm particle size (Mo et al., 2021). The stability of Av-NDs was three to four times higher in comparison to Av, and Av-NDs showed controlled release of the insecticide. The Av-NDs displayed enhanced insecticidal activity compared with Av.

Gao et al. (2020) synthesized a temperature-responsive release formulation for the controlled release of thiamethoxam, a broad-spectrum systemic insecticide. The nanoformulation was prepared by coating hollow mesoporous silica (HMS core) with the thermoresponsive copolymer, poly-(N-isopropylacrylamide-co-methacrylic acid) (P(NIPAMMAA)) (outer shell). The THI@HMS@P (NIPAM-MAA) formulation protected THI against UV degradation and showed good leaf adherence. The insecticide was released by nanoformulation in a temperature-sensitive manner, and release was 25% at 28°C and 50% at 34°C. The insecticide release and mortality of brown planthoppers were positively correlated with temperature. The nanoformulation was active for 14 days and did not affect the rice seedling's growth. This approach has the potential for improved pesticide efficacy and reduced pesticide usage.

The lepidopteron insect pest *Plutella xylostella* infests several vegetables and crops. *Plutella* has become resistant to insecticides, and innovative strategies for pest control are highly desirable. Bilal et al. (2020) synthesized fluorescent mesoporous silica nanoparticles by embedding carbon dots (FL-SiO_2NPs) and loading them with indoxacarb (IN@FL-SiO_2NPs, 24% loading). The IN@FL-SiO_2NPs displayed pH-sensitive release of indoxacarb and better insecticidal activity against *P. xylostella* than indoxacarb. Interestingly, the particles suppressed the activities of *P. xylostella* detoxification enzymes, such as GST, CarE, and P450. These results indicated that nano-mediated delivery may have applications for the management of insecticide resistance.

Lambda-cyhalothrin (LC) is a broad-spectrum pyrethroid insecticide that mimics *Chrysanthemum cinerariaefolium* natural pyrethroid molecules. However, LC is a hydrophobic pesticide with the capacity to bind to soil components and was banned in the EU in 2021 due to the risk posed to aquatic organisms. Thus, the encapsulation of LC was attempted by Ganilho et al. (2022) to reduce its toxicity for sustainable agriculture. Lipid nanoparticles showed high encapsulation of LC (98% entrapment) and stability for 4 months. The LN–LC nanoparticles were 165 nm in size and had a –38 mV charge. The environmental safety of LN–LC nanoparticles was evaluated with the crops *Solanum lycopersicum* (dicotyledonous) and *Zea mays* (monocotyledonous), soil invertebrates, and soil microbial factors. Crop growth and invertebrate reproduction were not inhibited by the LN–LC nanoparticles. However, the effect of the particles on soil pH at the highest concentration can inhibit or stimulate microbial parameters. Encapsulation of pesticides by lipid nanoparticles can contribute to decreasing their release into the environment.

The effect of nanoencapsulation on the pesticidal activity of imidacloprid was studied by Kumar et al. (2014). The nanoencapsulation was done using sodium alginate and calcium chloride. An emulsion of sodium alginate was prepared, and to it, calcium chloride was added. For pesticide encapsulation, the pesticide was dissolved in the alginate solution before adding calcium chloride. The cytotoxicity of imidacloprid on Vero cell lines was reduced by nanoencapsulation. The pesticidal activity was checked on Okra leafhoppers by spraying the conventional formulation and nanoformulation. After 15 days, the population of leafhoppers on control plants was 15 per three leaves, 1 per three leaves on plants sprayed with nanoformulation, and 5 per three leaves on plants sprayed with normal pesticide. Nanoencapsulation increased the pesticidal activity of nanoformulation, which lasted for 15 days (Kumar et al., 2017).

Pandey et al. (2020) studied the pesticidal and antifungal activity of ready-to-use, oil-in-water-based nanoemulsions of diacyl and sulphonyl acyl hydrazines. Administration of the nanoemulsion by topical application and diet incorporation showed its improved pesticidal activity as compared with conventional formulations. Both the topical application method and the diet incorporation method showed that the sulphonyl acyl derivative displayed the highest pesticidal activity with growth inhibition (0.010–0.015 mg L^{-1}

GI_{90}). The nanoformulations also showed enhanced antifungal activity as compared with conventional formulations. All the nanoformulations showed moderate to good antifungal activity (Pandey et al., 2020).

Prochloraz (N-propyl-N-[2-(2,4,6-trichlorophenoxy) ethyl]-imidazole-1-carboxamide) is a broad-spectrum chemical fungicide used against different pathogenic fungi. However, low absorption and distribution in plants hampers its efficiency. Therefore, Abdelrahman et al. (2021) studied the translocation and distribution of prochloraz-loaded mesoporous silica nanoparticles cross-linked with pectin (Pro@MSN-Pec) and their effectiveness against *Magnaporthe oryzae*, the rice blast disease fungus. The Pro@MSN-Pec (70 nm) displayed a controlled release of prochloraz when it encountered pathogen-secreted pectinase enzyme. The Pro@MSN-Pec had an average size of 70 nm. It was reported that the addition of pectinase enzyme induced higher fungicide release from Pro@MSN-Pec as compared with the control. Pro@MSN-Pec had a higher fungicidal activity (effective concentration, 0.248 mg L^{-1} EC_{50}) as compared with prochloraz (EC_{50} value, 0.725 mg L^{-1}). The application of nanoformulation on leaves and roots showed that fluorescently labeled mesoporous silica nanoparticles and prochloraz were translocated to the whole plant. Application of Pro@MSN-Pec in plants showed slightly higher residual prochloraz levels than other treatments, although they were lower than the maximum allowable concentrations (Abdelrahman et al., 2021).

Majumder et al. (2020) studied the effectiveness of amphiphilic polymer-based nanoformulations of mancozeb on the early blight of tomatoes. Mancozeb is a dithiocarbamate fungicide that inhibits spore germination and is used for controlling different fungal diseases. Amphiphilic copolymers of dimethyl-5-hydroxyisophthalate and polyethylene glycols of different molecular weights (600, 1000, 1500, and 2000) were synthesized and used for the encapsulation of mancozeb. It was observed that the formulations were effective in controlling the disease as compared with mancozeb. The field efficacy of mancozeb nanoformulations was checked at 50 mg L^{-1} and 100 mg L^{-1} doses, and a single application of 60 g a.i./ha was effective in comparison to the commercial formulation (650 g a.i./ha) (Majumder et al., 2020). A controlled release of mancozeb was reported, with a period of optimum availability of up to 35 days.

Metsulfuron methyl is an herbicide that inhibits the synthesis of certain amino acids and is used against weed growth during crop cultivation. The effectiveness of metsulfuron-methyl-loaded pectin nanocapsules against the weed *Chenopodium album* was checked by Kumar et al. (2017). The herbicide-loaded nanocapsules had a particle size of 164 nm, and the herbicide loading was 6.3%. Cytotoxicity studies showed that the pectin-loaded nanocapsules had lower toxicity than the original herbicide. In field trials, it was observed that the nanocapsulated herbicide was more effective in inhibiting the growth of weeds (dry biomass 5 g/m^2) even at lower concentrations as compared with bare herbicide (Kumar et al., 2017).

12.3 NANO-MEDIATED BIOPESTICIDES DELIVERY

Although the immense contribution of synthetic pesticide application against crop pests and the high productivity of farm yields are undeniable, their harmful environmental effects are considerable. Apart from health challenges to humans, these pesticides damage the ecosystem, causing environmental pollution, the development of pest resistance, and the loss of biodiversity. It is pertinent to reduce the dependence on synthetic pesticides and adopt bio-based pesticides. Alternative methods of pest control are sought due to the harmful effects of pesticides. Botanical compounds consist of a rich source of bioactive chemicals that display a variety of insect-deterrent activities. Essential oils are bioactive compounds of plant origin that have toxicant, growth-regulating, repellent, and phagodeterrant properties (Werdin-Gonzalez et al., 2014). However, essential oils face the problem of stability due to rapid evaporation and degradation upon exposure to high temperatures, light, and air. Nanoformulations are reported as an important tool in improving the stability profiles of these compounds by protecting the AI of essential oils from degradation and evaporation. Campos et al. (2018) have described β-cyclodextrin functionalized chitosan nanoparticles (CNPs) for the entrapment of carvacrol and linalool. The nanoparticulate system displayed good insecticidal activity with high mortality against the lepidopteran *Helicoverpa armigera* (86% mortality) and a sixfold reduction in oviposition of the spider mite, *Tetranychus urticae*. The nanoencapsulation compounds showed reduced cytotoxicity as compared with the essential oils. Therefore, a nanocarrier system for the two botanical insecticides (carvacrol and linalool) showed enhanced efficacy and reduced toxicity as an environmentally safe system for the control of agricultural pests.

The effect of citrus peel essential oil-loaded polyethylene glycol nanoparticles was studied against *Tuta absoluta*, an invasive tomato pest (Campolo et al., 2017). The citrus oil nanoformulation showed significant insecticidal activity with higher mortality as compared with the oil emulsion.

In addition, the nanoformulated citrus oil significantly reduced the phytotoxicity, suggesting that such nanoformulations have the potential for integrated pest management. Further, the nanoformulation of *Rosmarinus officinalis* essential oil was studied for the management of *Tribolium castaneum*, the red flour beetle (Khoobdel et al., 2017). The authors used the nanoprecipitation method to synthesize rosemary oil-loaded nanocapsules. The EO nanoformulation showed increased fumigant and contact toxicity with a slow release of active components and was more effective against *T. castaneum* than the bare oil.

Ziaee et al. (2014) demonstrated that cumin oil-loaded nanogels were more effective than pure oil against two stored product pests, *Tribolium confusum* and *Sitophilus granarius*. They reported that the nanoformulation-stabilized cumin oil lost 15% fumigant toxicity against *T. confusum* and 60% for *S. granarius* after 12 days, whereas the bare oil completely lost its fumigant toxicity. Similarly, Yang et al. (2009) reported that polyethylene glycol-coated nanocapsules displayed 80% control efficacy after 5 months, whereas free garlic at a similar concentration achieved 11% control efficacy.

The contact toxicity of *Mentha longifolia* L. essential oil on *Ephestia kuehniella* Zeller, the Mediterranean flour moth was enhanced by its nanoemulsion, showing enhanced contact toxicity and stability (Louni et al., 2018). The nanoemulsion protected the EO and showed a slow-release profile. Evaluation of geranium essential oil nanoformulation showed high biological efficacy against the first instar of *Phthorimaea operculella,* the potato tuber moth (Ibrahim, 2017). The nanoformulation comprising geranium oil-loaded solid lipid nanoparticles led to significantly increased larval stage mortality, reduced adult progeny, reduced female fecundity, and egg hatchability. Application of the nanoformulated oil on potato crops under field conditions exhibited improved efficacy compared with free essential oil. This suggests that the application of EO nanoformulation can contribute to reduced insecticide applications for insect control.

Gayathri et al. (2018) carried out a field evaluation of nanoparticle-enriched biopesticides. The bacterial and fungal entomopathogens *Bacillus thuringiensis* (Bt), *Nomuraea rileyi,* and *Beauveria bassiana* were grown in nanoparticle-enriched media and sprayed against an armyworm, *Spodoptera litura,* infestation in the peanut crop. The *Bt* with calcium oxide (CaO NPs) in barley-based solid formulations against *S. litura* larvae resulted in the highest mean percentage reduction (77%), followed by CaO NPs-based *B. bassiana* talc formulations (73.72%), and CaO NPs-based *N. rileyi* broken rice formulations (66.44%). The Bt with CaO NPs was an effective insecticide, having the highest pod and haulm yields of 1993 kg ha^{-1}, and 3899 kg ha^{-1}, respectively (Gayathri et al., 2018).

Spinosad, a new generation of eco-friendly pesticides, faces limitations due to its short-duration effect and low potency in field conditions. Zhang et al. (2020) loaded spinosad on silica nanoparticles with spiky nanotopography for protection and adhesion. Exposure to solar radiation revealed that the nanospinosad was >10 times more photostable than the commercial product. Topically applied nanospinosad formulations displayed improved adhesion to both cattle hair and pest surfaces, contributing significantly to the higher mortality of ticks (*Rhipicephalus microplus*, an ectoparasite). Nanotechnology-inspired nanobiopesticides can contribute to sustainable agricultural applications.

A variety of agricultural insect pests have been successfully controlled using surface-modified hydrophobic nanosilica by physisorption. Properly functionalized lipophilic nanosilica is absorbed into insect cuticular lipids, causing damage to the protective wax layer (made of a variety of fatty acids and lipids that prevent water from evaporating) and death by desiccation. It has been reported that porous hollow silica nanoparticles with a shell thickness of nearly 15 nm and a pore diameter of 4–5 nm can shield pesticides from UV light degradation (Nair et al., 2010). Nanoencapsulation of botanical insecticides imparts greater selectivity and specificity in pest control and contributes to a reduced impact on other organisms and the ecosystem. An enzyme-responsive nanoencapsulation botanical insecticide was devised by loading a mixture of botanical insecticides in zein nanoparticles with release triggered by proteolytic enzymes (Camara et al., 2021). The nanoformulations displayed >95% encapsulation efficiency and stabilized the essential oils. The trypsinization of NPs increased the release of geraniol from 15 to 56%. Upon consumption of the NPs, the proteolytic enzyme in the midguts of insects and larvae reacted with the proteinaceous zein NPs to release the insecticides. This approach displays a significant perspective on the control of crop insects.

Dikbaş and Cinisli (2019) suggested that bacterial-origin chitinase enzymes immobilized on zinc oxide nanoparticles may be applied as nanobiopesticides against harmful insects in agricultural production. The nanoparticles can contribute to enzyme stabilization. They proposed that the nano-bio formulation could cause insect pest mortality through the breakdown of chitin-containing insect intestinal cells and suppression of intestinal enzymes. Thus, nanobiopesticides based on new nanobiotechnological methods can reduce the risk of pollution due to traditional pesticides and emerge as alternatives to conventional pesticides. Copper and silver nanoparticles (AgNPs) synthesized by the green synthetic method and crude fermented neem extract (FNE) were applied for pest management of the fruit fly, *Bactrocera dorsalis, using the* insecticidal properties of neem (*Azadirachta indica*) (Paragas et al., 2021). Application of 100% FNE, 20% FNE-copper nanoparticles, and 20% FNE-AgNPs caused the maximum death of adult male fruit flies (83–100%) and exhibited ovicidal properties. FNE can be used to formulate biopesticides enhanced with nanoparticles for controlling *B. dorsalis.*

12.4 NANOTECHNOLOGY-ENABLED DNA AND RNA DELIVERY FOR IMPROVED PLANT PROTECTION

Delivery of DNA and siRNA is important for the development of transgenic pest-resistant crops through altered gene expression. Conventional methods for plant genetic transformation like *Agrobacterium-mediated* transformations, microprojectile bombardment, and viral gene delivery face the problem of low efficiency. *Agrobacterium-mediated* transformations are limited to suitable host plants, while particle bombardment can harm plant tissues and cause random insertions. Viral gene delivery techniques for plant transformation display a narrow host range, have limited size for inserted genetic material, and face the possibility of viral infection.

Nanotechnology can contribute to the development of efficient gene delivery to overcome such hurdles during genetic transformation. Particularly, nanomaterials provide the ideal platform for the delivery of biomolecules due to their unique physiochemical properties and capacity to cross the difficult barrier of the plant cell wall (Kumar et al., 2020). The current methods of crop management can greatly benefit from the addition of nanotechnology. Numerous nanoparticles, including chitosan, liposomes, cationic dendrimers, and others, have emerged as a result of the rapid advancement of nanotechnology over the past 10 years and have been constructed and designed for the delivery of dsRNA and small interfering RNA (siRNA) (Yan et al., 2021).

Different biomolecules, such as DNA, miRNA, siRNA, and ribonucleoproteins, can be delivered by the nanomaterials. These systems are biocompatible and regarded as non-cytotoxic vectors. Delivery systems for genetic material face difficulties with limited host range, transport across the cellular membrane, and targeting to the nuclear contents. By introducing new genes that correspond to crop improvement, nanotechnology has demonstrated its value in the genetic modification of plants. When compared with traditional and conventional approaches to gene transformation, this system offers significant advantages (Ghormade et al., 2011). First, nanoparticle methods can be used on any plant, monocot, or dicot, regardless of the type of tissue or organ. Second, by regulating the DNA copies in conjunction with nanoparticles, they can be utilized to overcome transgenic silencing. Thirdly, if necessary, nanoparticles can be easily functionalized to boost transformation efficiency further. Lastly, nanoparticle-mediated multigene transformation can be accomplished without the use of conventional approaches that necessitate complex carriers (Kashyap et al., 2015). For instance, the "gene-gun" system uses DNA-coated Gold NPs (5–25 nm) embedded in carbon matrices as bullets to bombard plant cells and tissues to transfer genes (Vijayakumar et al., 2010). Consequently, it was suggested that the biolistic co-transformation method using gold nanoparticles for gene delivery into rice varieties was successful (Mortazavi et al., 2018). Hao et al. (2013) synthesized magnetic gold nanoparticles (mGNPs) by sonication treatment and bound them covalently with the fluorescent dye fluorescein isothiocyanate (FITC). The FITC-labeled mGNPs were bound covalently to plasmids expressing the GUS (β glucuronidase) gene. The FITC-labeled nanoparticles achieved high delivery efficiency and showed high expression of the GUS gene in transformed canola plant cells with and without cell walls. In addition, Torney et al. (2007) exhibited that a honeycomb surface-functionalized mesoporous silica NP framework with 3 nm pores covered with disulfide bond-held gold NPs (10–15 nm) moved DNA and synthetics into isolated plant cells and leaves aided by biolistic method. Recently, Hajiahmadi et al. (2019) delivered the Cry1Ab gene loaded on mesoporous nanoparticles in tomato plants by foliar spray or by shoot injection. The CRY proteins are a family of toxins produced by the bacterium *Bacillus thuringiensis*. The MSNs successfully delivered the Cry1Ab gene to plant cells and achieved transient gene expression and protection against tomato leafminer *Tuta absoluta* larvae. These studies disclose the potential of MSNs for non-transgenic DNA delivery to plants.

The utility of carbon nanotubes (CNTs) is demonstrated for the unaided delivery of plasmid DNA (pDNA) and siRNA into a wide range of both model and non-model plant species (Kwak et al., 2019; Demirer et al., 2019). CNTs are nano-scale structures made of single-layer carbon atoms, like graphene sheets. Giraldo et al. (2014) reported that coating SWCNTs with chitosan facilitated their passive passage through the lipid bilayer membrane of chloroplasts. Kwak et al. (2019) functionalized SWCNTs by covalent binding with chitosan molecules (CS-SWCNTs). The CS-SWCNTs selectively delivered the pDNA with the reporter YFP gene to the chloroplasts of tobacco, spinach, watercress, and arugula plants. Binding with SWCNTs protected the pDNA from nuclease degradation. While the mild pH conditions in cell cytosol helped the binding of pDNA to chitosan-complexed SWCNTs, the acidic pH of chloroplast stroma caused the release of the plasmid from SWCNTs.

Golestanipour et al. (2018) reported DNA delivery to tobacco root cells with chimeric peptides and arginine-functionalized SWCNTs (Arg-SWCNTs). Recently, Ghaghelestany et al. (2020) employed ultrasound for CNTs-mediated gene transfer to German chamomile cells. Functionalization of CNTs with polyethyleneimine rendered them cationic, followed by attachment of ssDNA-FITC (fluorescein isothiocyanate) conjugate to the CNT surface by electrostatic interactions. Cationic polymers such as chitosan are used to condense and deliver DNA in mammalian cell lines and in animal models (Kim et al., 2007). For the purpose of delivering pDNA, researchers have looked into nanoparticles (NS) made with the biodegradable polymer poly (d,l-lactide-co-glycolide) (PLGA). Chitosan (CHS), a cationic polymer, was incorporated into the PLGA matrix to boost cellular uptake and pDNA loading efficiency (Zeng et al., 2011).

Thagun et al. (2022) developed peptide nanocarriers as a high-throughput delivery platform for nucleic acids to plants for application as a foliar spray. The cell-penetrating peptide (CPP)-based foliar spraying showed efficient DNA delivery into plants. Furthermore, they achieved gene silencing in chloroplasts by delivery of siRNA by chloroplasts targeting CPP complexes. Upon spraying, the uptake of the nucleic acid-bound CPP complexes in plant cells depended on peptide characteristics and a stomatal-dependent mechanism. Such technology has the potential for non-transgenic transformation of plants and engineering economically important traits. Recently, a nanoparticle-mediated plant transformation approach was reported using biocompatible hydroxyapatitie nanoparticles (nHAs) in model and crop plants (Izuegbunam et al., 2021). Transgenes were successfully introduced by arginine-functionalized nHAs for transient expression into Arabidopsis, tobacco (*Nicotiana benthamiana*), and common ice plant (*Mesembryanthemum crystallinum*) leaves. Moreover, germinating seeds of *Arabidopsis*, field mustard (*Brassica rapa*), wheat (*Triticum aestivum*), and barley (*Hordeum vulgare* L.) were successfully transformed. Hydroxyapatite (HA) nanoparticles were successful in the passive, efficient, passive, and harmless delivery of genetic cargo.

Insect pest control is a serious challenge in agriculture. The rapid development of resistance to chemical insecticides and Bt-transformed plants has exacerbated the situation. The emergence of non-target pests and environmental concerns necessitate new strategies for pest management. The incorporation of a novel RNAi approach for pest management has given pest management a new outlook. In the fields of pest management and functional gene analysis, RNAi has emerged as an essential method. Since its discovery in the nematode *Caenorhabditis elegans*, double-stranded RNA (dsRNA)-triggered RNAi has been extensively utilized in the functional analysis of insect genes and the management of pests (Fire et al., 1998; Bellés, 2010; Zhu and Palli, 2020). RNAi-based pest control is a sustainable, eco-friendly strategy with the potential to target pest species with great specificity. However, insects are extremely diverse and have demonstrated wide ranges of RNAi responses and susceptibilities. Effective delivery of dsRNA molecules to insects is still a serious challenge. The largest order of insects is Lepidoptera, which includes butterflies, skippers, and moths. Some of them are significant crop pests. Lepidopteran adults play an important role in pollination, and some species, like the silkworm (*Bombyx mori*), can even be beneficial to humans. Although Lepidopteran larvae cause significant damage to important crops, an effective strategy is required to control them. Lepidopterans are highly adaptable, and some of these adaptations can protect them from food, the environment, and climate change. Lepidopterans have developed a variety of survival mechanisms thanks to their high adaptive value and environment-specific physiological conditions (Dawkar et al., 2013).

The discovery of RNA-based silencing has given a new vision to crop improvement by allowing specific control of insect pests that feed upon the dsRNA-producing plants (Auer and Frederick, 2009; Price and Gatehouse, 2008). The CNP-dsRNA complex has been used in different insects, and the target genes were found to be effectively knocked down in the species. Targeted *chitin synthase* genes *AgCHS1* and *AgCHS2* and the axon guidance gene *semaphorin-1a* (*sema1a)* were knocked down successfully in *A. gambiae* and *A. aegypti* (Zhang et al., 2015). The Asian corn borer, *Ostrinia furnacalis,* was fed dsRNA-loaded cationic core-shell fluorescent nanoparticles (which consist of fluorescent perylene-3,4,9,10-tetracarboxydiimide chromophore (PDI) in the center and two different types of outer shells) to knock down key CHT10 developmental gene expression (He et al., 2013). CHT10 is a chitinase gene present in the midgut periotrophic membrane of *O. furnacalis* and has a role in insect hatching, larval molting, pupation, and metamorphosis. The dsRNA FNPs caused serious defects during larval growth and molting, leading to insect death (He et al., 2013).

Guanylated nanoparticle-coated dsRNA was used to target chitin synthase in the lepidopteran species *Spodoptera exigua* with increased mortality (53%) in comparison to the bare dsRNA control (Christiaens et al., 2018). Application of CNPs for dsRNA delivery against *H. armigera* chitinase and lipase genes improved dsRNA uptake and protected it against degradation. The oral administration of CNPs-dsRNA with artificial or leaf feed to *H. armigera* larvae resulted in efficient gene downregulation, reduced enzyme activity, and increased insect mortalities (Kolge et al., 2023). The dsRNA is probably assimilated by the cell through endocytosis following restriction to the cell layer (Li et al., 2013). However, the precise mechanism for the nanoparticle-coated dsRNA's movement is unclear. The transport of nanoparticles was traced in *Tribolium castaneum* insect whole body with fluorescently labeled dsRNA complexed with branched amphiphilic peptide capsules (BAPC) delivery agent. As opposed to naked dsRNA, which did not exhibit any such fluorescence, it was demonstrated that dsRNA complexed BAPC was localized in the midgut, fat bodies, and the malphigian tubules (Avila et al., 2018). Studies conducted with silica, chitosan, and carbon quantum dot (CQD) nanoparticles complexed with dsRNA to target two mosquito genes (SNF7 and SRC) for controlling *Aedes aegypti* larvae showed 15%, 46%, and 75% mortalities, respectively (Das et al., 2015).

It is simple to perform high-efficiency gene knockdowns on a variety of genes in coleopteran insects (Palli et al., 2014). Lepidopteran insects, on the other hand, are not affected by dsRNA, and the main obstacles to RNAi success are the substantial dsRNA degradation by dsRNAases, the accumulation of dsRNA in acidic granules, and the absence of the core RNAi machinery (Huvenne and Smagghe, 2010; Yoon et al., 2017; Cooper et al., 2019). As a result, in order to increase the effectiveness of RNAi for the purpose of identifying functional genes in lepidopteran insects, a powerful yet simple strategy for dsRNA protection and effective delivery is urgently required. The degradation of dsRNA by nucleases greatly influences variability in the insect RNAi response. The insect digestive system contains various dsRNA nucleases as well as varying pH conditions. dsRNA is unstable in very alkaline conditions. Lepidopterans with a high alkaline gut pH are less sensitive to RNAi, while Coleopterans are reported to show robust RNAi responses (Adeyinka et al., 2020).

Recently, Mitter et al. (2017) developed biodegradable BioClay nanocarrier for dsRNA delivery to plants as a foliar spray. The dsRNA was bound to cationic layered double hydroxide (LDH) by electrostatic forces and was protected from nuclease activity. The application of BioClay to Arabidopsis leaf surfaces led to the generation of carbonic acid from atmospheric CO_2 and humidity, which led to BioClay degradation and dsRNA release. The RNAi-mediated systemic protection was achieved by BioClay's delivery of dsRNA to plant cells and protected the plants against virus infection.

Kolge et al. (2021) used foliar spray with CNPs for RNAi targeting *H. armigera* larvae. The CNPs protected dsRNA from nuclease activity and improved its gut uptake in *H. armigera* larvae. The CNPs-dsRNA were efficient, and their oral administration with artificial or leaf feed to *H. armigera* larvae in bioassays resulted in gene silencing, suppression of enzyme activity, and high insect mortalities (Kolge et al., 2023). Field application showed that a single spray of CNPs-dsRNA was as effective as the conventional chemical control. The modern CRISPR–Cas genome-editing technology received the Nobel Prize in 2020. CRISPR–Cas technology has immense potential in the areas of plant biology and agriculture for transformation in plant functional genomics research and crop improvement (Demirer et al., 2021). CRISPR–Cas genome editing has been extensively studied in animal systems, and the first clinical trials involved patient T4 cells. Delivery methods for CRISPR–Cas in animal cells include electroporation or viruses. Currently, cationic lipid-based and gold nanoparticles have emerged as alternative vehicles for delivering CRISPR editing components to mammalian cell lines and mice.

The application of CRISPR technologies in plants needs to address the challenges of CRISPR cargo delivery to plant tissue or cell culture and the need for common methods applicable across plant species (Demirer et al., 2021). These challenges can be addressed by nanoparticles as carriers for the delivery of Cas9 RNPs via cleavable chemical bonds for targeted release or their direct delivery into plant cells. The controlled delivery of Cas protein, sgRNA, and donor DNA with nanoparticles can be tailored to orchestrate the maximum cellular expression levels of Cas protein. Nano-scale carriers can contribute to the transformation of plant germline cells as they have the potential to infiltrate the pollen through its apertures. Further, negatively charged nanoparticles can be utilized for Cas-sgRNA complex stabilization and as a vehicle to carry the donor DNA to the nucleus. Mahmoud et al. (2022) used lipofectamine and cationic lipid nanoparticles to deliver donor DNA into citrus protoplasts. Polyethylene glycol improved the transfection of a Cas9 construct containing a gRNA targeting the *CsNPR3* gene by 51% as compared with bare lipofectamine (30%). Successful genome editing resulted in nine edited plants that downregulated *CsNPR3* expression and upregulated *CsPR1* expression.

Shen et al. (2023) employed magnetic nanoparticles (MNPs) as vehicles for the transfection of the CRISPR/Cas9 expression vector with the cargo gene. The rice pollen was magnetofected and was used to obtain T1 plants by pollination and T2 plants by self-pollination. The mutations obtained from T2 generation demonstrated the early development of mutant plants in a tissue culture-independent manner, which shortens the breeding cycle. This approach has significance for the acceleration of the rice breeding process for newer varieties.

12.5 NANO-MEDIATED FERTILIZER DELIVERY

Fertilizers provide nutrition to plants. The addition of fertilizers to soil contributes to increased soil fertility, improved crop yields, and nutrient-enriched crops. In general, either natural or chemical fertilizers can be used for improved yields. Fertilizers are categorized as bio-fertilizers (of biological origin) and chemical fertilizers (of synthetic origin). Chemical fertilizers or synthetic fertilizers containing nitrogen, phosphorus, and potassium are added to soils for improved fertility and enhanced crop yields. Although, the extensive fertilizer applications increase the crop productivity, it is harmful in the long run as they can disturb the natural soil fertility, beneficial microflora, and organic matter contents. Generally, there is an application of excess fertilizer. The plants use 40% of the nutrients from the applied fertilizers, while the rest is washed away (Kumaraswamy et al., 2021). Tilman et al. (2002) reported that excess use of fertilizers and pesticides leads to pathogen and pest resistance, reduces beneficial soil microbial organisms, lowers nitrogen fixation, leads to pesticide bioaccumulation, and destroys habitats for birds.

Nanoparticles have potential applications in agriculture systems, viz., for controlled delivery of fertilizers, nutrients, and genetic material (Ghormade et al., 2011). Usually, the nutrients are loaded on nanoparticles by absorption, ligand-mediated attachment, encapsulation, and entrapment (Torney et al., 2007; Tsuji et al., 2001; Boehm et al., 2003). Another method is to synthesize nano-sized nutrient particles. The nutrient use efficiency for conventional fertilizers like nitrogen, phosphorus, and potassium is 30–35%, 18–20%, and 35–40%, respectively. Nanofertilizers can contribute to improved nutrient utilization, lower cost, and lower vol/wt due to their nanosize-related properties (Azam et al., 2022).

Various strategies are emerging for nanofertilizer applications. Briefly, improved yields are reported by the slow release of chemical compounds for plant growth entrapped in double-layered zinc-aluminum hydroxide nanocomposites. Moreover, the absorption of silica nanoparticles by roots enhances the plant's resistance to stress and contributes to improved yield. The composition of nanoparticles plays a significant role in their application. For example, polymer-coated nanoparticles are used as agrochemical carriers due to their controlled-release ability, whereas metal nanoparticles show size-dependent properties, such as magnetism, fluorescence, and photocatalytic degradation, which have applications in development of detection devices, degradation of pesticides, removal of soil contaminants. Surface coatings of nanomaterials on fertilizer particles have a strong holding capacity, thus helping in controlled release. Delivery of nanofertilizers for supplying macro- and micronutrients shows a controlled release of agrochemicals, targeted delivery, amelioration of toxicity, and improvement of nutrient utilization (Cui et al., 2010). Nanotechnology has improved the scale-up process of various metal nanoparticles production for use in fertilizer formulations. Such an approach will contribute to their increased nutrient uptake in plants with minimal losses (Solanki et al., 2015).

Globally, nitrogen (N) is a critical element that affects agricultural production and faces the limitation of low N use (< 50% efficiency) (NUE). N is commonly applied in granular form to the soil, which results in losses via leaching, ammonia volatilization, and oxide (N_2O, NO, and NOx) emissions. The incorporation of innovative nanomaterial-based strategies for N fertilizer is expected to improve the NUE through effective N delivery to plants and the reduction of environmental losses (Mejias et al., 2021). The impact of nanofertilizers was investigated on the yield and milled grain quality of two rice cultivars in Iran (Taghi et al., 2021). Nano-nitrogen (nN), nano-phosphorus (nP), nano-potassium (nK), nano-NPK (nNPK), and its combination (NPK þ nNPK) treatment were compared with the conventional NPK application in Tarom and Shiroudi rice varieties for 2 years. An increase in grain yield and milled rice yield was reported for NPK, nNPK, and their combination (NPK þ nNPK). The treated grains displayed improved characteristics, such as enhanced milling quality, elongation ratio, gelatinization temperature, reduced gel consistency, and amylose content. While nNPK and NPK þ nNPK applications enhanced grain and milled rice production, N, nN, nNPK, and NPK þ nNPK fertilizers yielded the highest grain quality. Therefore, nanofertilizer applications can improve fertilizer efficiency for improved rice yield and quality.

Recently, the patented nanourea liquid fertilizer was released by the government of India and included under the Fertilizer Control Order (Kumar et al., 2021). Nanourea spray enters the leaf through the stomata and is easily assimilated by cells. Small particle sizes (20–50 nm) improve availability by 80%. Thus, nano-scale urea leads to more efficient absorption and penetration of nitrogen for improved plant growth (Kumar et al., 2021).

Carmona et al. (2020) demonstrated the potential use of nitrate-doped calcium phosphate nanoparticles as nanofertilizers. This approach added to the knowledge of efficient macronutrient delivery to plants and the solubilization of sparingly soluble components. The role of NPs morphology was highlighted, wherein the spherically fully amorphous NPs (ACPs) showed a rapid nitrate loss in comparison to the apatite-amorphous nanoplatelets (NPLs). NPLs exhibited a slower release similar to that of Ca^{2+} ions. Further, the authors synthesized an efficient N-nanofertilizer by modification of ACP NPs with urea, which has three times more N than was obtained by equivalent one-pot synthesis (Carmona et al., 2021). Application of the nanofertilizer to cucumber plants in hydroponic conditions showed the development of an equivalent root and shoot biomass without nitrogen depletion compared with the conventional urea treatment. The nanofertilizer showed high nitrogen use efficiency (69%) and supported sustainable real-world usage.

A nanofertilizer was synthesized by the sorption of phosphorus and potassium into zeolite as a carrier material (Rajonee et al., 2017). Zeolite is a microporous aluminum silicate solid. X-ray diffraction (XRD) and chemical analysis confirmed the sorption of fertilizer material into zeolite. The *in vitro* incubation study at field moisture conditions for 30 days showed that the nano fertilizer has a higher release of P and K nutrients in comparison to conventional fertilizer. Nanofertilizer-treated *Ipomoea aquatica* (Kalmi) plants showed a higher accumulation of P and K than conventional fertilizer (Rajonee et al., 2017). Corradini et al. (2010) demonstrated the interaction and stability of CNPs colloidal mixture with N, P, and K fertilizers, which can be useful for agricultural applications (Corradini et al., 2010). Urea-modified HA nanoparticles and urea have a strong interaction that contributes to sustained urea release. These nanofertilizers showed an initial burst and subsequently slow release of nitrogen up to 60 days of plant growth compared with commercial fertilizer, which shows release of only up to 30 days (Kottegoda et al., 2011). The introduction of nanotechnology for plant nutrition has led to the development of delivery systems designed for prolonged fertilizer release. The triple-layered electrospun nanofibers were produced by electrospun coating of polyhydroxybutyrate (PHB) nanofibers as an external layer on the urea fertilizer incorporating PLLA nanofibers (Javazmi et al., 2021). The triple-layered urea-nanofertilizer exhibited a slow-release rate of cumulative nitrogen for >3 months. Therefore, prolonged fertilizer release by the nanofibrous structures displays potential for application in agriculture for improved plant growth.

Tarafder et al. (2020) proposed a hybrid nanofertilizer (HNF) composed of urea-modified HA, a rich source of nitrogen, calcium, and phosphate, incorporating copper, iron, and zinc nanoparticles for enhanced efficiency. The HNF showed a slow, sustainable release of nutrients into the soil and water. Field application of HNF to *Abelmoschus esculentus* (okra) by foliar and soil routes confirmed the availability of Ca^{2+}, PO_4^{3-}, NO_2^-, NO_3^-, Cu^{2+}, Fe^{2+}, and Zn^{2+}due to the slow release of HNF. The hybrid fertilizer exhibited considerable water absorption and retention capabilities that compared favorably to the traditional fertilizer. The improved characteristics of HNF permitted low-dose application at 50 mg/week, while traditional fertilizer was required at 5 g/week.

The ability of multiwalled carbon nanotubes (MWCNTs) is demonstrated to enhance the seed germination and growth of tomato plants. Based on relative root elongation, oxidized MWCNTs were more effective at lower concentrations than non-oxidized MWCNTs (Khodakovskaya et al., 2009). Inorganic silica nanoparticles can perform as effective carrier systems for agrochemical compounds, which improves efficiency and economic utilization. Mesoporous silica nanoparticles (150 nm) with pore sizes of 2.5 nm were reported to entrap urea. High loading of urea was observed by physiabsorption (80%, w/w) within the nanoparticle pores and demonstrated a controlled urea release profile in soil and water (Wanyika et al., 2012). Fluorescence-labeled monodispersed mesoporous silica nanoparticles were employed to understand the mode of nanoparticle uptake and their distribution in plants. The fluorescent NPs were found to penetrate the roots via symplastic and apoplastic pathways and translocate via xylem tissue to the aerial parts of the plants, including the stems and leaves (Sun et al., 2014). Suriyaprabha et al. (2012) reported that treatment with SiO_2 nanoparticles in maize plants significantly enhanced the plant's dry weight and enhanced the levels of proteins, chlorophyll, and phenols.

Kumaraswamy et al. (2021) reported a chitosan-silicon nanofertilizer (CS-Si NF) as a foliar spray and seed priming for the improvement of plant growth. They encapsulate silicon (Si) in a synthesized chitosan-tripolyphosphate (TPP) nanomatrix that enables its slow release of Si and promotes growth and yield in maize crops. Seed priming with CS-Si NF displayed increased seedling vigor as compared with SiO_2. The foliar nanofertilizer spray increased the total chlorophyll content and leaf area to increase the photosynthesis in comparison to SiO_2. Field application of foliar nanofertilizer resulted in a higher yield per plot, indicating its potential for sustainable agriculture.

Fertilizer application has an important effect on fruit yield and quality. The effect of fertilizer application was monitored on strawberry cultivar, quality, and nutritional parameters (Weber et al., 2021). Nitrogen and calcium altered the sugar, organic acid, volatile, and phenolic contents of strawberry fruits. Higher nitrogen and calcium doses increased the presence of hexanal aldehydes (3.8-fold) and (E)-2-hexen-1- (3.7-fold), giving an unpleasant aroma and decreasing the ketone and terpenoids fruity aroma. Lithovit is a nanofertilizer made from limestone composed of high calcium and magnesium contents. Strawberries fertilized with the nanofertilizer Lithovit exhibited higher fruity esters content (up to 2.3-fold), total phenols, and individual hydroxycinnamic or hydroxybenzoic acid derivatives. While nitrogen and calcium fertilizers displayed a negative impact on strawberry flavor, the nanofertilizer enhanced the phenolic contents and aroma of strawberry.

Zinc solubility and dissolution kinetics of ZnO nanoparticles and bulk ZnO particles coated on macronutrient fertilizers (urea and monoammonium phosphate) have been compared (Milani et al., 2012). Burman et al. (2013) reported that ZnO nanoparticles at 1.5 ppm had a positive effect on growth and antioxidant system chickpea seedling. Azam et al. (2022) chemically synthesized a zinc oxide nanofertilizer for application to maize cultivars, and responses reported enhanced growth, photosynthetic pigments, and antioxidant activity. Foliar and soil application of ZnO nanofertilizer to plants enhanced the growth of photosynthetic pigments and antioxidant activity versus control maize plants. Therefore, ZnO nanofertilizers have the potential to enhance crop growth and yield and boost zinc-deficient soil quality.

Zinc deficiency in plants is caused by zinc-deficient soils. Khanm et al. (2018) studied the effect of seed priming and foliar spraying with ZnO nanoparticles on improving the yield and zinc contents of tomato plants. Application of ZnO NPs fertilizers through both methods indicated that the zinc content increase was higher than that of granular zinc sulfate. Nano-sized particles display a high rate of penetration through the plant cell membrane to boost crop nutrition. The nanofertilizer-treated plants showed enhanced yield as compared with the granular zinc sulfate. Nanofertilizer with enhanced uptake displayed potential avenues for nano-scale fertilizers for improved crop productivity.

Yuvaraj and Subramanian (2021) synthesized a zinc nanofertilizer by loading zinc sulfate on carbon nanospheres. The zinc nanofertilizer showed extended nutrient release up to 600 h in percolation reactor studies, while zinc release halted after 216 h for the conventional zinc sulfate. The zinc nanofertilizer has the potential to improve zinc use efficiency and crop yield in agriculture. The application of Zn and Cu NPs significantly improved the leaf chlorophyll and carotenoid contents of basil plants (Abbasifar et al., 2020). The Zn and Cu NPs were synthesized by green synthesis after exposure to the extract of the basil plant. The application improved the antioxidant activity in the plant.

Nanosilver was better than silver nitrate at preventing leaf abscission and improving the seed yield of *Borago officinalis* plants. The foliar spray of nanosilver and silver nitrate demonstrated that nanosilver (60 ppm) was effective at a lower concentration than silver nitrate (100 ppm) (Sahandi et al., 2011). Namasivayam and Chitrakala (2011) reported that green synthesized AgNPs did not alter the seedling emergence, chlorophyll content, and phyllosphere microbial presence in cowpea, black gram, green gram, horse gram, and sorghum. They reported that Watermelon (*Citrullus lanatus*) is a high-value specialty triploid crop that faces problems with low seed germination. Acharya et al. (2020) synthesized AgNPs from agro-industrial byproducts and turmeric oil nanoemulsions (TNE) as nanopriming agents for diploid (Riverside) and triploid (Maxima) watermelon seeds. AgNPs-treated triploid seeds showed a higher seedling emergence rate at 14 days in comparison to the untreated control. Enhanced soluble sugar contents for the AgNPs-treated seeds were observed during germination at 96 h. Transplantation of greenhouse-grown seedlings showed higher yields for the AgNPs-treated Riverside and Maxima watermelons.

Application of nano-TiO_2 promoted photosynthesis and growth in spinach. Treatment with 2000 ppm TiO_2 nanoparticles led to improved seed germination and seedling vigor in rapeseed plant (Lei et al., 2007; Mahmoodzadeh et al., 2013). Treatment with gold nanoparticles increased plant stress tolerance that was associated with increased redox status in the plants. It is reported that foliar spraying of gold nanoparticles resulted in increased plant growth, pod formation, and yield of *Brassica juncea* plants in field (Arora et al., 2012). Lee et al. (2008) reported that copper nanoparticles can cross the cell membrane of mung bean and wheat and agglomerate in the cell.

Elfeky et al. (2013) studied the effect of iron nanoparticles on levels of active essential oil constituents and the yield of basil crops. Iron is an essential element for plant growth and serves as a cofactor for several enzymes. It plays an important role in photosynthesis and the development of chloroplasts. Application of iron (Fe_3O_4 NPs) nanoparticles (12.6 nm) to basil plants by foliar spray or soil addition led to improved carbohydrate, chlorophyll, essential oil, and iron contents. The foliar spray was more effective and showed enhanced plant growth and biomass. An inverse relationship was observed between the total carbohydrate content and the percentage of essential oil in the leaves. This study demonstrated the potential of nanofertilizer for improving plant growth and yields.

Encapsulation of agrochemicals for sustainable agriculture is developed via nano-based systems, but few nanocarriers are reported for plant growth regulators (PGRs) (Santo Pereira et al., 2019). Seed priming treatments improve the germination rate and contribute to increased plant growth and yield. Alginate/chitosan (*nano*ALG/CS) and chitosan/tripolyphosphate (*nano*CS/TPP) nanocarriers containing gibberellic acid (GA_3) were developed for seed priming of *Solanum lycopersicum*. The nanocarrier systems improved plant growth and yield, with a fourfold increase with *nano*ALG/CS-GA_3. The slow release of GA_3 from seed coating and the possible uptake in growing tissues were attributed to the enhanced effect of the GA_3 nanosystems. This innovative application of nanocarrier systems with PGRs is important for sustainable agriculture.

12.6 NANOCARRIERS FOR BIOLOGICAL FERTILIZER DELIVERY

Nanoparticles can affect plant growth-promoting bacteria and plants by directly improving nutrient availability or indirectly stimulating the PGPB, thereby improving the growth-promoting effects (Prado de Moraes et al., 2021). Nanocarriers protect the PGPB inoculants in the field and activate various mechanisms beneficial to the plant–microbe association. Application of NPs leads to increased cell viability and bacterial growth rate, nitrogen fixation, and the secretion of specific microbial compounds to protect plants from adverse conditions (Ghalamboran and Ramsden, 2012; Karunakaran et al., 2013; Abdel Latif et al., 2018). The application of copper oxide NPs (CuO NPs) increased indole acetic acid synthesis in *Pseudomonas chlororaphis,* while ZnO NPs increased siderophore production (Dimkpa et al., 2012). Nitrogen fixation potential in *Nitrosomonas europaea* was enhanced by AgNPs by upregulation of amoA1 and amoC2 gene expression for nitrification (Yang et al., 2013).

The promotion of PGPB growth by the NPs association can contribute to phytopathogen suppression due to competition for niches and nutrients. Additionally, nutrient uptake and increased secondary metabolite production by PGPB make plants more resistant to phytopathogen attacks. However, the effects of NPs on bacteria are not well understood, and further studies are necessary.

A study of the interaction of nanotitania (TNs) and colonization of growth-promoting rhizobacteria (PGPR) was carried out by Timmusk et al. (2018). Importantly, these experiments were conducted with peat and sandy soil to consider the behavior of PGPRs in the field. They reported that the PGPR inoculants with TNs attached stably to wheat seedling roots and showed enhanced biomass production. The plants were also resistant to the wilt pathogen, *Fusarium culmorum*. MoradiPour et al. (2019) explored the beneficial effects of nanoparticles in the enhancement of PGPR using silica nanoparticles and CNTs for improving UCB1 commercial pistachio rootstock micropropagation. The use of nanoencapsulation by *Pseudomonas fluorescens* VUPF5 and *Bacillus subtilis* VRU1 significantly improved root length and plant proliferation.

12.7 REGULATORY FRAMEWORK FOR THE DEVELOPMENT OF NANOPESTICIDES AND NANOFERTILIZERS

The use of pesticides and fertilizer potentially presents a risk to humans and environmental health. In the case of excessive use of nitrogen fertilizer, serious threats to the environment and human health are identified. The rate of nitrogen fertilizer application is closely related to the accumulation of nitrate in the environment, groundwater, and leafy and root vegetables. The high nitrate contents in plants contribute to endogenous nitrosation, leading to thyroid conditions, human cancers, diabetes, and defects during fetal development. Regulations are defined for controlled-release fertilizers containing polymers. The polymer would undergo a plant growth acute toxicity test, an earthworm acute toxicity test, and a nitrification inhibition test with soil microorganisms.

Regulatory frameworks provide the requirements for toxicity risk data for AI for the assessment of human health risk (Kah et al., 2021). Conventional pesticides undergo rigorous safety testing based on AI under defined regulatory frameworks to prevent unacceptable risks. Although nanopesticides reduce the risk of AI by reducing the required applied dose, they can also pose a health risk by increasing the bioavailability of the AI. In the case of nanopesticides, the contribution of coformulants and excipients to pesticide effectiveness needs to be considered. Additionally, nanopesticides can exist in different states in the environment (such as single particle, aggregation, agglomeration, partly dissolved or degraded, and fully dissolved or degraded), and that can affect AI performance (Skjolding et al., 2016). Walker et al. (2018) mentioned that the release characteristics of AI from nanopesticides (e.g., fast, moderate, slow, or no-release) should be evaluated.

Conventional pesticides are evaluated by regulatory agencies primarily on the basis of AI and its formulations. In contrast to the AI, other components of the formulation (coformulants or adjuvants) have generally been considered "inerts." However, some regulatory agencies, like the European Union and Canada, require safety data for the coformulants. Currently known modes of nanopesticide action involve enhancing the effect of already known pesticides by nano-scale carriers or on dissolution from metallic nanoparticulate biocides such as copper and silver (Zhao et al., 2018; Grillo et al., 2015). The concerning issue will be whether the nanopesticide's behavior is similar to the original pesticide's or follows a pathway governed more by the nanoform, or will these pathways be interrupted?

The risk assessment guideline for nanomaterials in food and feed by the European Food Safety Authority (EFSA) outlines that surfactants, solvents, carriers, or wetting agents that contribute to the formulation must be considered. Moreover, the safety of the nanopesticide entity comprising AI+ coformulants needs to be evaluated, irrespective of whether the AI or coformulants were separately evaluated as safe.

There is a lack of consensus regarding the definition of nanoparticle size; some agencies define the size as up to 1000 nm, while others restrict it to 1–100 nm. The US Federal Insecticide, Fungicide, and Rodenticide Act (FIFRA) Scientific Advisory Panel (SAP) has mentioned that the current methods may be inadequate to assess the eco-toxicology of pesticide nanomaterials (FIFRA-SAP, 2010). The European regulatory frameworks have defined the approaches for assessment and control of the risk of nanomaterials to humans and the ecosystem. In Asia and Africa, the regulation of agricultural nanoproducts is treated heterogeneously in a country-specific manner (Subramanian and Rajkishore, 2019). In India, although the guidelines for the evaluation of nanofertilizers and nanopesticides have been published, the policy and regulatory frameworks remain to be developed (India, Government, 2019). The Australian government has published the "regulatory considerations for nanopesticides and veterinary nanomedicines," to address the definitions and risk recommendations with the regulatory frameworks (Australian Government, 2014). As this is a new area, several countries are still formulating the regulations. Human exposure to pesticides by dermal, inhalation, and ingestion may occur at several stages, that is, mixing-loading, transportation, application, and post-application. Different forms of pesticide (for example, concentrated, diluted, and transformed in the field) and exposed populations (professional, residential) are involved at each stage. Adequate personal protection equipment can alleviate occupational hazards, while options for non-occupational hazards (including bystanders, residents, and consumers) are limited.

Regulatory requirements for AI toxicokinetics include a battery of tests depending on the route of exposure, like oral, dermal, and inhalation, intravenous and oral absorption (distribution, metabolism, and excretion), acute systemic toxicity, skin and eye irritation, skin sensitization, and short-term and long-term toxicity (Kah et al., 2021). Other tests include genotoxicity (*in vitro* and *in vivo*), carcinogenicity, and reproductive and developmental toxicity. Currently, the toxicity data requirements for formulations (AI + coformulants) include acute systemic toxicity (oral, dermal, inhalation), skin and eye irritation, skin sensitization, and sometimes dermal absorption. For example, new OECD test guidelines for nano-relevant properties are currently being developed (Rasmussen et al., 2019) to address the two broad classes of nanopesticide products: polymer nanoparticles (AI complexes) and inorganic metal or metal oxide NPs. Nanopesticides composed of AI complexed with a polymer will require the characterization of the AI contents incorporated into the carrier and its release kinetics. Other issues include the possibility of the presence of inert silica nanoforms in several nanoformulations which may affect pesticide bioavailability. Further, the interaction of the biological environment is not well understood for a range of nanopesticides related to alteration in nanomaterial characteristics that require the development of validated methods.

Additionally, test methods could include longer-term exposure tests because of their high persistence in the environment, which are not incorporated in the standard framework for risk assessment (Kah et al., 2021). Degradation products or the aging of nanopesticides may manifest over longer exposures in the environment and need to be addressed. The tests could possibly involve full life cycle, multigenerational, metabolomics, and multispecies or mesocosm-level tests (Grillo et al., 2020). The development of new tests or modifications of existing protocols should consider the recommendations for nanopesticides to address the various concerns. Currently, there is no internationally accepted definition of nanopesticides, and thus different regulatory agencies may adopt different size ranges and limits. Finally, computational approaches are proposed for the risk assessment of nanopesticides (Villaverde et al., 2018). The proposed *in silico* models for nanomaterials include nano-QSAR (Quantitative Structure-Activity Relationship for nanomaterials) and nano-QSPR (Quantitative Structure-Property Relationship for nanomaterials) models (Gajewicz et al., 2012; Abramenko et al., 2020). These models are based on quantitative and qualitative physicochemical and nanodescriptors and the ecotoxicological effects of the nanomaterials (Burk et al., 2018; Villaverde et al., 2018). The implementation of computational tools for nanopesticide risk assessment will ease the decision path as more experiments can be carried out with reduced costs. However, the current nano-QSAR and QSPR models are basic and need to be developed for application to nanopesticides. To summarize, the regulatory framework needs to develop an unambiguous definition of nanopesticide to set clear outlines and avoid undue clubbing of conventional (non-nano) formulations with nanopesticides. The standardization of the nanopesticide testing guidelines is necessary to consider the properties and behavior of the nano-scale materials. Long-term testing of nanopesticides is necessary over longer exposures and against appropriate organisms in consideration of their interactions and persistance in the environment. An improved knowledge of the hazards posed by nanopesticides or their degradation products is required. These guidelines will permit the design of more efficient, environmentally safe, and sustainable nanomaterials and formulations.

12.8 CONCLUSION

Due to its ability to more effectively and precisely target the delivery of pesticides and fertilizers, nanotechnology has shown significant promise in a variety of industries, including agriculture. The use of nanotechnology in agriculture aims to increase pesticide effectiveness while reducing environmental impact. There are numerous ways that nanotechnology interventions can enhance the delivery of pesticides and fertilizers, including nanoencapsulating pesticides and fertilizers, nano-scale slow-release fertilizers, nanopesticides, nanobased formulations, smart delivery systems, precision agriculture, diminished environmental impact, and improved nutrient uptake. The use of nanotechnology in agriculture is not without difficulties and potential drawbacks, despite the fact that it offers a lot of promise. These include potential toxicity to organisms other than the ones being targeted, the persistence of nanoparticles in the environment, and regulatory issues. To ensure the safety and efficacy of agricultural products based on nanotechnology, extensive testing and evaluation are necessary. Overall, the use of nanotechnology in the delivery of fertilizer and pesticides has the potential to completely transform modern agriculture by making it more effective, sustainable, and ecologically benign.

ACKNOWLEDGMENTS

VG thanks the Department of Biotechnology (BT/PR36430/NNT/28/1713/2020) and DST-SERB (EMR 004243), Government of India, for the funding. GM is grateful to DST-INSPIRE for the grant of fellowship, and DC is thankful to UGC-CSIR for the grant of fellowship. SS and VB are thankful to DBT for the grant of a research studentship.

KEYWORDS

- **Biopesticides**
- **fertilizer**
- **fungicide**
- **nanomaterials**
- **pesticide**
- **siRNA**

REFERENCES

Abbasifar, A., Shahrabadi, F., & ValizadehKaji, B. (2020). Effects of green synthesized zinc and copper nano-fertilizers on the morphological and biochemical attributes of basil plant. *Journal of Plant Nutrition*, *43*, 1104–1118.

Abdel Latef, A. A. H., Srivastava, A. K., El-sadek, M. S. A., Kordrostami, M., & Tran, L. S. P. (2018). Titanium dioxide nanoparticles improve growth and enhance tolerance of broad bean plants under saline soil conditions. *Land Degradation and Development*, *29*, 1065–1073.

Abdelrahman, T. M., Qin, X., Li, D., Senosy, I. A., Mmby, M., Wan, H., Li, J., & He, S. (2021). Pectinase-responsive carriers based on mesoporous silica nanoparticles for improving the translocation and fungicidal activity of prochloraz in rice plants. *Chemical Engineering Journal*, *404*, 126440.

Abramenko, N., Kustov, L., Metelytsia, L., Kovalishyn, V., Tetko, I., & Peijnenburg, W. (2020). A review of recent advances towards the development of QSAR models for toxicity assessment of ionic liquids. *Journal of Hazardous Materials*, *384*, 121429.

Acharya, P., Jayaprakasha, G. K., Crosby, K. M., Jifon, J. L., & Patil, B. S. (2020). Nanoparticle-mediated seed priming improves germination, growth, yield, and quality of watermelons (*Citrullus lanatus*) at multi-locations in Texas. *Scientific Report*, *10*, 1–16.

Adeyinka, O. S., Riaz, S., Toufiq, N., Yousaf, I., Bhatti, M. U., Batcho, A., Olajide, A. A., Nasir, I. A., & Tabassum, B. (2020). Advances in exogenous RNA delivery techniques for RNAi-mediated pest control. *Molecular Biology Report*, *47*, 6309–6319.

Arora, S., Sharma, P., Kumar, S., Nayan, R., Khanna, P. K., & Zaidi, M. G. H. (2012). Gold-nanoparticle induced enhancement in growth and seed yield of Brassica juncea. *Plant Growth Regulation*, *66*, 303–310.

Auer, C., & Frederick, R. (2009). Crop improvement using small RNAs: Applications and predictive ecological risk assessments. *Trends Biotechnology*, *27*, 644–651.

Australian Government. (2014). Regulatory considerations for nanopesticides and veterinary nanomedicines, 1. Australian Pesticides and Veterinary Medicines Authority, Australia.

Avila, L. A., Chandrasekar, R., Wilkinson, K. E., Balthazor, J., Heerman, M., & Bechard, J. (2018). Delivery of lethal dsRNAs in insect diets by branched amphiphilic peptide capsules. *Journal of the Controlled Release*, *273*, 139–146.

Azam, M., Bhatti, H. N., Khan, A., Zafar, L., & Iqbal, M. (2022). Zinc oxide nano-fertilizer application (foliar and soil) effect on the growth, photosynthetic pigments and antioxidant system of maize cultivar. *Biocatalogy*, *42*, 102343.

Bellés, X. (2010). Beyond drosophila: RNAi in vivo and functional genomics in insects. *Annual Review of Entomology*, *55*, 111–128.

Bilal, M., Xu, C., Cao, L., Zhao, P., Cao, C., Li, F., & Huang, Q. (2020). Indoxacarb-loaded fluorescent mesoporous silica nanoparticles for effective control of *Plutella xylostella* L. with decreased detoxification enzymes activities. *Pest Management Science*, *76*, 3749–3758.

Boehm, A. L., Martinon, I., Zerrouk, R., Rump, E., & Fessi, H. (2003). Nanoprecipitation technique for the encapsulation of agrochemical active ingredients. *Journal of Microencapsulation*, *20*, 433–441.

Burk, J., Sikk, L., Burk, P., Manshian, B. B., Soenen, S. J., Scott-Fordsmand, J. J., Tamm, T., & Tamm, K. (2018). Fe-doped ZnO nanoparticle toxicity: assessment by a new generation of nanodescriptors. *Nanoscale*, *10*, 21985–21993.

Burman, U., Saini, M., & Kumar, P. (2013). Effect of zinc oxide nanoparticles on growth and antioxidant system of chickpea seedlings. *Toxicological and Environmental Chemistry*, *95*, 605–616.

Camara, M. C., Monteiro, R. A., Carvalho, L. B., Oliveira, J. L., & Fraceto, L. F. (2021). Enzyme stimuli–responsive nanoparticles for bioinsecticides: An emerging approach for uses in crop protection. *ACS Sustainable Chemistry and Engineering*, *9*, 106–112.

Campos, E. V. R., Proença, P. L., Oliveira, J. L., Pereira, A. E. S., Ribeiro, L. N. M., Fernandes, F. O., Gonçalves, K. C., Polanczyk, R. A., Pasquoto-Stigliani, T., Lima, R., Melville, C. C., Vechia, F. F. D., Andrade, D. J., & Fraceto, L. F. (2018). 1,2 Carvacrol and linalool co-loaded in β-cyclodextrin-grafted chitosan nanoparticles as sustainable biopesticide aiming pest control. *Scientific Rep*ort, *8*, 7623.

Campolo, O., Cherif, A., Ricupero, M., Siscaro, G., Grissa-Lebdi, K., Russo, A., Cucci, L. M., Pietro, P. D., Satriano, C., Desnuex, N., Biondi, A., Zappala, L., & Palmeri, V. (2017). Citrus peel essential oil nanoformulations to control the tomato borer, *Tuta absoluta*: Chemical properties and biological activity. *Scientific Report*, *7*, 13036.

Carmona, F. J., Sasso, G. D., Bertolotti, F., Ramirez-Rodriguez, G., Delgado-Lopez, J. M., Pedersen, J. S., Masciocchi, N., & Guagliardi, A. (2020). The role of nanoparticle structure and morphology in the dissolution kinetics and nutrient release of nitrate-doped calcium phosphate nanofertilizers. *Scientific Report*, *10*, 12396.

Carmona, F. J., Sasso, G. D., Bertolotti, F., Ramirez-Rodriguez, G. B., Pii, Y., Delgado-Lopez, J. M., Guagliardi, A., & Masciocchi, N. (2021). Urea-functionalized amorphous calcium phosphate nanofertilizers: Optimizing the synthetic strategy towards environmental sustainability and manufacturing costs. *Scientific Report*, *11*, 3419.

Christiaens, O., Tardajos, M. G., Reyna, Z. L. M., Dash, M., Dubruel, P., & Smagghe, G. (2018). Increased RNAi efficacy in *Spodoptera exigua* via the formulation of dsRNA with guanylated polymers. *Frontiers in Physiology*, *9*, 316.

Cooper, A. M., Silver, K., Zhang, J., Park, Y., & Zhu, K. Y. (2019). Molecular mechanisms influencing efficiency of RNA interference in insects. *Pest Management Science*, *75*, 18–28.

Corradini, E., De-Moura, M. R., & Mattoso, L. H. C. (2010). A preliminary study of the incorporation of NPK fertilizer into chitosan nanoparticles. *Express Polymer Letters*, *4*, 509–515.

Cui, H. X., Sun, C. J., Liu, Q., Jiang, J., & Gu, W. (2010). Applications of nanotechnology in agrochemical formulation, perspectives, challenges and strategies. In *International Conferences on Nanoagriculture, Sao Pedro, Brazil* (pp. 28–33).

Das, S., Debnath, N., Cui, Y., Unrine, J., & Palli, S. R. (2015). Chitosan, carbon quantum dot, and silica nanoparticle mediated dsRNA delivery for gene silencing in Aedes aegypti: A comparative analysis. *ACS Applied Materials and Interfaces*, *7*, 19530–19535.

Dawkar, V. V., Chikate, Y. R., Lomate, P. R., Dholakia, B. B., Gupta, V. S., & Giri, A. P. (2013). Molecular insights into resistance mechanisms of lepidopteran insect pests against toxicants. *Journal of Proteome Research*, *12*, 4727–4737.

Demirer, G. S., Silva, T. N., Jackson, C. T., Thomas, J. B., Ehrhardt, D. W., Rhee, S. Y., Mortimer, J. C., & Landry, M. P. (2021). Nanotechnology to advance CRISPR–Cas genetic engineering of plants. *Nature Nanotechnology*, *16*, 243–250.

Demirer, G. S., Zhang, H., Goh, N. S., González-Grandío, E., & Landry, M. P. (2019). Carbon nanotube–mediated DNA delivery without transgene integration in intact plants. *Nature Protocols*, *14*, 2954–2971.

Dikbaş, N., & Cinisli, K. T. (2019). Microbial metabolites powered by nanoparticles could be used as pesticides in future? (NanoBioPesdicides). *Biotechnology Journal International*, 23, 1–4.

Dimkpa, C. O., Mclean, J. E., Britt, D. W., & Anderson, A. J. (2012). CuO and ZnO nanoparticles differently affect the secretion of fluorescent siderophores in the beneficial root colonizer, *Pseudomonas chlororaphis* O6. *Nanotoxicology*, *6*, 635–642.

Elfeky, S. A., Mohammed, M. A., Khater, M. S., Osman, Y. A. H., & Elsherbini, E. (2013). Effect of magnetite nano-fertilizer on growth and yield of *Ocimum basilicum* L. *International Journal of Indigenous Medicinal Plants*, *46*, 1286–1293.

FIFRA-SAP. (2010). Evaluation of the hazard and exposure associated with nanosilver and other nanometal pesticide products. U.S. Federal Insecticide, Fungicide, and Rodenticide

Act (FIFRA) Scientific Advisory Panel (SAP); SAP Minutes 2010–01. http://www.epa.gov/scipoly/sap/meetings/2009/november/110309ameetingmi nutes.pdf

Fire, A., Xu, S., Montgomery, M. K., Kostas,S. A., Driver, S. E., & Mello, C. C. (1998). Potent and specific genetic interference by double-stranded RNA in *Caenorhabditis elegans*. *Nature*, *391*, 806–811.

Gajewicz, A., Rasulev, B., Dinadayalane, T. C., Urbaszek, P., Puzyn, T., Leszczynska, D., & Leszczynski, J. (2012). Advancing risk assessment of engineered nanomaterials: Application of computational approaches. *Advanced Drug Delivery Reviews*, *64*, 1663–1693.

Ganilho, C., da Silva, M. B., Paiva, C., de Menezes, T. I., dos Santos, M. R., Pereira, C. M., Pereira, R., & Andreani, T. (2022). Environmental safety assessments of lipid nanoparticles loaded with Lambda-Cyhalothrin. *Nanomaterials*, *12*, 2576.

Gao, Y., Xiao, Y., Mao, K., Qin, X., Zhang, Y., Li, D., Zhang, Y., Li, J., Wan, H., & He, S. (2020). Thermoresponsive polymer-encapsulated hollow mesoporous silica nanoparticles and their application in insecticide delivery. *Chemical Engineering Journal*, *383*, 123169.

Gayathri, M., Venkateswarlu, N. C., & MuraliKrishna, T. (2018). Evaluation of nano based biopesticides against *S. litura* on groundnut. *International Journal of Current Microbiology and Applied Science*, *7*, 1614–1625.

Ghaghelestany, A. B., Jahanbakhshi, A., & Taghinezhad, E. (2020). Gene transfer to German chamomile (*L chamomilla M*) using cationic carbon nanotubes. *Scientia Horticulturae*, *263*, 109106.

Ghalamboran, M. R., & Ramsden, J. J. (2012). Viability of *Bradyrhizobium japanicum* on soybean seeds enhanced by magnetite nanoparticles during desiccation. *International Journal of Biological Life Sci*ence, *8*, 228–233.

Ghormade, V., Deshpande, M. V., & Paknikar, K. M. (2011). Perspectives for nano-biotechnology enabled protection and nutrition of plants. *Biotechnology Advance*, *29*, 792–803.

Giraldo, J. P., Landry, M. P., Faltermeier, S. M., McNicholas, T. P., Iverson, N. M., Boghossian, A. A., Reuel, N. F., Hilmer, A. J., Sen, F., Brew, J. A., & Strano, M. S. (2014). Plant nanobionics approach to augment photosynthesis and biochemical sensing *Nature Materials*, *13*, 400.

Golestanipour, A., Nikkhah, M., Aalami, A., & Hosseinkhani, S. (2018). Gene delivery to tobacco root cells with single-walled carbon nanotubes and cell-penetrating fusogenic peptides. *Molecular Biotechnology*, *60*, 863–878.

Grillo, R., Clemente, Z., de Oliveira, J. L., Campos, E. V. R., Chalupe, V. C., Jonsson, C. M., de Lima, R., Sanches, G., Nishisaka, C. S., Rosa, A. H., Oehlke, K., Greiner, R., & Fraceto, L. F. (2015). Chitosan nanoparticles loaded the herbicide paraquat: the influence of the aquatic humic substances on the colloidal stability and toxicity. *Journal of Hazardous Materials*, *286*, 562–572.

Grillo, R., Fraceto, L. F., Amorim, M. J. B., Scott-Fordsmand, J. J., Schoonjans, R., & Chaudhry, Q. (2020). Ecotoxicological and regulatory aspects of environmental sustainability of nanopesticides. *Journal of Hazardous Materials*, *404*(Part A), 124148.

Hajiahmadi, Z., Shirzadian-Khorramabad, R., Kazemzad, M., & Sohani, M. M. (2019). Enhancement of tomato resistance to *Tuta absoluta* using a new efficient mesoporous silica nanoparticle-mediated plant transient gene expression approach. *Scientia Horticulturae*, *243*, 367–375.

Hao, Y., Yang, X., Shi, Y., Song, S., Xing, J., Marowitch, J., Chen, J., & Chen, J. (2013). Magnetic gold nanoparticles as a vehicle for fluorescein isothiocyanate and DNA delivery into plant cells. *Botany*, *91,* 457–466.

He, B., Chu, Y., Yin, M., Müllen, K., An, C., & Shen, J. (2013). Fluorescent nanoparticle delivered dsRNA toward genetic control of insect pests. *Advanced Materials*, *25*, 4580–4584.

Huvenne, H., & Smagghe, G. (2010). Mechanisms of dsRNA uptake in insects and potential of RNAi for pest control: A review. *Journal of Insect Physiology*, *56*, 227–235.

Ibrahim, S. S. (2017). *Applications of some biotechnological methods for controlling the most important pests of potatoes* (Thesis). Faculty of Science Al-Azhar University, Egypt.

India, Government. (2019). Guidelines for evaluation of nano-agriinput products and nano- agriproducts, 1. Department of Biotechnology Ministry of Science & Technology Government of India, India.

Izuegbunam, C. L., Wijewantha, N., Wone, B., Ariyarathne, M. A., Sereda, G., & Wone, B. W.M. (2021). A nano-biomimetic transformation system enables in planta expression of a reporter gene in mature plants and seeds. *Nanoscale Advanced*, *3*, 3240.

Javazmi, L., Young, A., Ash, G. J., & Low, T. (2021). Kinetics of slow release of nitrogen fertiliser from multi-layered nanofibrous structures. *Scientific Report*, *11*, 1–8.

Kah, M., Johnston, L. H., Kookana, R. S., Bruce, W., Haase, A., Ritz, V., Dinglasan, J., Doak, S., Garelick, H., & Gubala, V. (2021). Comprehensive framework for human health risk assessment of nanopesticides. *Nature Nanotechnology*, *16*, 955–964.

Karunakaran, G., Suriyaprabha, R., Manivasakan, P., Yuvakkumar, R., Rajendran, V., Prabu, P., & Kannan, N. (2013). Effect of nanosilica and silicon sources on plant growth promoting rhizobacteria, soil nutrients and maize seed germination. *IET Nanobiotechnol*ogy, *7,* 70–77.

Kashyap, P. L., Xiang, X., & Heiden, P. (2015). Chitosan nanoparticle-based delivery systems for sustainable agriculture. *International Journal of Biological Macromolecules*, *77*, 36–51.

Khanm, H., Vaishnavi, B. A., & Shankar, A. G. (2018). Raise of nano-fertilizer era: Effect of nano scale zinc oxide particles on the germination, growth and yield of tomato (*Solanum lycopersicum*). *International Journal of Current Microbiology and Applied Science*, *7*, 1861–1871.

Khodakovskaya, M., Dervishi, E., Mahmood, M., Xu, Y., Li, Z., Watanbe, F., & Biris, A. S. (2009). Carbon nanotubes are able to penetrate plant seed coat and dramatically affect seed germination and plant growth. *ACS Nano*, *3*, 3221–3227.

Khoobdel, M., Ahsael, S. M., & Farzane, M. (2017). Insecticidal activity of polycaprolactone nanocapsules loaded with Rosmarinus officinalis essential oil in Tribolium castaneum(Herbst). *Entomology Research*, *47*, 175–184.

Kim, T. H., Jiang, H. L., Jere, D., Park, I. K., Cho, M. H., Nah, J. W., Choi, Y. J., Akaike, T., & Cho, C. S. (2007). Chemical modification of chitosan as a gene carrier in vitro and in vivo. *Progress in Polymer Science*, *32*, 726–753.

Kolge, H., Kadam, K., Galande, S., Lanjekar, V., & Ghormade, V. (2021). New frontiers in pest control: Chitosan nanoparticles- shielded dsRNA as an effective topical RNAi Spray for gram podborer biocontrol. *ACS Applied Biomaterials*, *4*, 5145–5157

Kolge, H., Kadam, K., & Ghormade, V. (2023). Chitosan nanocarriers mediated dsRNA delivery in gene silencing for *Helicoverpa armigera* biocontrol. *Pesticide Biochemistry and Physiology*, *189*, 105292.

Kottegoda, N., Munaweera, I., Madusanka., & Karunaratne, V. (2011). A green slow-release fertilizer composition based on urea-modified hydroxyapatite nanoparticles encapsulated wood. *Current Science, 101*, 73–78.

Kumar, S., Bhanjana, G., Sharma, A., Dilbaghi, N., Sidhu, M. C., & Kim, K.-H. (2017). Development of nanoformulation approaches for the control of weeds. *Science of the Total Environment, 586*, 1272–1278.

Kumar, S., Bhanjana, G., Sharma, A., Sidhu, M. C., & Dilbaghi, N. (2014). Synthesis, characterization and on field evaluation of pesticide loaded sodium alginate nanoparticles. *Carbohydrate Polymers, 101*, 1061– 1067.

Kumar, S., Nehra, M., Dilbaghi, N., Marrazza, G., Tuteja, S. K., & Kim, K. H. (2020). Nanovehicles for plant modifications towards pest-and disease-resistance traits. *Trends Plant Science, 25*, 198–212.

Kumar, Y., Tiwari, K. N., Singh, T., & Raliya, R. (2021). Nanofertilizers and their role in sustainable agriculture. *Annals of Plant Soil Research, 23*, 238–255.

Kumaraswamy, R. V., Saharan, V., Kumari, S., Choudhary, R. C., Pal, A., Sharma, S. S., Rakshit, S., Raliya, R., & Biswas, P. (2021). Chitosan-silicon nanofertilizer to enhance plant growth and yield in maize (*Zea mays* L.) *Plant Physiology and Biochemistry, 159*, 53–66.

Kwak, S. Y., Lew, T. T. S., Sweeney, C. J., Koman, V. B., Wong, M. H., Bohmert-Tatarev, K., Snell, K. D., Seo, J. S., Chua, N. H., & Strano, M. S. (2019). Chloroplast-selective gene delivery and expression in planta using chitosan-complexed single-walled carbon nanotube carriers. *Nature Nanotechnology, 14*, 447–455.

Lee, W., An, Y. J., Yoon, H., & Kweon, H. S. (2008). Toxicity and bioavailability of copper nanoparticles to the terrestrial plants mung bean (*Phaseolus radiatus*) and wheat (*Triticum aestivum*): Plant agar test for water-insoluble nanoparticles. *Environmental Toxicology and Chemistry, 27*, 1915–1921.

Lei, Z., Mingyu, S., Chao, L., Liang, C., Hao, H., Xiao, W., Xiaoqing, L., Fan, Y., Fengqing, G., & Fashui, H. (2007). Effects of nanoanatase TiO2 on photosynthesis of spinach chloroplasts under different light illumination. *Biological Trace Element Research, 119*, 68–76.

Li, J., Wang, X. P., Wang, M. Q., Ma, W. H., & Hua, H. X. (2013). Advances in the use of the RNA interference technique in Hemiptera. *Insect Science, 20*, 31–39.

Liu, J., Luo, Y., Jiang, X., Sun, G., Song, S., Yang, M., & Shen, J. (2022). Enhanced and sustained pesticidal activity of a graphene-based pesticide delivery system against the diamondback moth *Plutella xylostella. Pest Management Science, 78*, 5358–5365.

Louni, M., Shakarami, J., & Negahban, M. (2018). Insecticidal efficacy of nanoemulsion containing *Mentha longifolia* essential oil against *Ephestia kuehniella* (Lepidoptera: Pyralidae). *Journal of Crop Protection, 7*, 171–182.

Mahmoodzadeh, H., Nabavi, M., & Kashefi, H. (2013). Effect of nanoscale titanium dioxide particles on the germination and growth of canola (Brassica napus). *Journal of Ornam Horticulture Plants, 3*, 25–32.

Mahmoud, L. M., Kaur, P., Stanton, D., Grosser, J. W., & Dutt, M. (2022). A cationic lipid mediated CRISPR/Cas9 technique for the production of stable genome edited citrus plants. *Plant Methods, 18*, 33.

Majumder, S., Kaushik, P., Rana, V. S., Sinha, P., & Shakil, N. A. (2020). Amphiphilic polymer based nanoformulations of mancozeb for management of early blight in tomato. *Journal of Environmental Science and Health B, 55*, 501–507.

Mali, S. C., Raj, S., & Trivedi, R. (2020). Nanotechnology a novel approach to enhance crop productivity. *Biochemistry and Biophysics Reports*, *24*, 100821.

Mejias, J. H., Salazar, F., Pérez Amaro, L., Hube, S., Rodriguez, M., & Alfaro, M. (2021). Nanofertilizers: A cutting-edge approach to increase nitrogen use efficiency in grasslands. *Frontiers in Environental Science*, *9*, 52.

Milani, N., McLaughlin, M. J., Stacey, S. P., Kirkby, J. K., Hettiarachchi, G. M., Beak, D. G., & Cornelis, G. (2012). Dissolution kinetics of macronutrient fertilizers coated with manufactured zinc oxide nanoparticles. *Journal of Agricultural and Food Chemistry*, *60*, 3991–3998.

Mitter, N., Worrall, E. A., Robinson, K. E., Li, P ., Jain, R. G., Taochy, C., Fletcher, S. J., Carroll, B. J., Lu, G. Q. M., & Xu, Z. P. (2017). Clay nanosheets for topical delivery of RNAi for sustained protection against plant viruses. *Nat Plants*, *3*,16207.

Mo, D., Li, X., Chen, Y., Jiang, Y., Gan, C., Zhang, Y., Li, W., Huang, Y., & Cui, J. (2021). Fabrication and evaluation of slow-release lignin-based avermectin nano-delivery system with UV-shielding property. *Scientific Report*, *11*, 23248.

MoradiPour, M., Saberi-Riseh, R., Mohammadinejad, R., & Hosseini, A. (2019). Nano-encapsulation of plant growth-promoting rhizobacteria and their metabolites using alginate-silica nanoparticles and carbon nanotube improves UCB1 pistachio micropropagation. *Journal of Microbiology and Biotechnol*ogy, *29*, 1096–1103.

Mortazavi, S. E., & Zohrabi, Z. (2018). Biolistic co-transformation of rice using gold nanoparticles. *Iran Agricultural Research*, *37*, 75–82.

Nair, R., Varghese, S. H., Nair, B. G., Maekawa, T., Yoshida, Y., & Kumar, D. S. (2010). Nanoparticulate material delivery to plants. *Plant Science*, *179,* 154–163.

Namasivayam, S. K. R.; & Chitrakala, K. (2011). Ecotoxicological effect of *Lecanicillium lecanii* (Ascomycota: *Hypocreales*) based silver nanoparticles on growth parameters of economically important plants. *Journal of Biopesticides*, *4*, 97–101.

Nicolopoulou-Stamati, P., Maipas, S., Kotampasi, C., Stamatis, P., & Hens, L. (2016). Chemical pesticides and human health: The urgent need for a new concept in agriculture. *Frontiers in Public Health*, *4*, 148. DOI: 10.3389/fpubh.2016.00148.

Nuruzzaman, M., Rahman, M. M., Liu, Y., & Naidu, R. (2016). Nanoencapsulation, nano-guard for pesticides: A new window for safe application. *Journal of Agricultural and Food Chemistry*, *64*(7), 1447–1483.

Palli, S. R. (2014). RNA interference in Colorado potato beetle: Steps toward development of dsRNA as a commercial insecticide. *Current Opinion in Insect Science*, *6*, 1–8.

Pandey, A., Srivastava, S., Aggarwal, N., Srivastava, C., Adholeya, A., & Kochar, M. (2020). Assessment of the pesticidal behaviour of diacyl hydrazine-based ready-to-use nanoformulations. *Chemical and Biological Technologies in Agriculture*, *7*, 10.

Paragas, D. S., Cruz, K. D. C., & Fiegalan, E. R. (2021). Fermented neem (*Azadirachta indica*) leaves-metal nanoparticles and their insecticidal properties against *Bactrocera dorsalis* (Hendel). *Indian Journal of Science and Technology*, *14*, 1338–1346.

Pouci, F. (2008). Polymer in agriculture: A review. *American Journal of Agriculture and Biological Science*, *3*, 299–314.

Prado de Moraes, A. C., da Silva Ribeiro, L., Rodrigues de Camargo, E., & *Teixeira Lacava*, P. (2021). The potential of nanomaterials associated with plant growth-promoting bacteria in agriculture. *3 Biotech*, *11*, 318.

Pretty, J. (2018). Intensification for redesigned and sustainable agricultural systems. *Science*, *362*, eaav0294.

Price, D. R., & Gatehouse, J. A. (2008). RNAi-mediated crop protection against insects. *Trends Biotechnology*, *26*, 393–400.

Rajonee, A. A., Zaman, S., & Huq, S. M. I. (2017). Preparation, characterization and evaluation of efficacy of phosphorus and potassium incorporated nano fertilizer. *Advances in Nanoparticles*, *6*, 62–74.

Rasmussen, K., Rauscher, H., Kearns, P., González, M., & Riego Sintes, J. (2019). Developing OECD test guidelines for regulatory testing of nanomaterials to ensure mutual acceptance of test data. *Regulatory Toxicology and Pharmacology*, *104*, 74–83.

Sahandi, S. M., Sorooshzadeh, A., Rezazadeh, H. S., & Naghdibadi, H. A. (2011). Effect of nanosilver and silver nitrate on seed yield of borage. *Journal of Medicinal Plants Research*, *5*, 706–710.

Santo Pereira, A. E., Caixeta Oliveira, H., & Fernandes Fraceto, L. (2019). Polymeric nanoparticles as an alternative for application of gibberellic acid in sustainable agriculture: a field study. *Scientific Report*, *9*, 7135.

Shen, R., Peng, Z., Zhao, L., Chen, C., Wang, H., Chen, Z., Wang, J., & Guo, T. (2023). Creation of fragrant rice by targeted editing of fgr gene using magnetic nanoparticle-mediated pollen magnetofection in Rice. Research Square; Preprint 1–21.

Shoaib, A., Waqas, M., Elabasy, A., Cheng, X., Zhang, Q., & Shi, Z. (2018). Preparation and characterization of emamectin benzoate nanoformulations based on colloidal delivery systems and use in controlling *Plutella xylostella* (L.) (Lepidoptera: Plutellidae). *RSC Advances*, *8*, 15687–15697.

Skjolding, L. M., Sorensen, S. N., Hartmann, N. B., Hjorth, R., Hansen, S. F., & Baun, A. (2016). Aquatic ecotoxicity testing of nanoparticles-the quest to disclose nanoparticle effects. *Angewandte Chemie International Edition English*, *55*, 15224–15239.

Solanki, P., Bhargava, A., Chhipa, H., Jain, N., & Panwar, J. (2015). Nano-fertilizers and their smart delivery system. In *Nanotechnologies in food and agriculture* (pp. 81–101). Springer Switzerland.

Subramanian, K. S., & Rajkishore, S. K. (2019). Regulatory framework for nanomaterials in agri-food systems. In M. Rai, J. K. Biswas (Eds.), *Nanomaterials: Ecotoxicity, safety, and public perception* (pp. 319-342). Springer

Suriyaprabha, R., Karunakaran, G., Yuvakkumar, R., Prabu, P., Rajendran, V., & Kannan, N. (2012). Growth and physiological responses of maize (Zea mays L.) to porous silica nanoparticles in soil. *Journal of Nanoparticle Research*, *14*, 1294–1308.

Sun, D., Hussain, H., Yi, Z., Siegele, R., Cresswell, T., Kong, L., & Cahill, D. (2014). Uptake and cellular distribution, in four plant species, of fluorescently labeled mesoporous silica nanoparticles. *Plant Cell Reports*, *33*, 1389–1402.

Taghi, S., Valojai, S., Niknejad, Y., Amoli, H. F., & Tari, D. B. (2021). Response of rice yield and quality to nano-fertilizers in comparison with conventional fertilizers. *Journal of Plant Nutrition*, *44*, 1971–1981.

Tarafder, C., Daizy, M., Alam, M. M., Ali, M. R., Islam, M. J., Islam, R., Ahommed, M. S., Aly, M. A. S., & Khan, M. Z. H. (2020). Formulation of a hybrid nanofertilizer for slow and sustainable release of micronutrients. *ACS Omega*, *5*, 23960–23966.

Thagun, C., Horii, Y., Mori, M., Fujita, S., Ohtani, M., Tsuchiya, K., Kodama, Y., Odahara, M., & Numata, K. (2022). Non-transgenic gene modulation via spray delivery of nucleic acid/peptide complexes into plant nuclei and chloroplasts. *ACS Nano*, *16*, 3506–3521.

Tilman, D., Knops, J., Wedin, D., & Reich, P. (2002). Plant diversity and composition: Effects on productivity and nutrient dynamics of experimental grasslands. In M. Loreau, S. Naeem,

P. Inchausti (Eds.), *Biodiversity and ecosystem functioning* (Vol. 21, pp. 21–35). O. U. P. Oxford.

Timmusk, S., Seisenbaeva, G., & Behers, L. (2018). Titania (TiO2) Nanoparticles Enhance the Performance of Growth-Promoting Rhizobacteria. *Scientific Report*, *8*, 617.

Torney, F., Trewyn, B. G., Lin, V. S-Y., & Wang, K. (2007). Mesoporous silica nanoparticles deliver DNA and chemicals into plants. *Nature Nanotechnology*, *2*, 295–300.

Tsuji, K. (2001). Microencapsulation of pesticides and their improved handling safety. *Journal of Microencapsulation*, *18*, 137–147.

Vijayakumar, P. S., Abhilash, O. U., Khan, B. M., & Prasad, B. L. (2010). Nanogold-loaded sharp-edged carbon bullets as plant-gene carriers. *Advanced Functional Mater*ials, *20*, 2416–2423.

Villaverde, J. J., Sevilla-Moran, B., Lopez-Goti, C., Alonso-Prados, J. L., & Sandin-Espana, P. (2018). Considerations of nano-QSAR/QSPR models for nanopesticide risk assessment within the European legislative framework. *Science of the Total Environment*, *634*, 1530–1539.

Walker, G. W., Kookana, R. S., Smith, N. E., Kah, M., Doolette, C. L., Reeves, P. T., Lovell, W., Anderson, D. J., Turney, T. W., & Navarro, D. A. (2018). Ecological risk assessment of nano-enabled pesticides: A perspective on problem formulation. *Journal of Agricultural and Food Chemistry*, *66*, 6480–6486

Wanyika, H., Gatebe, E., Kioni, P., Tang, Z., & Gao, Y. (2012). Mesoporous silica nanoparticles carrier for urea: Potential applications in agrochemical delivery systems. *Journal of Nanoscience and Nanotechnology*, *12*, 2221–2228.

Weber, C. N., Koron, D., Jakopic, J., Veberic, R., Hudina, M., & Baša Cesnik, H. (2021). Influence of nitrogen, calcium and nano-fertilizer on strawberry (Fragaria _ ananassa Duch.) Fruit Inner and Outer Quality. *Agronomy*, *11*, 997.

Yan, S., Ren, B. Y., & Shen, J. (2021). Nanoparticle-mediated double-stranded RNA delivery system: A promising approach for sustainable pest management. *Insect Science*, *28*, 21–34.

Yang, F. L., Li, X. G., Zhu, F., & Lei, C. L. (2009). Structural characterization of nanoparticles loaded with garlic essential oil and their insecticidal activity against Tribolium castaneum (Herbst) (Coleoptera: Tenebrionidae). *Journal of Agricultural and Food Chemlstry*, *57*, 10156–10162.

Yang, Y., Wang, J., Xiu, Z., & Alvarez, P. J. J. (2013). Impacts of silver nanoparticles on cellular and transcriptional activity of nitrogen-cycling bacteria. *Environmental Toxicology and Chemlstry*, *32*, 1488–1494.

Yoon, J. S., Gurusamy, D., & Palli, S. R. (2017). Accumulation of dsRNA in endosomes contributes to inefficient RNA interference in the fall armyworm, *Spodoptera frugiperda. Insect Biochemistry and Molecular Biology*, *90*, 53–60.

Yuvaraj, M., & Subramanian, K. S. (2021). Carbon sphere-zinc sulphate nanohybrids for smart delivery of zinc in rice (*Oryza sativa* L). *Scientific Report*, *11*, 9508.

Werdin-Gonzalez, J. O., Gutiérrez, M. M., Ferrero, A. A., & Band, B. F. (2014). Essential oils nanoformulations for stored-product pest control—Characterization and biological properties. *Chemosphere*, *100*, 130–138.

Zeng, P., Xu, Y., Zeng, C., Ren, H., & Peng, M. (2011). Chitosan-modified poly (D, L-lactide-co-glycolide) nanospheres for plasmid DNA delivery and HBV gene-silencing. *International Journal of Pharmaceuticals, 415*, 259–266.

Zhang, W. (2018). Global pesticide use: Profile, trend, cost/benefit and more. *Proceedings of the International Academy and Ecology and Environmental Science*, *8*, 1–27.

Zhang, J., Brown, G., Fu, G., James, P., Mukandiwa, L., Huang, X., & Yu, C. (2020). Nanobiopesticides: Silica nanoparticles with spiky surfaces enable dual adhesion and enhanced performance. *EcoMat, 2*, e12028.

Zhang, X., Mysore, K., Flannery, E., Michel, K., Severson, D. W., Zhu, K. Y., & Duman-Scheel, M. (2015). Chitosan/interfering RNA nanoparticle mediated gene silencing in disease vector mosquito larvae. *JoVE (Journal of Visualized Experiments)*, *97*, 52523.

Zhao, X., Cui, H., Wang, Y., Sun, C., Cui, B., & Zeng, Z. (2018). Development strategies and prospects of nano-based smart pesticide formulation. *Journal of Agricultural and Food Chemistry*, *66*, 6504–6512.

Zhu, K. Y., & Palli, S. R. (2020). Mechanisms, applications, and challenges of insect RNA interference. *Annual Reviews of Entomology*, *65*, 293.

Ziaee, M., Moharramipour, S., & Mohsenifar, A. (2014). MA-chitosan nanogel loaded with Cuminum cyminum essential oil for efficient management of two stored product beetle pests. *Journal of Pest Science*, *87*, 691–699.

PART III

Biochar and Nanobiochar Production and Applications

CHAPTER 13

Fabrication, Characteristics, and Application of Biochar and Engineered Biochar in the Agriculture Sector

RAHUL KUMAR DHAKA[1], SARITA DHAKA[2], ANUJ RANA[3], and KRISHNA PAL SINGH[4,5]

[1]*Department of Chemistry (COBS&H), CCS Haryana Agricultural University, Hisar, India*

[2]*Department of Chemistry, Sanatan Dharm (PG) College, Muzaffarnagar, Affiliated to Maa Shakumbhari University, Saharanpur, India*

[3]*Department of Microbiology (COBS&H), CCS Haryana Agricultural University, Hisar, India*

[4]*Biophysics Unit, College of Basic Sciences and Humanities, G. B. Pant University of Agriculture and Technology, Pantnagar, India*

[5]*Vice-Chancellor's Secretariat, Mahatma Jyotiba Phule Rohilkhand University, Bareilly, India*

ABSTRACT

Biochar and engineered biochar-based nanomaterials such as nanobiochar and nanocomposites thereof have been widely used in the energy, materials, environment, and agriculture sectors. These materials have improved tunable surface, chemical, and physical properties. The potential applications of biochar and engineered biochar in the agriculture sector are enormous, including as soil conditioners, plant growth boosters, carriers for controlled delivery of agrochemicals, agents for plant disease management, microbial community enrichment in soil, pesticide remediation, support for enzymes in

Nanobiotechnology for Agricultural Sciences: Nano-Agri-Input Products for Crop Production and Environmental Protection. J. C. Tarafdar, Deepak Kumar, Sanjay K. Singh, and Krishna Pal Singh (Eds.)

sensing, and others. Further, these biochar-based engineered materials, being sustainable, help in carbon sequestration and climate change mitigation, making them an ideal candidate for agricultural applications as discussed above. In comparison to biochar, engineered biochars, nanobiochar, and nanocomposites outperform in immobilizing nutrients and contaminants in soil by influencing their dynamics. This chapter covers the methodologies for the production of biochar and biochar-based nanocomposites, followed by their potential application in the agriculture sector.

13.1 INTRODUCTION

Biochar-based materials have a wide range of applications in the environment, energy, and materials sectors, including the agriculture sector (Chausali et al., 2021). These materials have a significant role in boosting plant growth, carbon sequestration, and mitigate global warming (Chen et al., 2017). Surface characteristics of these materials like high microporosity, surface area, sorption capacities, abundance in the environment, and structural properties make them suitable candidates for environmental remediation applications. Biochar-based nanocomposites with reductive/oxidative and catalytic potential (such as graphitic C_3N_4 and nanoscale zero-valent iron, etc.) find application environmental remediation (Tang et al., 2022). Biochar and its composites have frequently been used for pesticide remediation in water. Biochar's effectiveness can be improved on a nanoscale and in composite form. Biochar nanocomposites derived from agriwaste with unique characteristics may improve the removal of various pollutants from wastewater (Noreen and Abd-Elsalam, 2021). Simultaneous production of biofuels and syngas in thermochemical treatment of biomass during the biochar preparation process makes it an economically feasible process versus the utilization of biomass through combustion.

Biochar is a carbon-negative material that finds a plethora of uses in the agriculture sector, such as enhancement of plant growth, soil health, and fertility. The potential of biochar may be boosted using its engineered forms. Biochar acts as a precursor while preparing biochar-based nanocomposites with improved surface area, functional group intensity, mechanical stability, and thermal stability (Noreen and Abd-Elsalam, 2021). The nanosize of particles may be controlled by operational pyrolysis temperature or exfoliation, thus resulting into better nutrient-holding capacity. Hence, nanobiochar inherits the benefits of biochar and nanotechnology (Naghdi et al., 2017).

Biochar-based nanomaterials are suitable inputs for the agriculture sector; they improve soil characteristics, enhance plant growth, control plant diseases, facilitate the bioremediation of pesticides and pollutants emitted from water-contaminated surface run-off from agricultural fields, and are employed as support materials for enzymes and other agrochemicals. Low cost and sustainability make these materials an ideal alternative to traditional agri-inputs, which play a role in carbon sequestration and climate change mitigation. Biochar-based nanomaterials retain nutrients firmly versus bulk biochar, leading to their controlled delivery for improved use efficiency and minimum environmental impact (Chausali et al., 2021). Porosity makes nanobiochar a promising choice for soil amendment, enzyme support, and as an adsorbent. The nanobiochar (NBC) demonstrates enhanced absorption capacity for a variety of contaminants due to its increased specific surface area, microporosity, and surface hydrophobicity. NBC counteracts soil acidity due to its basic nature. NBC presents a viable, cost-effective, and long-term solution to a number of problems in the agricultural industry.

Nanobiochar can be utilized to fabricate nanocomposites with potential applications in the production of energy, wastewater treatment, and agriculture sectors. NBC and composites thereof find various applications in agriculture, including soil amendment and as a carrier for smart delivery of agrochemicals, among others. Nanocomposites are effective supporting materials to immobilize enzymes for catalysis (Naghdi et al., 2017). Desorption, followed by long-distance transport of biochar-derived nanocomposites and leaching of metals polluting soil and water, is a challenge. This chapter provides information about the production of nanobiochar and nanocomposites thereof and the potential application of these materials in agriculture. Further, the research gaps and challenges for future studies on biochar and biochar-based nanocomposites have been highlighted.

13.2 BIOCHAR

Biochar is a sustainable carbonaceous material that can be prepared by pyrolyzing agricultural and forest residues (like wheat residue, paddy residue, manure, wood, and leaves) under nil (pyrolysis) or a limited supply of oxygen (gasification). Biochar has been used in diverse areas, namely, but not limited to soil conditioning, crop quality improvement, climate change, as an adsorbent in water purification and catalysis, waste utilization, and waste management. This section summarizes the pyrolysis process, methods for the synthesis of biochar, and the influence of modification methods on biochar properties.

13.2.1 BIOCHAR SYNTHESIS

Biochar can be synthesized via the thermochemical decomposition of biomass under controlled conditions (pyrolysis, torrefaction, hydrothermal carbonization, and gasification). The pyrolysis process involves biomass heating at 300–900°C in an oxygen-limited atmosphere. Pyrolysis processes can be classified as slow pyrolysis, fast pyrolysis, and flash pyrolysis. Pyrolysis of biomass gives solid (biochar), liquid (bio-oil), and gaseous (syngas) products. The process parameters (carrier gas, heating rate, temperature, reaction time, and reactors) govern biochar production, including its yield. Biochar yield is also dependent on the characteristics of biomass precursor and the continuous gas phase reaction.

The pyrolysis temperature is similar in cases of fast and slow pyrolysis, except the heating rate (>200°C min^{-1}) and the very short residence time (~2–5 s) in cases of fast pyrolysis. Fast pyrolysis is used when the intended product is bio-oil. The temperature is even higher in the case of flash pyrolysis (~1000°C for 1 s). Flash pyrolysis is used when the preferred product is bio-oil. Torrefaction involves a low heating rate and a mild temperature (~200–300°C), giving solid torrefied biomass under anoxic conditions. The torrefied biomass inherits the properties of biomass precursors and biochar. Torrefied biomass is not true biochar. Therefore, torrefaction is normally considered a pretreatment step of the pyrolysis process. Recently, microwave-assisted pyrolysis has also been used for biomass carbonization.

Hydrothermal carbonization is used for hydrochar production from agricultural residues, including lignocellulosic biomass. The hydrothermal method is preferred for animal and human excreta, municipal waste, and sewage sludge (with a high amount of water). The drying of these biomasses for better yield requires process energy. The hydrothermal processes are classified as hydrothermal carbonization (<250°C, main product: biochar), hydrothermal liquefaction (250–400°C, bio-oil), and hydrothermal gasification (>400°C, gaseous product). Gasification (~800°C) of biomass gives syngas, tars, and char (5–10 wt% of the feedstock mass) through drying, pyrolysis, partial oxidation, and reduction in the presence of CO_2, O_2, N_2, air, steam, or a gas mixture. The carbon content in gasification-derived chars is low (20–60 wt%), and the abundance of surface functional groups is poor.

On the other hand, their stability (against chemical oxidation and microbial mineralization) is higher due to condensed aromatic rings. A combined process of pyrolysis and gasification may produce char of high porosity and surface area (Cha et al., 2016). In summary, thermochemical technologies are effective for biochar, bio-oil, solid fuel, and syngas production using a number of precursor biomasses.

13.2.2 PROPERTIES OF BIOCHAR

Application of biochar depends upon its properties, which are further determined by precursor biomass and physical, chemical, and molecular changes during pyrolysis. Hemicellulose completely decomposes from 250 to 350°C, while cellulose and lignin decompose partially. Biochar quality may be customized by pretreatment of biomass with magnetic precursors or other agents, for example alkali (Liu et al., 2015). The following chapters discuss the physical and chemical properties of biochar, their importance, and the factors influencing them.

13.2.2.1 PHYSICAL PROPERTIES

Pyrolysis temperature and biomass influence yield, elemental content, surface area, pore volume, and microstructure of biochar; higher surface area and porosities improve water-holding capacity and cation exchange capacity of biochar. Lignin content in biomass improves biochar yield. Animal manure and sludge offer high biochar yields versus crop residues and wood biomass biochar because of the relatively higher inorganic content in animal manure and sludge. The mineral ash in higher amounts increases yield but simultaneously reduces the surface structure of biochar.

Complete and ordered micro or mesoporous structures are developed in biochar with improved specific surface area and active sites due to volatile release. Generally, high pyrolysis temperatures and forest wood give biochar with a higher surface area. The increase in surface area with temperature is relatively less in the case of agricultural residues, aquatic plants, industrial and sewage sludge, and animal waste. Loss of volatile substances hinders up to some extent in the presence of metal, rendering biochar less porous. The configuration of biochar also affects its porosity, mechanical stability, and grindability. Increasing the pyrolysis temperatures induces porosity in biochar, which may fall at 800–1000°C due to the destruction of cell structure in precursor biomass. The emission of volatile matter, dehydration, and rise in pyrolysis temperature help to form pore volumes in biochar. The surface area of biochar decreases at high temperatures of 800–1000°C due to shrinkage of the solid matrix, followed by an initial increase.

13.2.2.2 CHEMICAL PROPERTIES

The content (%) of C, H, N, S, and O, anions like HCO_3^-, CO_3^{2-}type biomass precursors, and pyrolysis conditions (heating rate at a particular temperature and residence time) influence the chemical properties of biochar. Chemical properties are further affected by the presence of C–C, C=C, –OH, –COOH, C–O–H, C=O, C–H linkages (in cellulose, hemicellulose, and lignin). The O, H, C, N, and S escape due to biomass pyrolysis, giving condensable and non-condensable gases and carbon-rich biochar. The abundance of Oxygen containing functional groups (OCFGs) at the biochar surface generally depends upon biomass feedstocks, pyrolytic temperature, and pyrolytic duration. Phenol or quinone moieties in biochar are contributed by lignin. These moieties transfer electrons to the metals (generally transition metals) to generate persistent free radicals (PFRs) on biochar, imparting catalytic properties for the activation of oxidants. The PFR's formation is mainly governed by the pyrolytic temperature and residence time. The functional groups on the biochar surface (double-bond, O-, and N-containing functional groups) play a major role in biochar activation.

The aromaticity, hydrophobicity, and hydrophilicity of biochar depend on different elemental ratios (O/C, H/C, (O+N)/C, and (O+N+S)/C). The pyrolysis at high temperatures leads to more aromatic content in biochar due to its thermodynamic stability. The extent of aromaticity in biochar improves its stability and simultaneously enhances surface (adsorption and catalytic) properties. The high-temperature pyrolysis enhances the hydrophobic nature of biochar due to the decomposition of lignocellulose structure and the dehydration of oxygen-containing functional groups, lowering polar functional groups. Biochar prepared at low and high temperatures behaves as a π acceptors and π- donors, respectively.

As far as the pH is concerned, raw biomass is slightly acidic, whereas biochar is alkaline in nature. This is attributed to the decomposition of acidic carboxyl and hydroxyl groups during pyrolysis, followed by the removal of CO_2, H_2, and a rise in pH with increasing pyrolysis temperature. In addition to this, a rise in pyrolysis temperature increases biochar ash content, which further enhances its alkalinity. Biomass ash contains SiO_2, Al_2O_3, K_2O, P_2O_5, CaO, and MgO. Low-temperature pyrolysis of biomass renders it with an adequate amount of functional groups, which provide a negative charge to biochar with high cation exchange capacity (Weber et al., 2018).

13.3 BIOCHAR-BASED NANOCOMPOSITES

Biochar and nanoparticles form composite materials with improved properties and potential applications (Wang et al., 2017). Incorporating nanomaterials enhances the sorption properties, mechanical properties, electrical properties, thermal stability, and ion exchange capabilities of the porous biochar. Biochar–nanocomposite not only enhances the properties of biochar but also provides the combined benefits of biochar and associated nanoparticles having an intrinsically large specific surface area, making them more efficient.

13.3.1 FABRICATION OF BIOCHAR-BASED NANOCOMPOSITE

A variety of biochar-based nanocomposites can be prepared via biomass pretreatment, biochar post-treatment, and target element enrichment methods (Rodriguez et al., 2019).

13.3.1.1 PRETREATMENT OF BIOMASS

The pre-pyrolysis treatment approach involves biomass pretreatment with metal oxide/salts or functional nanoparticles, and nanomaterial integration takes place concurrently in one process. Biomass is immersed in a solution of desired metal salt for a predetermined amount of time, allowing the metal ions to adhere to the surface or penetrate the biomass's core. A biomass matrix with metal ions embedded or biochar nanocomposites is formed by pyrolysis. This process increases the functional group density, surface area, and ion exchange capacity of biochar. Ca–Mg/biochar nanocomposites were prepared by immersing corncob biomass in $MgCl_2$ and $CaCl_2$ solutions, followed by pyrolysis. These nanocomposites comprise CaO and MgO nanoparticles and a lot of organic functional groups (Fang et al., 2015). The pre-pyrolysis approach is, therefore, a simple approach to develop biochar-based nanocomposite inherited with the catalytic activity of metal ions using only one pyrolysis step.

13.3.1.2 POST-TREATMENT OF BIOMASS

Post-pyrolysis preparation of nanocomposites is accomplished in a two-step process involving biochar fabrication via pyrolysis of biomass followed by the attachment of nanoparticles to the biochar matrix. The most commonly used techniques to produce post-pyrolysis synthesis of biochar nanocomposites are impregnation, encapsulation, and chemical coprecipitation. Lai et al. (2019) prepared a $MnFe_2O_4$/biochar composite via the coprecipitation method for improved catalytic activity versus pure $MnFe_2O_4$ for tetracycline (TC) elimination in water (Lai et al., 2019).

13.3.1.3 TARGET ELEMENT ENRICHMENT METHOD

Biochar-based nanocomposites can also be prepared via bioaccumulation to enrich the biomass feedstock used to produce biochar with the desired metal elements. However, this approach is less popular because the metal-enriched plant is not intended for human food and can contaminate the soil (Rodriguez et al., 2019).

Biochar-based nanocomposites may be divided into three groups: functional nanoparticles coated with biochar; oxide/hydroxide biochar nanocomposites; and magnetic biochar nanocomposites. Functional nanoparticle-coated biochar composites are prepared by coating the surface of biochar with functional nanoparticles, such as chitosan, carbon nanotubes, grapheme, graphene oxide (GO), ZnS nanocrystals, nanoscale zero-valent iron, etc. These functional materials enhance porosity, surface area, thermal stability, and surface activity. These are prepared by either post-pyrolysis infusion of functional nanoparticles or pretreatment of biomass with functional nanomaterials. The biochar–carbon nanotube composite, prepared by dip-coating biochar in a suspension of carbon nanotubes while using sodium dodecylbenzenesulfonate, was more effective than native biochar in water remediation. The wheat straw biochar–graphene composite had an improved surface area, smaller pores, more functional groups, and greater thermal stability (Tan et al., 2016). Oxide/hydroxide biochar nanocomposites can be prepared by primarily using three techniques: bioaccumulation of target elements, biomass pretreatment with intended metal salts, and post-pyrolysis insertion of metal oxide.

13.4 APPLICATION OF BIOCHAR-BASED NANOCOMPOSITES FOR THE AGRICULTURE SECTOR

Soil nutrient depletion is primarily caused by anthropogenic practices (including industrial activity and deforestation). The same has led to low yields and slower plant development, which ultimately have an impact on food security. The sustainable agriculture sector has enormous potential for nanotechnology usage. GO, fullerene, carbon nanotubes, carbon nanotubes, carbon nanofibers, carbon nanohorns, carbon nanodots, carbon nanoonions, and nanodiamonds are a few examples of carbon nanomaterials (CNMs) to be utilized in the agriculture sector in various capacities. The first seven are sp^2 carbon, while the latter seven are sp^3 carbon. The use of carbon nanomaterials (CNMs) in agriculture has recently made headlines. The CNMs have demonstrated varying performance in agro-activities depending on the choice of allotrope, morphology, size, and the potential to combine to generate hybrids (Chandel et al., 2022). Biological processes, namely, photosynthesis, energy production, fatty acid metabolism, lignin biosynthesis, and glucose metabolism were activated when copper hydroxide nanowires were applied topically in a dose-dependent manner. The activation of the amino acid metabolism and the oxidative stress response were elevated as a result of treatments with commercially available copper-based fungicides (Majumdar et al., 2021). Specifically, biochar finds application in the agriculture sector for fertilization, pesticide remediation, plant growth enhancement, and support for microorganisms (Helal et al., 2019; Wang et al., 2017) (Figure 13.1, Tables 13.1–13.4).

13.4.1 SOIL CONDITIONING VIA BIOCHAR AND COMPOSITES

Chemical fertilizers have been overused in an attempt to retain soil nutrients. In contrast, nitrogen loss and runoff reduced their bioavailability in soil and encouraged chemical-based fertilizers. In addition to improving soil fertility, permeability, and water-holding capacity, biochar also contains a significant quantity of nutrients, which increases crop yield (Liu et al., 2016; Wang et al., 2016). The addition of biochar to soil affected soil characteristics, such as texture, porosity, particle size distribution,

FIGURE 13.1 Applications of biochar or engineered biochar in the agriculture sector.

density, and salinity. The presence of organic material enhances soil's pH, CEC, EC, and levels of nutrients like C, N, and P, improving soil's ability to exchange cations and increasing electrical conductivity, which in turn improves the efficiency of metal sorption. It significantly changes the soil's microbiological and nutritional health within the rhizosphere, air, and oxygen content, water-holding capacity, the potential for compaction by diminishing tensile strength, and other factors. Further, applying tailored biochar (together with cutting-edge technology and nanomaterial anchoring) may produce greater results.

13.4.2 PLANT GROWTH ENHANCEMENT VIA BIOCHAR AND COMPOSITES

A crucial ingredient for plant growth is phosphorus (P). Phosphorus can be obtained by plants as a result of P adsorption, desorption, and the transition between soil particles and solution (Helal et al., 2019; Debicka et al., 2016).

Wheat plants treated with biochar containing water-soluble carbon nanoparticles showed a noticeable improvement in plant development (Saxena et al., 2014). Alkaline biochar significantly boosted root development and root biomass as compared with acidic biochar (Bopp et al., 2016). Biochar with non-pyrogenic organic amendments supports the growth of plants, but this combination is not widely accepted due to unique shifting behaviors. Furthermore, rice seedlings exposed to nanobiochar showed reduced metal toxicity (Cd^{2+}) and oxidative stress (Yue et al., 2019). P dynamics in soil regulate its uptake and usage by plants; therefore, biochar can be used to amend soil to increase P and N availability while decreasing P fixation, nitrogen leaching, and N_2O emission. As a result, biochar enhances plant growth and is crucial for managing soil nitrogen levels for enhanced agricultural output.

TABLE 13.1 Biochar/Engineered Biochar for Plant Growth Enhancement.

Biochar/Engineered Biochar	Crop/ Plant	Effect	References
Decorated (MgO) magnetic sugarcane harvest residue-derived biochar	Ryegrass	Plant growth and yield rise due to phosphorous adsorption	Li et al. (2016)
MgO-modified peanut shell biochar	Rice	Yields improvement in saline-alkaline soil due to phosphate retention	Wu et al. (2019)
Wheat straw Biochar-based fertilizer	Rice	Yield and pH increase, Nutrient enrichment (%) (N-40, P-46), Plant biomass (67%)	Chew et al. (2020)
Bamboo biochar inoculated *Pseudomonas putida*	Grape	Increase fruit weight, soluble protein content, and hardness by 7.6%, 28.6%, and 10.8%, respectively	Wei et al. (2020)
Iron-rich sludge biochar	Grass	Phosphorous adsorption and enhanced the seedling growth	Wang et al. (2020)
MgO-peanut shell biochar composite	Rice	Increase rice shoot biomass by 6.0%, increase shoot phosphorus content by 1.0%	Wu et al. (2019)
Palm branches and rice straw Fe(II) biochar composite	Rice and wheat	Reduce Cd accumulation in wheat and rice grain by 16% and 57%, respectively; reduce as accumulation in wheat and rice grain by 48% and 44%, respectively	Tang et al. (2020)

13.4.3 BIOCHAR AND BIOCHAR COMPOSITES AS FERTILIZER

Biochar enriched with micronutrients like Cu, Mn, Fe, Zn, and ash, as well as other macronutrients like N, P, Mg, K, Ca, and S, is a natural, organic fertilizer (Hossain et al., 2011). Plant biomass-derived biochar (wood and green waste) contains less P and N than biochar made from animal sources (broiler litter and sewage sludge) (Bopp et al., 2018). While N intake was adequately decreased, production from biochar and urea was considerably improved (Joseph et al., 2013). The advantages of fertilizer use can also be increased by using biochar composites made of minerals and organic components. The use of biochar-derived nanocarbons as nanofertilizers has received a lot of attention due to their capacity to retain water and nutrients as well as their simplicity (Manya, 2012; Qian et al., 2014). The growth and productivity of the wheat crop were considerably improved by the application of biochar–carbon nanoparticles (Zhang et al., 2020). Nanocarbons and nanobiochar, along with their advantageous effects on fostering plant growth, are tabulated in Table 13.1. Corn cob-derived biochar nanocomposite has reduced traditional fertilizers' leaching issues and helped with pollution control by turning trash into useful products (Lateef et al., 2019). Biochar in agricultural soil may reduce the demand for nitrogen fertilizer to safeguard the environment from pollution due to seepage into groundwater. However, more research is required to determine whether biochar-based fertilizer can effectively provide nitrogen while immobilizing heavy metals.

13.4.4 DISEASE MANAGEMENT VIA BIOCHAR/ENGINEERED BIOCHAR

Biochar can significantly control plant diseases caused by pathogens and nematodes. The same has been true as a result of interactions between specific soil microorganisms like *Bacillus, Trichoderma*, and *Pseudomonas* species and plant roots. Jaiswal and colleagues observed correlations between biochar-induced changes in rhizosphere microbial community structure, taxonomic and functional diversity, and activity. Further, they noticed reduced soilborne diseases like *Fusarium* crown and root rot in tomatoes, along with enhanced plant growth. Biochar reduced *Fusarium* root colonization and survival in the soil while boosting the culturable counts of numerous biocontrol agents. High rhizosphere microbial diversity and activity have previously been linked to disease suppression caused by soilborne pathogens and plant growth promotion, which may explain the significant reduction in disease and improvement in plant performance observed in the presence of biochar (Jaiswal et al., 2017, Table 13.2).

Fusarium root rot was studied using natural recovery as a control, and the influence of vermicompost or biochar application after soil sterilization on soilborne diseases was studied. After one growing season, the disease index of root rot, phenolic acids, and microbial communities in American ginseng rhizosphere soil, as well as the disease index, decreased versus control. A study advises a combined application of biochar and vermicompost to the rhizosphere soil to prevent *Fusarium* root rot in replanted American ginseng (Tian et al., 2021).

TABLE 13.2 Biochar/Engineered Biochar for Disease Management in Plants.

Biochar	Crop/Plant	Effects	References
Eucalyptus wood chip biochar (EUC600) and greenhouse pepper plant waste biochar (GHW350)	Tomato	EUC600 and GHW350 reduced the disease severity by 44% and 61%, respectively	Jaiswal et al. (2017)
Corn straw biochar	American Ginseng	Vermicompost, biochar, and biochar + vermicompost lowered disease index by 33.32%, 19.03%, and 80.96%, respectively	Tian et al. (2021)
Garden waste biochar (GWB) and wood biochar (WB)	Tomato	GWB alone or with *Bacillus subtilis* (PGPR) was most effective in suppressing *Alternaria solani* (fungus causing early blight disease)	Rasool et al. (2021)
Eucalyptus wood chip biochar	Cucumber	14 days of pre-conditioning through biochar lowers pre-emergence damping-off by *P. aphanidermatum* up to 71% in cucumber seedlings	Jaiswal et al. (2019)
Tobacco stem biochar	Tobacco	Biochar reduced *Ralstonia solanacearum* population in rhizosphere, consequently suppressing bacterial wilt	Li et al. (2022)
Jerusalem artichoke biochar	Tomato	Biochar reduced *Fusarium oxysporum* f. sp. *Lycopersici*, ultimately suppressing *Fusarium* wilt disease	Jin et al. (2023)

13.4.5 SOIL MICROBIAL COMMUNITY ENRICHMENT VIA BIOCHAR/ ENGINEERED BIOCHAR

Microbial communities improve plant–soil ecosystems, preserve soil health, maintain nutrient cycling, suppress disease, and improve crop yields. (Thies et al., 2015). Microorganisms in soil need nutrients for their survival and the proliferation of communities; they shift according to the availability of nutrients. Most microorganisms grow in the plant rhizosphere region, which contains sugar, amino acids, and organic compounds for better growth (Lata et al., 2011). Soil needs organic carbon as a platform to maintain microbial growth. Biochar (a rich carbon dioxide source) provides support to soil microorganisms. Additionally, it alters the surface and other physicochemical properties of soil, eventually enriching soil microbial communities (Lehmann et al., 2011). Nanobiochar has also been reported to enhance bioremediation by increasing the microbial diversity and biomass of Actinobacteria and Bacteroidetes in cadmium-contaminated soil to mitigate toxicity. The co-application of biochar and microorganisms can improve soil conditions by removing pollutants (Zheng et al., 2022). Biochar can provide shelter for microorganisms and stimulate soil enzymatic activity (Table 13.3). Biochar-based soil amendments have improved the biodegradation efficiency of microorganisms versus individual microorganisms (Chen et al., 2017). Biochar in soil enriches the microbial community and simultaneously improves overall soil health, nutrient cycling, and fertility (Thies et al., 2015; Liao et al., 2019). For example, Calcium-based magnetic biochar increased the bacterial population, drastically modified the relative abundance of microbial taxa, and shifted its composition.

Biochar's surface functional properties have a great impact on the influence of soil microbial enzymatic activity. Biochar prepared from switchgrass (*Panicum virgatum L.*), corn stover (*Zea mays L.*), and Ponderosa pine wood residue (*Pinus ponderosa Lawson* and *C. Lawson*) has diverse effects on the mineralization of soil C, N, and S-related enzymatic activity (Chintala et al., 2014). Biochar prepared under different gases changes its functional properties and negatively impacts soil dehydrogenase, fluorescein di acetate, beta-glucosidase, and protease activity (Chintala et al., 2014). Dissolved organic matter from biochar plays key roles in the soil ecosystem by altering the soil's biological and physicochemical properties and providing nutrients to the soil microbiome. DOM properties depend upon biomass, pyrolysis temperatures, retention time, and doping (Ajem et al., 2023). Soil fertility and carbon sequestration are directly enhanced by the addition of biochar to the soil. Biochar amendment in terrestrial ecosystems significantly changed the functional diversity of soil microbial communities and influenced soil microbial enzymatic activities (Rutigliano et al., 2014). Biochar applications in acidic soil significantly induced alpha diversity of soil microorganisms and altered their relative abundance via increased soil total carbon, total nitrogen, and C/N ratios under pot trials. It also decreased N_2O emissions by reducing them to N_2 (Xu et al., 2014). Biochar treatments significantly increased microbial enzymatic activity via nitrogen reduction to enhance soil health under field conditions in rapeseed crops (Tian et al., 2022). Yuan et al. (2023) studied the influence of three nitrogen fertilizers and two biochar on the soil microbiome using high-throughput sequencing. A significant decrease was detected in bacterial and fungal diversity in the top soil. Moreover, the soil total microbial community composition was also changed by the application of biochar and nitrogen. Furthermore, keystone microbial species of nitrogen, phosphorus, and potassium networks were also changed by biochar (Yuan et al., 2023). In a study, biochar addition to soil significantly increased the maize yield and altered the beta diversity of the soil microbial community; however, the alpha diversity remained the same. Amendment of 31.5–126.0 Mg/ha biochar reduced bacterial communities in the rhizosphere but had no significant effect on the fungal communities in aeolian soil (Hu et al., 2023).

Liao and others supported the same and showed that biochar amendment altered the soil microbial community by assimilating plant-derived carbon. Further, biochar amendment increased the relative densities of *Alicyclobacillus*, *Bacillus*, *Clostridium*, *Sporomusa*, and *Desulfosporosinus* (Firmicutes and Bacteroidetes members) while decreasing the loads of plant-derived carbon. A calcium-based magnetic biochar has considerably boosted the quantity and diversity of bacterial populations, altered the relative abundance of microbial taxa, and caused a composition shift (Chausali et al., 2021).

13.4.6 PESTICIDE REMEDIATION VIA BIOCHAR/ENGINEERED BIOCHAR

Biochar effectively sequesters pesticides due to its superior porosity and rich surface chemistry with abundant adsorption sites. It may lessen pesticide leaching and bioavailability in soil, leading to the remediation of pesticide-contaminated soils (Nag et al., 2011; Khorram et al., 2016). Biochar may have aided in a decrease in pesticide content in soil due to its strong pesticide-binding capacity (Jones et al., 2011, Table 13.4). A further investigation

TABLE 13.3 Biochar/Engineered Biochar for Microbial Enrichment in Soil.

Biochar	Plant	Effect	References
Fruit trees residue biochar	Soil samples taken from tomato crop sown area	Increase MBC and MBN Significant increase in the number of amoA-type nitrifiers and decreased the number of nirS-type denitrifies, at biochar rate ≤30 t ha^{-1} Increase in plant height, stem thickness, and biomass under biochar treatment Maximum yield of tomato and vitamin C were observed under application rate 50 t ha^{-1}	Guo et al. (2021)
Residual sulfur-enhanced biochar-microorganisms	*Capsicum annuum* L.	Increased macro- and micro-nutrient concentration, plant growth, productivity, dehydration tolerance, and irrigation use efficiency Decreased Na^+ and Cd^{2+} uptake with a concomitant increase in N, P, K^+, Ca^{2+}, Fe, Mn, Cu, and Zn uptake	Abd El-Mageed et al. (2020)
Corn straw biochar	–	Increase in soil microbes and enzymes activity	Lin et al. (2019)
Walnut shell biochar	One genotype of *Tritica secale Wittmack*; and two genotype of *Pisum sativum* sp. *arvense* L., namely, Taşkent and Özkaynak	Enhance in seed germination, seed vigor index, seedling dry and fresh weight, radicle and plumule length, at 40 Mg ha^{-1} and 80 Mg ha^{-1} application rates	Uslu et al. (2020)
Bamboo feedstock biochar	Tomato	Increased plant growth index, glucose, fructose, soluble solids, ascorbic acid, and sugar-to-acid ratios under biochar prepared at 300°C (3.0%) and 450°C (1.0%)	Suthar et al. (2018)
Coconut shells	Maize	Improved the plant height, stem diameter, and leaf area Increased yield by 4.4% and 5.7% at 0.7% and 1% nanobiochar treatment, respectively	Yang et al. (2020)
Phosphorus-modified biochar	*Pseudostellaria heterophylla* (medicinal plant)	Increased leaf area, root length density, dry biomass of tuber and yield	Ng et al. (2022)

revealed that plants' uptake of the herbicide fomesafen was reduced due to biochar's greater sorption capacity (Khorram et al., 2016). Biochar is a suitable habitat for microorganisms due to its nutrients and ash, which in turn benefit the soil in many ways.

TABLE 13.4 Biochar/Engineered Biochar for Pesticide Remediation.

Biochar	Effect	References
Maize straw/pig manure biochar	The sorption of three common neonicotinoids (imidacloprid, clothianidin, and thiacloprid) was greater in acid-deashed biochar than in untreated biochar	Zhang et al. (2018)
Red gum wood biochar	The biochar amendment (especially for BC850) lowers the bioavailability of chlorantraniliprole (CAP) to earthworms in the soils	Wang et al. (2012)
Rice straw biochar, wheat straw biochar and swine manure biochar	Compared with plant residue-derived biochar, swine waste-derived biochar had lesser enhancement of sorption capacity toward imidacloprid, isoproturon, and atrazine at a given biochar application rate	Jin et al. (2016)
Wood chip biochar, macadamia nut shell biochar, and hardwood biochar	Wood chip biochar had maximum sorption of the pesticides (aminocyclopyrachlor and bentazone) from soil	Carbera et al. (2014)
Cotton straw biochar	Adding 1% biochar (BC850) to the soil reduced chlorpyrifos and fipronil in plants by 81% and 52%, respectively	Yang et al. (2010)

13.4.7 OTHER USEFUL CARBON-BASED NANOMATERIALS EMPLOYED IN THE AGRICULTURE SECTOR

Agrochemicals in agriculture undergo degradation and cause irreparable damage to the environment. Bio-reduced graphene oxide (RGO) from microbial sources was used as a nanocarrier for the efficient delivery of plant growth regulator indole acetic acid to maize plants. RGO applied to maize seeds by a seed priming approach showed a slow, sustained release of indole acetic acid to increase the root length, shoot length, and biomass of maize plants (Bhattacharya et al., 2023). A carbon dot-based novel slow-release zinc fertilizer influenced yield in wheat (*Triticum aestivum*) depending upon zinc concentration and growth indices (Alikhani et al., 2023). In comparison to $ZnSO_4$ treatment, Zn (carbon dot nanocarrier) increased grain yield, fertile spikelet yield, and root-shoot biomass by 20, 44, 16, and 43%, respectively. Concentrations of zinc and nitrogen in the grain increased by 19% and 118%, respectively, versus ZnSO4 treatment, whereas phytic acid decreased by 18%. Through vascular bundles, wheat plants were able to absorb and transmit Zn-NCDs from the roots to the stems and leaves, according to microscopic examinations. It is essential for biological imaging technology that Zn-NCDs have a good *in vivo* fluorescence imaging effect (Alikhani et al., 2023).

13.5 SUMMARY AND FUTURE RECOMMENDATIONS

The sustainable agriculture sector has enormous potential for nanotechnology usage. Biochar, being porous and having a large surface area with high carbon and mineral content, is suitable for agriculture and water and soil remediation. The practical application of functional nanoparticle-coated biochar composites is nonetheless constrained by the high cost of components like carbon nanotubes and graphene; hence, low-cost nanomaterials and protocols must be devised. Biochar and engineered biochar have been used as soil conditioners, plant growth boosters, carriers for controlled delivery of agrochemicals, agents for plant disease management, agents for microbial community enrichment in soil, pesticide remediation, support for enzymes in sensing, and others. Further, these biochar-based engineered materials help in carbon sequestration and climate change mitigation. In comparison to biochar, engineered biochar, nanobiochar, and nanocomposites outperform in immobilizing nutrients and contaminants in soil by influencing their dynamics. More research is required to determine whether biochar-based fertilizer can effectively provide nitrogen while immobilizing toxicants such as heavy metals. These CNMs can be a sustainable solution to pesticide-induced soil contamination and may be a possibility for pesticide remediation due to their increased affinity for pesticides. Biochar nanocomposites-related work should target sorption capability for various pesticides and herbicides.

ACKNOWLEDGMENT

RKD and AR acknowledge the Department of Science and Technology, Government of Haryana, for providing financial support (Grant No. HSCIT/R&D/2022-23/4497) and HSCSIT/R&D/2024/517, dated 16/02/2024.

KEYWORDS

- **Biochar**
- **controlled delivery of agrochemicals**
- **engineered biochar**
- **fertilizer**
- **plant health**
- **pesticide remediation**
- **soil health**

REFERENCES

Abd El-Mageed, T. A., Rady, M. M., Taha, R. S., Abd El Azeam, S., Simpson, C. R., & Semida, W. M. (2020). Effects of integrated use of residual sulfur-enhanced biochar with effective microorganisms on soil properties, plant growth and short-term productivity of Capsicum annuum under salt stress. *Scientia Horticulturae, 261*, 108930.

Alikhani, M., Mirbolook, A., Sadeghi, J., & Lakzian, A. (2023). Effect of a new slow-release zinc fertilizer based on carbon dots on the zinc concentration, growth indices, and yield in wheat (Triticum aestivum). *Plant Physiology and Biochemistry, 200*, 107783.

Bhattacharya, N., Kochar, M., Bohidar, H. B., Yang, W., & Cahill, D. M. (2023). Biologically synthesized and indole acetic acid-loaded graphene as biostimulants for maize growth enhancement. *ACS Agricultural Science & Technology, 3*(5), 432–444.

Bopp, C., Christl, I., Schulin, R., & Evangelou, M. W. (2016). Biochar as possible long-term soil amendment for phytostabilisation of TE-contaminated soils. *Environmental Science and Pollution Research, 23*, 17449–17458.

Cha, J. S., Park, S. H., Jung, S. C., Ryu, C., Jeon, J. K., Shin, M. C., & Park, Y. K. (2016). Production and utilization of biochar: A review. *Journal of Industrial and Engineering Chemistry, 40*, 1–15.

Chandel, M., Kaur, K., Sahu, B. K., Sharma, S., Panneerselvam, R., & Shanmugam, V. (2022). Promise of nano-carbon to the next generation sustainable agriculture. *Carbon, 188*, 461–481.

Chausali, N., Saxena, J., & Prasad, R. (2021). Nanobiochar and biochar-based nanocomposites: Advances and applications. *Journal of Agriculture and Food Research, 5*, 100191.

Chen, L., Chen, X. L., Zhou, C. H., Yang, H. M., Ji, S. F., Tong, D. S., Zhong, Z. K., Yu, W. H., & Chu, M. Q. (2017). Environmental-friendly montmorillonite-biochar composites: Facile production and tunable adsorption-release of ammonium and phosphate. *Journal of Cleaner Production, 156*, 648–659.

Chew, J., Zhu, L., Nielsen, S., Graber, E., Mitchell, D. R., Horvat, J., Mohammed, M., Liu, M., van Zwieten, L., Donne, S., & Munroe, P. (2020). Biochar-based fertilizer: Supercharging root membrane potential and biomass yield of rice. *Science of the Total Environment, 713*, 136431.

Chintala, R., Schumacher, T. E., Kumar, S., Malo, D. D., Rice, J. A., Bleakley, B., Chilom, G., Clay, D. E., Julson, J. L., Papiernik, S. K., & Gu, Z. R. (2014). Molecular characterization of biochars and their influence on microbiological properties of soil. *Journal of Hazardous Materials, 279*, 244–256.

Debicka, M., Kocowicz, A., Weber, J., & Jamroz, E. (2016). Organic matter effects on phosphorus sorption in sandy soils. *Archives of Agronomy and Soil Science, 62*(6), 840–855.

Fang, C., Zhang, T., Li, P., Jiang, R., Wu, S., Nie, H., & Wang, Y. (2015). Phosphorus recovery from biogas fermentation liquid by Ca–Mg loaded biochar. *Journal of Environmental Sciences, 29*, 106–114.

Guo, L., Yu, H., Kharbach, M., Zhang, W., Wang, J., & Niu, W. (2021). Biochar improves soil-tomato plant, tomato production, and economic benefits under reduced nitrogen application in northwestern China. *Plants, 10*(4), 759.

Helal, M. I., Husein, M. E., Walaa, G., & Mostafa, E. D. (2019). Characterization of agricultural residues-based nano biochar and its efficiency in adsorption/desorption of nutrients. *International Journal of Environmental Research and Public Health, 8*(2), 130–141.

Hu, Y., Cong, M., Yan, H., Sun, X., Yang, Z., Tang, G., Xu, W., Zhu, X., & Jia, H. (2023). Effects of biochar addition on aeolian soil microbial community assembly and structure. *Applied Microbiology and Biotechnology, 107*(11), 3829–3845.

Hossain, M. K., Strezov, V., Chan, K. Y., Ziolkowski, A., & Nelson, P. F. (2011). Influence of pyrolysis temperature on production and nutrient properties of wastewater sludge biochar. *Journal of Environmental Management, 92*(1), 223–228.

Jaiswal, A. K., Elad, Y., Paudel, I., Graber, E. R., Cytryn, E., & Frenkel, O. (2017). Linking the belowground microbial composition, diversity and activity to soilborne disease suppression and growth promotion of tomato amended with biochar. *Scientific Reports, 7*(1), 44382.

Jaiswal, A. K., Graber, E. R., Elad, Y., & Frenkel, O. (2019). Biochar as a management tool for soilborne diseases affecting early-stage nursery seedling production. *Crop Protection, 120*, 34–42.

Jin, J., Kang, M., Sun, K., Pan, Z., Wu, F., & Xing, B. (2016). Properties of biochar-amended soils and their sorption of imidacloprid, isoproturon, and atrazine. *Science of the Total Environment, 550*, 504–513.

Jin, X., Zhou, X., Wu, F., Xiang, W., & Pan, K. (2023). Biochar amendment suppressed fusarium wilt and altered the rhizosphere microbial composition of tomatoes. *Agronomy, 13*(7), 1811.

Jones, D., Edwards-Jones, G., & Murphy, D. (2011). Biochar mediated alterations in herbicide breakdown and leaching in soil. *Soil biology and Biochemistry, 43*(4), 804–813.

Joseph, S., Graber, E., Chia, C., Munroe, P., Donne, S., Thomas, T., Nielsen, S., Marjo, C., Rutlidge, H., Pan, & G.-X. (2013). Shifting paradigms: Development of high-efficiency biochar fertilizers based on nano-structures and soluble components. *Carbon Management, 4*(3), 323–343.

Khorram, M. S., Zhang, Q., Lin, D., Zheng, Y., Fang, H., & Yu, Y. (2016). Biochar: A review of its impact on pesticide behavior in soil environments and its potential applications. *Journal of Environmental Sciences, 44*, 269–279.

Lai, C., Huang, F., Zeng, G., Huang, D., Qin, L., Cheng, M., Zhang, C., Li, B., Yi, H., Liu, S., & Chen, L. (2019). Fabrication of novel magnetic MnFe2O4/bio-char composite and heterogeneous photo-Fenton degradation of tetracycline in near neutral pH. *Chemosphere, 224*, 910–921.

Lin, Z., Li, X. Liu, et al., (2019). Effects of Fe-Mn modified biochar composite treatment on the properties of As-polluted paddy soil, *Environmental Pollution, 244*, 600–607 https://doi.org/10.1016/j.envpol.2018.10.011

Liu, W. J., Jiang, H., & Yu, H. Q. (2015). Development of biochar-based functional materials: Toward a sustainable platform carbon material. *Chemical Reviews, 115*(22), 12251–12285.

Liu, S. B., Tan, X. F., Liu, Y. G., Gu, Y. L., Zeng, G. M., Hu, X. J., Wang, H., Zhou, L., Jiang, L. H., & Zhao, B. B., (2016). Production of biochars from Ca impregnated ramie biomass (*Boehmeria nivea* (L.) Gaud.) and their phosphate removal potential. *Rsc Advances, 6*(7), 5871–5880.

Li, R., Wang, J. J., Zhou, B., Awasthi, M. K., Ali, A., Zhang, Z., Lahori, A. H., & Mahar, A. (2016). Recovery of phosphate from aqueous solution by magnesium oxide decorated magnetic biochar and its potential as phosphate-based fertilizer substitute. *Bioresource Technology, 215*, 209–214.

Li, C., Ahmed, W., Li, D., Yu, L., Xu, L., Xu, T., & Zhao, Z. (2022). Biochar suppresses bacterial wilt disease of flue-cured tobacco by improving soil health and functional diversity of rhizosphere microorganisms. *Applied Soil Ecology, 171*, 104314.

Majumdar, S., Long, R. W., Kirkwood, J. S., Minakova, A. S., & Keller, A. A. (2021). Unravelling metabolic and proteomic features in soybean plants in response to copper hydroxide nanowires compared to a commercial fertilizer. *Environmental Science & Technology, 55*(20), 13477–13489.

Nag, S. K., Kookana, R., Smith, L., Krull, E., Macdonald, L. M., & Gill, G. (2011). Poor efficacy of herbicides in biochar-amended soils as affected by their chemistry and mode of action. *Chemosphere, 84* (11), 1572–1577.

Naghdi, M., Taheran, M., Brar, S. K., Kermanshahi-Pour, A., Verma, M., & Surampalli, R. Y. (2017). Immobilized laccase on oxygen functionalized nanobiochars through mineral acids treatment for removal of carbamazepine. *Science of the Total Environment, 584*, 393–401.

Ng, C. W. W., Wang, Y. C., Ni, J. J., & So, P. S. (2022). Effects of phosphorus-modified biochar as a soil amendment on the growth and quality of Pseudostellaria heterophylla. *Scientific Reports, 12*(1), 7268.

Noreen, S., & Abd-Elsalam, K. A. (2021). Biochar-based nanocomposites: A sustainable tool in wastewater bioremediation. In *Aquananotechnology* (pp. 185–200). Elsevier.

Rasool, M., Akhter, A., Soja, G., & Haider, M. S. (2021). Role of biochar, compost and plant growth promoting rhizobacteria in the management of tomato early blight disease. *Scientific Reports, 11*(1), 6092.

Rodriguez-Narvaez, O. M., Peralta-Hernandez, J. M., Goonetilleke, A., & Bandala, E. R. (2019). Biochar-supported nanomaterials for environmental applications. *Journal of Industrial and Engineering Chemistry, 78*, 21–33.

Rutigliano, F. A., Romano, M., Marzaioli, R., Baglivo, I., Baronti, S., Miglietta, F., & Castaldi, S. (2014). Effect of biochar addition on soil microbial community in a wheat crop. *European Journal of Soil Biology, 60*, 9–15.

Saxena, M., Maity, S., & Sarkar, S. (2014). Carbon nanoparticles in 'biochar' boost wheat (Triticum aestivum) plant growth. *Rsc Advances, 4*(75), 39948–39954.

Suthar, R. G., Wang, C., Nunes, M. C. N., Chen, J., Sargent, S. A., Bucklin, R. A., & Gao, B. (2018). Bamboo biochar pyrolyzed at low temperature improves tomato plant growth and fruit quality. *Agriculture, 8*(10), 153.

Tan, X. F., Liu, Y. G., Gu, Y. L., Xu, Y., Zeng, G. M., Hu, X. J., Liu, S. B., Wang, X., Liu, S. M., & Li, J. (2016). Biochar-based nano-composites for the decontamination of wastewater: A review. *Bioresource Technology, 212*, 318–333.

Tang, X., Shen, H., Chen, M., Yang, X., Yang, D., Wang, F., Chen, Z., Liu, X., Wang, H., & Xu, J. (2020). Achieving the safe use of Cd-and As-contaminated agricultural land with an Fe-based biochar: A field study. *Science of the Total Environment, 706*, 135898.

Tang, R., Gong, D., Deng, Y., Xiong, S., Zheng, J., Li, L., Zhou, Z., Su, L., & Zhao, J. (2022). π-π stacking derived from graphene-like biochar/g-C3N4 with tunable band structure for photocatalytic antibiotics degradation via peroxymonosulfate activation. *Journal of Hazardous Materials, 423*, 126944.

Tian, G. L., Bi, Y. M., Jiao, X. L., Zhang, X. M., Li, J. F., Niu, F. B., & Gao, W. W. (2021). Application of vermicompost and biochar suppresses Fusarium root rot of replanted American ginseng. *Applied Microbiology and Biotechnology, 105*(18), 6977–6991.

Thies, J. E., Rillig, M. C., & Graber, E. R. (2015). Biochar effects on the abundance, activity and diversity of the soil biota. In *Biochar for Environmental Management: Science, Technology and Implementation*, (Vol. 2, pp. 327–389).

Tian, X., Li, Z., Wang, Y., Li, B., Xie, X., & Wang, L. (2023). Effects of biochar combined with nitrogen reduction on the soil microbial community of rapeseed. *Archives of Agronomy and Soil Science, 69*(11), 2038–2052.

Uslu, O. S., Babur, E., Alma, M. H., & Solaiman, Z. M. (2020). Walnut shell biochar increases seed germination and early growth of seedlings of fodder crops. *Agriculture, 10*(10), 427.

Wang, B., Gao, B., & Fang, J. (2017). Recent advances in engineered biochar productions and applications. *Critical Reviews in Environmental Science and Technology, 47*(22), 2158–2207.

Wang, S., Dai, G., Yang, H., & Luo, Z. (2017). Lignocellulosic biomass pyrolysis mechanism: A state-of-the-art review. *Progress in Energy and Combustion Science, 62*, 33–86.

Wang, Y. Y., Lu, H. H., Liu, Y. X., & Yang, S. M. (2016). Removal of phosphate from aqueous solution by SiO_2–biochar nanocomposites prepared by pyrolysis of vermiculite treated algal biomass. *Rsc Advances, 6*(87), 83534–83546.

Weber, K., & Quicker, P. (2018). Properties of biochar. *Fuel, 217*, 240–261.

Wang, H., Xiao, K., Yang, J., Yu, Z., Yu, W., Xu, Q., Wu, Q., Liang, S., Hu, J., Hou, H., & Liu, B. (2020). Phosphorus recovery from the liquid phase of anaerobic digestate using biochar derived from iron− rich sludge: A potential phosphorus fertilizer. *Water Research, 174*, 115629.

Wang, T. T., Cheng, J., Liu, X. J., Jiang, W., Zhang, C. L., & Yu, X. Y. (2012). Effect of biochar amendment on the bioavailability of pesticide chlorantraniliprole in soil to earthworm. *Ecotoxicology and Environmental Safety, 83*, 96–101.

Wu, L., Wei, C., Zhang, S., Wang, Y., Kuzyakov, Y., & Ding, X. (2019). MgO-modified biochar increases phosphate retention and rice yields in saline-alkaline soil. *Journal of Cleaner Production, 235*, 901–909.

Wei, M., Liu, X., He, Y., Xu, X., Wu, Z., Yu, K., & Zheng, X. (2020). Biochar inoculated with Pseudomonas putida improves grape (Vitis vinifera L.) fruit quality and alters bacterial diversity. *Rhizosphere, 16*, 100261.

Xu, H. J., Wang, X. H., Li, H., Yao, H. Y., Su, J. Q., & Zhu, Y. G. (2014). Biochar impacts soil microbial community composition and nitrogen cycling in an acidic soil planted with rape. *Environmental Science & Technology, 48*(16), 9391–9399.

Yang, X. B., Ying, G. G., Peng, P. A., Wang, L., Zhao, J. L., Zhang, L. J., Zhang, L. J., Yuan, P., & He, H. P. (2010). Influence of biochars on plant uptake and dissipation of two pesticides in an agricultural soil. *Journal of Agricultural and Food Chemistry, 58*(13), 7915–7921.
Yang, Y., Zhou, B., Hu, Z., & Lin, H. (2020). The effects of nano-biochar on maize growth in Northern Shahnxi province on the loess plateau. *Applied Ecology & Environmental Research, 18*(2), 2863–2877.
Yue, L., Lian, F., Han, Y., Bao, Q., Wang, Z., & Xing, B. (2019). The effect of biochar nanoparticles on rice plant growth and the uptake of heavy metals: Implications for agronomic benefits and potential risk. *Science of the Total Environment, 656*, 9–18.
Yuan, M., Zhu, X., Sun, H., Song, J., Li, C., Shen, Y., & Li, S. (2023). The addition of biochar and nitrogen alters the microbial community and their cooccurrence network by affecting soil properties. *Chemosphere, 312*, 137101.
Zhang, P., Sun, H., Ren, C., Min, L., & Zhang, H. (2018). Sorption mechanisms of neonicotinoids on biochars and the impact of deashing treatments on biochar structure and neonicotinoids sorption. *Environmental Pollution, 234*, 812–820.
Zheng, X., Xu, W., Dong, J., Yang, T., Shangguan, Z., Qu, J., Li, X., & Tan, X. (2022). The effects of biochar and its applications in the microbial remediation of contaminated soil: A review. *Journal of Hazardous Materials, 438*, 129557.

CHAPTER 14

Assessment of Immobilization of *Bacillus* sp. on Biochar on the Rate of Cellulose Decomposition and the Ecological State of Haplic Chernozem

TATIANA MINNIKOVA, SERGEY KOLESNIKOV, NIKITA MININ, VLADIMIR CHISTYAKOV, TATIANA MINKINA, SAGLARA MANDJIEVA, and VISHNU D. RAJPUT

Southern Federal University, Academy of Biology and Biotechnology named by D.I. Ivanovsky, Rostov on Done, Russia

ABSTRACT

The area of fields occupied by grain crops in the world reaches tens of millions of hectares, which are dominated by crops of winter wheat, barley, maize, and sorghum. According to the FAO forecast (2022), the world's crop harvest in the 2022–2023 season will be at least 2.774 million tons. Although this figure has decreased compared to 2021–2022, the amount of plant residues from grain crops remaining in the fields is steadily increasing by 5 million tons. The crop residues left on the fields after harvesting are a source of soil carbon, which is important for the formation of humus. The yield of crops grown on such soil directly depends on the rate of decomposition and the chemical composition of crop residues. However, the process of natural decomposition of crop residues compared to vegetable crops in the soil is quite long. To accelerate the decomposition of plant residues from grain crops in the soil, various biostimulants are used, such as microbiological preparations and mineral and organic fertilizers. In addition to the independent introduction of strains of various microorganisms, it is advisable to combine them with

Nanobiotechnology for Agricultural Sciences: Nano-Agri-Input Products for Crop Production and Environmental Protection. J. C. Tarafdar, Deepak Kumar, Sanjay K. Singh, and Krishna Pal Singh (Eds.)

organic carriers having a porous structure and a high specific surface area, such as biochar. Immobilization of *Bacillus* sp. on biochar is an innovative tool in the process of increasing the number and activity of soil biota, mainly through the decomposition of crop residues from grain crops. The purpose of this chapter is to assess the rate of decomposition of winter wheat crop residues and the ecological state of Haplic Chernozem under the influence of the immobilization of *Bacillus* sp. on biochar. Biochar and *Bacillus* sp. were added to the prepared soil samples to assess the rate of decomposition of crop residues in the soil and cotton web samples. Experimental options included independent and combined exposure to biochar and *Bacillus* sp., including immobilization of *Bacillus* sp. on the biochar. The main sensitive and informative biological indicators reflecting the ecological state of the soil were analyzed: catalase and dehydrogenase activity, total number of soil bacteria, emission of CO_2, and intensity of initial growth and development of barley (*Hordéum vulgáre* L.). Haplic Chernozem incubation with cellulose cotton, biochar, and *Bacillus* sp. was carried out within 30 days. The introduction of biochar into the soil with crop residues did not affect the rate of fiber decomposition while stimulating the indicators of the intensity of the initial growth and development of barley seeds. *Bacillus* sp., at a dose recommended for the treatment of fungal diseases of plants, inhibited the roots and shoots of wheat but stimulated the activity of dehydrogenases, indicators of the intensity of the initial growth of barley. *Bacillus* sp. immobilized on biochar contributed to the decomposition of cotton in the soil by 91% of the initial rate and stimulated the intensity of the initial growth and development of barley. The results of the study should be used to stimulate the decomposition of plant residues from grain crops, to increase soil fertility and the quality of agricultural products, as well as for biomonitoring of the soil condition on agricultural land.

14.1 INTRODUCTION

14.1.1 THE CONTENT OF CROP RESIDUES OF GRAIN CROPS IN THE SOILS OF AGROPHYTOCENOSIS

According to the report of the FAO (2022), the lack of moisture in the northern hemisphere soils caused a significant reduction in the forecast level of cereal production in 2022 by 38.9 million tons compared with the same period last year (1.4%). World cereal production in 2022 was previously estimated at 2,774 million tons, down 17.2 million tons from the previously published July forecast (FAO, 2022). Under such conditions, maintaining the soil fertility of agrophytocenoses becomes a priority. The UN FAO recommends that the countries of Eastern Europe and Central Asia switch to conservation agriculture based on three principles: no-till and direct seeding, diversified crop rotation, and ensuring a permanent organic cover on the soil surface (https://www.fao.org/conservation-agriculture/en/). According to UN FAO calculations, the soil should be 100% covered with crop residues, and their volume and thickness depend on climatic conditions: in conditions of low average annual temperatures and low air humidity (a cold climate), the layer of residues should be thinner compared to agricultural lands in a warmer climate. The UN FAO estimates that about 3–3.5 tons of crop residues per hectare are needed to completely shade the land with crop residues. However, sometimes half of the specified volume is enough. The minimum amount of crop residue is as follows: 1 t/ha on moderate soils, 1.5 t/ha on medium soils, 2.5 t/ha on sandy soils.

Often, crop residues of small grains after harvesting are either removed from the field and disposed of at domestic waste landfills or burned directly in the field. All these technological operations lead to the loss of soil carbon, which is so necessary for the formation of the humus layer and soil fertility. Crop residues on the soil surface have a direct impact on soil fertility by stabilizing the content of soil organic carbon (SOC) and nitrogen (N), phosphorus (P), and potassium (K) in the soil (Novikov and Kisarov, 2012; Thapa et al., 2021; Torma et al., 2018). SOC is a vital indicator of soil health. The use of long-term conventional tillage for many years has led to a deterioration of the physical, chemical, and biological condition of the soil and, consequently, a marked decrease in soil carbon (Minnikova et al., 2022; Mokrikov et al., 2019). With no-till farming, crop residues are ploughed into the soil for subsequent decomposition and SOC preservation. Direct seeding compared to conventional technology increases crop yields and reduces the cost of major crops (wheat and sunflower) (Mokrikov et al., 2019). Direct seeding technology was found to have a favorable effect on nutrients in the topsoil (0–10 cm) during the active growing season (Minnikova et al., 2019; Kravtsova et al., 2019). It was revealed that in the soils of winter wheat agrocenoses, the use of direct seeding led to an increase in mobile phosphorus and potassium. Crop residues affect such indicators of fertility as the enzymatic activity of the soil. A favorable effect of binary sunflower with sweet clover was found on the activity of dehydrogenases of Haplic Chernozem within the Rostov Region compared with mono-cultured sunflower by 11–111%. Protease in crops with sweet clover and vetch increased up to 7.6 times compared with the control. The stimulation of protease during the transition from the wet (June) to the cold (September) period in sunflower crops with sweet clover was established. In addition, the rate of cellulose decomposition affects the soil bacteria's abundance. In soil with cellulose and straw, taxa whose numbers both increased (*Chthoniobacteraceae, Xanthomonadaceae, and Chitinophagaceae*) and decreased (*Gaiellaceae*) were found (Awad et al., 2012). Among the former were classical cellulolytic bacteria (*Chitinophaga* and representatives of the *Streptosporangiaceae* and *Micromonosporaceae* families) and bacteria not yet seen in biodegradation (*Chthoniobacter and Chitinophaga*).

14.1.2 BIOLOGICAL METHODS OF DECOMPOSITION OF CROP RESIDUES AND ASSESSMENT OF THE ECOLOGICAL SOILS UNDER AGROPHYTOCENOSIS

The load on agricultural soils is associated not only with declining fertility due to soil depletion but also with chemical pollution by pesticides and fertilizers (Bokade et al., 2023). Often, microorganisms such as algae are used to restore the ecological and agricultural functions of the soil. Algae are the most common microorganisms on Earth, with potential applications in agriculture as biofertilizers and means to improve soil conditions and increase plant productivity (Chapman, 2013; Duarte et al., 2018). Most algae, in particular cyanobacteria, are able to fix nitrogen from the atmosphere, and several species, including *Anabaena* sp., *Nostoc* sp., and *Oscillatoria angustissima*, are effective biostimulants. *Acutodesmus dimorphus*, *Spirulina platensis*, *Chlorella vulgaris*, *Scenedesmus dimorphus*, *Anabaena azolla*, and *Nostoc* sp. are some of the types of green microalgae and cyanobacteria that have been successfully used as bio-fertilizers to enhance crop growth. In addition, *Chlorella vulgaris* is one of the most commonly used microalgae in biofertilizer research. Adding species of seaweed, *Sargassum* sp. and *Gracilaria verrucosa,* leads to chemical changes as an indicator of soil fertility on clay and sandy soils: an increase in organic matter and normalization of soil acidity (pH) (Shinde et al., 2022). Among various soil remediation strategies, microbial bioremediation faces inherent challenges when applied in the field because the microbe interaction with polluted soil varies greatly depending on climatic conditions (Bokade et al., 2023). A number of studies have found that combining microalgae and fungi has a synergistic effect on soil properties (Wu et al., 2022).

Biochar has a high carbon concentration (up to 90%), increases cation exchange and water holding capacity, reduces nutrient leaching, and maintains moderate soil acidity (Ahmad et al., 2014; El-Bassi et al., 2021; Jones et al., 2012; Lehmann et al., 2011). Biochar is used for the remediation of soils polluted with pesticides using the following stages: bioremediation, phytoremediation, ionization, hydrolysis, redox restoration, photocatalysis, and photodegradation (Marican et al., 2018; Morillo et al., 2017). The use of biochar for the restoration of pesticide-polluted soil is quite common (Khalid et al., 2020; Liang et al., 2021; Ogura et al., 2021). The main factor affecting the biochar's properties and pesticide sorption ability is the pyrolysis temperature. Soils enriched with biochar have a greater sorption capacity for pesticides, which limits potential leaching and running into groundwater. Biochar reduces the bioavailability of pesticides, reducing their toxicity to soil organisms and increasing soil fertility and crop yields. Currently, it is necessary to study the potential desorption of biochar associated with pesticide pollution. Biochar production with the utilization of crop residues should be applied to restore the sustainability of agricultural systems (Ogura et al., 2021). The positive biochar effect on the productivity of legumes with symbiotic microorganisms has been established (Liu et al., 2022).

The development of fertilizers with a slow release of nutrients to the soil is needed (Tang et al., 2019; Liang et al., 2021). There are three reasons for obtaining such biochar-based fertilizers (Wang et al., 2022; Wu et al., 2022). (a) Biochar is rich in carbon, porous, and contains many functional groups, so it can be used to create fertilizers, (b) biochar is useful to improve soil cation exchange capacity and water-holding capacity, and (c) biochar has the advantages of low cost, environmental friendliness, and renewability.

Burning dry agricultural crop residues is considered one of the most environmentally friendly approaches to the safe disposal of plant residues (Mohan et al., 2018). Obtaining biochar from plant residues reduces the area of crop residues subject to uncontrolled burning and helps restore soil properties (Ahmad et al., 2014). Biochar is a stable form of carbon, and it takes from 20 to more than 1000 years for its decay. Thus, burning plant residues under anaerobic conditions to produce biochar is an important tool for carbon sequestration and reducing greenhouse gas emissions.

The synergistic effect of inoculation with arbuscular mycorrhiza and organic substances, such as biochar, on soil condition and increased plant productivity during pollution has been proven. The effectiveness of remediation with inoculated substances depends on the chemical nature of the polluting substance. When the soil is polluted with cadmium (Cd), the addition of biochar inoculated with arbuscular mycorrhiza has not been studied much yet. Cadmium accumulates easily in the soil because, unlike organic pollutants, it is not biodegradable (Liu et al., 2022). Inoculation with arbuscular mycorrhiza and biochar introduction improve plant growth, stimulate the activity of soil microbes to reduce the Cd phytoavailability at elevated Cd concentrations in the soil and have a synergistic effect on reducing Cd uptake by corn. Inoculation of phosphate-solubilizing bacteria is also a sustainable approach to increasing the available phosphorus content in soils for crop production (Wang et al., 2022). This application is limited by the low survival rate of phosphate-solubilizing bacteria in the field. The effectiveness of biochar as a phosphate-solubilizing bacteria carrier is determined by its physicochemical properties, which depend on the starting crude and pyrolysis temperatures. The purpose is to evaluate the effect of *Bacillus* sp. immobilization on biochar on the rate of cellulose decomposition and the ecological state of Haplic Chernozem.

14.2 MATERIAL AND METHODS

14.2.1 SOIL

The object of the study was Haplic Chernozem (WRB, 2022). The sampling point was the Botanical Garden of the Southern Federal University, Rostov-on-Don. Haplic Chernozem belongs to one of the most fertile soils in the world. The zone of Chernozem occurrence extends from the Central Chernozem zone of Russia to the South of the European part of Russia (Lukin and Zhuikov, 2022). Chernozem physic and chemical properties: pH = 7.5–7.8, organic matter = 5.5–6.0%, carbonate = 0.3–0.5%, total absorbed bases = 34.0–36.0 mg-eq/100 g of soil.

14.2.2 IMMOBILIZATION OF BACILLUS SP. ON BIOCHAR

A microbiological preparation based on a consortium of aerobic spore-forming *Bacillus* sp. strains with antagonistic activity against phytopathogenic fungi was used. The total recommended dose of the preparation in a vegetative vessel is 20 ml per 1 ha of soil. The recommended dose of *Bacillus* sp. was analyzed, and the dose was 100 times more than the recommended one (*Bacillus* sp. ×100). The biochar was pre-reduced to a fraction size of 0.2–0.5 cm and filled with a solution of a microbiological preparation for 48 h. After that, the biochar, together with a solution containing bacterial strains, was introduced into the soil.

14.2.3 MODEL EXPERIMENT

To assess the effect of *Bacillus* sp. and biochar on the decomposition of winter wheat crop residues in the laboratory environment, winter wheat crop residues in the soil together with *Bacillus* sp. and biochar were simulated (Figure 14.1).

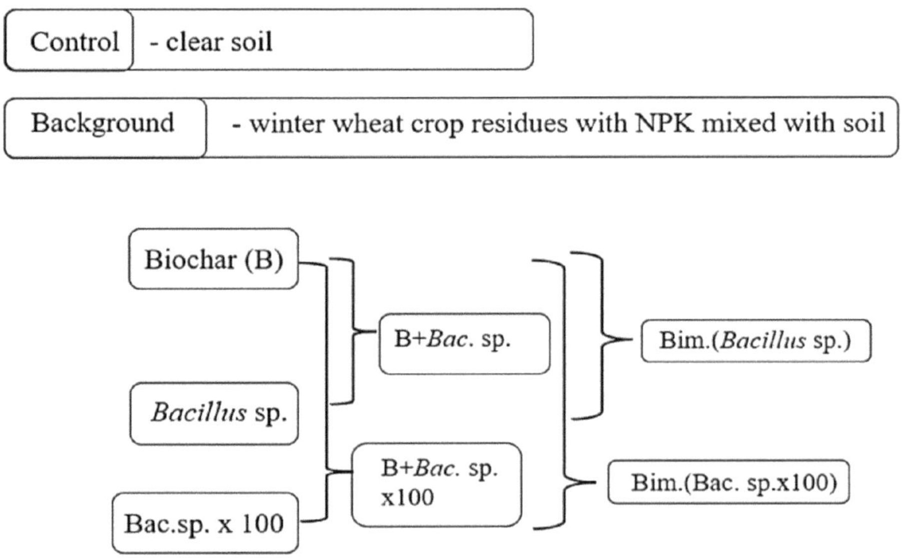

FIGURE 14.1 Scheme of the experiment.

The independent effects of *Bacillus* sp. and biochar on the decomposition of winter wheat crop residues, as well as the combined effects of *Bacillus* sp. and biochar, were compared in the experiment. The effect of biochar immobilized with *Bacillus* sp. was evaluated based on the biological properties of Haplic Chernozem. It was assumed that adding *Bacillus* sp. and biochar would increase the decomposition rate of chaff crop residues and stimulate the biological indicators of common chernozem.

14.2.4 MEASUREMENT OF BIOLOGICAL INDICATORS

The biological activity and ecological state of the soil after the introduction of biochar and *Bacillus* sp. were determined according to the methods in Table 14.1.

TABLE 14.1 Methods for Determining the Biological Activity of the Haplic Chernozem

Sl. No.	Indicator	Methodology
1	Cellulolytic activity	Determination of the cellulolytic activity of the soil according to the method of E. N. Mishustin, I. S. Vostrova, and A. N. Petrova (1993)
2	CO_2 emission	Measurement by gas analyzer EGM-5 PP Systems (USA) every 3–5 days during the entire period of incubation
3	Catalase activity	Catalase activity according to A. Sh. Galstyan (1978)
4	Dehydrogenases activity	The activity of dehydrogenases according to the method of A. Sh. Galstyan in the modification of F. Kh. Khaziev (1990)
5	Total number of bacteria	Bacteria enumeration method on a luminescent microscope in incident light in 20 fields of view
6	Barley germination rates:	Changes in soil phytotoxicity were judged by changes in the germination rates of barley seeds (*Hordeum Vulgare* L.) of the variety "Leon," %: germination, energy, friendliness, and germination rate
6.1	Germination	
6.2	Germination energy	
6.3	Friendliness of germination	
6.4	Germination rate	
7	Indicators of the intensity of the initial growth of barley:	It was estimated by changes in the length of shoots and roots of barley (*Hordeum Vulgare* L.) of the variety "Leon," mm; the weight of shoots and roots, g
7.1	Length of shoots	
7.2	Length of roots	
7.3	Weight of shoots	
7.4	Weight of roots	

14.2.5 CELLULOSE DECOMPOSITION

Cellulose decomposition in the soil or cellulolytic activity in the soil was determined by E. N. Mishustin, I. S. Vostrova, and A. N. Petrova (1993). In each sample, cloth fabrics were placed in a pre-made profile pit in the container filled with soil. After the exposure period, the surviving remnants of the cloth strips were carefully removed from the profile pit, dried, cleaned of soil, and weighed.

14.2.6 ENZYME ACTIVITY

Catalase activity was determined in air-dry soil samples by A. Sh. Galstyan (1978). A sample weight of 1 g was added to one of the compartments of the double flask. In another compartment of the flask, 5 ml of a 3% hydrogen peroxide solution was added, after which the released oxygen was recorded in the containers with the soil; the activity was expressed in ml O_2/min/g of soil.

Hydrogenase activity was determined in air-dry soil samples by A. Sh. Galstyan in the modification of Khaziev (1990) and (Khaziev, 2005). After adding 1 ml of a 0.1 M glucose solution and 1 ml of a 1% TTC solution, the tubes were placed in a vacuum desiccator for 24 h, after which 10 ml of ethyl alcohol was added, and then the resulting stained TPP solution was filtered and calorimetrically analysed; activity was expressed in mg TPP per 10 g of soil for 24 h.

14.2.7 EMISSION OF CO_2

The intensity of CO_2 emissions was assessed using an EGM PP Systems (USA) gas analyzer by measuring each container with a soil sample. The assessment of CO_2 emissions from the soil was calculated by increasing CO_2.

14.2.8 TOTAL NUMBER OF SOIL BACTERIA

The total number of bacteria was calculated by luminescent microscopy. Preparations with bacteria were stained with 0.1% acridine orange, after which, by fluorescence of live bacterial and dead cells stained with fluorochromes in microscopic preparations on a Carl Zeiss Axio luminescent microscope, bacteria were counted in billions/g of soil.

14.2.9 THE INTENSITY OF INITIAL GROWTH AND DEVELOPMENT OF BARLEY

Phytotoxicity was measured by indicators of seed germination (*Hordeum Vulgare* L.) of the Leon variety (germination friendliness, germinative power, seedling vigor, and germination rate) and indicators of the intensity of initial seed growth (length of shoots and roots, weight of shoots and roots). Changes in these characteristics were recorded daily for seven days after sowing the seeds.

Based on the above informative and sensitive biological indicators (catalase and dehydrogenase activity, total bacterial count, root length, and germination), a soil total biological state index (IIBS) was calculated. For this purpose, the maximum value of each of the indicators was taken as 100%, and the same indicator value in relation to it was expressed as a percentage in the other samples (Equation 14.1):

$$B_1 = \left(\frac{B_x}{B_{\max}}\right) \times 100\% \qquad 14.1$$

where B_1 is the relative score of the indicator; Bx is the actual value of the indicator; B_{max} is the maximum value of the indicator.

After that, the average evaluation score of the studied indicators was calculated for the sample (variant), the absolute values of which cannot be summed up, since they have different units of measurement. The soil IIBS was calculated similarly to the Equations 14.1 and 14.2.

$$IIBS = \left(\frac{B_{av}}{B_{\max}}\right) \times 100\% \qquad 14.2$$

where B_{av} is the average evaluation score of all indicators; $B_{av\ max}$ is the maximum evaluation score of all indicators.

14.2.10 STATISTICAL PROCESSING

Statistical processing and multiple correlation analyses were performed to obtain reliable data. The sensitivity of the indicators was evaluated according to their decrease depending on the biostimulant used, and the informativeness was evaluated according to the tightness of the relationship between straw decomposition and biological indicators.

14.3 RESULTS AND DISCUSSION

14.3.1 CHANGES IN CELLULOSE DECOMPOSITION

The cellulolytic activity of soils reflects the rate of decomposition of the cellulose fabric in the soil and the activity of the soil biota. The lower the weight of the fabric after incubation, the more effective the activity of cellulose-destroying microorganisms. Figure 14.2 shows the percentage of decomposition of the fabric when adding biochar and *Bacillus* sp. In soil with a background (winter wheat crop residues with soil) without biochar and *Bacillus* sp., the fabric decomposition rate is only 22% lower than the initial fabric weight. It was found that, in comparison with the background content, when adding 1% biochar to the background, there was a 35% decrease in activity compared to the initial fabric weight.

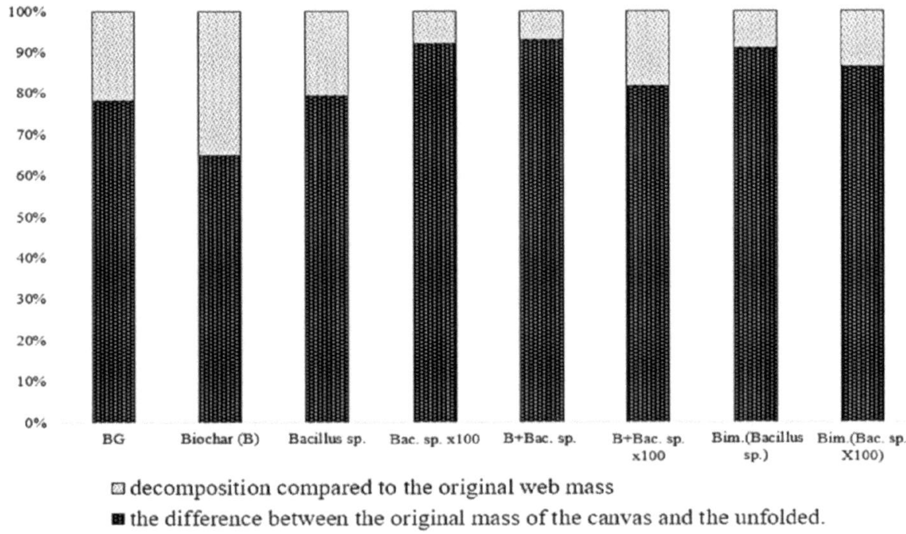

FIGURE 14.2 Changes in the cellulolytic activity of Haplic Chernozem with the addition of biochar and *Bacillus* sp.

When adding *Bacillus* sp. and *Bacillus* sp. ×100, cellulolytic was reduced by 20 and 8%, respectively, compared to the initial fabric weight. When adding the bacterial preparation *Bacillus* sp. with biochar and *Bacillus* sp. ×100 with biochar, cellulolytic activity was reduced by 7 and 18% compared to the initial fabric weight, respectively. When adding biochar inoculated with *Bacillus* sp. and biochar inoculated with *Bacillus* sp. ×100, cellulolytic activity was reduced by 9 and 14% compared to the initial fabric weight, respectively. Based on the results obtained, it is obvious that the use *of* *Bacillus* sp. ×100, *Bacillus* sp. with biochar, and biochar inoculated with *Bacillus* sp. stimulates the fabric decomposition by more than 90% compared to the initial weight.

14.3.2 CHANGE IN CO2 EMISSIONS

The CO_2 emission from Haplic Chernozem reflects by-products of carbon-containing material decomposition, including oil and cellulose (Minnikova et al., 2019; Kudeyarov, 2015). Figure 14.3 shows that biochar had no effect on CO_2 emissions compared to the background. When adding *Bacillus* sp. and *Bacillus* sp. ×100, biochar + *Bacillus* sp., the inhibition of CO_2 was 11, 10, and 7% compared to the background.

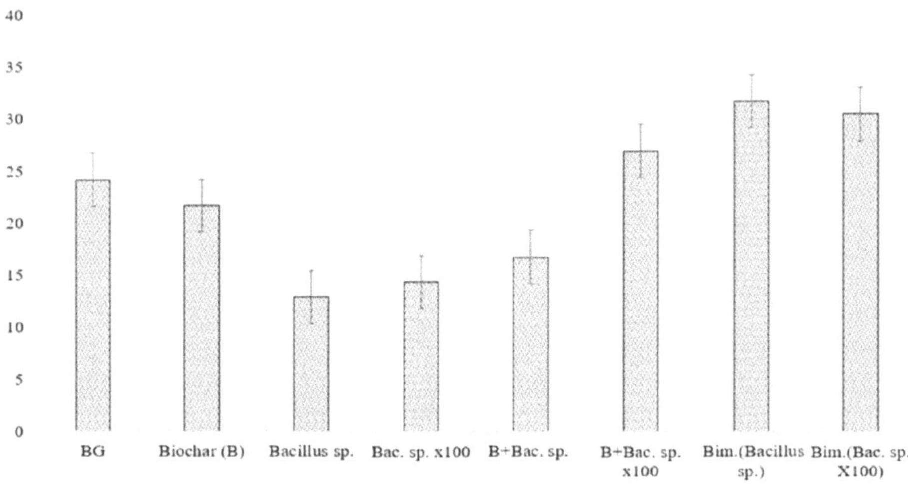

FIGURE 14.3 Changes in CO_2 emission of Haplic Chernozem with the addition of biochar and *Bacillus* sp.

When adding *Bacillus* sp. ×100 and biochar, CO_2 emissions did not significantly differ from the background. When adding both biochar inoculated with *Bacillus* sp. and biochar inoculated with *Bacillus* sp. ×100, total Ibs were increased by 8 and 7%, respectively, compared to the background. Thus, it was established that the biochar inoculated with *Bacillus* sp. was the initiator of an increase in CO_2 in the soil air and, therefore, the decomposition of cellulose fabric at a higher rate.

14.3.3 CHANGES IN THE TOTAL NUMBER OF SOIL BACTERIA

The total soil bacterial count is a very sensitive biological indicator for various types of anthropogenic impact. When adding biochar to the soil with crop residues, an inhibition of the number by 48% was detected (Figure 14.4).

When adding *Bacillus* sp. and biochar, no significant differences from the background were found. When adding both biochar inoculated with *Bacillus* sp. and biochar inoculated with *Bacillus* sp. ×100, the total number of bacteria was inhibited by 39 and 52% compared to the background, respectively.

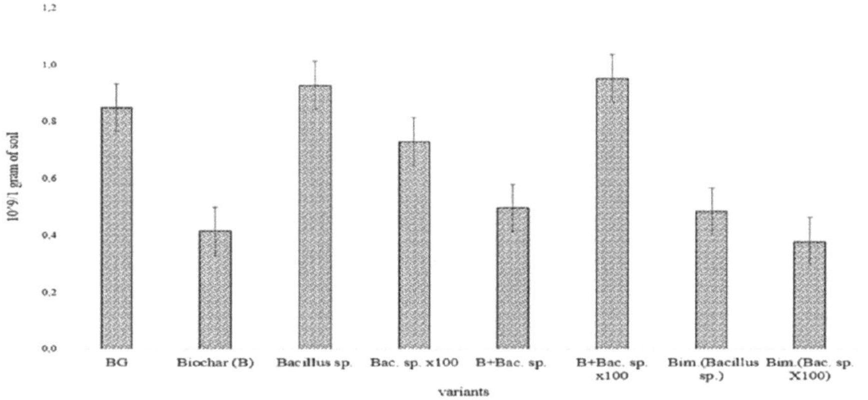

FIGURE 14.4 Changes in the total number of soil bacteria in Haplic Chernozem with the addition of biochar and *Bacillus* sp.

14.3.4 CHANGES IN CATALASE AND DEHYDROGENASES ACTIVITY

Figure 14.5 shows changes in catalase activity, where, with the addition of 1% biochar, catalase activity was stimulated by 31% compared to the background. When adding *Bacillus* sp. and *Bacillus* sp. ×100, catalase activity decreased by 12 and 21%, respectively, compared to the background.

When adding *Bacillus* sp. and biochar, *Bacillus* sp. ×100 and biochar inhibited catalase activity by 14 and 17% compared to the background. When adding biochar inoculated with *Bacillus* sp., catalase activity was not significantly different from the background. With an increase in biochar inoculated with *Bacillus* sp. ×100, inhibition was 42% compared to the background.

The dehydrogenases activity under the impact of biochar and *Bacillus* sp. changed differently. When adding biochar, the inhibition of activity was found to be 16% compared to the background. When adding *Bacillus* sp. and *Bacillus* sp. ×100, dehydrogenase activity was stimulated by 51 and 109%, respectively, compared to the background. When adding *Bacillus* sp. and biochar, *Bacillus* sp. ×100, and biochar, dehydrogenases activity was inhibited by 11 and 56% compared to the background. When adding biochar inoculated with *Bacillus* sp. and biochar inoculated with *Bacillus* sp. ×100, dehydrogenases activity was increased by 29 and 104%, respectively, compared to the background.

Thus, catalase activity was inhibited in most variants of biochar and *Bacillus* sp. addition, but dehydrogenases activity was stimulated when adding *Bacillus* sp. ×100 and biochar inoculated with *Bacillus* sp. ×100.

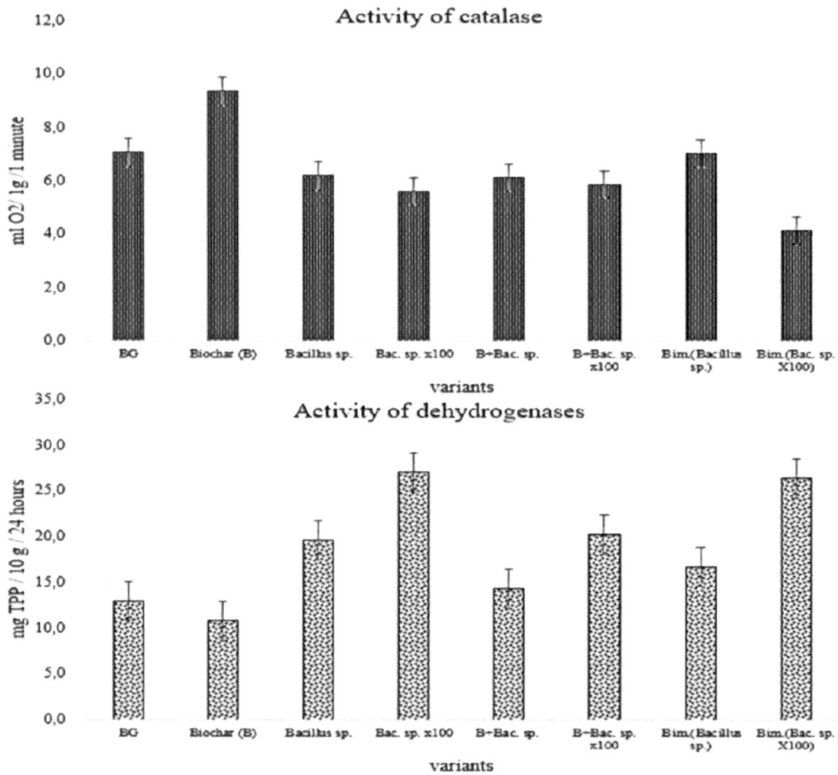

FIGURE 14.5 Changes in catalase activity in Haplic Chernozem with the addition of biochar and *Bacillus* sp.

14.3.5 CHANGE IN THE INTENSITY OF THE INITIAL GROWTH AND DEVELOPMENT OF BARLEY

The change in intensity of the initial growth of barley with the addition of biochar and *Bacillus* sp. is shown in Figure 14.6. The addition of biochar did not have a significant effect on the germination friendliness, germination, germinative energy, or germination rate of barley. The addition of *Bacillus* sp. resulted in the inhibition of germination friendliness, germination, germinative energy, and the germination rate of barley by 20–25% relative to the background. With an increase in the concentration of *Bacillus* sp. ×100, germination friendliness and germination were less inhibited by 16 and 15%, respectively, and the germinative power and germination rate were at the level recommended by *Bacillus* sp. at 21 and 25%, respectively, relative to the background.

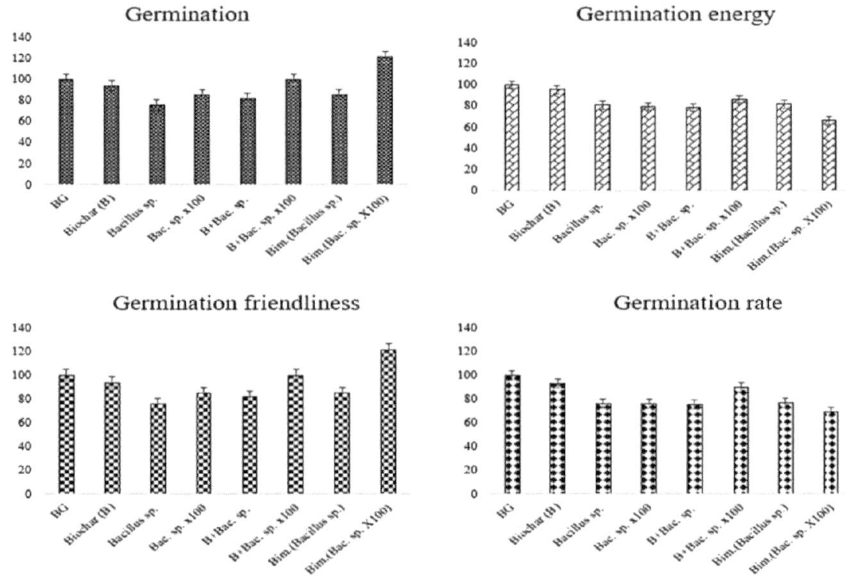

FIGURE 14.6 Changes in the intensity of the initial growth of barley seeds on Haplic Chernozem with the addition of biochar and *Bacillus* sp.

When adding biochar with *Bacillus* sp., a decrease in the intensity of initial growth was found by 19–25% relative to the background. With a higher dose of *Bacillus* sp. ×100, germination friendliness and seedling vigor reached the background level, and the germinative power and germination rate were reduced by 14 and 16%, respectively, relative to the background. The use of biochar inoculated with *Bacillus* sp. resulted in a decrease in germination friendliness, seedling vigor, and germinative power of barley by 16–19% relative to the background, and the germination rate was reduced by 24%. When adding biochar inoculated with *Bacillus* sp. ×100, the stimulation of germination friendliness and seedling vigor was observed by 20 and 21%, respectively, compared with the background, and the germinative power and germination rate decreased by 34 and 32%, respectively.

Indicators of barley development after sprouting were evaluated by changes in shoot and root length as well as in the weight of dry shoots and roots (Figure 14.7). Biochar had no significant effect on shoot width, inhibiting root width, shoot length, and root length by 11, 26, and 15%, respectively, relative to the background.

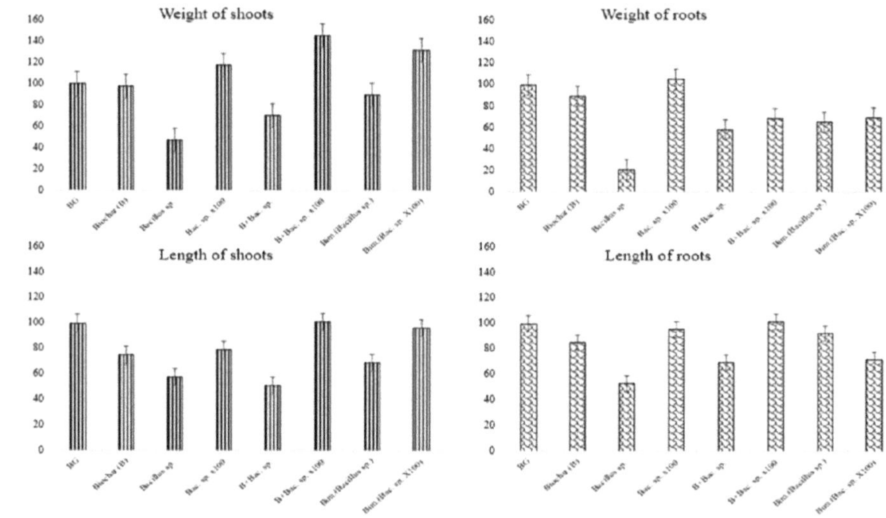

FIGURE 14.7 Changes in the growth rates of barley on Haplic Chernozem with the addition of biochar and *Bacillus* sp.

When adding *Bacillus* sp., a decrease in the weight of shoots and roots was observed by 53 and 80% relative to the background, respectively. At the same time, the length of shoots and roots after *Bacillus* sp. was inhibited only by 43 and 48% relative to the background, respectively. With an increase in the dose of *Bacillus* sp. ×100, the weight of roots and the length of shoots did not differ from the control, and the weight of shoots was stimulated by 17% relative to the background. The length of barley shoots with the addition of *Bacillus* sp. ×100 was inhibited by 22% relative to the background.

Biochar with *Bacillus* sp. inhibited the weight of shoots and roots and the length of shoots and roots by 49, 31, 30, and 42% relative to the background, respectively. When increasing the dose of biochar with *Bacillus* sp. ×100, the length of shoots and roots did not differ from the control; the width of shoots was stimulated by 46%, and the width of roots was inhibited by 32% relative to the background. Inoculation of *Bacillus* sp. on biochar resulted in the inhibition of shoot length and root width by 32 and 35%, respectively. Root length and shoot weight were inhibited by only 8 and 11%, respectively. With increasing the dose of *Bacillus* sp. ×100 in the biochar, stimulation of the weight of shoots by 32% and inhibition of the length and weight of roots by 29 and 31% of the background, respectively, were established.

14.3.6 INTEGRATING INDICATOR OF BIOLOGICAL SOIL HEALTH

The Integral indicator of the biological state of the Haplic Chernozem (IIBS) after the introduction of biochar and *Bacillus* sp. is shown in Figure 14.8. The introduction of all substances stimulated IIBS by 26–47% compared to the background. The addition of biochar increased IIBS by 36% compared to the background.

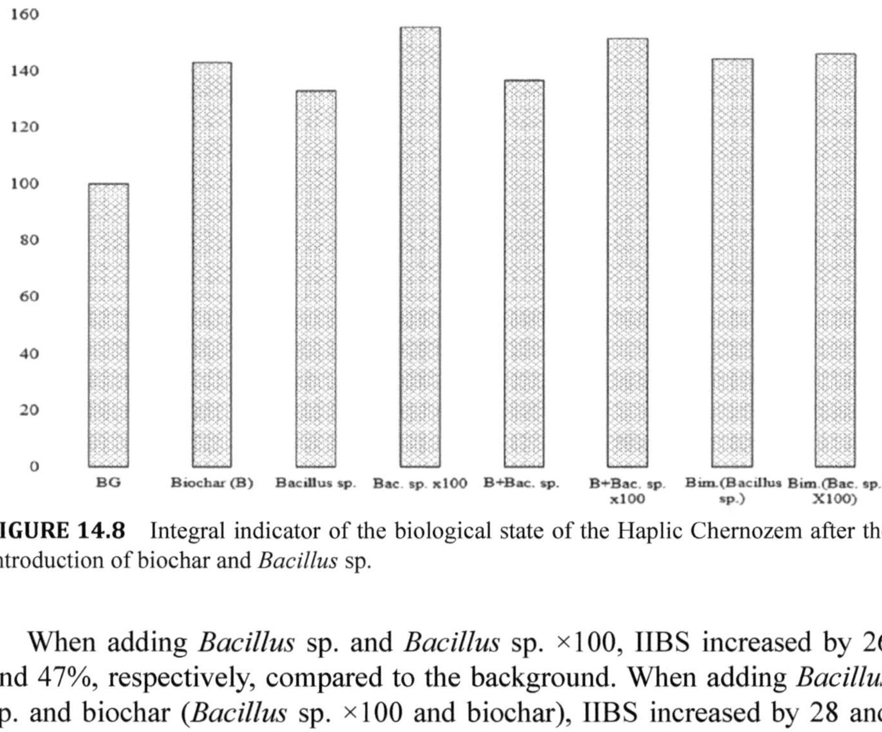

FIGURE 14.8 Integral indicator of the biological state of the Haplic Chernozem after the introduction of biochar and *Bacillus* sp.

When adding *Bacillus* sp. and *Bacillus* sp. ×100, IIBS increased by 26 and 47%, respectively, compared to the background. When adding *Bacillus* sp. and biochar (*Bacillus* sp. ×100 and biochar), IIBS increased by 28 and 44%, respectively, compared with the background. When adding biochar inoculated with *Bacillus* sp. and biochar inoculated with *Bacillus* sp. ×100, total Ibs increased by 36 and 38%, respectively, compared to the background. Thus, the use of biochar and *Bacillus* sp. separately is not as effective as the combined use of biochar, *Bacillus* sp., and biochar inoculated with *Bacillus* sp., confirming the positive synergistic effect.

14.3.7 ASSESSMENT OF SENSITIVE AND INFORMATIVE BIOLOGICAL INDICATORS OF SOIL

The sensitivity of biologicals was evaluated by the change in the index relative to the background variant in %. The more the index decreases, the greater the sensitivity to the introduction of biochar, *Bacillus* sp., or a combination of biochar and *Bacillus* sp. this indicator has (Table 14.2).

When biochar is applied to the soil with winter wheat crop residues, biological indicators are arranged in the following order by their sensitivity: catalase activity (70) > dehydrogenases activity (74) > total number of bacteria (88) > CO_2 emission (104) > germination energy (141) > germination friendliness (143) = germination (143) > weight of shoots (151) > germination rate (158) = length of roots (158) > weight of roots (196) > length of shoots (209).

When applying *Bacillus* sp. in the soil with winter wheat crop residues, biological indicators are arranged in the following order according to their sensitivity: weight of roots (40) > catalase activity (61) > weight of shoots (71) > length of roots (84) > total number of bacteria (96) > CO_2 emission (102) > germination friendliness (109) > germination (109) > dehydrogenases activity (111) > germinative energy (114) > germination rate (120) > length of shoots (121).

When *Bacillus* sp. ×100 is introduced into the soil with winter wheat crop residues, biological indicators are arranged in the following order according to their sensitivity: catalase activity (56) > total number of bacteria (75) > CO_2 emission (102) > germinative energy (112) > germination rate (120) > germination friendliness (122) > germination (122) > length of roots (152) > dehydrogenases activity (154) > shoot length (165) > weight of shoots (178) > weight of roots (206).

In the case of joint application of biochar with *Bacillus* sp. in the soil with winter wheat crop residues, biological indicators are arranged in the following order according to their sensitivity: total number of bacteria (51) > catalase activity (61) > dehydrogenases activity (81) > CO_2 emission (103) > shoot length (107) > weight of shoots (107) > germinative energy (111) > length of roots (111) > weight of roots (115) > germination friendliness (117) > germination (117) > germination rate (119).

When increasing the dose of *Bacillus* sp. ×100 with biochar into the soil with winter wheat crop residues, biological indicators are arranged in the following order according to their sensitivity: catalase activity (58) > total number of bacteria (98) > CO_2 emission (105) > dehydrogenases activity (115) > germinative energy (122) > weight of roots (134) > germination rate (142) > germination friendliness (143) > germination (143) > length of roots (162) > length of shoots (212) > weight of shoots (220).

In the case of inoculation of *Bacillus* sp. in the biochar in the soil with winter wheat crop residues, biological indicators are arranged in the following order according to their sensitivity: total number of bacteria (50) >

TABLE 14.2 Sensitivity of Biological Indicators of Haplic Chernozem to the Decomposition of Crop Residues When Biochar and Bacillus sp. are Introduced, % of the Background Value.

Biological Indicators	BG + Biochar	BG + Bacillus sp.	BG + Bacillus sp. ×00	BG + Bacillus sp. + Biochar	BG + Bacillus sp. ×100 + Biochar	BG + Bacillus sp. + Biochar (im)	BG + Bacillus sp. ×100 + Biochar (im)
CO_2 emission	104	102	102	103	105	105	105
Germination	143	109	122	117	143	122	174
Germination energy	141	114	112	111	122	115	93
Germination friendliness	143	109	122	117	143	122	174
Germination rate	158	120	120	119	142	121	109
Length of shoots	209	121	165	107	212	144	202
Length of roots	158	84	152	111	162	147	114
Weight of shoots	151	71	178	107	220	136	200
Weight of roots	196	40	206	115	134	128	136
Total number of bacteria	88	96	75	51	98	50	39
Catalase activity	70	61	56	61	58	70	41
Dehydrogenases activity	74	111	154	81	115	95	150

BG, background.

catalase activity (70) > dehydrogenases activity (95) > CO_2 emission (105) > germinative energy (115) > germination rate (121) > germination friendliness (122) > germination (122) > weight of roots (128) > weight of shoots (136) > length of shoots (144) > length of roots (147).

In the case of biochar inoculated with *Bacillus* sp. ×100 into the soil with winter wheat crop residues, according to their sensitivity, biological indicators are arranged in the following order: total number of bacteria (39) > catalase activity (41) > germination energy (93) > CO_2 emission (105) > germination rate (109) > length of roots (114) > weight of roots (136) > dehydrogenases activity (150) > germination friendliness (174) > germination (174) > weight of shoots (200) > length of shoots (202).

The most sensitive biological indicators when applying biochar and *Bacillus* sp. are catalase activity and the total bacterial count. The informativeness was assessed by the strength of relationship of the correlation coefficients with the percentage of cellulose decomposition: the higher the percentage of fabric decomposition, the higher the biological indicators: CO_2 emission (0.78) < shoot length (0.60) < germination friendliness (0.50) = germination (0.50) < weight of shoots (0.49) < length of root (0.48) < germination rate (0.41) < weight of roots (0.34) = dehydrogenases activity (0.34) < germinative energy (0.26) < total number of bacteria (−0.60) < catalase activity (−0.85).

According to the constructed series, the CO_2 emission and the length of barley shoots have the closest relationship to cellulose fabric decomposition in the soil after the addition of biochar and *Bacillus* sp. At the same time, the least informative indicators are catalase activity and the total number of bacteria. The introduction of biochar and *Bacillus* sp. at all concentrations stimulated the emission of CO_2 and cellulose fabric decomposition. It is known that dehydrogenases characterize the soil microbial pool, participate in redox processes, and allow determining the intensity of soil organic matter transformation processes (Wolińska and Stępniewska, 2012; Minnikova et al., 2017). Based on the results obtained, the introduction of *Bacillus* sp., *Bacillus* sp. ×100, *Bacillus* sp. ×100 and biochar and *Bacillus* sp., and the addition of biochar inoculated with *Bacillus* sp. stimulated the enzymatic activity of dehydrogenases compared to the background, confirming the synergistic effect of the use of biochar and *Bacillus* sp. (Lehmann et al., 2011; Harindintwali et al., 2020). Application of wheat straw biochar (0.2%) inoculated with *Bacillus* sp. ZM20 proved to be the most effective treatment for improving soil biological properties, plant growth, yield, and yield quality of maize compared to all other treatments (Ahmad et al., 2020). When adding biochar inoculated with *Bacillus* sp. ×100, maximum stimulation of germination friendliness and germination was observed. Such results partially support the hypothesis that biochar application to the soil improves the studied Haplic Chernozem fertility, which was noted in other studies (Romero et al., 2021).

14.4 CONCLUSION

The study of the impact of biochar and *Bacillus* sp. on the process of cellulose decomposition in soil found that the independent application of biochar and *Bacillus* sp. had a stimulating effect on most biological indicators, but there was no effect on the rate of cellulose fabric decomposition in the soil. The introduction of *Bacillus* sp. ×100 and the combined use of biochar and *Bacillus* sp. stimulated cellulose decomposition and biological indicators characterizing the ecological state of the soil. Inoculation of *Bacillus* sp. on biochar was effective already at the recommended dose for both cellulose fabric decomposition and soil remediation, demonstrating a synergistic effect. The results obtained must be used in carrying out measures to improve agricultural soil fertility and environmental monitoring of soil conditions.

ACKNOWLEDGMENT

The study was financially supported by the project of the Ministry of Science and Higher Education of Russia (FENW-2024-0001) and the Strategic Academic Leadership Program of the Southern Federal University ("Priority 2030") (No. SP-12-23-01).

KEYWORDS

- **Biochar**
- **biological indicators**
- **decomposition of crop residues**
- **decomposition of plant residues**
- **fertility**
- **Haplic Chernozem incubation**
- **immobilization of Bacillus sp**
- **microbiological preparations**

REFERENCES

Ahmad, M., Rajapaksha, A. U., Lim, J. E., Zhang, M., Bolan, N., Mohan, D., Vithanage, M., Lee, S. S., & Ok, Y. S. (2014). Biochar as a sorbent for contaminant management in soil and water: A review. *Chemosphere, 99*, 19–33. https://doi.org/10.1016/j.chemosphere.2013.10.071.

Ahmad, M., Wang, X., Hilger, T. H., Luqman, M., Nazli, F., Hussain, A., Zahir, Z. A., Latif, M., Saeed, Q., Malik, H. A., & Mustafa, A. (2020). Evaluating biochar-microbe synergies for improved growth, yield of maize, and post-harvest soil characteristics in a semi-arid climate. *Agronomy*, *10*(7), 1055. https://doi.org/10.3390/agronomy10071055.

Awad, Y. M., Blagodatskaya, E., Kuzyakov, Y., & Ok, Y. S. (2012). Effects of polyacrylamide, biopolymer, and biochar on decomposition of soil organic matter and plant residues as determined by ^{14}C and enzyme activities. *European Journal of Soil Biology*, *48*, 1–10. https://doi.org/10.1016/j.ejsobi.2011.09.005.

Bokade, P., Gaur, V. K., Tripathi, V., Bobate, S., Manickam, N., & Bajaj, A. (2023). Bacterial remediation of pesticide polluted soils: Exploring the feasibility of site restoration. *Journal of Hazardous Materials*, *441*, 129906. https://doi.org/10.1016/j.jhazmat.2022.129906.

Chapman, R. L. (2013). Algae: The world's most important "plants"-an introduction. *Mitigation and Adaptation Strategies for Global Change*, *18*, 5–12. https://doi.org/10.1007/s11027-010-9255-9.

Chirak, E. L., Orlova, O. V., Aksenova, T. S., Kichko, A. A., Chirak, E. R., Provorov, N. A., & Andronov E. E. (2017). Dynamics of the microbial community of a typical chernozem during the biodegradation of cellulose and barley straw. *Agricultural Biology*, *52*, 3:588-596. https://doi.org/10.15389/agrobiology.2017.3.588rus.

Duarte, I., Hernández, S., Ibañez, A., & Canto, A. (2018). Macroalgae as soil conditioners or growth promoters of *Pisum sativum* (L). *Annual Review of Plant Biology*, 27, 1–8. https://doi.org/10.9734/ARRB/2018/43272.

El-Bassi, L., Azzaz, A. A., Jellali, S., Akrout, H., Marks, E. A. N., Ghimbeu, C. M., & Jeguirim, M. (2021). Application of olive mill waste-based biochars in agriculture: Impact on soil properties, enzymatic activities and tomato growth. *Science of the Total Environment*, *755*, 142531. https://doi.org/10.1016/j.scitotenv.2020.142531.

FAO. (2022). *Crop Prospects and Food Situation – Quarterly Global Report No. 2, July 2022, Rome*. https://doi.org/10.4060/cc0868en.

Harindintwali, J. D., Zhou, J., Yang, W., Gu, Q., & Yu, X. (2020). Biochar-bacteria-plant partnerships: Eco-solutions for tackling heavy metal pollution. *Ecotoxicology and Environmental Safety*, *204*, 111020. https://doi.org/10.1016/j.ecoenv.2020.111020.

Jones, D. L., Rousk, J., Edwards-Jones, G., DeLuca, T. H., & Murphy, D. V. (2012). Biochar-mediated changes in soil quality and plant growth in a three-year field trial. *Soil Biology and Biochemistry*, 45, 113–124. https://doi.org/10.1016/j.soilbio.2011.10.012.

Khalid, S., Shahid, M., Murtaza, B., Bibi, I., Natasha, M., Naeem, A., & Niazi, N. K. (2020). A critical review of different factors governing the fate of pesticides in soil under biochar application. *Science of the Total Environment*, *711*, 134645. https://doi.org/ 10.1016/j.scitotenv.2019.134645.

Khaziev, F. K. (2005). *Methods of soil enzymology* (p. 252 c). Institute of Biology of the UFA Scientific Center.

Kravtsova, N. E., Mokrikov, G. V., Kazeev, K. Sh., Minnikova, T. V., & Kolesnikov, S. I. (2019). Influence of tillage practices on the dynamics of the content of nutrients in

Haplic Chernozems of the Rostov region. *Agrochemical Bulletin, 1*, 33–36. https://doi.org/10.24411/0235-2516-2019-10008.

Kudeyarov, V. N. (2015). Current state of the carbon budget and the capacity of Russian soils for carbon sequestration. *Eurasian Soil Science*, *48*(9), 923–933. https://doi.org/10.7868/S0032180X15090087.

Lehmann, J., Rillig, M. C., Thies, J. E., Masiello, C.A., Hockaday, W. C., & Crowley, D. E. (2011). Biochar effects on soil biota – A review. *Soil Biology & Biochemistry*, *43*, 1812–1836. https://doi.org/10.1016/J.SOILBIO.2011.04.022.

Liang, L., Xi, F., Tan, W., Meng, X., Hu, B., & Wang, X. (2021). Review of organic and inorganic pollutants removal by biochar and biochar-based composites. *Biochar*, *33*(3), 255–281. https://doi.org/10.1007/s42773-021-00101-6.

Liu, M., Ke, X., Joseph, S., Kadambot Siddique, H. M., Pan, G., & Solaiman Z. M. (2022). Interaction of rhizobia with native AM fungi shaped biochar effect on soybean growth. *Industrial Crops and Products Part B, 187*, 115508. https://doi.org/10.1016/j.indcrop.2022.115508.

Lukin, S. V., & Zhuikov, D. V. (2022). Content and balance of trace elements (Co, Mn, Zn) in agroecosystems of the central chernozemic region of russia. *Agriculture, 12*(2), 154. https://doi.org/10.3390/agriculture12020154.

Marican, A., & Durán-Lara, E. F. (2018). A review on pesticide removal through different processes. *Environmental Science and Pollution Research*, *25*(3), 2051–2064. https://doi.org/10.1007/s11356-017-0796-2.

Minnikova, T. V., Kazeev, K. Sh., Medvedeva, A. M., Biryukova, O. A., Keswani, C., Minkina, T. M., Sushkova, S. N., Elgendy, H., Kolesnikov, S., & Mokrikov, G. (2022). Soil organic carbon dynamics in response to tillage practices in the steppe zone of Southern Russia. *Processes*, *10*(2), 244. https://doi.org/10.3390/pr10020244.

Minnikova, T. V., Kolesnikov, S. I., & Denisova, T. V. (2019). Effect of nitrogen and humic fertilizers on the biochemical state of oil-contaminated chernozem. *South of Russia: Ecology, Development*, *14*(2), 189–201 (In Russ.). https://doi.org/10.18470/1992-1098-2019-2-189-201.

Minnikova, T. V., Kravtsova, N. E., Mokrikov, G. V., Kazeev, K. Sh., & Kolesnikov, S. I. (2019). Influence of direct sowing of winter wheat on the content of nutrients in the chernozem. *Agrochemistry*, *10*, 64–71. https://doi.org/10.1134/S0002188119100119.

Minnikova, T. V., Mokrikov, G. V., Kazeev, K. Sh., Akimenko, Yu. V., & Kolesnikov, S. I. (2017). Evaluation of the enzymatic activity of the chernozems of the Rostov region under binary crops of sunflower. *News of the Timiryazev Agricultural Academy*, *6*, 141–155. https://doi.org/10.26897/0021-342X-2017-6-141-155.

Mohan, D., Abhishek, K., Sarswat, A., Patel, M., & Singh, P., Pittman CU. (2018). Biochar production and applications in soil fertility and carbon sequestration-a sustainable solution to crop-residue burning in India. *RSC Advances*, *8*, 508–520. https://doi.org/10.1039/C7RA10353K

Mokrikov, G. V., Minnikova, T. V., Kazeev, K. Sh., & Kolesnikov, S. I. (2019). Influence of precipitation and moisture reserves on the yield of crops under different tillage. *Agronomy Research*, *17*(6), 2350–2358. . https://doi.org/10.15159/AR.19.202

Mokrikov, G. V., Minnikova, T. V., Kazeev, K. Sh., & Kolesnikov, S. I. (2019). The influence of productive moisture reserves and the amount of precipitation on productivity under the condition of direct sowing of agricultural crops in the Rostov region. *Samara Scientific Bulletin* *1*(26), 69–75. https://doi.org/10.24411/2309-4370-2019-11111.

Morillo, E., & Villaverde J. (2017). Advanced technologies for the remediation of pesticide-contaminated soils. *Science of the Total Environment, 586*, 576–597. https://doi.org/10.1016/j.scitotenv.2017.02.020.

Novikov, A. A, & Kisarov, O. P. (2012). Substantiation of the role of root and crop residues in agrocenoses. *Polythematic Network Electronic Scientific Journal of the Kuban State Agrarian University,* (78), 643–652.

Ogura, A. P., Lima, J. Z., Marques, J. P., Sousa, L. M., Guimarães, V., Rodrigues, S., & Gaeta Espíndola, E. L. (2021). A review of pesticides sorption in biochar from maize, rice, and wheat residues: Current status and challenges for soil application. *Journal of Environmental Management, 300*, 113753. https://doi.org/10.1016/j.jenvman.2021.113753.

Romero, C., Hao, X., Li, C., Owens, J., Schwinghamer, T., McAllister, T. A., & Okine, E. (2021). Nutrient retention, availability and greenhouse gas emissions from biochar-fertilized chernozems. *CATENA, 198*, 10504. https://doi.org/10.1016/j.catena.2020.105046

Shinde, R., Shahi, D. K., Mahapatra, P., Shekhar Singh, C., Kumar Naik, S., Thombare, N., & Singh, A. K. (2022). Management of crop residues with special reference to the on-farm utilization methods: A review. *Industrial Crops and Products, 181*, 114772. https://doi.org/10.1016/j.indcrop.2022.114772.

Tang, Y., Yang, Y., Cheng, D., Gao, B., Wan, Y., Li, Y. C., Yao, Y., Xie, J., & Liu, L. (2019). Multifunctional slow-release fertilizer prepared from lignite activated by a 3D-molybdate-sulfur hierarchical hollow nanosphere catalyst. *ACS Sustainable Chemistry & Engineering, 7*, 10533–10543. https://doi.org/10.1021/jf1029306.

Thapa, V. R., Ghimire, R., & Marsalis, M. A. (2021). Cover crops for resilience of a limited-irrigation winter wheat–sorghum–fallow rotation: Soil carbon, nitrogen, and sorghum yield responses. *Agronomy, 11*(4), 762. https://doi.org/10.3390/agronomy11040762.

FAO, Three principles of conservation agriculture. https://www.fao.org/conservation-agriculture/en/. (accessed October 18, 2022).

Torma, S., Vilček, J., Lošák, T., Kužel, S., & Mårtensson, A. (2018). Residual plant nutrient in crop residues – an important resource. *Acta Agriculturae Scandinavica, Section B — Soil & Plant Science, 68*, 358–366. https://doi.org/10.1080/09064710.2017.1406134.

Wang, Z., Chen, H., Zhu, Z., Fang, X. S, Guang, W. S., & Chen B. (2022). Low-temperature straw biochar: Sustainable approach for sustaining higher survival of B. megaterium and managing phosphorus deficiency in the soil. *Science of The Total Environment, 830*, 154790. https://doi.org/10.1016/j.scitotenv.2022.154790.

Wolińska, A., & Stępniewska, Z. (2012). Dehydrogenase activity in the soil environment. *Dehydrogenases* (Vol. 8, pp. 183–210). INTECH Publisher. https://doi.org/10.5772/48294..

World Reference Base for Soil Resources. (2022). *International soil classification system for naming soils and creating legends for soil maps* (p. 234). 4th edition published in 2022 by the International Union of Soil Sciences (IUSS), Vienna, Austria.

Wu, W., Ke, T., Zhou, X., Li, Q., Tao, Y., Zhang, Y., Zeng, Y., Cao, J., & Chen L. (2022). Synergistic remediation of copper mine tailing sand by microalgae and fungi. *Applied Soil Ecology, 175*, 104453. https://doi.org/10.1016/j.apsoil.2022.104453.

PART IV

Nanomaterials in Soil Remediation and Environment Protection

CHAPTER 15

Exploration of Nanomaterials in the Remediation of Soil and Environment Pollution

PALLAVI JOSHI[1], PRIYANK VYAS[2], and PRERNAA SHARMA[1]

[1]*Graphic Era Hill University, Dehradun, Uttarakhand, India*

[2]*Graphic Era Deemed to be University, Dehradun, Uttarakhand, India*

ABSTRACT

Multiple pollutants have entered the food chain as a result of the rapid growth of the human population, industrialization, urbanization, and intensive agriculture. The exceptional reactivity of nanomaterials (NMs) has made nanoremediation one of the most exciting new approaches to cleaning up polluted environments. Polymers, clay minerals, zeolites, activated carbon, and biochar can all be added to NMs to increase their ability to filter out contaminants, and there is a lot of literature on the effectiveness of NMs in this regard. In order to achieve sustainable success in soil and water remediation, this chapter gives an overview of contemporary NMs and their applications in innovative remediation approaches, such as adsorption/filtration, catalysis, photodegradation, electronano-remediation, and nano-bioremediation. Based on what we know about the social and economic effects of using NMs to clean up contaminated ground and water, it is clear that the underlying science and engineering behind these technologies need to advance significantly before they can be widely adopted. New studies involving ecofriendly nanotechnology, nano-bioremediation, electro-nanoremediation, risk assessment of NMs, and outreach efforts are desired to realize the successful functions of nanoremediation at regional and global scales.

Nanobiotechnology for Agricultural Sciences: Nano-Agri-Input Products for Crop Production and Environmental Protection. J. C. Tarafdar, Deepak Kumar, Sanjay K. Singh, and Krishna Pal Singh (Eds.)

15.1 INTRODUCTION

15.1.1 THE ROLE OF NANOTECHNOLOGY AND NANOMATERIALS IN THE ELIMINATION OF CONTAMINANTS

In the realm of nanotechnology, "materials" are fabricated at the atomic and molecular levels of structure. These proportions contradict the known principles of physics and chemistry. Between the nano- and macro-scales, materials undergo dramatic changes in their appearance, hardness, conductivity, and sensitivity. Carbon "nanotubes," for instance, are a hundred times stronger and six times lighter than steel. This overview explains the different kinds of nanoparticles, how to characterize them, and how they can be used in fields as diverse as medicine and environmental clean up. There are several laboratory and field (pilot and full-scale) studies that support the use of nanotechnology in pollution control and water purification. Though there are numerous overviews of these nanotechnology-based applications, it is clear that a comparison between current treatment methods and emerging nanotechnology-based treatments is necessary to demonstrate their significance and guide their further development (Singh and Naveen, 2014; Figure 15.1).

The dangers of outdoor air pollution include both organic and inorganic toxins that are released into the air. Human activity and the natural environment both contribute to pollution. To name a few, we have NOx (nitrogen oxides), CO (carbon monoxide), SO*x* (sulfur oxides), O_3 (ozone), and VOCs (volatile organic compounds). The combustion of fossil fuels, including coal, petroleum, and natural gas, is a primary source of these pollutants. Chemical processes involving nitrogen oxides (NO_2), VOCs, and ultraviolet (UV) light produce ozone in the atmosphere. Coal combustion produces carbon monoxide (CO), a colorless, odorless gas, during the combustion process. Fuel-burning vehicles, generators, and boilers are the most common sources of CO pollution. Carbon monoxide, or CO, is produced when tobacco is burned, and it is generally agreed that this is a dangerous byproduct. Sulfur dioxide (SO_2), a very reactive gas, is created during the refining of sulfur-containing commodities such as crude oil and coal (Alptekin and Celiktas, 2022).

Wastewater production, groundwater contamination, and land degradation all contribute to a decline in environmental quality, making it one of the most important problems facing most countries today (Li et al., 2021). The bioaccumulation of toxic trace elements is exacerbated by repeated exposure to polluted soil and water. Cadmium (Cd), lead (Pb), arsenic (As), and fluoride (F) exposure in humans due to contaminated water and food has been linked

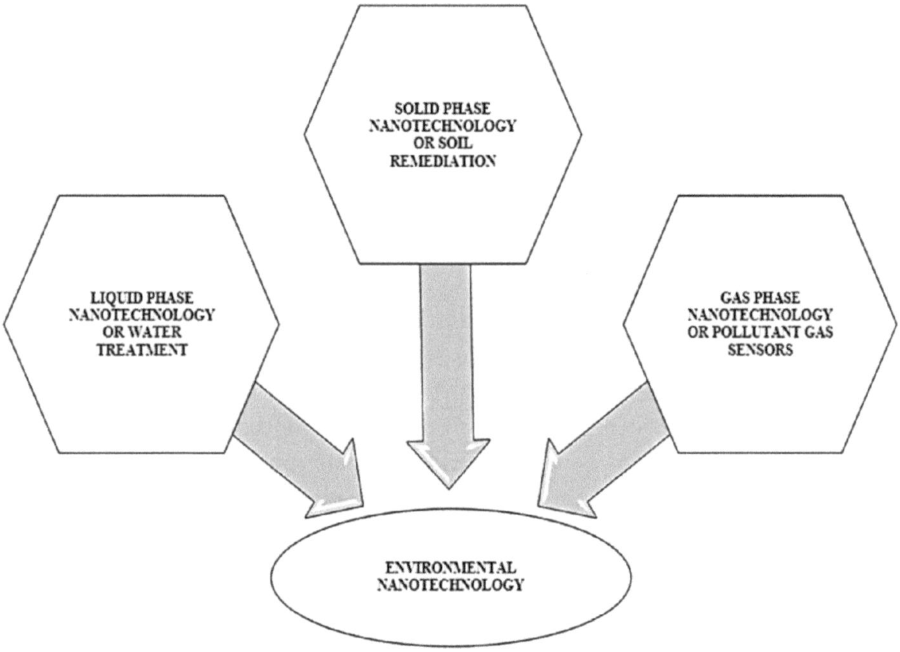

FIGURE 15.1 Role of nanotechnology in the environment.

Source: Adapted from Singh and Naveen (2014).

to negative effects on the skin, lungs, kidneys, and brain. Industrial dyes and other organic pollutants are challenging to handle because of their persistence and dependence on environmental physicochemical parameters for degradation. Human health is endangered by the availability of heavy metals in food (Lin et al., 2022). Vegetable contamination has become a major issue since it exposes people to hazardous heavy metals and can also cause them to accumulate in the body. Industrial processes and vehicle emissions are two major contributors to heavy metal contamination in the environment. Sewage from both cities and factories is a significant contributor to those materials.

Due to their many applications, nanomaterials can be broken down into three broad classes: adsorbents (used to soak up contaminants), catalysts (used to break down and change pollutants), and membranes (used to filter out impurities from wastewater and seawater using pressure) (Anjum et al., 2019). Even so, the vast majority of published works deal with the cleanup of aquatic systems, and most of these investigations are conducted on a laboratory scale. Less emphasis has been placed on the use of NMs for cleaning up polluted soils.

Surface assimilation, absorption, synthetic reactions, photocatalysis, and filtration are just a few of the environmental remediation techniques used to get rid of pollutants in the environment. Due to their low volume and large surface area, nanomaterials are often more reactive on a biogeological and chemical level than their macroscale counterparts. In particular, they are well-suited for this kind of work due to the enhancements made to their design and performance. Nanoscale materials are finding applications in many different disciplines, from medicine and the life sciences to environmental protection and manufacturing. The manufacture and application of nanoscale products with remediation functions for the environment have increased in recent years (Roy et al., 2021).

Sites contaminated by chlorinated solvents and oil spills are only two examples of the various types of hazardous waste that nanomaterials have been used to clean up. Since manufactured nanoparticles have their own unique physicochemical, surface, and optical-electronic properties, they can solve problems that conventional approaches could not. However, most of the published papers focus on cleaning up aquatic systems, and most of the studies are undertaken on a small scale in the lab. Using NMs for the remediation of polluted soil has received scant research interest. The widespread use of nanotechnologies for ecological restoration has been stymied by a number of technical, social, and economic obstacles (such as the substantial investment required for NM planning and deployment and problems with removing contaminants in situ that are not acceptable). There is currently no comprehensive critical review that addresses all of the issues related to water and soil remediation utilizing different nanomaterials, such as their fates and concurrent effects on living organisms, feasibility in field-level applications, environmentally benign, and inexpensive synthesis methods, and scientific, social, and economic bottlenecks preventing widespread adoption (Sánchez-Castro et al., 2023; Roy et al., 2021).

15.2 USES OF NANOMATERIALS IN ENVIRONMENTAL POLLUTION

15.2.1 NANOMATERIALS IN THE REMEDIATION OF WATER POLLUTION

Sewage sludge and synthetic chemicals from homes, in addition to industrial and commercial wastewater, constitute a threat to ecosystems and human health. Streams of any size can be rendered unfit for human consumption by the presence of heavy metal ions, chemical compounds, and oils. There is now a wide variety of state-of-the-art technologies based on nanomaterials, all made possible by advances in nanotechnology. The production weight of nanoparticles is substantially lower than that of activated carbon, but their superiority as adsorbent systems due to their capacity to remove contaminants and their huge surface area for reactions makes up for this. Several different types of nanomaterials have been studied as potential adsorbents in this field. These include zero-dimensional nanoparticles, one-dimensional nanofibers and nanotubes, two-dimensional nanosheets, and three-dimensional nanoflowers. Carbon-based and noncarbon-based nanomaterials, among others, can be used to create water filtration systems. The purification process is highly dependent on the physical and chemical interface between contaminants as well as the surface of the nanomaterial. The advancement of the permeable segregator unit can be greatly aided by the use of nanomaterials, either as a nanocomposite ultimate filter material or even as a thin film addition to improving the membrane's aqueous solubility and thermal-mechanical capabilities. Methods of remediation include lysis, filtration, adsorption, and oxidation. Zinc and titanium oxides, ceramic, nanowire and polymeric membranes, carbon nanotubes, and submicron particles are all materials that are used regularly.

Because of its carcinogenic properties, dye pollution poses a particularly high risk. Since ancient times, humans have employed dyes to add color to textiles, artworks, and other materials. As of right now, businesses all around the world are mass-producing somewhere around 0.1 million distinct synthetic colors. Each year, we go through more than 1.6 million metric tons of dye. Fifteen percent is thrown away as waste. Magnetic nanoparticles with a sugar coating of chitosan are very effective in adsorbing Crocein orange G and acid green. It is worth noting that a magnetic field might be used to conveniently retrieve the adsorbent nanoparticles. Roy et al. (2021) used this idea to develop Fe_3O_4/activated carbon nanoparticles that efficiently partition dyes at concentrations from 138 to 166.6 mg/g (methylene blue and brilliant green, respectively).

Filtering water with methylene blue dye can be done with the help of electrospun polyether sulfone nanofibers doped with vanadium nanoparticles. Due to their low isoelectric point, the nanofibers can be employed under alkaline conditions to generate a huge, highly hydroxylated surface area, which can subsequently be used to adsorb cationic methylene blue molecules. Although the process of adsorption can lessen the intensity of tainted hues, it cannot completely remove or neutralize them. Instead, it actively seeks out the dye's constituent parts, collecting and cataloging them before deliberately rearranging their molecular phases. The accumulation of dye-based sewage on the infrequently reused membrane could make it difficult to reach this capacity. Decolorizing textile wastewater may be possible through the use of hydroxyl radicals (OH) produced by various progressive oxidation methods. Due to its lack of electron pairs, the hydroxyl radical is very reactive and can oxidize even the most stable organic pollutants. Photocatalysis is one of the most sophisticated oxidative remediation strategies and has advanced in recent years to become a practical means of removing organic dyes and pesticides (Figure 15.2).

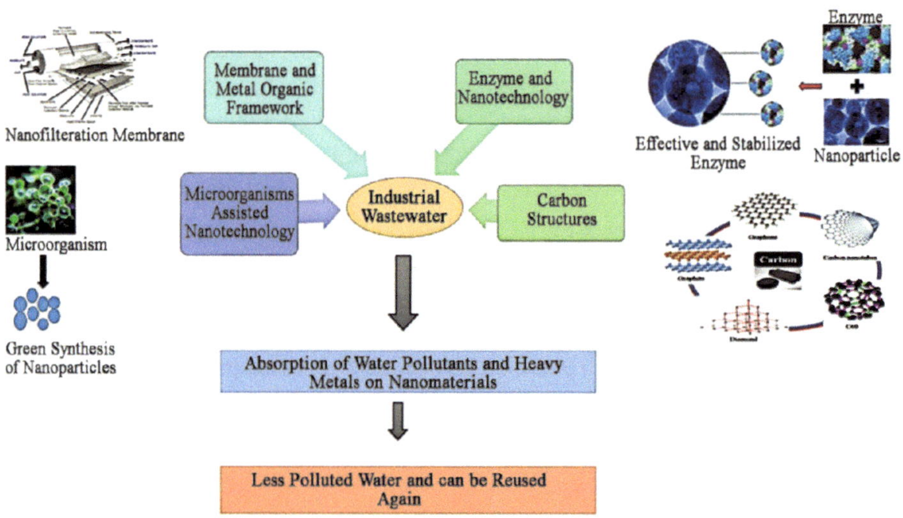

FIGURE 15.2 Nanotechnology approach for bioremediation of industrial wastewater treatment.

Source: Adapted from Shukla (2020).

The metal oxides zinc oxide and titanium oxide, both of which can function as semiconductors, showed remarkable efficacy in the degradation of dye pollution. Semiconductors are vulnerable to degradation when exposed to the light of different dyes (organic pollutants), but semiconductors with energy gaps in the visible light spectrum (groups II–VI) are not very stable or efficient in these settings. Nanoparticles that are inorganic or made of metals in general, there are three types of nanomaterials: (a) nanoparticles based on carbon, (b) nanomaterials based on polymers, and (c) composite nanomaterials. Below is a simplified depiction of the wide variety of NMs that might be employed to purify water and air (Figure 15.3). Due to their enormous surface area, high chemical reactivity, and selectivity, NMs are rapidly gaining appeal as a tool for soil remediation. However, geochemical conditions may alter their reactivity. The bioavailability of both single and multimetalloids (As, Cd, Cr, Pb, and Zn) in contaminated soils was measured using nZVI. At 10% nZVI (w/w), the availability of As, Cr, and Pb was reduced by 82%, while the availability of Zn and Cd was reduced by 31–75% and 13–42%, respectively (Baragaño et al., 2020).

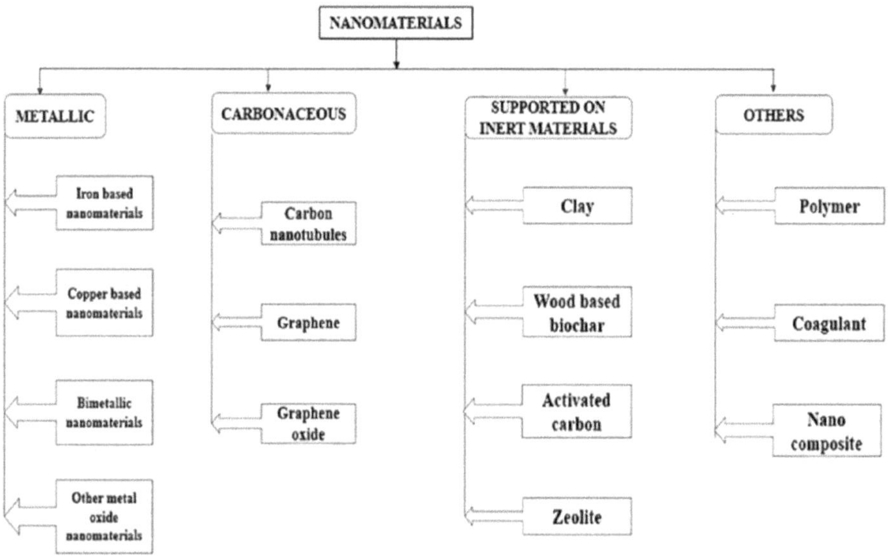

FIGURE 15.3 A description of the wide variety of nanomaterials (NMs) that might be employed to purify water and air.

Source: Adapted from Mukhopadhyay et al. (2021).

15.2.2 NANOMATERIALS IN THE REMEDIATION OF SOIL POLLUTION

There is a worldwide focus on remediating soil contamination and deterioration, a serious environmental issue that shows no signs of abating. We must act quickly to prevent further damage to agricultural output and food security as a result of the worldwide depletion of soil reservoirs. Contamination of soil is already an issue, but the introduction of heavy metals (toxic residues), herbicides, and POPs makes the situation much worse (persistent organic pollutants). More of those compounds in the soil are likely to biomagnify and make their way into the food chain. There is a direct correlation between agricultural output, the pressing need to increase food production in order to feed a growing global population, and the difficulty of halting the degradation of agricultural land. In medicine, immobilization, or adsorption, is frequently used. The advantages of this sort of remediation method are its efficiency, low cost, and lack of environmental impact in excluding metal contaminants from the soils. Some examples of synthetic nanomaterials used to immobilize soil contaminants include carbon-based nanomaterials like carbon nanotubes, metal oxides including ferric oxide and titanium oxide, and a wide range of nanocomposites. Nanoparticles of ferric oxide, for instance, can be used to remove heavy metals like Cd (cadmium) and As (arsenic) from various media samples. As a result of the oily waste, the aquatic environment in the area is suffering. Iron nanoparticles are able to remove 88.34% more Total Petroleum Hydrocarbons (TPHs) from water than conventional methods. Nanotechnology-based treatments are highly effective since they can effectively remove heavy metals and other impurities while also reducing hazardous substances. Nanotechnology and biological treatments can lessen the impact of environmental contaminants. The use of nanoparticles in industrial processes has proven to be an effective method for removing color from organic dyes. Gold and silver nanoparticles can act as catalysts in the two-step process of dye degradation. In order to reduce a particle, first electrons must be concentrated on its surface, and then the dye must diffuse to the particle's interior.

Adsorption and catalytic pollution cleanup are two of several areas where magnetic nanoparticles might be useful. Beneficial effects on soil erosion and arsenate leaching can be achieved by applying polyacrylamide-modified magnetite nanoparticles to the ground. Decontaminating polluted soil through on-site immobilization of pollutants is gaining popularity as a practical and cost-effective solution. Scientists have investigated numerous NM-based full-form alterations in their quest to find an ideal material that is cost-effective, efficient, stable, environmentally benign, and reusable. Metal concentrations in contaminated sediments and soils were successfully reduced after being immobilized in nanohydroxyapatite particles. The authors claim that using nZVI that had been stabilized with sodium carboxymethyl cellulose was able to extract 80% of the soil-bound Cr(VI). Abiotic reductive reactions, whether used alone or in conjunction with immobilization techniques, can greatly decrease the mobility and toxicity of redox-sensitive components that are environmental contaminants. Soil polluted with pyrene, for instance, was treated with nZVI to establish strong reducing conditions and facilitate the reduction and subsequent removal of the contaminant. Nanomaterials are used in advanced oxidation processes (AOPs) for the degradation of organic contaminants in polluted soil, which also employs a wide range of oxidants.

When soluble Fe(II) is mixed with persulfate or hydrogen peroxide, highly reactive radicals are produced, which can be employed for the effective oxidative elimination of organic pollutants (water radical, sulfur dioxide radical). The need for an acidic pH (to prevent Fe(II) precipitation) and the related disadvantages, such as the high expense of early acidification and the harmful effects on soil quality and microorganisms, make these endeavors challenging. The use of insoluble iron phases has been proposed as an alternative to soluble Fe for chemical oxidation since it does not change the pH(II). The process is enhanced by the addition of chelating ligands to the iron particles. When it comes to cleaning up areas that have been contaminated by metal ions, green nanoparticles have a wide range of possible applications. The use of these materials in waste site cleaning and rehabilitation is possible. Ferrous NP can be used to purify water contaminated with heavy metals and soil contaminated with heavy metals, in addition to being an effective, environmentally friendly fertilizer. Because they do not break down into chemical or gaseous forms that the developing plant cannot use, nano fertilizers help synchronize plant nutrient uptake and lessen groundwater contamination. To create optical coatings for solar energy applications, scientists have combined bacteria-produced siliceous material with magnetite (Fe_3O_4) and greigite (Fe_3S).

Many studies have demonstrated that NMs can be used to bind toxic metalloids in soils (Baragaño et al., 2020; Matos et al., 2020; Gil-Díaz et al., 2019). However, it is not well understood what happens to NMs after they are applied to soils to remove contaminants. Soil contamination remediation treatments can either reduce pollutant concentrations to well below a threshold level or stabilize contaminants within the soil to reduce their immediate risk to environmental receptors (Sánchez-Castro et al., 2023; Floris et al., 2017).

15.2.3 NANOMATERIALS IN THE REMEDIATION OF AIR POLLUTION

Nanotechnology has numerous applications that can help enhance air quality. For gaseous processes, one method involves the use of nanocatalysts, which have a much higher surface area. Chemical reactions that neutralize toxic exhaust emissions from vehicles and industry can be sped up with the use of catalysts. Getting rid of the filtered chemicals was hard because of poor air waste collection. Filters are useful tools to have on hand in situations where air quality is compromised. Filters are tiny, porous devices that enable gas to flow through while trapping any particles that may be in the air. Toxins are removed from the gas stream by the membrane due to the particles' direct contact with the filter structure, the inertial force created by the gas's re-direction, and the electric charge created between the particles and the filter. A major downside of filters is the pressure drop they cause, which necessitates considerably more power than is otherwise needed. Because their pores are only 1–10 nm in size, nanofilters are highly effective at filtering out microorganisms and organic pollutants of all kinds. Carbon nanotube membranes are superior at separating CO_2 from other gases. Carbon nanotubes are one hundred times faster at entrapping gases than other gas separation technologies. This makes them applicable in a variety of settings. Carbon nanotube membranes overcome the paradoxical link between the efficiency of gas separation and the volume traveling across traditional layers (Bakshi and Abhilash, 2020).

The separation and purification of gases and contaminated vapors, as well as the prevention of their discharge into the environment, are two examples of the potential applications of nanomembrane technology. The ability to detect and react to physical changes on the nanometer scale is what makes a nanosensor such a cutting-edge device. Nanoparticles, which are tiny, spherical, three-dimensional molecules, are used as sensors to detect the presence of poisons in the environment. In the business environment, spills may be both a nuisance and a safety hazard. Recent developments in commercial sensing technology have enabled the detection of these gases at extremely low concentrations (Dhall et al., 2021). Because of this, we need to take advantage of modern technology to develop sensors that are quick and precise. This biosensor, which consists of one or more membrane nanotubes, can identify and collect noxious gas molecules. Smart Dust is the popular name for this innovative nanosensor (Masuda, 2022). The ultimate aim is to assemble a fleet of ultra-portable nanocomputers that can perform sophisticated sensing tasks. Silicon-based ultrafine particles have been designed to wirelessly send data to a centralized server. A data transfer rate of 1 kbps or more is necessary for the proper operation of this sort of sensor. Solar power allows these nanosensors to remain airborne for extended periods of time. Climate conditions, humidity, and the number of contaminants can all be conveyed by smart dust, allowing for continuous management of air pollution in a specific area (Beni and Jabbari, 2022).

15.2.4 NANOMATERIALS IN THE REMEDIATION OF HEAVY METALS

An increase in human activity is primarily responsible for the presence of HMs in the natural environment. However, their release as pollutants is also attributable to disruptions in biogeochemical cycles. Some metals, like lead, mercury, arsenic, mercury, and cadmium, serve no biological purpose. Major inorganic contaminants include heavy metals, which have profoundly harmful effects on biota even at low concentrations. The bioavailability and absorption of HMs are also crucial to their toxicity (Vázquez-Núñez et al., 2020). The toxicity of HMs is exacerbated by an acidic environment, which is especially true in nutrient-poor and structurally weak soil (e.g., mining sites). Because they stimulate the production of reactive oxygen species (ROS), which in turn damage macromolecules like proteins and nucleic acids, heavy metals are particularly toxic to plants and soil organisms. Crop and plant growth, nutrient quality, and environmental circumstances are all impacted by heavy metals in the soil. The ways in which HMs affect plant growth depend on factors such as the type of crop grown, the soil's physicochemical characteristics, and the HMs themselves. Crop plants are sensitive to HMs because of the way they cause ROS to be produced, which in turn damages cellular organelles, macromolecules such as nucleic acids and proteins, and other parts of a plant's structure and function. The quantity, activity, variety, and genetic makeup of beneficial soil microorganisms have all been documented to be affected. Processes including breathing and photosynthesis, enzyme functions, oxidative stress, biomass formation, and agricultural output all fall under this category. Because site stabilization stops the contaminants from migrating in the soil or being washed away by rain, it is a useful strategy for minimizing heavy metals. The use of biogenic and other nanoparticles for the elimination of heavy metals has garnered a lot of attention as of late. Synthetic biological nanoparticles are called "biogenic nanoparticles." *Morganella psychrotolerans* is responsible for the production of the most well-known biogenic nanoparticles, including Ag nanoparticles (Cao et al., 2020).

Using a Gram-negative bacteria, *Halomonas* sp., to bioremediate Pb and Cd-contaminated soil, the addition of FeO nanoparticles coated with polyvinylpyrrolidone (PVP) led to better results. Pb was removed at over 100% after 24 h and Cd after 48 h, making this technique vastly superior to removal by bacteria or simply NPs. A biosorbent made of *S. aureus*-treated magnetic Fe_3O_4 NPs, with the surface encapsulated with phthalic acid (as an *n*-Fe_3O_4-Phth-S complex), was able to adsorb 795, 1355, and 985 mol g^{-1} of

Cu, Pb, and Ni, respectively. Recovery rates for Cu^{2+}, Pb^{2+}, and Ni^{2+} varied from 83.5 to 89.5%, 99.4 to 100%, and 92.7 to 7.5%, respectively. The study comparing n-Fe_3O_4-Phth-S to dried *S. aureus* for HM removal revealed that the n-Fe_3O_4-Phth-S core of the NPs was essential for the removal of HMs, in addition to the functional groups present on the microbial surface. Thus, this investigation showed that the functional groups on the microbial surface, in addition to the core of the NPs, considerably influenced the efficiency with which the pollutants were removed (Aliyari Rad et al., 2023).

15.3 TECHNIQUES USED IN REMEDIATION OF ENVIRONMENTAL POLLUTION

Adsorption, photodegradation, heterogeneous catalysis, microbiological engagement (nano-bioremediation), and the use of electrical fields are only some of the many technological approaches used in nanoremediation (electro-nanoremediation). Therefore, it is essential to understand the operation of the aforementioned methodologies prior to implementing nanomaterials for repair (Gobinath et al., 2021).

15.3.1 ADSORPTION

Contaminants are "adsorbed" onto the surface of an adsorbent at the solid-liquid contact. Solid-liquid intermolecular forces attract the adsorbent and the solute in solution (Sadegh et al., 2017). The constituent atoms of the adsorbent form bonds with other atoms in the substance, making the adsorbent very efficient (including ionic, covalent, and metallic bonds). Adsorbate is taken up by an adsorbent because the adsorbent's surface has unbalanced atomic bonds. However, the particular adsorption mechanism is species-specific and involves a lot of intricacies. Adsorption occurs primarily due to Van der Waals forces, electrostatic attraction, and chemisorption, as determined by a number of studies (Gupta et al., 2016; Sadegh et al., 2017). The solute mass is transported to the solid surface and bonded thereby both chemisorption and physical adsorption. With a high specific surface area, an adsorbent's uptake of an adsorbate can be more precisely regulated (Sadegh et al., 2017).

15.3.2 PHOTODEGRADATION

One of the most efficient ways to clean up contaminated areas, photo-catalysis is also relatively cheap and kind to the surrounding ecosystem

(Raizada et al., 2019). When a catalyst is added to a chemical reaction that is stimulated by light, it is called a photocatalytic reaction. Depending on the catalyst, electron–hole pairs may be formed during photocatalytic activity. These electron–hole pairs combine to form superoxide radicals, which break down environmental contaminants. Titanium dioxide nanoparticles (NPs) have found practical applications as catalysts. An abundance of carbon-based nanomaterials, including graphene composites doped with NPs, showed enhanced photocatalytic activity (e.g., TiO_2 NPs). Because the energy of the photons was greater than the band gap, ultraviolet light vacated electrons from the valence band and displaced holes from the valence band in carbon materials. Organic contaminants were decomposed by superoxide radicals produced when holes were present, while heavy metal contaminants were destroyed by electrons generated when superoxide radicals were present (Raizada et al., 2019).

Both UV and visible light can facilitate the photocatalytic decomposition of MO by Cu-doped TiO_2 NPs, as described by Khairy and Zakaria (2014). Metal doping (with Cu, for example) had little effect on the crystal structure but drastically impacted particle size and the rate of photodegradation. Degradation of MO by Cu-doped TiO_2 NPs was highest (73%) when exposed to UV light compared to visible light (50%). The metals inserted into the pure oxide matrix served as electron–hole separation centers, hastening MO's breakdown (Khairy and Zakaria, 2014). The addition of doping metal ions to TiO_2 NPs makes them more photocatalytic.

15.3.3 NANOREMEDIATION FOR THE REMOVAL OF HEAVY METALS

There are three types of technological approaches for the removal of heavy metals, which are based on three methods: primary, secondary, and tertiary. Physical methods include precipitation, ion exchange, electrowinning, electrocoagulation, membrane filtration, electrodialysis, and cementation. In order to precipitate metal salts, the right type of anion must be added to the mixture. Chemicals like manganese sulfate, copper sulfate, ammonium sulfate, alum, ferric salts, etc. are frequently used in this procedure. Reduced productivity occurs when the pH is low or there are salts (ions) present, and sludge treatment increases expenses. Precipitation methods, including bisulfide, lime, and ion exchange, tend to be inefficient and unsuccessful when used to remove metal ions at lower concentrations (Ahmed et al., 2021).

Second, ion exchange is the gold standard in the water treatment industry for eliminating metals. This technique makes use of a solid-phase ion exchange substance that can swap cations for anions. Many ion exchange matrices rely on synthetic ion exchange resin. It is costly and time-consuming to get ppb results, even when treating a very large volume. Processing high metal concentrations is not possible due to foul growth in the matrix caused by the presence of particles and organic compounds in the effluent. Not only it is insensitive to the pH of the solution, but it also lacks any selectivity. Like electrolysis, electrocoagulation uses an electric current to separate metal ions from a solution. To recap, an external voltage is used in this procedure to keep the contaminants suspended in the wastewater solution. By combining them with ions of the opposite electrical charge, the electrocoagulation mechanism renders the ions and charged particles unstable (Ahmed et al., 2022).

Metal transformation, acid mine drainage, the electronics and electrical industries, and heap leaching are just some of the many places where electrowinning is employed to recover and remove heavy metals (Thenepalli et al.,2019). On the other hand, Cementation, number five, is an electrochemical kind of precipitation that incorporates metals with greater oxidation capability into the solution. Copper, lead, gallium, gold, silver, antimony, cadmium, tin, and arsenic are only some of the metals that are regularly extracted using cementation. The use of this method facilitates the retrieval of arsenic (As). Metal species bond to the adsorbent surface via physisorption and chemisorption during the adsorption process (Sun et al., 2022). The effectiveness with which metal ions are eliminated is dependent on a number of factors, such as pH, adsorbent surface area, energy, and others. Granular ferric hydroxide, iron oxide-coated sand, copper–zinc granules, glauconite coated with potassium permanganate ($KMnO_4$), and so forth.

Removal of contaminants by means of a membrane: The procedure involves pressurizing water and then forcing it over a semipermeable membrane to filter out the metals. The precipitation of Fe^{2+} and Mn^{2+} ions in the water is the main source of fouling when using this method. The high price tag is largely attributable to factors like water preparation and pressure difference monitoring (Carmona and Abejón, 2023). Electrodialysis uses a semipermeable membrane and an electric field to remove metal ions from polluted water, much like the reverse osmosis (RO) method. Several factors, such as porosity, pH, groundwater flow velocity, texture, ionic conductivity, and water content, are associated with increased efficiency in the removal of heavy metals from groundwater. Reactive zone treatment, membrane filtration, surfactant flushing, reactive zonal treatment, permeable reactive barriers (PRBs), and bioaugmentation are only some of the additional methods that can be employed to achieve the desired remediation results (Carmona and Abejón, 2023).

In chemical methods, since heavy metal contaminants are so dispersed in the groundwater, removing those using conventional methods is an enormous challenge. We will discuss the pros and cons of some different chemical therapies below. At first, areas with high permeability and an alkaline soil pH are pumped full of reductants like gaseous hydrogen sulfide and dithionites. These contaminated areas break down or immobilize toxins. One drawback of the reduction process is that it results in potentially harmful intermediates. Colloidal zerovalent ion (ZVI) is one such material; it can be injected far down into the aquifer before decomposing into harmful chemicals. Chemical washing, in which strong extractants like acids are used to dissolve metal ions and then wash them away, is one technique for accomplishing this. This method is harmful because it degrades soil quality. There are significant risks involved with the *ex vivo* treatment of polluted soil, and managing hazardous waste is a complex and difficult process (Ahmed et al., 2022).

The technique of extracting significant quantities of heavy metals is called "chelate flushing," and it involves the regeneration and recycling of the active agents used in the process (Hashim et al., 2011). PRBs use solvent-loaded resins because they are highly efficient and can be reused indefinitely. Chelating agents are one solution, although some of them, like EDTA and DTPA, can be rather expensive and even carcinogenic.

Biological methods include bioremediation, the broad application of biotechnology in the realm of environmental science, namely for water purification. Water treatment methods that are based on biotechnology are addressed below. Today, bioremediation is seen as the most environmentally benign and cost-effective method for cleaning up metal contaminants in water and soil. It is ideal to use bioremediation; however, sometimes the toxins are so strong that the microorganisms can not handle them (Ayilara and Babalola, 2023). Since heavy metal contaminants are so dispersed in the groundwater, removing them using conventional methods is an enormous challenge.

Bioremediation is a method that uses beneficial organisms, including bacteria, fungi, algae, and some plants, to remove or reduce toxic environmental toxins to safe levels (Ayilara and Babalola, 2023). Common bacteria employed in bioremediation include *Escherichia coli*, *Citrobacteria chrysogenum*, *Klebsiella spp.*, *Rhodococcus spp.*, *Staphylococcus aureus*,

Alcaligenes spp., *Bacillus spp.*, and *Pseudomonas spp.* Several different remediation strategies can be used in bioremediation, including bioaugmentation (the natural attenuation process using indigenous microorganisms), biostimulation (the stimulation of the process by adding nutrients), the use of genetically modified organisms (GMOs), phytoremediation (the use of some plants), and biomineralization (the complete biodegradation of organic substances into inorganic components) (Bala et al., 2022). For its part, the biofilter is a porous medium that can contain both water and bacteria. This method makes use of the fact that pollutants and organic compounds in water are capable of forming complexes. After the porous media take in metabolic waste, biomass, carbon dioxide, and water, they undergo biotransformation to become the final products. A biofilter, in its most basic definition, is any device that aids in the adhesion, development, breakdown, and removal of microbes.

Third on our list is biosorption, a well-known biological technique that employs materials with a biological origin, such as biomass generated from non-living or decomposing microbes. The process is passive and does not necessitate any additional energy to complete, yet the benefits include efficient biosorbent regeneration, metal recovery, minimum sludge formation, and low cost. Biomass can function as an ion exchange matrix, binding the metals and exhibiting their natural properties, allowing for their removal from much-diluted water solutions. The cell walls of certain algae, fungi, and bacteria serve as examples of biosorbent materials. For these reasons, biosorption is being looked at more and more as the best option for cleaning up metal-rich wastewater (Abo-Alkasem et al., 2023).

Phytoremediation uses a variety of plant species to clean up polluted areas by absorbing and decomposing harmful chemicals in the soil, water, ground, and sediments (Mocek-Płóciniak et al., 2023; Nedjimi, 2021). Phytofiltration is a technique for purifying water that takes advantage of plants' ability to collect and show tolerant behavior toward toxic metals like arsenic. Phytofiltration often makes use of hydrilla plants; by growing these plants in realistic field conditions while subjecting them to contaminated water, improvements to the process can be explored (Hanus-Fajerska and Koźmińska, 2016). Phytoextraction is the process by which plants absorb and then move harmful substances through their systems via their roots. Although this method is effective for cleaning up metal-contaminated soil, it does come with some downsides, such as having to get rid of the plant biomass that is removed along with the reclaimed metal and dealing with the metals' phototoxic effect (Kafle et al., 2022). The term "phytodegradation," which is identical to "phytotransformation," refers to the process by which plants remove pollutants from the environment, either by metabolizing them within the plant itself or by influencing the environment through plant-produced enzymes. The degradation of the products and the generation of harmful intermediates are the main problems with this process. Another way in which plants can help remove pollutants from the environment is called phytovolatilization. This process occurs when plants take in pollutants and then release them into the air through transpiration (Yan et al., 2020).

15.4 APPLICATIONS AND SCOPE OF NANOMATERIALS IN THE FUTURE

These days, polymeric materials are widely used because of their many advantages over more conventional building supplies. Because of their superior physical and chemical qualities, polymeric materials are most commonly employed in the transportation and plastics industries. There is currently a disposal problem with the massive waste produced by industries thanks to the extensive use of plastic in packaging for consumables like food, cosmetics, detergents, chemicals, pharmaceuticals, and so on.

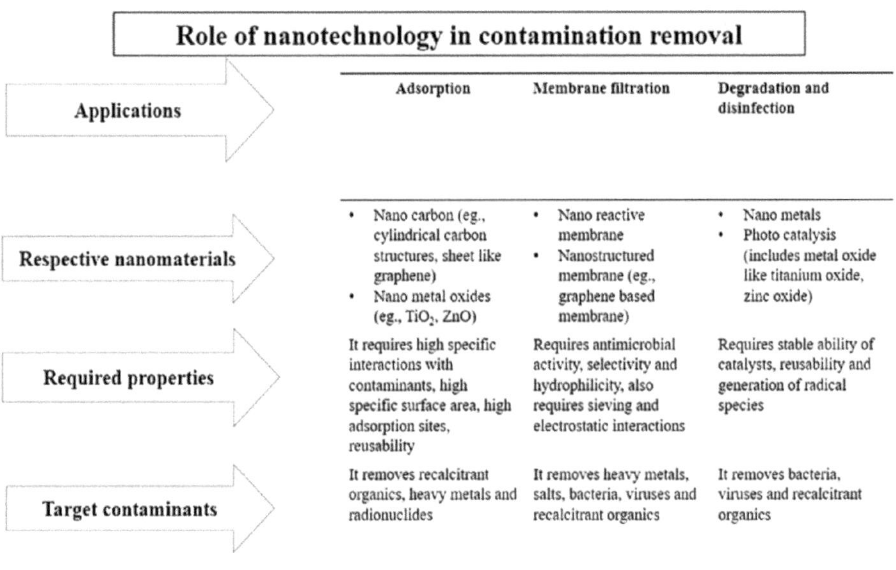

FIGURE 15.4 Applications of nanotechnology.

Source: Adapted from Divyapriya et al. (2016).

Unlike synthetic polymers, natural polymers can decompose in the environment without leaving any residue (Attia and Elsheery, 2020). Starch, a macromolecule consisting primarily of glucose, is the second most abundant natural polymer that can be extracted. At the start of the day, plants use carbon dioxide and water in a reaction called photosynthesis to create starch, which they then use as an energy source (Divyapriya et al., 2016). In an effort to reduce reliance on finite petrochemical resources, scientists have been working on developing starch-based polymers. These materials have several advantages over their petrochemical counterparts, including being completely biodegradable, easily accessible, and less likely to cause.

Recent publications detail the use of NPs in the remediation of contaminated soils, including polycyclic aromatic hydrocarbons (PAHs) of high molecular weight (Karnchanasest and Santisukkasaeom, 2007). Amphiphilic polymer nanoparticles (NPs) have been used by scientists as a means of purifying water. Titanate nanotubes have been shown to be useful as solid-phase extraction adsorbents for seven PAHs found in environmental water samples by Kefi et al. (2011). These PAHs are acenaphthylene, acenaphthylene, anthracene, fluorene, phenanthrene, fluoranthene, and pyrene. PAHs had a recovery rate between 90 and 100%.

It has been suggested that the NPs' increased affinity for the xenobiotics is responsible for their enhanced scavenging capacity for PAH and other pollutants (Karnchanasest and Santisukkasaem, 2007). Recently, additives made from titanate nanosheets and titanate nanotubes (TNT) have been developed to purify cigarette smoke by absorbing harmful chemicals such as nicotine, tar, ammonia, hydrogen cyanide, and phenolic compounds (Qixin et al., 2011). The adsorption capacity of TNT for many toxins is quite high. This could be due to an inherent property of NPs (Qixin et al., 2011).

As the primary and most frequent site of nano-based remediation, groundwater is a common and regular focus (Bardos et al., 2015; US-EPA, 2014). Innovative and potentially fruitful is the concept of using zerovalent metals (ZVMs) to clean up polluted groundwater (Phenrat et al., 2020; Caliman et al., 2011). Nanoremediation is a technique for cleaning up polluted areas that does not require extracting groundwater or digging up vegetation. For this method to work, a slurry of NPs must be injected into an injection well at a depth that corresponds to the probe's vertical range. After being transported to the site of action, small particles (NPs) remove contaminants through adsorption, immobilization, and conversion to less dangerous compounds.

The degradation of pollutants primarily occurs through redox reactions (Phenrat et al., 2020; Galdames et al., 2020). Toxins in the environment can only be cleaned up to a certain extent if NPs are evenly dispersed. Direct-pump wells are less expensive than the drilling and packing method (Pandey and Fulekar, 2012). Surface water can be cleaned, disinfected, and desalinized with the use of carbon nanotubes and ultrafine TiO_2, according to *in vitro* research. In membrane filtration, nanoparticles (NPs) serve as sorbents, reactive agents (photocatalysts, redox agents), and nanofilters (Theron et al., 2008).

Nanoparticles (NPs) have the potential to be used in the detection and removal of even minute quantities of harmful substances in polluted settings. Handheld equipment and instruments are usually insufficient for identifying trace pollutants outside of a laboratory setting. Nanoparticles (NPs) are useful sorbents because of their high surface area and potent reactivity, allowing them to remove the targeted traces of pollutants. They achieve this by using a solid-phase micro-extraction technique that relies on monolayers of self-assembled material supported by a mesoporous matrix. Mercury (Hg), lead (Pb), cadmium (Cd), chromium (Cr), and arsenic (As) are examples of heavy/trace metals that may be easily absorbed by this matrix (Balali-Mood et al., 2021; Ray and Shipley, 2015). Cancer-causing radionuclides are not limited to those two. Organic and inorganic substances, as well as radioactive materials like 99Tc and 137CS, are among these. Saturated magnetite nanoparticles (Fe_3O_4) are effective at removing a wide variety of harmful pollutants. Nanostructured silica can be used to purify wastewater by absorbing metal ions (Sheet et al., 2014).

15.5 CONCLUSION

Various conventional treatments are discussed, including physical, chemical, physiochemical, and biological approaches, specifically nanotechnology's potential applications and the mechanisms behind bioremediation. Nanoparticles made of metal oxides can be employed as efficient adsorbents for cleaning water of toxic metals. Adsorption is an electrochemical process that involves attaching metal species to the adsorbent surface in order to precipitate them out of the solution. Metals in water can be filtered out with the help of a semipermeable membrane, and the process is known as membrane filtration. When it comes to cleaning up heavy metal contamination, traditional methods often fail. The benefits and drawbacks of several chemical therapy options are mentioned below. Bioremediation describes the broad use of biotechnology in the environmental field for water purification. Attenuation occurs when native organisms are used; stimulation occurs when nutrients are added (biostimulation); and biomineralization occurs when organic chemicals are biodegraded thoroughly into inorganic components; all three processes are involved in bioremediation. Effluents containing metals can be cleaned up in a number of different ways.

KEYWORDS

- **Environmental hazards**
- **environmental remediation**
- **nanoremediation**
- **soil contamination**
- **wastewater treatment**

REFERENCES

Abo-Alkasem, M. I., Hassan, N. H., & Abo Elsoud, M. M. (2023). Microbial bioremediation as a tool for the removal of heavy metals. *Bulletin of the National Research Centre*, *47*, 31. https://doi.org/10.1186/s42269-023-01006-z.

Ahmed, M., Mavukkandy, M. O., Giwa, A., Elektorowicz, M., Katsou, E., Khelifi, O., Naddeo, V., & Hasan, S. W. (2022). Recent developments in hazardous pollutants removal from wastewater and water reuse within a circular economy. *npj Clean Water*, *5*, 12. https://doi.org/10.1038/s41545-022-00154-5.

Ahmed, S. F., Kumar, P. S., Rozbu, M. R., Chowdhury, A. T., Nuzhat, S., Rafa, N., Mahli, T. M. I., Ong, H. W., & Mofijur, M. (2022). Heavy metal toxicity, sources, and remediation techniques for contaminated water and soil. *Environmental Technology & Innovation*, *25*, 102114. https://doi.org/10.1016/j.eti.2021.102114Hashim.

Ahmed, T., Noman, M., Ijaz, M., Ali, S., Rizwan, M., Ijaz, U., Hameed, A., Ahmad, U., Wang, Y., Sun, G., & Li, B. (2021). Current trends and future prospective in nanoremediation of heavy metals contaminated soils: A way forward towards sustainable agriculture. *Ecotoxicology and Environmental Safety*, *227*, 112888. https://doi.org/10.1016/j.ecoenv.2021.112888.

Aliyari Rad, S., Nobaharan, K., Pashapoor, N., Pandey, J., Dehghanian, Z., Senapathi, V., Minkina, T., Ren, W., Rajput, V. D., & Asgari Lajayer, B. Nano-microbial remediation of polluted soil: A brief insight. *Sustainability*, *15*, 876. https://doi.org/10.3390/su15010876.

Alptekin, F. M., & Celiktas, M. S. (2022). Review on catalytic biomass gasification for hydrogen production as a sustainable energy form and social, technological, economic, environmental, and political analysis of catalysts, *ACS Omega*, *7*, 29, 24918–24941. https://doi.org/10.1021/acsomega.2c01538.

Anjum, M., Miandad, R., Waqas, M., Gehany, F., & Brakat, M.A. (2019). Remediation of wastewater using various nano-materials. *Arabian Journal of Chemistry*, 12(8), 4897–4919.

Attia, T. M. S., & Elsheery, N. I. (2020). Nanomaterials: Scope, applications, and challenges in agriculture and soil reclamation. In Hayat, S., Pichtel, J., Faizan, M., & Fariduddin, Q. (Eds.) *Sustainable agriculture reviews 41. Sustainable agriculture reviews* (Vol. 41). Springer. https://doi.org/10.1007/978-3-030-33996-8_1.

Ayilara, M. S, & Babalola O. O. (2023) Bioremediation of environmental wastes: The role of microorganisms. *Frontiers in Agronomy, 5*, 1183691. https://doi.org/10.3389/fagro.2023.1183691.

Bakshi, M., & Abhilash, P. C. (2020). Nanotechnology for soil remediation: Revitalizing the tarnished resource. In Singh, P., Borthakur, A., Mishra, P. K., & Tiwary (Eds.). *Nanomaterials as photocatalysts for degradation of environmental pollutants: challenges and possibilities* (pp. 345–370). https://doi.org/10.1016/B978-0-12-818598-8.00017-1

Bala, S., Garg, D., Thirumalesh, B. V., Sharma, M., Sridhar, K., Inbaraj, B. S., & Tripathi, M. (2022). Recent strategies for bioremediation of emerging pollutants: A review for a green and sustainable environment. *Toxics*, 10(8):484. https://doi.org/10.3390/toxics10080484.

Balali-Mood, M., Naseri, K., Tahergorabi, Z., Khazdair, M. R., & Sadeghi M. (2021). Toxic mechanisms of five heavy metals: Mercury, lead, chromium, cadmium, and arsenic. *Frontiers in Pharmacology*, *12*, 643972. doi: https://doi.org/10.3389/fphar.2021.643972.

Baragaño, D., Forján, R., Welte, L., & Gallego, J. L. R. (2020). Nanoremediation of As and metals polluted soils by means of graphene oxide nanoparticles. *Scientific Reports, 10*, 1896. https://doi.org/10.1038/s41598-020-58852-4

Bardos, P., Bone, B., Černík, M., Elliott, D. W., Jones, S., & Merly, C. (2015). Nanoremediation and international environmental restoration markets. *Remediation*, *25*(2), 83–94. https://doi.org/10.1002/rem.21426

Beni, A.A., & Jabbari, H. (2022). Nanomaterials for environmental applications. *Results in Engineering*, *15*, 100467. https://doi.org/10.1016/j.rineng.2022.100467.

Caliman, F. A., Robu, B. M., Smaranda, C. Pavel, V. L., & Gavrilescu, M. (2011). Soil and groundwater cleanup: Benefits and limits of emerging technologies. *Clean Technologies and Environmental Policy*, *13*, 241–268. https://doi.org/10.1007/s10098-010-0319-z.

Cao, X., Alabresm, A., Chen, Y. P., Decho, A. W., & Lead, J. (2020). Improved metal remediation using a combined bacterial and nanoscience approach. *Science of The Total Environment, 704*, 135378. https://doi.org/10.1016/j.scitotenv.2019.135378.

Carmona, B., & Abejón, R. (2023). Innovative membrane technologies for the treatment of wastewater polluted with heavy metals: Perspective of the potential of electrodialysis, membrane distillation, and forward osmosis from a bibliometric analysis. *Membranes*, *13*, 385. https://doi.org/10.3390/membranes13040385

Dhall, S., Mehta, B. R., Tyagi, A. K., & Sood, K. (2021). A review on environmental gas sensors: Materials and technologies, *Sensors International*, *2*, 100116. https://doi.org/10.1016/j.sintl.2021.100116.

Divyapriya, G., Nambi, I. M., & Senthilnathan, J. (2016). Nanocatalysts in Fenton based advanced oxidation process for water and wastewater treatment. *Journal of Bionanoscience*, *10*(5), 356–368.

Floris, F., Fornasari, L., Marini, A., Bellani, V., Banfi, F., Roddaro, S., Ercolani, D., Rocci, M., Beltram, F., Cecchini, M., Sorba, L., & Rossella, F. (2017). Self-Assembled InAs nanowires as optical reflectors. *Nanomaterials*, *7*, 400. https://doi.org/10.3390/nano7110400.

Galdames, A., Ruiz-Rubio, L., Orueta, M., Sánchez-Arzalluz, M., & Vilas-Vilela, J. L. (2020). Zero-valent iron nanoparticles for soil and groundwater remediation. *International Journal of Environmental Research and Public Health, 17*(16), 5817. https://doi.org/10.3390/ijerph17165817..

Gil-Díaz, M., Rodríguez-Valdés, E., Alonso, J., Baragaño, D., Gallego, J. R., and Lobo, M. C. (2019). Nanoremediation and long-term monitoring of brownfield soil highly polluted with As and Hg. *Science of the Total Environment,* 675, 165–175.

Gobinath, R., Manasa, V., Rajendiran, S., Kumar, K., Paul, R., & Basavaraj, K. (2021). Nanoparticle-mediated adsorption of pollutants: A way forward to mitigation of environmental pollution. *Microbial Rejuvenation of Polluted Environment* (pp. 317–348). Springer.

Gupta, V., Moradi, O., Tyagi, I., Agarwal, S., Sadegh, H., Shahryari-Ghoshekandi, R., Makhlouf, A., Goodarzi, M., & Garshasbi, A. (2016). Study on the removal of heavy metal ions from industry waste by carbon nanotubes: effect of the surface modification: a review. *Critical Reviews in Environmental Science and Technology, 46*(2), 93–118.

Hanus-Fajerska, E., & Koźmińska, A. (2016). The possibilities of water purification using phytofiltration methods: A review of recent progress. *BioTechnologia, 97*(4), 315–322. https://doi.org/10.5114/bta.2016.64549

Hashim, M. A., Mukhopadhyay, S., Sahu, J. N., & Sengupta, B. (2011). Remediation technologies for heavy metal contaminated groundwater. *Journal of Environmental Management, 92*(10), 2355–2388. https://doi.org/10.1016/j.jenvman.2011.06.009

Kafle, A., Timilsina, A., Gautam, A., Adhikari, K., Bhattarai, A., & Aryal, N. (2022). Phytoremediation: Mechanisms, plant selection and enhancement by natural and synthetic agents. *Environmental Advances, 8*, 100203. https://doi.org/10.1016/j.envadv.2022.100203.

Karnchanasest, B., & Santisukkasaem, O. (2007). A preliminary study for removing phenanthrene and benzo(a) pyrene from soil by nanoparticles. *Journal of Applied Sciences, 7*(21), 3317–3321.

Kefi, B. B., El-Atrache, L. L., Kochkar, H., & Ghorbel, A. (2011). TiO_2 nanotubes as solid-phase extraction adsorbent for the determination of polycyclic aromatic hydrocarbons in environmental water samples. *Journal of Environmental Sciences, 23*(5), 860–867. https://doi.org/10.1016/S1001-0742(10)60481-0

Khairy, M., & Zakaria, W. (2014). Effect of metal-doping of TiO_2 nanoparticles on their photocatalytic activities toward removal of organic dyes. *Egyptian Journal of Petroleum, 23*, 419–426. https://api.semanticscholar.org/CorpusID:95586296.

Li, P., Karunanidhi, D., Subramani, T., & Srinivasamoorthy, K. (2021). Sources and consequences of groundwater contamination. *Archives of Environmental Contamination and Toxicology, 80*, 1–10. https://doi.org/10.1007/s00244-020-00805-z.

Lin, L., Yang, H.,& Xu, X. (2022). Effects of water pollution on human health and disease heterogeneity: A review. *Frontiers in Environmental Science, 10*, 880246. https://doi.org/10.3389/fenvs.2022.880246.

Masuda, Y. (2022). Recent advances in SnO_2 nanostructure-based gas sensors. *Sensors and Actuators B: Chemical*, 364, 131876. https://doi.org/10.1016/j.snb.2022.131876

Matos, J. C., Laurenti, M., Vighetto, V., Pereira, L. C. J., Waerenborgh, J. C., Gonçalves, M. C., & Cauda, V. (2020). Biomimetic amorphous titania nanoparticles as ultrasound responding agents to improve cavitation and ROS production for sonodynamic therapy. *Applied Sciences, 10*, 8479. https://doi.org/10.3390/app10238479.

Mocek-Płóciniak, A., Mencel, J., Zakrzewski, W., & Roszkowski, S. (2023). Phytoremediation as an effective remedy for removing trace elements from ecosystems. *Plants*, *12*, 1653. https://doi.org/10.3390/plants12081653.

Mukhopadhyay, R., Sarkar, B., Khan, E., Alessi, D. S., Biswas, J. K., & Manjaiah, K. M., et al. (2021). Nanomaterials for sustainable remediation of chemical contaminants in water and soil, *Critical Reviews in Environmental Science and Technology*, *52*, 1–50. https://doi.org/10.1080/10643389.2021.1886891.

Nedjimi, B. (2021). Phytoremediation: A sustainable environmental technology for heavy metals decontamination. *SN Applied Sciences*, *3*, 286. https://doi.org/10.1007/s42452-021-04301-4

Pandey. B., & Fulekar, M. (2012). Bioremediation technology: A new horizon for environmental clean-up. *Biology and Medicine*, *4*(1), 51–59.

Phenrat, T., Skácelová, P., Petala, E., Velosa, A., & Filip, J. (2020). Nanoscale zero-valent iron particles for water treatment: From basic principles to field-scale applications. In: Filip, J., Cajthaml, T., Najmanová, P., Černík, M., & Zbořil, R. (Eds.), *Advanced nano-bio technologies for water and soil treatment. Applied environmental science and engineering for a sustainable future*. Springer. https://doi.org/10.1007/978-3-030-29840-1_2

Qixin, D., Chaozhang, H., Wei, X., Jianping, Z., Yiqiang, Z., Hong, Z., Pang, A., & Wei, M. (2011) Significant reduction of harmful compounds in tobacco smoke by the use of titanate nanosheets and nanotubes. *Chemical Communications*, *47*, 6153–6155.

Raizada, P., Sudhaik, A., & Singh, P. (2019). Photocatalytic water decontamination using graphene and ZnO coupled photocatalysts: A review. *Materials Science for Energy Technologies*, *2*(3), 509–525. https://doi.org/10.1016/j.mset.2019.04.007

Ray, P., & Shipley, H. (2015). Inorganic nano-adsorbents for the removal of heavy metals and arsenic: A review. *RSC Advances*, https://pubs.rsc.org/en/content/getauthorversionpdf/C5RA02714D

Roy, A., Sharma, A., Yadav, S., Jule, L. T., & Krishnaraj, R. (2021). Nanomaterials for remediation of environmental pollutants. *Bioinorganic Chemistry and Applications*. *28*, 1764647. https://doi.org/10.1155/2021/1764647.

Sadegh, H., Ali, G. A. M., Gupta, V. K., Makhlouf, A. S. H., Shahryari-ghoshekandi, R., Nadagouda, M. N., Sillanpää, M., & Megiel, E. (2017). The role of nanomaterials as effective adsorbents and their applications in wastewater treatment. *Journal of Nanostructure in Chemistry*, *7*, 1–14. https://doi.org/10.1007/s40097-017-0219-4.

Sánchez-Castro, P., Molina, L., Prieto-Fernández, M. A., & Segura, A. (2023). Past, present and future trends in the remediation of heavy-metal contaminated soil – Remediation techniques applied in real soil-contamination events. *Heliyon*, *9*(6), e16692. https://doi.org/10.1016/j.heliyon.2023.e16692.

Sheet, I., Kabbani, A., & Holail, H. (2014). Removal of heavy metals using nanostructured graphite oxide, silica nanoparticles and silica/graphite oxide composite. The international conference on technologies and materials for renewable energy, environment and sustainability, TMREES14. *Energy Procedia*, *50*(2014), 130–138. http://creativecommons.org/licenses/by-nc-nd/3.0/.

Shukla, P. (2020). Microbial nanotechnology for bioremediation of industrial wastewater. *Frontiers in Microbiology*, *11*, 590631.

Singh, M., & Naveen, B. P. (2014). Molecular nanotechnology: A new avenue for environment treatment. *Journal of Environmental Science, Toxicology and Food Technology*, *8*(1), 93–99.

Sun, Y., Yu, F., Han, C., Houda, C., Hao, M., & Wang, Q. (2022). Research progress on adsorption of arsenic from water by modified biochar and its mechanism: A review. *Water*, *14*, 1691. https://doi.org/10.3390/w14111691.

Thenepalli, T., Chilakala, R., Habte, L., Tuan, L. Q., & Kim, C. S. (2019). A brief note on the heap leaching technologies for the recovery of valuable metals. *Sustainability*, *11*, 3347. https://doi.org/10.3390/su11123347.

Theron, J., Walker, J. A., & Cloete, T. E. (2008). Nanotechnology and water treatment: Applications and emerging opportunities. *Critical Reviews in Microbiology*, 34(1), 43–69. http://dx.doi.org/10.1080/10408410701710442.

US-EPA. (2014). Climate change indicators in the United States. https://www.epa.gov/sites/default/files/2016-07/documents/climateindicators-full-2014.pdf.

Vázquez-Núñez, E., Molina-Guerrero, C. E., Peña-Castro, J. M., Fernández-Luqueño, F., de la Rosa-Álvarez, M. G. (2020). Use of nanotechnology for the bioremediation of contaminants: A review. *Processes*, *8*, 826. https://doi.org/10.3390/pr8070826.

Yan, A., Wang, Y., Tan, S. N., Mohd Yusof, M. L., Ghosh, S.,& Chen, Z. (2020). Phytoremediation: A promising approach for revegetation of heavy metal-polluted land. *Frontiers in Plant Science*, 11, 359. https://doi.org/10.3389/fpls.2020.00359.

CHAPTER 16

Novel Nanoscale Porphyrinic Inorganic Materials in Agricultural Development and Environment Sustainability

NIVEDITA CHAUDHRI[1], YATIN BATRA[1], and ASHISH KALKAL[2]

[1]*Department of Chemistry, Guru Nanak Dev University Amritsar, Punjab, India*

[2]*Department of Mechanical Engineering, University College London, London, UK*

ABSTRACT

Porphyrins are naturally occurring tetrapyrrolic biologically important molecules that can be also produced synthetically. The nanosized porphyrinic metal-organic frameworks are an emerging area of research interest due to their potential application in sustainable agriculture and environmental protection. Designing cost-effective, highly sensitive, and selective environmental pollutant real-time detectors is the fastest-growing demand at the present time. These porphyrin-based inorganic nanomaterials have a remarkable array of potential benefits for agricultural and environmental sustainability. In this chapter, we have discussed the conventional and green synthetic methods of preparation and their functionalization for various applications in maintaining agricultural sustainability via light-activated reactive oxygen species, controlled delivery of nutrients and pesticides for efficient cultivation, soil revitalization by binding contaminants with porphyrinic-MOFs, water purification by photocatalysis, and regulated plant growth for enhanced yields. As we continued our discussion of environmental sustainability, we also discussed the sensing of common environmental toxins like pesticides, heavy metals, and volatile organic compounds.

Nanobiotechnology for Agricultural Sciences: Nano-Agri-Input Products for Crop Production and Environmental Protection. J. C. Tarafdar, Deepak Kumar, Sanjay K. Singh, and Krishna Pal Singh (Eds.)

16.1 INTRODUCTION

The growing demand for food due to the increasing population has become the main concern and a matter of food insecurity. Thus, feeding a huge population with quality food becomes a global priority. Moreover, to supply enough and quality food, modern agriculture involves the usage of agrochemicals, phytohormones, fertilizers, and pesticides. Notwithstanding, the pathogen-free cultivation is still a big dream for the farmers. The early detection of these microorganisms is the most valuable task for the sustainable development of agriculture and concerned environmental conditions. The excess use of pesticides and artificial fertilizers is further a serious threat to soil conservation, human health, and a clean environment. Thus, the controlled delivery of these agrochemicals, their detection and removal from soil and water, and remediation are some of the parallel actions that need to be rigorously performed (Rojas et al., 2022).

Environmental Sustainability aims to conserve natural resources while developing alternative energy sources with minimal environmental impact. It involves identifying and addressing present environmental degradation scenarios, including issues like climate change, global warming, and glacier melting (Goodland, 1995; Morelli, 2011). In recent years, the concept of environmental sustainability has gained prominence, emphasizing the importance of harnessing the potential of chemistry while mitigating its negative impacts on the planet. Traditional processes have led to environmental challenges such as pollution, waste generation, and resource depletion (Marques and Machado, 2014).

Porphyrins, a group of naturally occurring compounds, hold immense value in promoting environmental sustainability by addressing various challenges and contributing to a more sustainable future. Porphyrin-based inorganic nanoparticles are an emerging area of research in sustainable agriculture and environment protection. Especially the metal-organic frameworks (MOFs) associated with porphyrin namely metal-porphyrinic frameworks are a subclass that is considered an important class among MOFs having many applications in the sustainable development of agriculture and related environmental considerations (Sajjadinezhad et al., 2022).

16.2 SYNTHESIS OF PORPHYRIN-BASED INORGANIC NANOPARTICLES: AN OVERVIEW

Porphyrins are a class of naturally occurring tetrapyrrolic pigments found in various biological systems, including haem in human blood and chlorophyll in plants (Montforts and Glasenapp-Breiling, 2002; Milgrom, 1997). In recent years, scientists have been exploring the potential of incorporating porphyrins into inorganic nanoparticles for various bioinspired and agricultural applications (Park et al., 2021; Liu et al., 2020). The most attractive feature of porphyrins which help them to become potential candidates in agrochemical sensing is they absorb light in visible window of electromagnetic radiations, nicely emissive and semi-conducting materials, and their ability to generate singlet oxygen. There is a wide variety of porphyrins that have been designed and isolated but among all few have real applications in agriculture development and environment sustainability (Nisa et al., 2023; Harvey, 2023). The synthesis of porphyrin-based inorganic nanoparticles involves the integration of porphyrin molecules into inorganic nanomaterials, such as MOFs or metal oxide nanoparticles. There are two following ways to synthesize the nanoscale porphyrinic architecture:

1. Chemical synthesis
2. Green synthesis

16.2.1 CHEMICAL SYNTHESIS OF PORPHYRIN-BASED INORGANIC NANOPARTICLES

Following are the general parameters and steps that need to be considered for the preparation of porphyrin-based inorganic nanoparticles:

16.2.1.1 COLLECTION OF PORPHYRIN AND INORGANIC PRECURSORS

Choose a specific porphyrin building block that has a proper coordination site and suits the desired application. Also, select the appropriate metal, metal cluster, or metal oxide precursors for the inorganic component. The construction of the framework totally depends on the coordination sites between the metal clusters and porphyrins while the structure of the constructed porphyrinic-MOFs depends on the geometry of porphyrin, the stoichiometric ratio of metal, and the porphyrin molecules, and lastly the solvent used. Thus, the change in variables leads to obtain different possible coordination sites, pore sizes of MOFs, different geometries, and sizes of the constructed porphyrin-containing MOF nanoparticles. Mostly, porphyrin-based MOFs contain anionic porphyrin as the primary building blocks the most suited example is tetrakis(4-carboxyphenyl) porphyrin ($TCPP^{4-}$) where the porphyrin is either in free state or accommodate different metal ions (Figure 16.1). While zirconium (IV) oxide ($Zr_6(\mu_3\text{-}O)_4$) cluster has been employed as the secondary building units (SBUs) where the coordination sphere is completed by oxygen-containing ligands such as water molecule, hydroxide, and carboxylate ions. The most essential condition for a porphyrinic-MOF which is being utilized in sustainable agriculture is that all the components should be either non-toxic or should exhibit minute toxicity. Porphyrin in free base form is non-toxic while metals like zinc, iron, or zirconium produce very low toxicity thus it is safe to use such metalloporphyrin in MOF synthesis similarly zirconium-based SBUs would be good complementary units as the inorganic component of the resulted architectures (Kebonye et al., 2022).

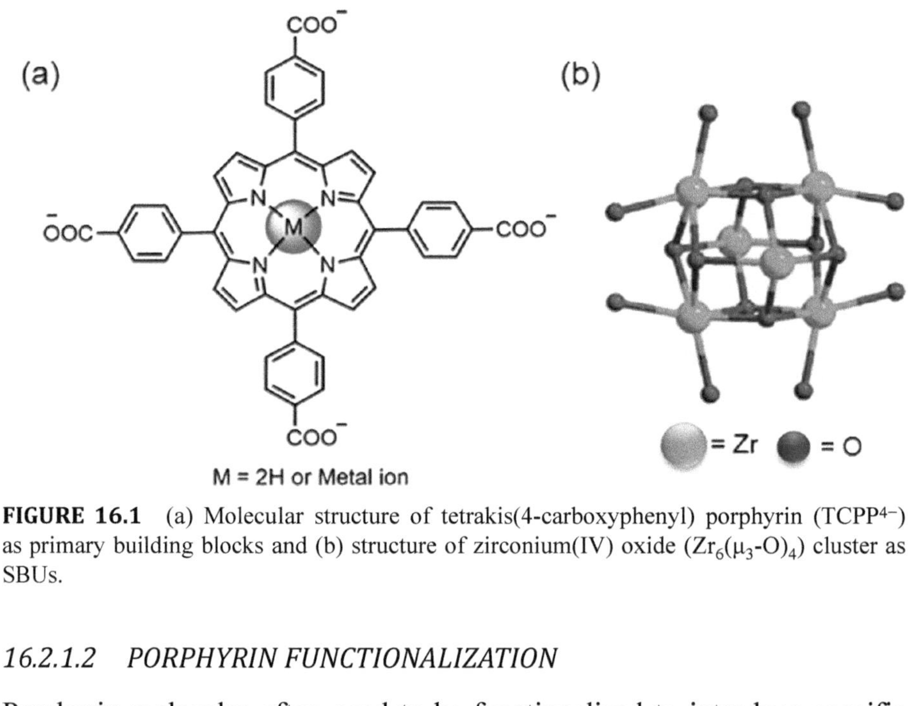

FIGURE 16.1 (a) Molecular structure of tetrakis(4-carboxyphenyl) porphyrin ($TCPP^{4-}$) as primary building blocks and (b) structure of zirconium(IV) oxide ($Zr_6(\mu_3\text{-}O)_4$) cluster as SBUs.

16.2.1.2 PORPHYRIN FUNCTIONALIZATION

Porphyrin molecules often need to be functionalized to introduce specific groups or functional units that can act as anchors or binding units to coordinate with inorganic nanoparticles. Functionalized porphyrins can be obtained by using two methods, first condensation of pre-functionalized precursors (pyrrole and aldehyde), second post-functionalization where we first synthesize the simple porphyrin molecule (tetraphenylporphyrin/tetraarylporphyrin) and then introduce various functional group using different chemical reactions (Hiroto et al., 2017).

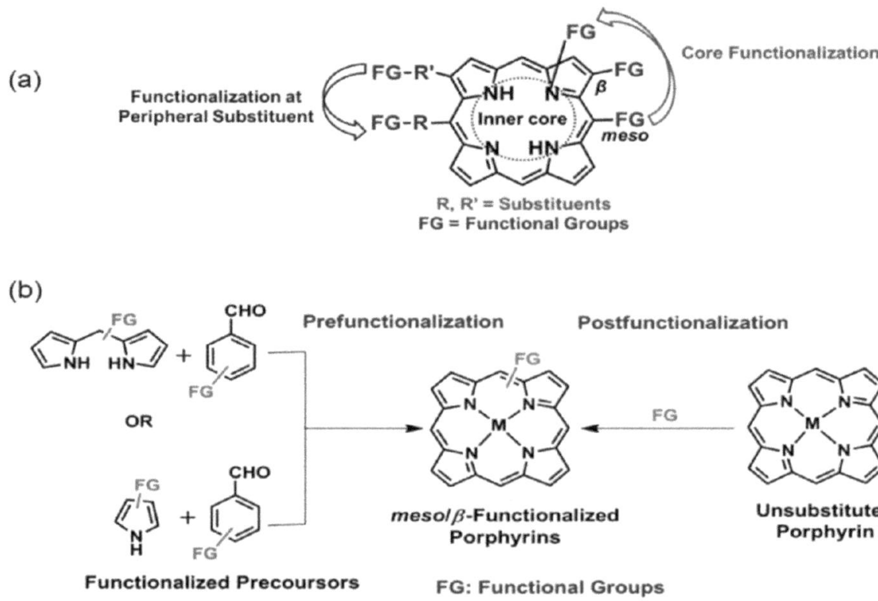

SCHEME 16.1 (a) Types of functionalization and (b) different methods to obtain functionalized porphyrins.

16.2.1.3 CONSTRUCTION OF METAL-PORPHYRINIC FRAMEWORKS (MPFS)

There are various methods available for the synthesis of metal-porphyrinic frameworks, but a few most efficient methods are discussed below:

1. **Conventional Method.** There are two most common reactions that are employed to synthesize porphyrin-based MOFs: (a) solvothermal method which takes place in a sealed tube at a very high temperature (temperature above boiling point), (b) nonsolvothermal which takes place in open vessels at temperature between ambient to the boiling point (Zhang et al., 2021).
2. **Sonochemical Method.** This method involves ultrasound radiations (20 kHz–10 MHz) as the energy source for the occurrence of chemical reactions between the reactants. This method has several advantages over others like less reaction time, room temperature synthesis, and ease of the process (Yu et al., 2021; Lee et al., 2022).
3. **Electrochemical Method.** This method includes the direct and indirect electrosynthesis of MOFs whereby a very thin film of the

MOF deposits on the surface of the substrates. Anodic dissolution and reductive electrosynthesis are the two best examples of this method in which MOFs get deposited on the surface of the electrodes through an electrochemical reaction whereas electrochemical reaction is just one of the steps in the indirect electrosynthesis of MOFs. Electrophoretic deposition, self-template synthesis, and galvanic displacement are well-known examples of indirect electrochemical synthesis (Varsha and Nageswaran, 2020).

4. **Microwave Synthesis.** Nowadays household microwave oven is being used to synthesize MOFs. This requires a magnetron irradiating microwave having a waveguide of 300 MHz to 300 GHz wavelength onto a reaction vessel (Jiang et al., 2020).

16.2.1.4 FORMATION OF PORPHYRIN-BASED NANOPARTICLES

There are various methods for synthesizing porphyrin-based inorganic nanoparticles. Some common techniques include:

1. **Coprecipitation:** This method involves the simultaneous precipitation of metal or metal oxide nanoparticles and porphyrin molecules by mixing appropriate precursor solutions under controlled conditions (La et al., 2022).
2. **Sol–Gel Method:** This method is widely used for the preparation of porphyrin–TiO_2 core–shell nanoparticles. The first step of this method is to prepare the self-organizing porphyrin dye and then use sol–gel process to form the core–shell hybrid nanomaterials. Further, during dialysis in water, the porphyrin molecules would self-assemble to form dispersed porphyrinic nanoparticles (Huang et al., 2014).
3. **Surface Modification:** Often, the pre-synthesized inorganic nanoparticles are functionalized with porphyrin molecules having some specific functional groups to modify their surface and utilize them for antimicrobial activities. For example, when the nanocrystalline TiO_2 materials assemble with sulfonated or amino porphyrins the resulting highly active surface-modified TiO_2 nanoparticles are found to be extremely useful in visible light antimicrobial therapy (Sułek et al., 2019; Saha et al., 2018).

16.2.2 GREEN SYNTHESIS OF PORPHYRIN-BASED INORGANIC NANOPARTICLES

Green synthesis refers to the environment friendly and sustainable approach to produce nanoparticles without using hazardous chemicals or harmful conditions. The green synthesis of porphyrin-based inorganic nanoparticles typically involves the use of natural and non-toxic materials as starting precursors and mild reaction conditions. Here is a general outline of the green synthesis or greener approach for the synthesis of porphyrinic-MOFs or hybrid porphyrin nanomaterials:

16.2.2.1 SELECTION PARAMETERS FOR GREEN SYNTHESIS

1. **Selection of Porphyrin and Inorganic Precursors:** Choose a suitable porphyrin molecule that will act as the building block for the inorganic nanoparticles. Porphyrins are naturally occurring compounds found in plants, algae, and certain bacteria thus in their metal-free state it is non-toxic in nature, similarly, the counterpart for the hybrid nanoparticle synthesis also must be environmentally friendly and non-toxic inorganic precursor. That will react with the porphyrin to form nanoparticles. Water-soluble porphyrins are most suited porphyrins for green synthesis. Commonly used inorganic precursors include metal salts or metal-organic compounds. Since few metals are highly toxic like mercury, lead, cadmium, and so forth. They produce acute toxicity (Balali-Mood et al., 2021). A trace amount of these metals is lethal to human life. Thus, the selection of non-toxic or less toxic metal is very important. For example, iron, zinc, magnesium, and manganese are examples of the less toxic metal.
2. **Solvent Selection:** Choose a green solvent or a benign medium that does not pose any harm to the environment or human health. Water and plant-based solvents are often preferred choices.
3. **Reaction Conditions:** Employ mild reaction conditions such as low temperatures and atmospheric pressure to minimize energy consumption and waste generation.
4. **Green Reducing Agents:** If the synthesis requires the reduction of metal ions, use green reducing agents derived from natural sources. For example, plant extracts can serve as eco-friendly reducing agents.

5. **Stabilization:** To prevent agglomeration and ensure stability, use natural stabilizing agents, such as plant extracts or biomolecules.

16.2.2.2 GREEN SYNTHESIS APPROACHES

Green syntheses or green approaches simply mean the development of synthetic procedures for the sustainable production of materials or compounds. At present, it has gained widespread attention because of its eco-friendly protocol and is being used worldwide for synthesizing wide range of materials/nanomaterials. The following are the most commonly used green synthetic approaches (Figure 16.2):

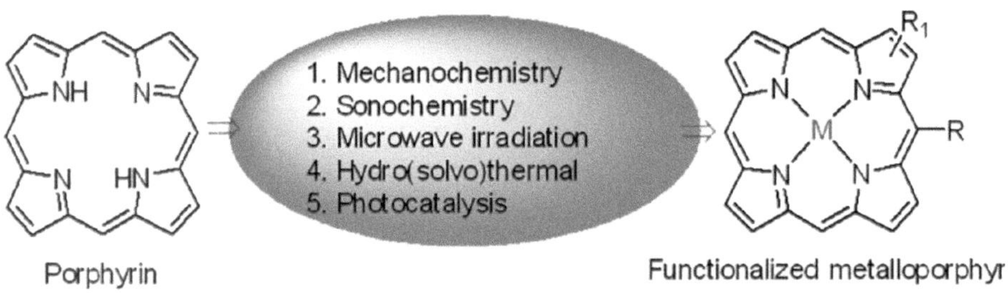

FIGURE 16.2 Different synthetic methods of green chemistry.

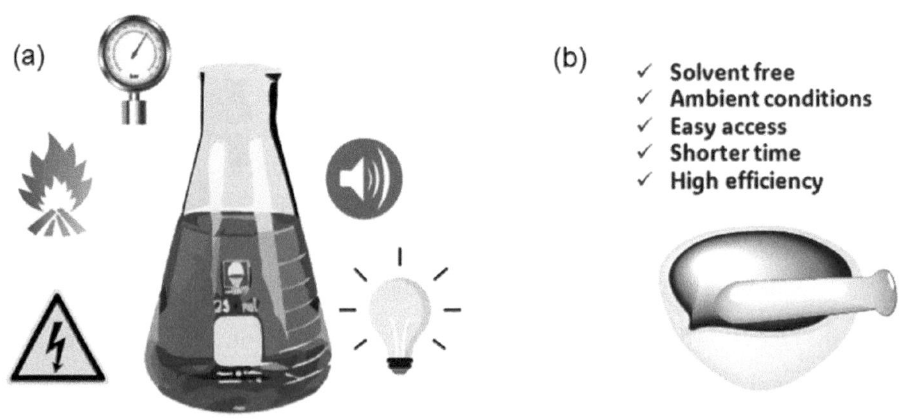

FIGURE 16.3 Solution chemistry versus mechanochemistry: (a) energy input requires in solution chemistry and (b) advantages of mechanochemistry.

1. **Mechanochemical Synthesis:** Mechanochemical synthesis is where the activation for the occurrence of a reaction is produced by mechanical force (Takacs, 2013). The probable advantages of mechanochemical synthesis over conventional solution-based synthesis are: the reaction process in solid state thus requires no solvent, takes a shorter time, and is easy handling process with large efficiency, on the other hand, solution-based reactions have access to different reaction pathways as well as require various energy sources for the occurrence of the reaction (Figure 16.3a). Over the past decades, porphyrin chemists have been actively engaged in synthesizing porphyrins using a ball mill mechanochemical process (Shy et al., 2014). Not only has synthesizing new molecules been proven effective but also the metal insertion inside the porphyrin using different metal sources with mechanochemical process is again one of the most suitable reactions of porphyrin metalation (Damunupola et al., 2020; Atoyebi et al., 2019).
2. **Sonochemical Synthesis:** As discussed above, sonochemical synthesis offers a simpler route to nanomaterials which requires ultrasound radiations for the occurrence of the reaction. This is the sustainable approach to obtain functionalized metalloporphyrins (Gomes et al., 2021). In recent years, sonochemistry has been successfully employed to make various nanomaterials. The fabrication of carbon nanotubes onto the porphyrin can also be achieved by a sonochemical approach. Additionally, sonochemical synthesis is commonly required for the preparation of porphyrin nanoparticles or nanosized hybrid materials (Kashani-Motlagh et al., 2009).
3. **Microwave Irradiation**: Recent advancements are being made to develop new methodologies to prepare eco-friendly porphyrin. Some microwave irradiation syntheses of porphyrins are reported where water beyond boiling point under 16 bar pressure acts as a catalyst without using any external solvent or oxidants (Henriques et al., 2014). Moreover, the controlled MOF synthesis via microwave irradiation is one of the most stabilized methods of MOF preparation (Jiang et al., 2020).
4. **Hydro(solvo)thermal Synthesis**: Hydrothermal synthesis is a green synthetic approach that includes high temperature and high vapor pressure conditions to make the porous material frameworks. Generally, water-soluble porphyrins which are stable at high temperatures, for example, sulfonated, pyridyl, and ionic porphyrins are being

used for this approach (Chen et al., 2014). The soluble porphyrins incorporate with microporous materials to make hybrid molecules that behave as potential catalysts in many catalytic reactions, ion exchangers, and absorbents.

5. **Photocatalytic Synthesis:** Photochemical synthesis is a different and diverse field of modern green chemistry that involves light as a source for the occurrence of chemical reaction. This can be considered a cleaner and more sustainable approach than other conventional syntheses since light is an ideal reagent for a greener environment due to its non-toxic nature. Moreover, it can be obtained from renewable sources and does not produce any waste. It is also a widely accepted method for the synthesis of porphyrin-based nanocomposites which are useful in many antimicrobial activities (Nejad et al., 2022).

16.2.3 CHARACTERIZATION

The synthesized porphyrin-based inorganic nanoparticles need to be thoroughly characterized using various analytical techniques. Common characterization techniques include scanning electron microscopy (SEM), transmission electron microscopy (TEM), X-ray diffraction (XRD), UV-visible spectroscopy, and fluorescence spectroscopy. Thoroughly characterize the synthesized porphyrin-based inorganic nanoparticles using various analytical techniques to confirm their size, shape, composition, and properties.

16.2.4 APPLICATIONS

Once synthesized and characterized, these hybrid nanoparticles and nanosized porphyrinic-MOFs can be explored for a wide range of applications, such as catalysis, photodynamic therapy, bioimaging, bio-sensing applications, and sustainable environment and agriculture development. Firstly, testing is required to evaluate their performance and efficacy. After evaluation, these nanosized systems can be potentially used in large-scale production.

16.3 APPLICATION OF PORPHYRIN-BASED INORGANIC NANOPARTICLES IN SUSTAINABLE AGRICULTURE

Here are some ways these nanoparticles could be used in sustainable agriculture given the following sections.

16.3.1 CROP PROTECTION

Bearing in mind the severe resistance of plant pathogens toward the most frequently used agricultural microbicides, together with the potential threats of agrochemicals to the green environment, there is a demanding need for developing antimicrobial approaches that can kill pathogens or inactivate them with good efficiency without risking the health of other living beings. Porphyrin-based inorganic nanoparticles can be engineered to have photocatalytic properties. When exposed to light, these nanoparticles can generate reactive oxygen species (ROS) or singlet oxygen ($^1O_2$), which have the potential to kill pathogens like bacteria, fungi, and viruses. By utilizing these nanoparticles as a form of environmentally friendly pesticide, it may be possible to reduce the reliance on traditional chemical pesticides that can have detrimental effects on the environment.

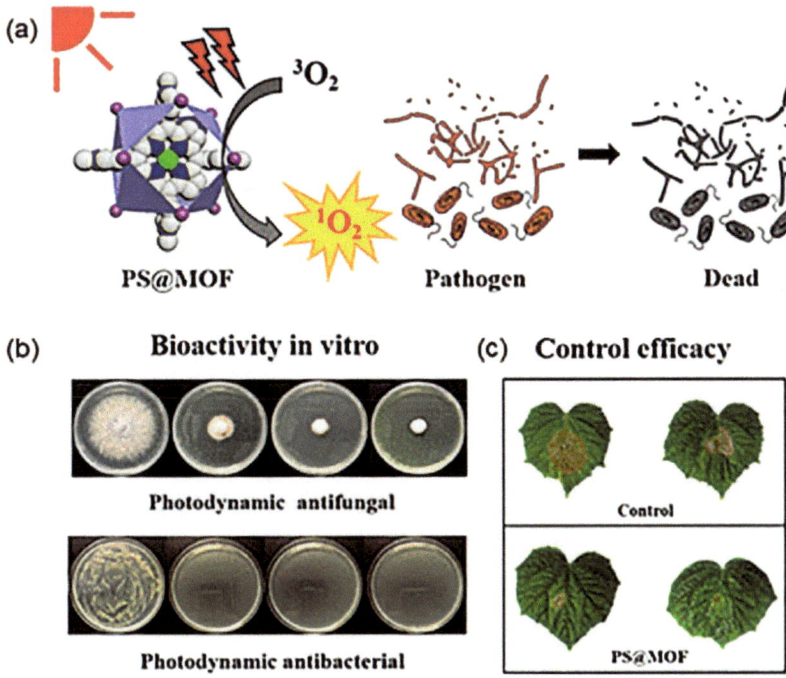

FIGURE 16.4 (a) Mechanism for killing the microbial pathogens, (b) bioactivity images for in vitro studies for photodynamic antifungal and antibacterial studies, and (c) the photos of disease spots on cucumber leaves treated with water (control), PS@MOF.

Source: Reprinted with permission from Tang et al., 2021, Copyright (2021) American Chemical Society.

Cao et al. (2017) synthesized PS@MOFs nanocomposites using 5,10,15,20-tetrakis(1-methyl-4-pyridinio) porphyrin tetra(*p*-toluenesulfonate) (TMPyP) as photosensitizer, benzene-1,3,5-tricarboxalate as linker and zinc acetate dihydrate as metal source. The resulting porphyrinic-MOF nanomaterials showed a broad-spectrum *in vivo* photodynamic antibacterial activity against two pathogenic bacteria and antifungal activity against three plant pathogenic fungi. Additionally, PS@MOF have shown a remarkable control efficacy against *Sclerotinia sclerotiorum* on cucumber leaves (Tang et al., 2021).

16.3.2 CONTROLLED AND EFFICIENT DELIVERY OF NUTRIENT AND PESTICIDES

Inorganic nanoparticles can serve as carriers for various agricultural inputs, including fertilizers and growth-promoting or plant growth regulator compounds. By functionalizing the surface of these nanoparticles with porphyrins, it may be possible to enhance the nutrient uptake in plants, improve fertilizer efficiency, and reduce nutrient leaching, thereby promoting more sustainable agriculture practices. The controlled release of nutrients and fertilizers has become routine in modern agriculture. The absorption and controlled release of the agrochemicals or plant nutrients can be achieved by designing the nanosized microporous porphyrinic-MOF. Since they have good porosity and due to the ease of controlling the size of nanoparticles. Cao and coworkers have developed the layered multi-composite dual-nature microbicide named Tebuc@PCN@pectin@chitosan nanocapsules (Tang et al., 2019). Where they synthesized PH-sensitive functionalized nano-MOF using PCN-224 and then covered them with pectin and chitosan layers which partially closed the pores but consisted of enough space for Tebuc loading. After loading, the pores were sealed as desired and the Nao capsules were pulverized onto the affected areas by fungi. The resulted composite Tebuc@PCN could absorb and release tebuconazole.

16.3.3 SOIL REMEDIATION

Due to different industrial and agricultural activities, soil fertility is getting lower. The use of excess chemicals severely pollutes the soil, and the concentration of pollutants increases tenfold at each trophic level. The presence of chemicals beyond a certain level increases the acidity or alkalinity (PH) of the soil and results in the degradation of the soil quality. The soil pollution leads to poor harvesting, climate change, and damage to health. Thus, looking into

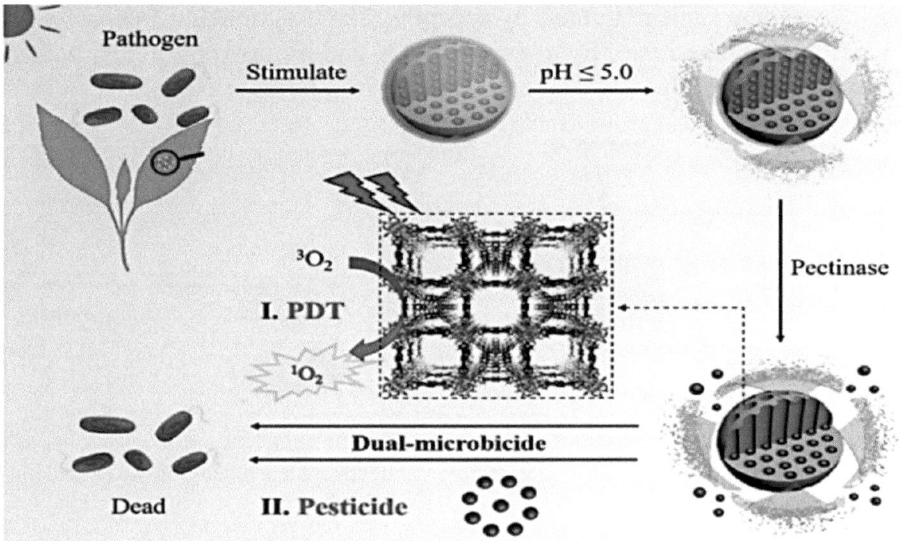

FIGURE 16.5 The schematic representation of the triggered release of tebuconazole and the dual microbicide effect of Tebuc@PCN@pectin@chitosan nanocapsules.

Source: Reprinted with permission from Tang et al., 2019. Copyright (2021), Elsevier.

the current scenario there is an urgent need for the development of non-toxic oxidative catalysts for soil remediation. Porphyrin-based inorganic nanoparticles may also be utilized in soil remediation efforts. These nanoparticles can be designed to bind with heavy metals and other contaminants in the soil, helping to immobilize or remove them from the environment. This could aid in the cleanup of polluted agricultural soils and contribute to restoring soil health. Humic acid plays an essential role in soil purification and very useful in context of agronomic parameters. Water-soluble porphyrins have been proven an outstanding class of compound possessing very high chemical reactivity and potential candidates in human health (Lim et al., 2006) and greener environmental applications (Buschmann et al., 1999). In this regard, few water-soluble Fe(III)-porphyrins have been synthesized and utilized as oxidative catalysts for the oligomerization of humic acid which further increase their conformational flexibility and reduce the amount of soil carbon dioxide released into the atmosphere (Gelsomino et al., 2010). The combined treatment by adding exogenous humic acid with oxidative iron porphyrin catalyst along with additional oxidant such as hydrogen peroxide (HA + FeP + H_2O_2) was found more significant in detecting and capturing the organic soil pollutants (Sannino et al., 2013). The addition of H_2O_2 results in the oxidation of organic pollutants by accepting electrons and thus reduces by its own, the combination of hydrogen peroxide and iron porphyrin involves the free radical oxidative reaction catalyzed by iron porphyrin.

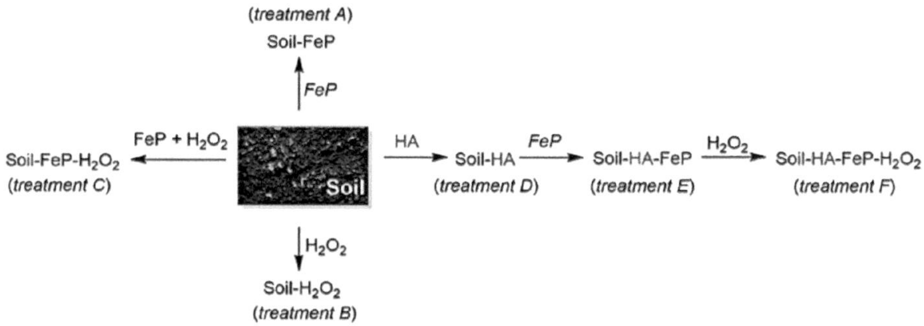

SCHEME 16.2 Schematic illustration of different treatment techniques for soil remediation.

Hydrogen peroxide is a single oxygen donor which upon reacting with iron porphyrin forms the highly reactive oxo-iron(IV)-porphyrin radical cations $(PFe(IV)=O^{\cdot +})$ which further reacts with organic pollutants and gets reduced to oxo-iron(IV)-porphyrin. The second electron transfer from organic pollutant (SH) to PFe(IV)=O transforms the highly oxidative porphyrin species back to the original iron(III) porphyrin PFe(III) as shown in Scheme 16.3.

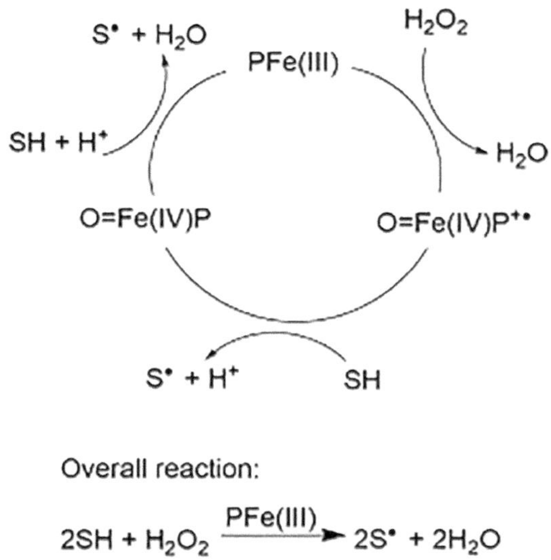

SCHEME 16.3 Schematic representation of oxidative catalysis of organic pollutants present in soil, SH = organic pollutants, and PFe = Iron(III) porphyrin.

16.3.4 WATER PURIFICATION

In addition to soil remediation, due to the enhanced photocatalytic activity porphyrin-based inorganic nanoparticles have potential applications in water purification for agricultural use. These nanoparticles could potentially be used to remove pollutants, organic compounds (Figure 16.6a), and even certain pathogens from water sources, making them safer for irrigation and other agricultural purposes. Over recent years, porphyrins have been extensively utilized as photocatalysts in oxidative reactions for the degradation of organic pollutants present in water. Nowadays the fabrication of porphyrin molecules onto the solid conducting materials is being considered as one of the most useful strategies to develop highly effective photocatalytic systems with advanced applicability. These conductive solid materials-based supported porphyrins form highly organized hybrid materials and perform great photochemical reactions (Figure 16.6b) (Silvestri et al., 2022).

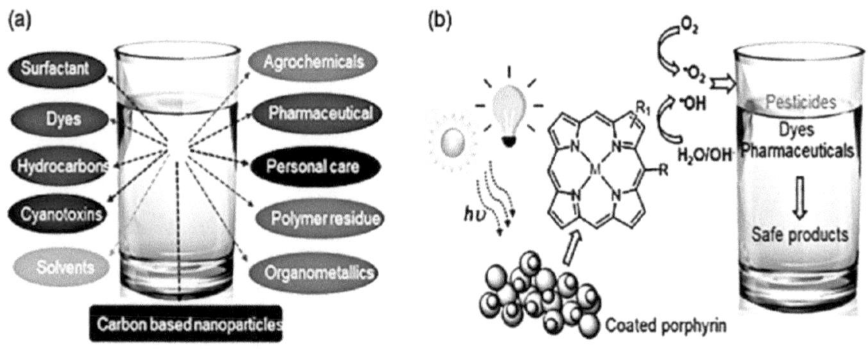

FIGURE 16.6 (a) Commonly found organic contaminants in water and (b) mechanism of photodegradation of water containing contaminants.

16.3.5 PLANT GROWTH REGULATION

Porphyrins have been known to be involved in various physiological processes in plants, including photosynthesis and hormone regulation. By incorporating porphyrin-based inorganic nanoparticles, it may be possible to influence plant growth and development in a controlled manner, potentially leading to improved crop yields and resource utilization. The fabrication of supramolecular nanovalves on nanosized porphyrinic-MOF is an emerging technique to develop a smart MOF-based supramolecular nano platform for the delivery of controlled plant hormones. Yang and coworkers have recently designed the smart multi-stimuli-responsive nano delivery system namely CLT6@PCN-Q via supramolecular interactions between quaternary ammonium stalks-functionalized porphyrinic nano-MOFs and water-soluble leaning tower[6]arene. The resulting hormone nano delivery system was used to regulate the plant growth of monocotyledonous wheat and dicotyledonous Chinese cabbage (Yang et al., 2021).

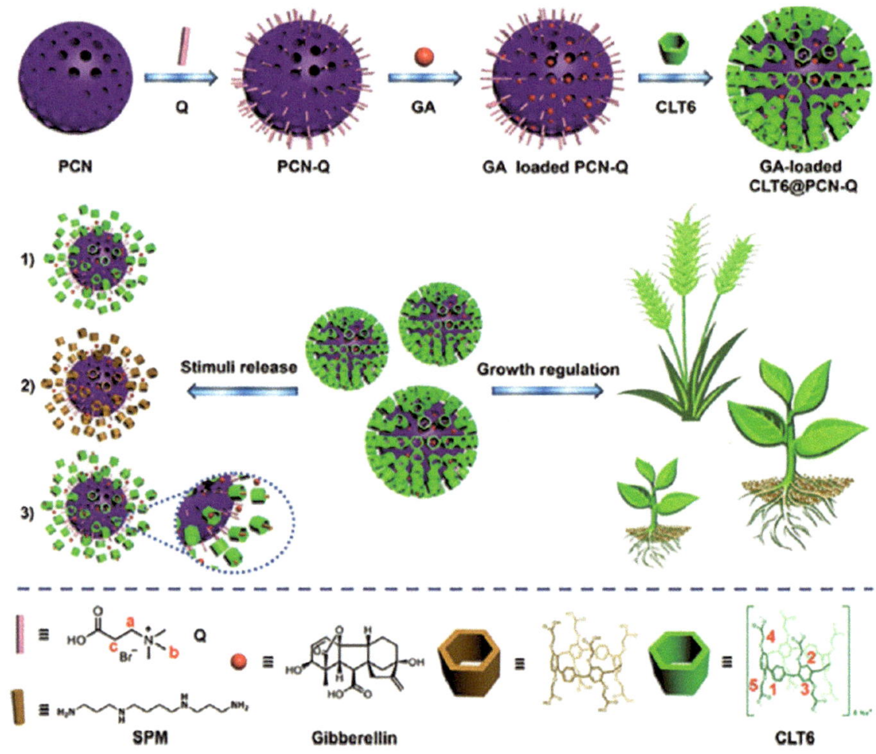

FIGURE 16.7 Schematic representation illustrating the formation of smart nano delivery gibberellin loaded multi-stimuli-responsive supramolecular platform namely GA-loaded CLT6@PCN-Q. Where GA = Gibberellin, CLT6 = Tower[6]arene, PCN = PCN-224, porphyrin, Q = Quaternary Ammonium Stalks, SPM = Spermine.

Source: Reprinted with permission from Yang et al., 2021, Copyright (2021) Elsevier.

It is important to note that while porphyrin-based inorganic nanoparticles hold promise in sustainable agriculture, their practical application and long-term effects on the environment and human health require further research. As with any novel technology, it is crucial to carefully assess potential risks and benefits before widespread adoption in agricultural practices.

16.4 ROLE OF PORPHYRINS IN ENVIRONMENT SUSTAINABILITY

Human activities such as agriculture, industrialization, transportation, urbanization, and mining are the main causes of global environmental pollution. Because the environment is getting more polluted each passing day, there is an urgent need to monitor the environmental contaminants with reliable devices such as effective sensors. In the past few decades, a variety of sensors have been scrutinized and employed as cost-effective, real-time, *in situ* detecting analytical tools for a healthy environment. For continuous monitoring of the environment, it is necessary that the detecting technique should be easy to perform, quick, and cost-effective. In the past few years, the porphyrin-based self-assembled nanostructures have attracted researchers' interest for their utilization in the detection and monitoring of environmental pollutants (La et al., 2023). Porphyrins can act as potential molecular sensors for environmental remediation. Functionalizing porphyrins with specific receptors or functional groups allows them to selectively bind to the target pollutants, leading to color changes (colorimetric detection), fluorescence emissions, or any other spectroscopic changes that help in determining the presence of the analyte. This property makes them valuable tools for real-time monitoring of environmental contaminants.

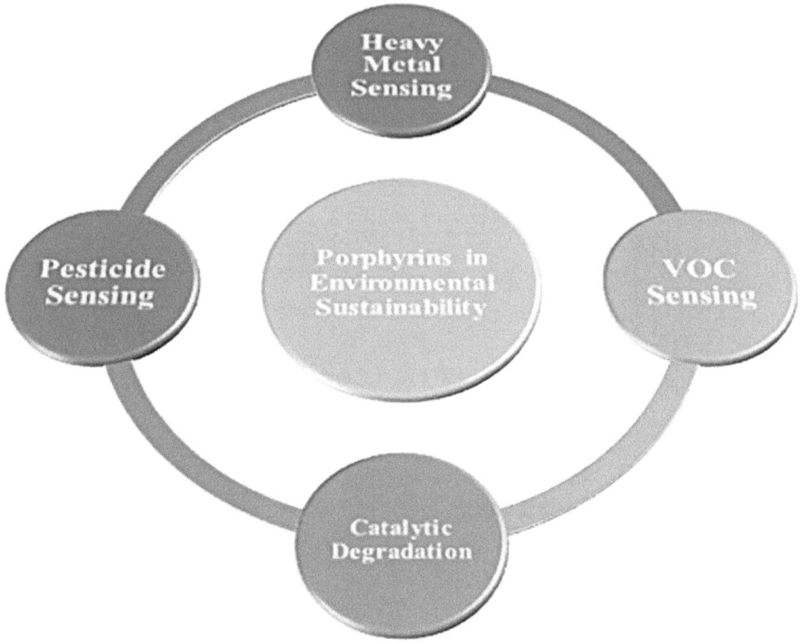

FIGURE 16. 8 Applications of porphyrins in environmental sustainability.

16.4.1 HEAVY METAL SENSING

Heavy metals, such as Pb, Hg, Ag, and Cu, harm crop health, human health, and soil quality. Porphyrins can be employed as heavy metal sensors, utilizing free-base porphyrins with varying substituents at the periphery (Qi et al., 2020). However, their water solubility remains a challenge to effectively utilize them in such applications (Ishihara et al., 2014). The most common approach to making water-soluble porphyrin is the substitution with the ionic functional groups (Sallam et al., 2020). Fluorescence-based techniques, such as Fluorescence Energy Transfer (FRET) (Figure 16.9) and Photo-induced Electron Transfer (PET), are widely used for metal sensing, contributing to fluorescence quenching (Paolesse et al., 2017). The Kinetics of metal complexation plays a significant role in porphyrin base sensing. Some metals may bind quickly and some may slow which can affect the detection rates of metal. Mercury is the most important target for sensing as it is the most toxic among all the heavy metals present in the environment, furthermore, a significant amount adds on to the agricultural land indirectly with fertilizers, sludge, and manures (Patra and Sharma, 2000).

The proper tuning of structural and optical properties of porphyrins leads to perturbing the electron clouds around the macrocyclic system and ultimately gives rise to unique properties. The electron-rich character enhances its affinity toward Hg(II) ions, making it an outstanding sensor for detecting Hg(II) among its counterparts (Yang et al., 2009).

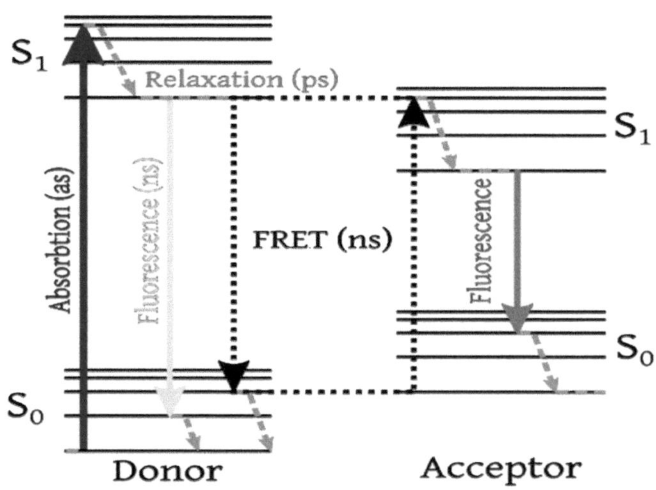

FIGURE 16.9 Representation of FRET event between donor and acceptor system.

A noteworthy development in the field of mercury sensing involves chemo-dosimetric sensing by water-soluble porphyrin-based sensors (Figure 16.10). This novel derivative enables the selective detection of mercury by capitalizing on mercury-promoted cyclization. The cyclization process leads to red-shifted absorption, facilitating the appearance of new absorption and fluorescence bands (Srivastava et al., 2013). Interestingly, in this case, mercury does not bind directly to the porphyrin core instead it selectively participates in the synthesis of an imidazole moiety. This specific interaction unveils the versatility of mercury in triggering distinct spectral responses, enabling precise and selective detection.

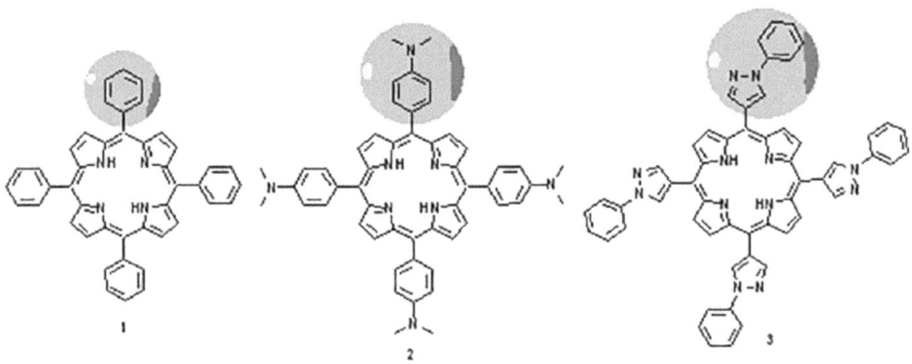

FIGURE 16.10 Molecular structures of porphyrin employed for Hg(II) sensing by tuning structural features.

Selective detection of Pb(II) ions was successfully achieved using phthalocyanine-porphyrin dyads, which create a donor-acceptor system in close proximity to each other. This spatial arrangement facilitates a Förster Resonance Energy Transfer (FRET) phenomenon between the Zn–porphyrin complex and phthalocyanine, owing to the significant overlap of their respective emission and absorption spectra. However, upon the interaction with Pb(II) ions, a fascinating transformation occurs. Pb(II) forms complexes with the phthalocyanine core, leading to a reduction in the spectral overlap between the two moieties and ultimately hindering the FRET event. This ingenious approach can be categorized as an "on–off" strategy, where the presence of the target analyte, Pb(II), acts as a molecular switch, effectively turning off the FRET mechanism (Zhang et al., 2017). The advanced "on–off" approach relies on ionic radii and simultaneously detects different metal ions such as sensing Ag(I), Cu(II), and Pb(II) ions. For simultaneous detection, the sensor is designed in different ways where porphyrins are linked to aza-crown ethers. Ag(I) ions exhibit a preference for binding to the aza-crown ether moiety rather than the porphyrinic core. As a result, Ag(I) selectively interacts with the aza-crown, leaving the porphyrin moiety free for other metal ions. Interestingly, if the Cu(II) is present among the mixture, it effectively displaces the Ag(I) and Pb(II) from the crown ring and fit nicely to the aza-crown moiety. This competitive binding phenomenon showcases the versatility of the aza-crown-porphyrin system in discerning and differentiating among various metal ions, thereby enabling specific and reliable detection of Cu(II) (Chen et al., 2013).

The remarkable selectivity and sensitivity demonstrated by these systems underscore their promising applications in environmental monitoring, industrial quality control, and analytical chemistry, where accurate and rapid detection of heavy metal ions is of utmost importance. Further exploration and optimization of such molecular recognition platforms hold the key to addressing current challenges in metal ion sensing and enhancing our understanding of these intriguing molecular interactions (Dar et al., 2020).

16.4.2 PESTICIDE SENSING

Pesticides are chemical and biological agents that are used to kill pests or any external microorganisms that are harmful to plants and destroy the crop. The unawareness about the requirement of the actual amount of pesticides needed as well as greed to increase crop production rapidly has led to an uncontrolled spray of pesticides hence extensive use of pesticides is of major concern as it is getting into the food chain causing ill and toxic effects. The commonly used pesticides comprise organophosphate and organochlorine which are synthetic. The excessive use of these pesticides leads to their entry into the air, water, and even our foods. Also, these organic pesticides are mostly non-biodegradable that is why they have more accumulation in soil or groundwater. In 2013, Patna, India witnessed the death of more than 30 students and many others severely infected by poisoning caused by organophosphate-based pesticides. Hence sensing these pesticides to check the toxicity level is of utmost importance these days.

The development of an effective and rapid pesticide sensor that can degrade pesticides and find realistic applications is a forthcoming challenge to warrant sustainable agriculture, environment development, and food safety. Nowadays the nanosized porphyrinic-MOF serves as bifunctional sensors that show fluorescence detection as well as the elimination of nitenpyram in agricultural environment (Liu et al., 2020). Pesticide sensing where porphyrins are being used as sensors is mostly based on the protonation and deprotonation of porphyrins where the pesticides bind to the central metal atom of the porphyrins which leads to changes in photophysical properties including UV-Vis. and fluorescence. 2,2-Dichloro vinyl dimethyl phosphate (DDVP), also known as dichlorvos is an organophosphate insecticide that is widely used in households to control pests. The detection of this pesticide by porphyrin leads to the color change of porphyrin from purple to green indicative of protonation of the porphyrin molecule. This type of detection is called colorimetric detection. Tetra-hydroxy-phenyporphyrin (THPP) and tetra-carboxy-phenylporphyrin (TCPP) have been used for the detection of DDVP (Murakami et al., 2015).

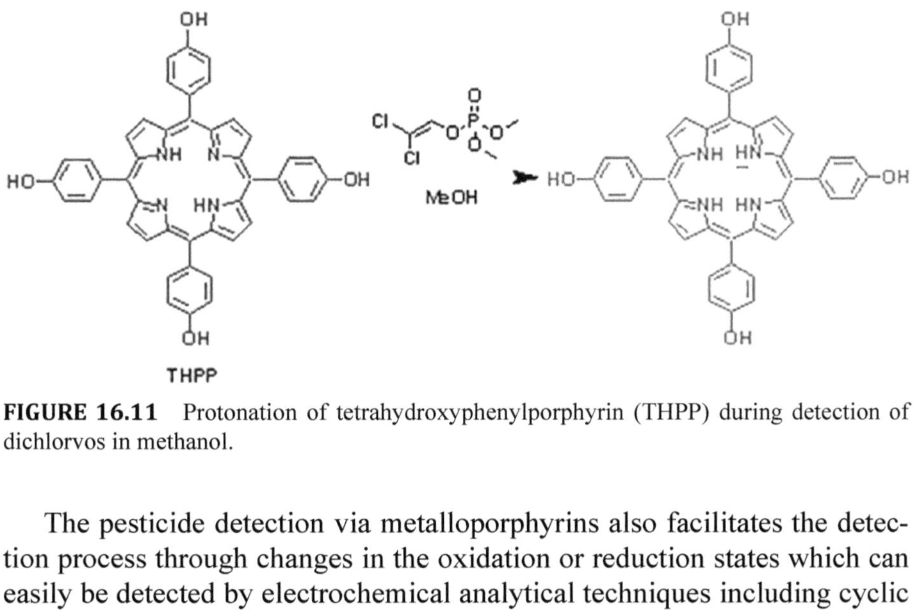

FIGURE 16.11 Protonation of tetrahydroxyphenylporphyrin (THPP) during detection of dichlorvos in methanol.

The pesticide detection via metalloporphyrins also facilitates the detection process through changes in the oxidation or reduction states which can easily be detected by electrochemical analytical techniques including cyclic voltammetry, however, this method requires electrochemical workstation and mostly non-colorimetric in nature (He et al., 2017; Jiang et al., 2022). There are several reports highlighting the use of porphyrins as electrochemical sensors for organochlorine and organophosphate agents. Zinc tetraphenylporphyrin has been used as organophosphate sensors in which the phosphoryl oxygen binds to the zinc leading to a change in the spectral behavior of the compound (Rompoti et al., 2015). These spectral changes quantify the concentrations of detected organophosphate thus providing valuable insights to find out the limit of detection (LOD) and limit of quantification (LOG) values using UV-Vis. spectroscopic data. Varying the spectral properties of porphyrins by either means discussed above paves the way for efficient and user-friendly pesticide detection technologies that contribute to safer agricultural practices and environmental protection.

16.4.3 VOLATILE ORGANIC COMPOUND (VOC) SENSING

VOCs as the name suggests are the compounds that vaporize at room temperature due to their high vapor pressure at room temperature. Long-term exposure to VOCs can lead to serious health hazards. Quantification of these gaseous toxins is of utmost importance these days due to their increased human exposure indoors as well as outdoors (Khatib and Haick, 2022). Non-reactivity of VOCs and low concentration is a major concern that limits their efficient detection. Thus, the development of highly efficient, selective, rapid, and sensitive detectors is needed for environmental protection and human health. The sensing mechanism to detect these compounds is referred to as electronic nose technology. In the present chapter, our main concern will be fluorometric and colorimetric detection of VOCs since these are cost-effective, rapid, and easy to perform detecting methods. The nanosized porphyrins-built functional MOFs have shown efficient sensing of gases and volatile compounds (Li et al., 2020).

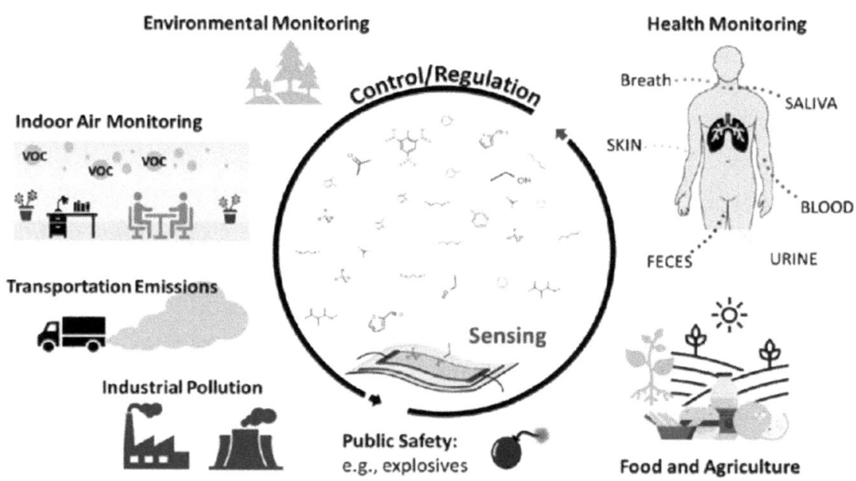

FIGURE 16.12 Volatile organic compounds control/regulation and sensing.

Source: Reprinted with permission from Khatib and Haick (2022), Copyright (2022) American Chemical Society.

Metalloporphyrins can detect volatile contaminants through axial coordination of the volatile ligands to the central metal ion which results in substantial color change resulting from the change in the wavelength as well as intensity shift upon coordination. By altering the binding sites, either by changing the metal center or through peripheral modification their chemical reactivity can be easily adjusted, thus creating a selective sensing platform (Li et al., 2019). Metalloporphyrins demonstrate a wide range of applications in sensing various ligating VOCs such as amines, carboxylic acids, phosphines, phosphites, and thiols. These VOCs trigger significant color changes in porphyrins when interacting with metalloporphyrin sensors. Various metalloporphyrins such as zinc, copper, magnesium, and so forth have been utilized for this purpose (Suslick et al. 2004; Dunbar et al., 2010).

16.5 REGULATORY FRAMEWORK FOR USAGE OF PORPHYRIN-BASED INORGANIC NANOPARTICLES-BASED PRODUCTS IN AGRICULTURE

The regulatory landscape for nanoparticles in agriculture can vary from country to country and may be subject to ongoing changes. When introducing new technologies like porphyrin-based inorganic nanoparticles or nanosized porphyrinic-MOFs into agriculture, it is crucial to consider potential environmental and health impacts as well as their efficacy in promoting sustainable practices. Here are some key aspects that may be relevant when considering the regulatory framework for porphyrin-based inorganic nanoparticles in sustainable agriculture and environment:

1. **Environment Impact Assessment (EIA)**: Regulatory authorities may require environmental impact assessments to evaluate the potential effects of these nanoparticles on soil health, water quality, and non-target organisms in the ecosystem before their bulk use.
2. **Toxicology and Safety Studies**: Safety evaluations might be necessary to understand the potential risks associated with the use of porphyrin-based inorganic nanoparticles on crops, farmers, and consumers. Handling guidelines may be required to properly address the issue concerning these porphyrinic-MOF nanosystems prior to their use.
3. **Efficacy and Performance Data:** Demonstrating the efficacy and performance of these nanoparticles in sustainable agriculture practices would be important for regulatory approval.

4. **Labeling and Usage Guidelines**: Approved nanoparticles might need specific labeling and usage guidelines to ensure proper application and to minimize unintended consequences.
5. **Product Registration**: Depending on the jurisdiction, products containing porphyrin-based inorganic nanoparticles might require registration before they can be legally sold or used in agriculture.
6. **International Harmonization**: For companies operating globally, compliance with international regulations and harmonization efforts may also be necessary.

16.6 CONCLUSION

Each passing day, industrial and agricultural activities severely pollute the soil, water, and environment resulting in food crop contamination and various human and plant diseases. Furthermore, providing food for the overcrowded population become a worldwide challenge. To achieve this goal modern agriculture involves the use of agrochemicals including fertilizers, plant hormones, and pesticides which increase the yield of agricultural production and growth of the crop. The excessive use of these agrochemicals becomes a serious threat to the environment. Thus, the detection and elimination of these environmental pollutants is the undertaken parallel action. The nanosized porphyrinic frameworks hold a special fit into this field. Their use comprises real-time detection and elimination of pesticides and related toxic elements, detection and elimination of toxic heavy metals, controlled delivery of plant nutrients and agrochemicals, microbial photodynamic therapy, food safety, and toxicity assessment. Detection can be done by absorbing, colorimetric, fluorescent, and electrochemical while the elimination processes involve catalytic hydrolysis, degradation, or inactivation of microbes and pathogens. 78% of investigations performed on the porphyrinic nanosized materials in the past, cover these two processes (detection and elimination) and the remaining 22% covers the other remediation techniques. However, the most prominent feature of porphyrin is that they absorbed in the visible region of electromagnetic radiations and readily photosensitize singlet oxygen $^1O_2$ resulting in the formation of ROS which is the prime source of photo-induced degradation of toxins and microbes. Nanosized porous porphyrinic-MOFs are also used in the controlled delivery of fertilizers, pesticides, and growth hormones. The photo-induced degradation process is not selective and can cause the premature decomposition of the porous nano-MOFs under irradiation conditions which affect the controlled release of a desired guest. There are several ways to obtain the porphyrin-based nanosized system including a conventional method of green synthesis and their hybrid porphyrinic-MOF can be synthesized by sonochemical, microwave, coprecipitation, and supramolecular self-assembly of porphyrin derivatives. In the past decade, the fabrication of porphyrin molecules with porous materials or self-assembled nanostructures gained researchers attention due to their remarkable contribution to sensing environmental pollutants and sustainable agricultural and environmental development. This chapter has outlined the various techniques for synthesizing the porphyrinic nanomaterials, and their use in soil, water, and environment remediation using various detection techniques.

16.7 FUTURE PERSPECTIVES

Sustainable agriculture is a question of how to feed a big population and reduce the impact of related climate change. The current scenario estimates that of the total greenhouse gas emissions, up to 30% is contributed by agricultural activities. Thus, there is an utmost need to reduce the impact of agriculture on the environment along with increased crop productivity to provide food for the global population. Thus, practicing sustainable agriculture is intended to protect the environment, restore or expand the natural resources of Earth, maintain soil quality, and improve soil fertility. Designing cost-effective, highly sensitive, and selective environmental pollutant detector is the prime need of the present time. Porphyrin in combination with inorganic nanoparticles can be an outstanding plethora for the sustainable development of the environment and agriculture. The desired outcomes of sustainable development are to satisfy human food insecurity, enhanced environmental quality, and remediation of the natural resources upon which the agriculture economy depends.

KEYWORDS

- **Controlled and efficient delivery**
- **environmental sustainability**
- **porphyrin-based inorganic nanomaterials**
- **sustainable agriculture**
- **volatile organic compounds**

REFERENCES

Atoyebi A. O. & Brückner, C. (2019). Observations on the mechanochemical insertion of zinc(II), copper(II), magnesium(II), and select other metal(II) ions into porphyrins. *Inorganic Chemistry*, *58*, 9631–9642.

Balali-Mood, M.,, Naseri, K., Tahergorabi, Z., Khazdair, M. R., & Sadeghi, M. (2021). Toxic mechanisms of five heavy metals: mercury, lead, chromium, cadmium, and arsenic. *Frontiers in Pharmacology*, *12*, 643972.

Buschmann, J., Angst, W., & Schwarzenbach, R. P. (1999). Iron Porphyrin and cysteine mediated reduction of ten polyhalogenated methanes in homogeneous solution: product analyses and mechanistic considerations. *Environmental Science & Technology*, *33*, 1015–1020

Cao, L., Lin, Z., Shi, W., Wang, Z., Zhang, C., Hu, X., Wang, C., & Lin, W. (2017). Exciton migration and amplified quenching on two-dimensional metal–organic layers. *Journal of the American Chemical Society*, 139, *20*, 7020–7029.

Chen, W.-T., Luo, Q.-Y., Xu, Y.-P., Dai, Y.-K., Huang, & S.-L., & Guo, P. Y. (2014). Hydrothermal synthesis, crystal structure and properties of a thermally stable dysprosium porphyrin with a three-dimensional porous open framework. *Inorganic Chemistry Communications*, *49*, 16-18.

Chen, Y. & Wang, K. (2013). Azacrown[N,S,O]-modified porphyrin sensor for detection of Ag^{+}, Pb^{2+}, and Cu^{2+}. *Photochemical and Photobiological Sciences*, *12*, 2001–2007.

Damunupola, D., Chaudhri, N., Atoyebi, A. O., & Brückner, C. (2020). Mechanochemical insertion of cobalt into porphyrinoids using $Co_2(CO)_8$ as a cobalt source. *Green Chemistry*, *22*, 3643–3652.

Dar, T. A. & Sankar, M. (2020). Fused Nickel(II) porphyrins-sensing of toxic anions and selected metal ions through supramolecular interactions. *Frontiers in Chemistry*, *8*, 595177.

Dunbar, A. D. F., Brittle, S., Richardson, T. H., Hutchinson, J., & Hunter, C. A. (2010). Detection of volatile organic compounds using porphyrin derivatives. *Journal of Physical Chemistry B*, *114*, 11697–11702.

Gelsomino, A., Tortorella, D., Cianci, V., Petrovicova, B., Sorgona, A. Piccolo, A., & Abenavoli, M. R. (2010). Effects of a biomimetic iron-porphyrin on soil respiration and maize root morphology as by a microcosm experiment. *Journal of Plant Nutrition and Soil Science*, *173*, 399–406.

Gomes, C., Peixoto, M., & Pineiro, M. (2021). Modern methods for the sustainable synthesis of metalloporphyrins. *Molecules*, *26*, 6652.

Goodland, R. (1995). The concept of environmental sustainability. *Annual Review of Ecology, Evolution, and Systematics*, *26*, 1–24.

Harvey, P. D. (2023). Sustainable development in the removal and photocatalytic reduction of heavy metals in wastewaters using environmentally friendly and health benign porphyrin-based metal-organic frameworks. *Separation and Purification Technology*, *322*, 124–214.

He, B., Mao, Y. L., Zhang, Y., Yin, W., Hou, C. J., Huo, D. Q., & Fa, H. B. (2017). A porphyrin molecularly imprinted biomimetic electrochemical sensor based on gold nanoparticles and carboxyl graphene composite for the highly efficient detection of methyl parathion. *Nano*, *12*, 1750046.

Henriques, C. A., Pinto, S. M. A., Aquino, G. L. B., Pineiro, M., Calvete, M. J. F., & Pereira, M. M. (2014). Ecofriendly porphyrin synthesis by using water under microwave irradiation. *ChemSusChem*, *10*, 2821–2824.

Hiroto, J. S., Miyake, Y., & Shinokubo, H. (2017). Synthesis and functionalization of porphyrins through organometallic methodologies. *Chemical Reviews*, *117*, 2910–3043.

Huang, C. C., Parasuraman, P. S., Tsai, H.-C., Jhu, J.-J., & Imae, T. (2014). Synthesis and characterization of porphyrin–TiO_2 core–shell nanoparticles as visible light photocatalyst. *RSC Advances*, *4*, 6540–6544.

Ishihara, S., Labuta, J., Van Rossom, W., Ishikawa, D., Minami, K., Hill, J. P., & Ariga, K. (2014). Porphyrin-based sensor nanoarchitectonics in diverse physical detection modes. *Physical Chemistry Chemical Physics*, *16*, 9713–9746.

Jiang, R., Pang, Y. H., Yang, Q. Y., Wan, C. Q., & Shen, X. F. (2022). Copper porphyrin metal-organic framework modified carbon paper for electrochemical sensing of glyphosate. *Sensors and Actuators B: Chemical*, *358*, 131492.

Jiang, Z. W., Zou, Y. C., Zhao, T. T., Zhen, S. J., Li, Y. F., & Huang, C. Z. (2020). Controllable synthesis of porphyrin-based 2D lanthanide metal organic frameworks with thickness- and metal node-dependent photocatalytic performances. *Angewandte Chemie International Edition*, *59*, 3300–3306.

Jiang, Z. W., Zou, Y. C., Zhao, T. T., Zhen, S. J., Li, Y. F.,, & Huang, C. Z. (2020). Controllable synthesis of porphyrin-based 2D lanthanide metal–organic frameworks with thickness and metal-node-dependent photocatalytic performance. *Angewandte Chemie International Edition*, *59*, 3300–3306.

Kashani-Motlagh, M. M., Rahimi, R., & Kachousangi, M. J. (2009). Ultrasonic method for The preparation of organic porphyrin nanoparticles. *Molecules*, *15*, 280–287.

Kebonye, N. M., Eze, P. N., John, K., Agyeman, P. C., Nemecek, K., & Boruvka, L. (2022). An in-depth human health risk assessment of potentially toxic elements in highly polluted riverine soils, Prıbram (Czech Republic). *Environmental Geochemistry and Health*, *44*, 369–385.

Khatib, M. & Haick, H. (2022). Sensors for volatile organic compounds. *ACS Nano*, *16*, 7080–7115.

La, D. D., Dang, T. D., Le, P. C., Bui, X. T., Chang, S. W., Chung, W. J., Kim, S. C., & Nguyen, D. D. (2023). Self-assembly of monomeric porphyrin molecules into nanostructures: Self-assembly pathways and applications for sensing and environmental treatment. *Environmental Technology & Innovation*, *29*, 103019.

La, D. D., Ngo, H. H., Nguyen, D. D., Tran, N. T., Vo, H. T., Nguyen, X. H., Chang, S. W., Chung, W. J., & Nguyen, M. D. (2022). Advances, and prospects of porphyrin-based nanomaterials via self-assembly for photocatalytic applications in environmental treatment, *Coordination Chemistry Reviews*, *463*, 214543.

Lee, J. H., Ahn, Y., & Kwak, S-Y. (2022). Facile sonochemical synthesis of flexible Fe-based metal–organic frameworks and their efficient removal of organic contaminants from aqueous solutions. *ACS Omega*, *7*, 23213–23222.

Li, H.-Y., Zhao, S.-N., Zang, S.-Q., & Li, J. (2020). Functional metal–organic frameworks as effective sensors of gases and volatile compounds. *Chemical Society Reviews*, *49*, 6364–6401.

Li, Z., Askim, J. R., & Suslick, K. S. (2019). The optoelectronic nose: Colorimetric and fluorometric sensor arrays. *Chemical Reviews*, *119*, 231–292.

Lim, D.-S., Ko, S.-H., Lee, C.-H., Ahn, W.-S., & Lee, W.-Y. (2006). DH-I-180-3-mediated photodynamic therapy: Biodistribution and tumour vascular damage. *Photochemistry and Photobiology*, *82*, 600–605.

Liu, J., Xiong, W. H., Ye, L. Y., Zhang, W. S., Yang, H. (2020). Developing a Novel Nanoscale Porphyrinic Metal−organic framework: A bifunctional platform with sensitive fluorescent detection and elimination of nitenpyram in agricultural environment, *Journal of Agricultural and Food Chemistry*, *68*(20), 5572–5578.

Marques, C. A. &, Machado, A. A. S. C. (2014). Environmental sustainability: Implications and limitations to the green chemistry. *Foundations of Chemistry*, *16*, 125–147.

Milgrom, L.R. (1997). *The colours of life: An introduction to the chemistry of porphyrins and related compounds* (p. 225). Oxford University Press.

Montforts, F.-P. &, Glasenapp-Breiling, M. (2002). Naturally occurring cyclic tetrapyrroles. *Fortschritte der Chemie organischer Naturstoffe*, *84*, 1–51.

Morelli, J. (2011). Environmental sustainability: A definition for environmental professionals. *Journal of Environmental Sustainability*, *1*, 1–9.

Murakami, T., Iwamuro, Y., Chinaka, S., Takayama, N., Komatsu, T. (2015). Highly sensitive detection of organophosphorus pesticides using 5,10,15,20-tetrakis(4-hydroxyphenyl) porphyrin. *Analytical Sciences*, *31*, 1325–1328.

Nejad, S. T., Rahimi, R., Rabbani, M., & Rostamnia, S. (2022). Zn (II)-porphyrin-based photochemically green synthesis of novel ZnTPP/Cu nanocomposites with antibacterial activities and cytotoxic features against breast cancer cells. *Scientific Reports*, *12*, 17121.

Nisa, K., Saxena, M,.,Lone, I. A., & Kumar, R. (2023). Supramolecular cages, and metal organic frameworks of porphyrins for a sustainable tomorrow: Challenges and applications, *sustain. Energy Fuels*, *7*, 2774–2801.

Paolesse, R., Nardis, S., Monti, D., Stefanelli, M., & Di Natale, C. (2017). Porphyrinoids for chemical sensor applications. *Chemical Reviews*, *117*, 2517–2583.

Park, J. M., Hong, K.-I., Lee, H., & Jang, W.-D. (2021). Bioinspired applications of porphyrin derivatives. *Accounts of Chemical Research*, *54*, 2249−2260.

Patra, M., & Sharma, A. (2000). Mercury toxicity in plants. *The Botanical Review*, *66*, 379-422.

Rojas, S., Rodriguez-Dieguez, A., & Horcajada, P. (2022). Metal–organic frameworks in agriculture. *ACS Applied Materials & Interfaces*, *14*, 16983–17007.

Rompoti, A., Dalal, N., Athanasopoulos, D., Rutan, S., & Helburn, R. (2015). Analysis of organophosphate-Zn metalloporphyrin interactions via UV-Vis spectroscopy and molecular modeling. *Spectrochimica Acta, Part A: Molecular and Biomolecular Spectroscopy*, *135*, 447–456.

Saha, U., Sharma, K., Chaudhri, N., Sankar, M., & Gopinath, P. (2018). Antimicrobial photodynamic therapy: Single-walled carbon nanotube (SWCNT)-porphyrin conjugate for visible light mediated inactivation of staphylococcus aureus. *Colloids and Surfaces B: Biointerfaces*, *162*, 108–117.

Sajjadinezhad, S. M., Tanner, K., & Harvey, P. D. (2022). Metal-porphyrinic framework nanotechnologies in modern agricultural management. *Journal of Materials Chemistry B*, *10*, 9054–9080.

Sallam, G., Shaban, S. Y., Nassar, A., & El-Khouly, M. E. (2020). Water soluble porphyrin as optical sensor for the toxic heavy metal ions in an aqueous medium. *Spectrochimica Acta Part A: Molecular and Biomolecular*, *241*, 118609.

Sannino, F., Spaccini, R., Savy, D., Piccolo, A. (2013). Remediation of highly contaminated soils from an industrial site by employing a combined treatment with exogeneous humic substances and oxidative biomimetic catalysis. *Journal of Hazardous Materials*, *261*, 55–62.

Shy, H., Mackin, P., Orvieto, A. S., Gharbharan, D., Peterson, G. R., Bampos, N., & Hamilton, T. D. (2014). The two-step mechanochemical synthesis of porphyrins. *Faraday Discussions, 170*, 59–69.

Silvestri, S., Fajardo, A. R., Iglesias, & B. A. (2022). Supported porphyrins for the photocatalytic degradation of organic contaminants in water: A review. *Environmental Chemistry Letters, 20*, 731–771.

Srivastava, P., Ali, R., Razi, S. S., Shahid, M., & Misra, A. (2013). Thiourea Based Molecular Dyad (ANTU): Fluorogenic Hg^{2+} Selective chemodosimeter exhibiting blue-green fluorescence in aqueous-ethanol environment. *Sensors and Actuators B: Chemical, 181*, 584–595.

Sułek, A.,, Pucelik, B.,, Kobielusz, M., Łabuz, P., Dubin, G, & D ą browski, J. M. (2019). Surface modification of nanocrystalline TiO_2 materials with sulfonated porphyrins for visible light antimicrobial therapy. *Catalysts, 9*, 821.

Suslick, K. S., Rakow, N. A., Sen, A. (2004). Colorimetric sensor arrays for molecular recognition. *Tetrahedron, 60*, 11133–11138.

Takacs, L. (2013). The historical development of mechanochemistry. *Chemical Society Reviews, 42*, 7649–7659.

Tang, J., Ding, G., Niu, J., Zhang, W., Tang, G., Liang, Y., Fan, C., Dong, H., Yang, J., Li, J., & Cao, Y. (2019). Preparation and characterization of tebuconazole metal-organic framework-based microcapsules with dual-microbicidal activity. *Chemical Engineering Journal, 359*, 225–232.

Tang, J., Tang, G., Niu, J., Yang, J., Zhou, Z., Gao, Y., Chen, X., Tian, Y., Li, Y., Li, J., & Cao, Y. (2021). Preparation of a porphyrin metal−organic framework with desirable photodynamic antimicrobial activity for sustainable plant disease management. *Journal of Agricultural and Food Chemistry, 69*, 2382−2391.

Varsha M. V., & Nageswaran, G. (2020). Review—direct electrochemical synthesis of metal organic frameworks. *Journal of the Electrochemical Society, 167*, 155527.

Yang, J., Dai, D., Cai, Z., Liu, Y.-Q., Qin, J.-C., Wang, & Yang, Y. W. (2021). MOF-based multi-stimuli-responsive supramolecular nanoplatform equipped with macrocycle Nanovalves for plant growth regulation. *Acta Biomaterialia, 134*, 664–673.

Yang, Y., Jiang, J., Shen, G., & Yu, R. (2009). An optical sensor for mercury ion based on the fluorescence quenching of tetra(*p*-dimethylaminophenyl) porphyrin. *Analytica Chimica Acta, 636*, 83–88.

Yu, K., Lee, Y.-R., Seo, J. Y., Baek, K.-Y., Chung, Y.-M,, & Ahn, W.-S. (2021). Sonochemical synthesis of Zr-based porphyrinic MOF-525 and MOF-545: Enhancement in catalytic and adsorption properties. *Microporous and Mesoporous Materials, 316*, 110985.

Zhang, D., Zhu, M., Zhao, L., Zhang, J., Wang, K., Qi, D., Zhou, Y., Bian, Y., & Jiang, J. (2017). Ratiometric fluorescent detection of Pb^{2+} by FRET-based phthalocyanine-porphyrin dyads. *Inorganic Chemistry, 56*, 14533–14539.

Zhang, L., Shi, X., Zhang, Z., Kuchel, R. P., Namivandi-Zangeneh, R., Corrigan, N., Jung, K., Liang, K. &,, Boyer, C. (2021). Porphyrinic zirconium metal–organic frameworks (MOFs) as heterogeneous photocatalysts for PET-RAFT polymerization and stereolithography. *Angewandte Chemie International Edition, 60*, 5489–5496.

Index

P

Q

R

S

T

V

W

Z